Animal Anatomy
and Physiology

Third Edition

Animal Anatomy and Physiology

Jesse F. Bone

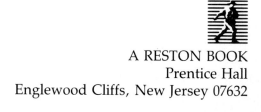

A RESTON BOOK
Prentice Hall
Englewood Cliffs, New Jersey 07632

Library of Congress Cataloging-in-Publication Data

Bone, Jesse Franklin
  Animal anatomy and physiology.

  Bibliography: p.
  Includes index.
    1. Veterinary anatomy.   2. Veterinary physiology.
I. Title.
SF761.B64   1988   636.089′2   87-7335
ISBN 0-8359-0099-1

Editorial/production supervision: Evalyn Schoppet
Manufacturing buyer: Peter Havens

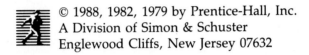
© 1988, 1982, 1979 by Prentice-Hall, Inc.
A Division of Simon & Schuster
Englewood Cliffs, New Jersey 07632

All rights reserved. No part of this book may be
reproduced, in any form or by any means,
without permission in writing from the publisher.

Printed in the United States of America

10 9 8 7 6 5 4 3 2 1

ISBN 0-8359-0099-1

Prentice-Hall International (UK) Limited, *London*
Prentice-Hall of Australia Pty. Limited, *Sydney*
Prentice-Hall Canada Inc., *Toronto*
Prentice-Hall Hispanoamericana, S.A., *Mexico*
Prentice-Hall of India Private Limited, *New Delhi*
Prentice-Hall of Japan, Inc., *Tokyo*
Simon & Schuster Asia Pte. Ltd., *Singapore*
Editora Prentice-Hall do Brasil, Ltda., *Rio de Janeiro*

*To my wife Faye,
without whose inspiration and encouragement
this book would not have been written*

Contents

Preface — xiii
Preface to the Third Edition — xv

Introduction — 1

Anatomy, 1
Physiology, 5
Taxonomy and Terminology, 6

Chapter 1 Cells — 11

The Vital Phenomena, 11
History of the Cell, 12
Gross Structure of the Cell, 12
Electron Microscopy, 13
Anatomy of the Mammalian Cell, 14
Protoplasm, 19
Physico-Chemical Factors Affecting Cells, 19
Organization of Cells, 20
Biological Rhythms, 21
Aging, 22

| Chapter 2 | **Bone and Cartilage** | **27** |

Bone, 27
Cartilage, 28
Bone Formation and Structure, 28

| Chapter 3 | **The Skeleton** | **33** |

The Appendicular Skeleton, 33
The Axial Skeleton, 77
The Visceral Skeleton, 95

| Chapter 4 | **Arthrology (Syndesmology)** | **97** |

Classification of Joints, 98

| Chapter 5 | **The Muscular System** | **101** |

Types of Muscle, 101
Muscle Physiology and Joint Movements, 113
Muscle Contraction, 114
Locomotion, 116
The "Stay Apparatus," 127

| Chapter 6 | **Teeth** | **129** |

Types and Anatomy of Teeth, 129
The Age of the Horse, 132
The Age of the Dog, 141
The Age of the Bovine, 142
The Age of the Sheep, 142
The Age of the Pig, 143
Dental Formula, 143
Physiology of the Teeth, 145

| Chapter 7 | **The Digestive System** | **149** |

Parts of the Alimentary Canal, 149
Exocrine Glands, 166
Accessory Structures of the Digestive System, 172
Physiology of Digestion, 174
Physiology of the Stomach, 187
Physiology of the Intestinal Tract, 193
Physiology of Rumination, 198

| Chapter 8 | **The Respiratory System** | **203** |

Gross Structures of the Mammalian External Respiratory System, 205
The Physiology of Respiration, 211
Air, Gaseous Exchange and Blood Transport, 223

| Chapter 9 | The Vascular System | 233 |

Blood, 234
The Fluid Element of the Blood, 248
Inflammation, 251
Other Defense Mechanisms, 252
Blood Types, 256
The Heart, 259
The Vascular Tree, 267
The Arterial System, 268
The Venous System, 272
The Lymphatic System, 279
Fetal Lymphatics, 285

| Chapter 10 | The Urinary System | 287 |

The Urinary Organs, 287
The Kidney, 288
The Secretory and Excretory Mechanism of the Kidney, 291
The Excretory Portion of the Kidney, 293
Renal Physiology, 298
Physical Aspects of Kidney Function, 300
Physiology of the Nephron, 303
Endocrine Aspects of the Kidney, 309

| Chapter 11 | The Nervous System | 311 |

Major Divisions of the Nervous System, 311
The Nerve Fiber or Neuron, 312
The Brain, 318
The Spinal Cord, 326
The Peripheral Nervous System, 329
The Autonomic Nervous System, 330
General Physiology of the Nervous System, 334
Central Nervous System, 338
Intelligence, Intellect and Memory, 342

| Chapter 12 | The Endocrine System | 345 |

General Considerations, 345
The Pituitary, 346
The Pineal Body, 350
The Thyroid, 351
The Parathyroid Glands, 352
The Thymus, 352
The Adrenal Glands, 353
The Pancreas, 355
The Testicles, 357

The Ovaries, 358
The Uterus, 358
The Stomach, 359
The Small Intestines, 359
The Kidney, 360
The Liver, 360
Prostaglandins, 360
Hormone Interrelationships, 361
Regulation of Reproductive and Accessory Sex Structures, 362
Regulation of the Intestinal Tract, 365
Regulation of Water and Electrolyte Balance, 366
Regulation of Metabolism, Body Temperature, and Hibernation, 366

Chapter 13   Intermediate Metabolism   369

Introduction, 369
The Dynamic State of Body Constituents, 369
Phases of Metabolism, 370
Carbohydrate Metabolism, 372
Fat Metabolism, 382
Protein Metabolism, 390
Physical Structure of Proteins, 396
Vitamins, 399

Chapter 14   The Reproductive System   401

Introduction, 401
The Anatomy of the Male Genital System, 402
Accessory Glands of the Male Genital System, 407
Anatomy of the Female Reproductive System, 411
Physiology of Reproduction, 418
Spermatogenesis, 423
Oogenesis, 427
The Estrous Cycle, 430
Copulation, 432
Sex Ratio, 438
Control of Pregnancy and Overpopulation, 443
Reproduction Patterns, 445

Chapter 15   The Common Integument and Its Derivatives   461

General Considerations, 461
Epithelium, Mesothelium, and Endothelium, 461
The Skin, 466

The Appendages of the Skin, 467
The Glands of the Skin, 473
The Bovine Udder, 475
Physiology of Lactation, 477

Chapter 16 The Organs of Special Sense 483

Light Reception, 483
The Eye, 484
Physiology of the Eye, 490
The Organ of Hearing and Equilibrium, 497
Physiology of the Ear, 502
The Organ of Smell, 505
Physiology of Smell, 506
The Organ of Taste, 507
Physiology of Taste, 508
Tactile Senses, 509

Chapter 17 Anatomy and Physiology of the Fowl 511

Cells, 512
Bone and Cartilage, 512
The Skeleton, 513
Arthrology, 517
Physiology of the Skeleton, 517
Musculature, 518
Digestive System, 521
Accessory Glands of the Digestive System, 525
Physiology of Digestion, 526
Respiratory System, 529
Blood Vascular System, 531
Urinary System, 533
Reproductive System, 533
Nervous System, 541
Endocrine System, 541
Common Integument, 542
The Organs of Special Sense, 547

A Glossary of Terms Used in Anatomy and Physiology, 551
Bibliography, 557
Index, 561

Preface

This is a basic book. It is my intention to present a fundamental understanding of animal structure and function. Advanced knowledge is more properly taught in medical and veterinary schools and in advanced courses in the life sciences.

The material contained herein is neither particularly complex nor difficult, nor is it intended to be. It is designed to open the gates to a broad field of knowledge that can be an endless fascination.

The text is presented from a systemic anatomical viewpoint, and consequently the organization of subject matter is different from works that are primarily physiologically oriented. This is the result of over a quarter century of teaching and conferring with students. For those who are not professionally involved with research or diagnostics, an anatomic approach appears to be more understandable, probably because it is based upon things about which students already have some knowledge. To initially expose nonprofessionally-oriented students to the complex acts of neural function, or the equally complex mechanisms of fluid and gaseous interchange, is one of the better ways to extinguish the light of comprehension and the excitement of learning.

In this text the horse will be used as the basic animal and comparison—where indicated—will be made with similar structures or functions in other species. While the cow and the dog have largely replaced the horse as the basic animal in professional schools, I feel that the relative simplicity and large size of equine structures make this animal an excellent subject for demonstration and study in a nonprofessional course. Additionally, there is a great bulk of relatively uncomplicated literature about horses which is available for outside reading by students who wish to travel beyond the restricted boundaries of this text.

Emphasis has been placed upon those portions of anatomy and physiology that have a practical function—which may explain the otherwise puzzling emphasis on skeletal anatomy, dentition, and locomotion. These subjects are important to people who have a functional interest in livestock and pets. It is of no great interest to the generalist or to animal science, agriculture, or wildlife students to have more than general knowledge of intermediate metabolism, ultrastructures, mathematical models, test procedures, or the histogenesis of organs and systems. Therefore, this and similar material is mainly left for the advanced texts designed for medical and veterinary students, specialists, and professionals; such specialist data that is included is simply for illustrative purposes.

This book will, I think, fill a gap that presently exists between the superficialities of survey courses in mammalian structure and function, and the complexities of the advanced works which assume a fundament of knowledge that often does not exist.

Preface to the Third Edition

I am reminded of an old rock-and-roll song that started with a hyperthyroid voice declaiming, "the beat goes on!" Somehow I seem to have gotten involved in a similar situation. What began a number of years ago as an attempt to keep a couple of dozen World War II veterans awake and reasonably interested in the structures and basic functions of life has now become *a book*, and moreover a book which has survived several years of use (by several thousand students), two editors, and a large number of headaches and heartaches over what can happen to a project between conception and publication.

This third edition is no exception. Frankly, I think it is better and more informative than either of the two previous editions, but I have that nagging certainty that I have either omitted, deleted, or failed to enter some new discovery of critical importance to students who want a brief but moderately comprehensive overview of the anatomy and physiology of animals.

I can console myself, however, that Gottfried Wilhelm Leibniz was the last of the universal geniuses (a man who knew *everything* worth knowing in his time), and he died in 1716. There hasn't been a universal genius since, and I lay no claims to that dubious honor. I do, however, believe that this edition is an improvement over the last. Illustrations and new information have been added. Terminology has been updated and is generally in conformance with the dicta of the *Nomina Anatomica Veterinaria*. Confusing passages have (I hope) been clarified. Old typographic and textual errors have been corrected (and doubtless replaced by new ones). Changes have been made, but I truly hope that the basic philosophy that prompted this book is still between its covers.

Learning is important, but it can also be fun. So . . . have fun!

Jesse F. Bone

Introduction

ANATOMY

Anatomy (Gr. *ana*—apart; *tomy*—cut) is the study of the structure of a body and the relation of its parts. The subject is usually studied by dissection and observation. It is one of the oldest branches of biological science, dating back at least to the fourth century, B.C., when the Greek philosopher Aristotle published some of his observations on the structure of fish and animals. For this work, Aristotle won the distinction of being called "the father of anatomy."

It has been said that, "It's a wise father who knows his own child," and as far as anatomy is concerned, the statement contains a considerable amount of truth. During the 23 centuries since it was "born," Aristotle's child has developed into one of the major branches of biological science. It is basic to any understanding of why and how an organism functions, and is important in introducing biological terminology and the locations and relationships of organs and systems that combine to form a functioning body. With the aid of the light microscope, the electron microscope, biochemistry, and the various subdivisions of the life sciences, anatomy has become so diverse and complex that Aristotle might well have difficulty recognizing it. Moreover, it has also produced a number of secondary or surrogate "fathers," including Claudius Galen and Andreas Vesalius. Galen, a Graeco-Roman physician living in Rome during the 2nd century, A.D., and Vesalius, a Belgian anatomist of the 16th century, did great work in advancing the knowledge of mammalian anatomy. Because of his extensive studies and dissections, Vesalius is often referred to as "the father of modern anatomy." In the field of microanatomy, Robert Hooke, Schleiden and Schwann, and Rudolf Virchow probably hold as much title to the term "father" as any others. However, the knowledge explosion in the 19th and 20th centuries made the term "father" applied to any discipline or subdiscipline in the biological sciences something of a misnomer; "pioneer" would be a more appropriate term.

In the Middle Ages, prior to the development and use of laboratory animals, the growth of anatomy as a branch of medicine depended a great deal upon cooperation of the local magistrates. Condemned criminals were in great demand and were often released into the custody of scientists for acute experimentation. This was very important for the advancement of anatomical studies, since cadavers tended to spoil rather quickly because of lack of refrigeration and/or embalming. A fresh cadaver was the best kind for making new and useful discoveries.

The 16th century anatomist Fallopius (of Fallopian tube fame) was once awarded two condemned criminals to "put to death in whatever way he pleases—and then anatomize them." Fallopius gave them opium. One survived, the other died. The magistrate pardoned the survivor, but Fallopius did not. He gave the unlucky man another 8 grains of opium—and thus helped establish the minimum lethal dose (MLD) for the drug. History, unfortunately, does not record the other results of Fallopius' studies upon these two cadavers.

Anatomy is generally considered to be a dead subject. With few exceptions, the work is performed upon carcasses or cadavers, although living animals (known as palpators) may be kept in or near the laboratory for reference. There is very little that is either new or dynamic in the anatomical field; the major disciplines are those of memory and correlation of various parts or structures into a complete organism. Yet the basics of any study are not new or dynamic. They are foundations upon which understanding is built. And it is in this sense that anatomy is important.

Planes of Reference

In describing an animal's body, anatomists have long used four arbitrary planes of reference to locate parts and structures. These planes are located in reference to the long axis of the body (which for practical purposes can be considered to be an imaginary line passing through the center of the spinal column) and can be applied regardless of the position or orientation of the animal.

The planes of reference are

1. Median
2. Sagittal
3. Transverse
4. Frontal

The median plane is the primary plane of reference. It is a single plane which passes through the center of the long axis of the body and divides the body into two equal halves. There is only one median plane. All the other planes are constructed in relationship to it.

Sagittal planes are those parallel to the median plane. Sagittal planes do not pass through the median axis of the body.

Transverse planes are located at right angles to the median plane. All transverse planes pass through the long axis of the body. The transverse plane divides the body into cranial (anterior) and caudal (posterior) portions. Transverse planes divide the body into cross sections.

Frontal planes are those which are located at right angles to both the median and transverse planes. They divide the body into dorsal (back) and ventral (belly) portions.

One must remember that these planes of reference, while used for both man and animals, have a slightly different meaning when applied to man or to primates. Man and the primates are considered anatomically to be upright bipedal animals with the long axis of the body vertical. Other animals are generally considered to have the long axes of their bodies horizontal. Consequently, the long axes of the limbs of man lie parallel to the long axis of the body, while in animals the long axes of the limbs lie generally at right angles to the long axis of the body.

Transverse planes in man and the primates will divide both trunk and limbs into cross sections, but will divide only the trunk of an animal into cross sections. Cross sections of most animals' limbs will be cut by frontal planes.

Anatomical Methods

The mammalian body is a complex structure composed of a number of interrelated systems combined into an integrated whole. To break the body down into parts that can be studied individually has been a difficult task and one which cannot be perfect since the body is an integrated unit and each structure is related to the others. Over the years a number of methods have been developed for the study of anatomy. Among these are included the following which are presently in use at various schools:

1. Systemic anatomy
2. Topographic anatomy
3. Regional anatomy
4. Special anatomy

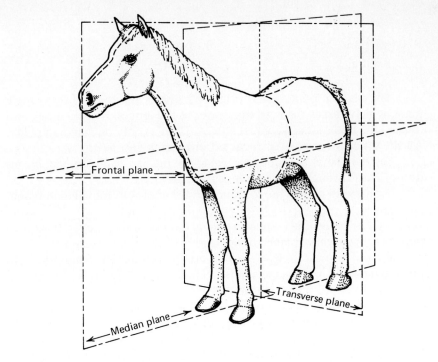

FIGURE A   Planes of reference.

Systematic or descriptive anatomy is the study of various organ groups or systems of the body. Each system is studied as a unit, and finally all the systems are correlated into a complete animal.

Topographic anatomy is the study of parts or systems in relation to their surrounding parts.

Regional anatomy is the study of limited portions or regions of the body. Under this method the body is normally broken down into four arbitrary regions which are studied topographically. These regions consist of the head and neck, thorax and forelimb, abdomen, and the pelvic region and hindlimb.

Special anatomy is the detailed study of certain limited regions of the body such as the teeth, the urogenital apparatus, the liver, stomach, lungs, or other specific structures.

Each of these methods has certain inherent advantages and disadvantages, but for a nontechnical course, systemic anatomy offers the best opportunity to gain the greatest amount of knowledge in a limited line.

Systems of the Body

The body falls more or less naturally into eleven major organ groups or systems which can be listed as follows:

1. Skeletal
2. Muscular
3. Vascular
4. Digestive
5. Respiratory
6. Urinary
7. Reproductive
8. Nervous
9. Endocrine
10. Common integument
11. Organs of special sense

Each of these systems is functionally related and physically connected with one or more of the others. Yet, insofar as separation can be accomplished, this method has stood the test of time with fewer changes than might be expected.

PHYSIOLOGY

Physiology (Gr. *physis*—nature; *logy*—study of) is the fountainhead of the "natural sciences" as they relate to living organisms. It deals with the function of living matter and includes a number of subsidiary disciplines such as behavior, biochemistry, and biophysics. Being a more recent discipline than anatomy it has fewer fathers, and like the fictional Topsy has "just growed." If any man could qualify as the "father of physiology" it would probably be Dr. William Harvey, a 17th century English physician who discovered that the blood circulates through a closed system, yet Harvey is usually considered to be an anatomist.

The time difference between Aristotle and Harvey is about 2,100 years, which indicates that physiology is a johnny-come-lately to the medical and paramedical field. However, physiology has grown to giant size and has built a towering edifice of discovery. Compared to anatomy, which is essentially pedestrian, physiology is airborne. Today the total knowledge accumulated in physiology is probably beyond the comprehension of any single man and the amount of knowledge about physiological processes constantly increases from year to year.

While physiology is a separate subject in veterinary, medical, and dental schools and is considered to be a separate branch of biological science, it is capable of being closely correlated with anatomy and particularly with systemic anatomy. This has considerable value, since an understanding of the functions of an animal is of more practical importance than a knowledge of its structures. It is through function that the usefulness of a species is expressed. And it is through adapting and developing function that animal husbandry has risen to a place of importance in agriculture. It is, of course, the duty of the breeder or the geneticist to develop useful traits among domestic animals, but those traits must be present in the animal in such a form that they will respond to development. A dairy cow, no matter how physically perfect she may be, is of no great value unless she produces milk. Racehorses are valueless unless they can run fast. Beef cattle, sheep, and swine are uneconomical unless they can turn a minimum of feed into a maximum of meat. These are physiological responses associated with anatomy but not entirely dependent upon it.

Similarly, the physiology presented in these pages will be associated with, but not entirely dependent upon, the anatomy. In certain areas where the data do not conform to anatomic classification or where the physiological aspects are important enough to require special consideration, the material will be given separately. However, in the main, anatomy and physiology will be included together in a potpourri, which should make both subjects more palatable.

TAXONOMY AND TERMINOLOGY

Taxonomy is that branch of biological science that attempts to classify living forms (or extinct life forms) into some sort of order and system which can be used to pigeonhole, relate, "hierarchize," or otherwise identify them and establish their relationships to each other and to the totality of life. In general, the taxonomic system is constructed according to the basic plan developed by Karl Von Linné (Linnaeus) in the middle part of the 18th century:

1. Kingdom
2. Phylum
3. Class
4. Order
5. Family
6. Genus
7. Species

For example, a human being belongs to the Animal Kingdom, Phylum Chordata, Class Mammalia, Order Primate, Family Hominidae, Genus *Homo*, Species *sapiens*. Note that only the specific designation commences with a lower case letter and that only the generic and specific identifications are italicized.

Linnaean classification was good enough as far as it went, but according to minutiae lovers, it didn't go far enough. The essential simplicity of the Linnaean system quickly became complicated as more and more minor details were discovered that separated or connected the various subdivisions of life forms until today we have such classification schemes that follow:

1. Kingdom
2. Phylum
   2a. Subphylum
3. Superclass
   3a. Class
   3b. Subclass
4. Superorder
   4a. Order
   4b. Suborder
5. Superfamily
   5a. Family
   5b. Subfamily
6. Genus
   6a. Subgenus
7. Species
   7a. Subspecies
8. Variety

This bewildering complexity is in some ways symptomatic of the problems involved in the knowledge explosion that makes life ever more complex. For practical purposes, most of this is unnecessary, but for academic purposes it is vital. At the same time, such a horde of categories reduces any taxonomic scheme to something best understood by a specialist. In this book we obviously cannot go into such detail, but we shall have a little taxonomy—mainly dealing with genera and species.

This is really not as difficult as it seems, and can, on occasion, be fun. I ran into a dinner menu one time which was written Linnaean style (probably for a taxonomists' meeting). I repeat it here as a small refresher course in taxonomy and nomenclature. I have also, mercifully, included a translation. You will note from the prices this is not new material.

DINNER MENU

Linnaean Style
APPETIZERS

Boiled Abdominal Musculature of *Penaeus setiferus* cocktail	$1.00
Fresh *Gallus domesticus* Hepatic Pate	.75
Pepos of *Cucumis melo* var. *cantalupensis* in Season	.40
Marinated *Clupea harengus* in Sour Cream—	
Served on Leaves of *Lactuca sativa*	.75
Backfin *Callinectes sapidus* Cocktail	1.50
Anchoa mitchilli canapes	.75
Chilled *Lycopersicum esculentum* Extract	.25
Venus mercenaria (Under 7.6 cm) On a Single Valve—	
Served with *Rorippa armoracia* Sauce	.90
Preserved Ova of *Acipenser oxyrhynchus*	2.50
Barbecued Intercostal Muscles with Attached Segments of Costal Bone and Cartilage of *Sus scrofa domestica*	2.00

SOUPS

Split Seeds of *Pisum sativum*	$ .35
Cream of *Agaricus campestris*	.45
French *Allium* spp. Au Grautin	.50
Chelydra serpentina Au Sherry	.60

ENTREE

Eviscerated *Salmo gairdneri*—Almondine	$2.75
Fresh Recently Molted *Callinectes sapidus*	4.00
Broiled Whole *Homarus americanus*	according to size
Pelvic Limbs of *Rana catesbeiana*—Saute *Allium sativum* Butter	3.75
Stuffed Musculus Pectoralis of *Gallus domesticus*	3.75
Hepatic Slices of Immature *Bos taurus* with *Allium* spp.— Saute—*Carya*—Smoked Sliced Lateral Thoracoabdominal Musculature of *Sus scrofa domestica*	4.00

Planked Chopped Gluteal Musculature of *Bos taurus*—*Agaricus campestris* Gravy	3.50
Fresh Thymus or Pancreas of *Bos* spp.—Saute	4.00
Adductor Muscles of *Pecten irradians*— Served with Carpels of *Citrus limon*	3.75
Sliced Roast *Meleagris gallopavo* with Gravy— Berries of *Vaccinium macrocarpon* Sauce	3.50
Broiled Caudal Muscles of *Panulirus argus*— Clarified Butter	4.50
Sliced Gluteal Musculature of Immature *Bos taurus*	3.25
U.S. Prime Thoracic Musculus Longissimus Dorsi of Castrated Male *Bos taurus* with Attached Halved Vertebra and Adjacent Costal Head	5.00

The Above Items Include Petioles of *Apium graveolens*, Black and Red Drupes of *Olea europea*, Pepos of *Cucumis sativus* (Previously Treated in 6% Acetic Acid), and Fleshy Taproots of *Raphanus sativus;* Choice of Two Vegetables; Tossed Chlorophyll-Pigmented Salad with Dressing; Bread, Butter, and Beverage.

VEGETABLES

Taproots of *Daucus carota* var. *sativa*, French-Fried Sliced Bulbs of *Allium* spp., Leafless Green Stems of *Asparagus officinalis*, Thick Undeveloped Inflorescences of *Brassica oleracea* var. *botrytis*, Taproots of *Beta vulgaris*, Fresh Leaves of *Spinacia oleracea*, Whipped Rhizomes of *Solanum tuberosum*, Swollen Tuberous Roots of *Ipomoea batatas*.

BEVERAGES

Fresh percolation of drupes of *Coffea* spp. in water	Lacteal secretion of *Bos taurus*	Terminal Leaf infusion of *Thea* spp. in water

DESSERTS

Sliced Hawaiian *Ananas comostus*	$ .60
Fresh *Cucurbita pepo* Pie	.40
Homemade Hot *Malus* spp. Pie	.40
A la Mode	.60
Oryza sativa Pudding Sprinkled with Milled Bark of *Cinnamonum zeylanicum*	.30

Your waiter will be most happy to suggest the proper 12% to 20% ethanol fermentation of the berries of various varieties of *Vitis vinifera* to complement your meal.

TRANSLATION—DINNER MENU

APPETIZERS

Shrimp Cocktail	$1.00
Fresh Chicken Liver Pate	.75
Fresh Cantaloupe in Season	.40
Marinated Herring in Sour Cream—Served on Lettuce	.75
Backfin Crabmeat Cocktail	1.50
Canape of Anchovies	.75
Chilled Tomato Juice	.25
Cherrystone Clams on the Halfshell— Served with Horseradish Sauce	.90
Caviar	2.50
Barbecued Spare Ribs	2.00

SOUPS

Split Pea	$ .35
Cream of Mushroom	.45
French Onion Au Grautin	.50
Snapper Au Sherry	.60

ENTREE

Rainbow Trout—Almondine	$2.75
Fresh Soft Shell Crabs	4.00
Broiled Whole Maine Lobster	according to size
Frog Legs—Saute, Garlic Butter	3.75
Stuffed Breast of Chicken	3.75
Fresh Calves Liver and Onions—Saute, Hickory Smoked Bacon	4.00
Planked Chopped Sirloin Steak—Mushroom Gravy	3.50
Fresh Sweetbreads—Saute	4.00
Fresh Sea Scallops—Served with Wedges of Lemon	3.75
Sliced Roast Turkey with Gravy—Cranberry Sauce	3.50
Broiled Lobster Tails—Drawn Butter	4.50
Veal Steak	3.25
U.S. Prime Rib of Steer Beef	5.00

The Above Items Include Celery, Olives, Pickles, and Radishes; Choice of Two Vegetables; Tossed Green Salad with Dressing; Bread, Butter, and Beverage.

VEGETABLES

Carrots, French-Fried Onion Rings, Asparagus, Broccoli, Beets, Spinach, Whipped Potatoes, Sweet Potatoes.

BEVERAGES

Coffee   Milk   Tea

DESSERTS

Sliced Hawaiian Pineapple	$ .60
Fresh Pumpkin Pie	.40
Homemade Hot Apple Pie	.40
A la Mode	.60
Rice Pudding with Cinnamon	.30

Your waiter will be most happy to suggest the proper wine to complement your meal.

No one should have trouble with *Bos taurus, Gallus domesticus, Sus scrofa* var, *domestica, Meleagris gallopavo,* or *Salmo gairdneri.* Most should easily recognize *Cucumis melo* var, *cantalupensis, Cucumis sativus, Asparagus officinalis, Spinacia oleracea,* and *Beta vulgaris* by inspection if not by name. The rest may be a mystery. Admittedly, it's not particularly relevant, except to illustrate the point that in biological science the terminology is the message, and that it is terminology that differentiates between *Equus caballus* and *Equus onager*, which is the difference between horse and ass.

    The taxonomy in this text will be confined mainly to genus and species and the terminology to common usage and to accepted veterinary terminology in Greek or Latin. In most instances the words will be explained and/or broken down into their components or roots. The others will be up to you to find in the glossary or in a medical dictionary.

chapter 1
Cells

The cell is the basic unit which possesses that indefinable thing called life. Life, until better knowledge is gained, is an organic means of converting one form of energy into another. In the process a number of acts are performed which are called the vital phenomena, or the attributes of life.

THE VITAL PHENOMENA

At our present state of knowledge, we cannot define life other than by telling what it does. We do not know what it is. All life, however, must fulfill four basic criteria; the so-called vital phenomena. These are Growth, Reproduction, Irritability, and Metabolism—and give rise to the aphorism that life is a "grim" proposition. Other criteria which are not basic, but which apply to higher life forms, include conductivity (the transmission of stimuli to other cells by direct or indirect contact) and organization (the grouping of cells and tissues into organs and parts, and the integration of these parts into a functioning whole).

HISTORY OF THE CELL

Like many other structures in living organisms, the basic structural and functional unit of living matter was named by mistake. When Robert Hooke applied the term "cell" to the compartments in cork in 1665, he really meant the chamber that encloses the living matter rather than the protoplasm itself. His description, however, was sufficiently unclear that later investigators confused the compartment with the contents and the word "cell" came to be applied to the protoplasm rather than to the walls around it.

Except in bone and cartilage, compartments containing cells are not found in animal bodies. Animal cells tend to be packed together in more or less specific ways to form tissues and organs. Compartments, however, are quite common in plants, and the word "cell" is very appropriate if one is a linguistic purist and wishes to make the word "cella" mean in English what it meant originally in Latin, e.g., "a small chamber."

It was not until 1772 that Corti (of inner ear fame) observed the jelly-like protoplasm within the cell, and it was even later that protoplasm came to be recognized as the life substance of plants and animals. In 1839 Schleiden and Schwann formulated the cell theory which states that "the elementary parts of all tissues are formed of cells in an analogous, though very diversified manner, so that it may be asserted that there is one universal principle of development for the elementary parts of organisms, however different, and that this principle is the formation of cells."

More simply, all living matter is composed of cells.

GROSS STRUCTURE OF THE CELL

The cell is a functional mass of protoplasm which contains, or has at one time contained, a nucleus. Classically, a cell is composed of five basic parts: cell membrane, cytoplasm, nuclear membrane, nucleus, and processes.

Most cells range in size from 10 to 100 microns (μ; 1 μ = 1/1000 mm. or about 1/25,000 inch) in diameter. The size limitation is necessary for efficient functioning since the rate at which substances reach all parts of a cell is inversely related to the square of the cell's diameter, assuming, of course, that the cell is spherical. Since a 10-fold increase in diameter would slow 100-fold the rate at which transportable substances could enter or leave the central parts of the cell, it is easy to see why size is important. Too large a cell probably could not transport waste products or nutrients or perform metabolic functions.

Large surface area in proportion to volume is another aspect of small cell size, and this can be very important in efficient functioning. The cell wall or cell membrane is very thin, about 70 to 110 Ångstroms in thickness. (1 Å = 1/10,000,000 mm.) It has, therefore, only very little tensile strength, which again means the cell must be small. This smallness is reflected in an even greater smallness of the cell organelles, the tiny structures found within the cytoplasm and nucleus that do the work of the cell. This has made cell study very difficult. It has not been until recently that we have known, except in a most general sort of way, what the fine structure of the cell is like, let alone what it does.

The cell membrane forms the outermost limits of the cell and separates it from other cells. The cytoplasm is the protoplasmic mass surrounding the nucleus. It contains the structures necessary for the specific functioning of the cell. The nucleus is the life center of the cell. It is usually located centrally, and is ordinarily separated from the cytoplasm by a nuclear membrane. It contains the genes, chromosomes, and other structures typical of the species and necessary for cell reproduction. The cell processes are extensions of the cell which are derived from the cytoplasm and which are covered by the cell membrane.

ELECTRON MICROSCOPY

Recently, with the development of electron microscopy, a fascinating new world of molecular biology has been opened to scientific investigation. This discipline deals with the structure and function of the nuclear and cytoplasmic components of the cell that could not previously be visualized because their size was smaller than a wavelength of visible light. Since the wavelength of light is the limiting factor in ordinary microscopic studies (which is one of the reasons for the use of blue light in light microscopy; the wavelength is shorter and the resolution or visibility of small objects is therefore better) objects less than 5 microns (1/5,000 inch) in size are fuzzy and unclear. The absolute limit of ordinary light microscope magnification is about 1,500 diameters—not because lenses cannot be made that will magnify even more, but because the object is so small that it will not interrupt the beam of light and produce an image that can be detected by the eye. The object literally becomes lost in the troughs of the waves of light.

The electron microscope, by using an electron stream that does not have a wave form, can theoretically resolve objects as small as the space between the electrons. Since electron beams are not suitable for direct observation, the specimen is photographed and the photograph studied at leisure. The focal field is extremely small and detailed, and one must know exactly where and at what one is looking. If the observer does not know, the preparation is useless.

The magnification can be as great as 400,000 diameters. Beyond that point, technical problems in photography and in fixing, staining, and preparing specimens interfere with further enlargement of objects and result in loss of clarity. However, this text does not intend to go into ultrastructures of the cell at any great length since this relatively unknown territory needs more exploration before its discoveries can be boiled down enough to be included in basic biological science. With ultramicroscopy as a tool, we shall undoubtedly learn a great deal more about the mechanisms of life, but it is equally probable that the definition of life itself is going to be a great deal more difficult.

ANATOMY OF THE MAMMALIAN CELL

The Cell Membrane

The cell membrane (sometimes called the plasma membrane) is the limiting structure of the cell and forms the outer boundary of the cytoplasm. It is approximately 100 Ångstrom units in thickness and is a fluid or semifluid structure in the living state. When observed through the light microscope it appears homogenous, but under electron microscopy it appears to be an irregularly three-layered structure consisting of lipids, proteins, and water. The usual explanation of the structure is that it is formed of an outer and inner layer of protein sandwiching a more or less central layer of lipid (mainly phospholipid). This is an acceptable structural explanation, but it fails functionally since it does not take into account that lipid molecules appear to be hydrophilic on one end and hydrophobic on the other. If this is true, a single layer of lipid would not permit the cell membrane to accomplish its function. The three-layered structure would have to be two layers of lipid sandwiching noncontinuous layers or aggregations of protein as described by Clinch in Phillis' *Veterinary Physiology*, and illustrated in the following sketch:

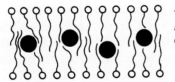

The wormlike structures represent the hydrophilic and hydrophobic ends of the lipoid molecules. The dark blobs represent protein molecules.

—after Phillis, *Veterinary Physiology*, p. 18.

FIGURE 1-1

The cell membrane supports structures such as desmosomes, pits, and microvilli. Its functional complexity is indicated by what it is capable of doing. In a red blood cell, for example, the cell membrane can distinguish between sodium and potassium ions even though these are virtually identical in size and electrical charge. In white blood cells the membrane is capable of passing entire foreign bodies into or out of the cytoplasm. The membrane can distinguish between nutrients and waste products and is capable, in some instances, of one-way transport that will pass materials from the bloodstream to surrounding tissues or to ducts.

The predominance of lipids in the cell membrane determines its permeability to specific substances. Ordinarily lipid soluble materials can cross cell membranes without excessive difficulty. Ionized substances, which are not lipid soluble, do not readily pass through cell membranes, but water, oxygen, and urea cross with minimal trouble. Nothing crosses the cell membrane easily; it is a barrier as well as being the limiting membrane of the cell.

The Cytoplasm

Contained within the cell membrane is the cytoplasm. At one time this was considered to be a homogenous substance called protoplasm, but gradually the structure and composition of the cytoplasm has been better recognized. A number of structures called organelles have been identified in animal cells. Among these are mitochondria, microbodies, lysosomes, centrosomes, Golgi apparatus, endoplasmic reticulum, vacuoles, and pinocytic vesicles. This is a far cry from Corti's description of protoplasm in 1772 that started this investigation.

1. The mitochondria are apparently the "power packs" of the cell and supply energy and substances to support cell function. They are the respiratory centers of the cell, and also extract energy from nutrients. The energy is packaged in the high energy phosphate bonds of ATP (adenosine triphosphate). ATP is furnished by the mitochondria to the energy-consuming structures of the cell.
2. The microbodies are apparently another form of a lysosome.
3. The lysosomes are similar in size and shape to mitochondria, but entirely different in composition and structure. They are the digestive apparatus of the cell and break down large molecules by enzymatic action, reducing their size and composition to units small enough for the mitochondria to handle. The enzymes in the lysosomes can quickly dissolve the cell if the protective membrane around the lysosome is ruptured.
4. The centrosomes (centrioles) are the oldest known organelles, and are clearly visible during cell division (mitosis) when they become the centers of the spindle apparatus that separates the chromosomes of the daughter cells. There is a similarity between the centrosomes and a special structure called the kinetosome which forms flagella or cilia. These latter processes occur on some mammalian cells, and in many one-celled animals.
5. Golgi apparatus has been known to exist almost as long as the centrosomes. Its function, once associated with cell reproduction, is now debatable. There seems to be no difference between Golgi

16
CELLS

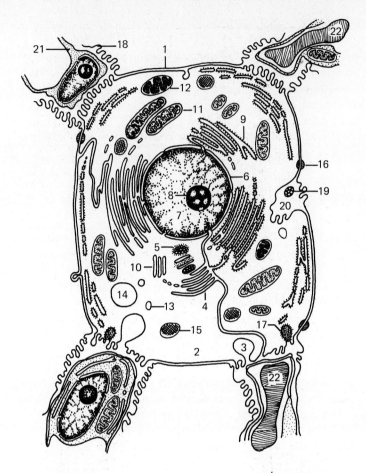

1. Cell membrane
2. Cytoplasm
3. Pinocytic vesicle
4. Golgi apparatus
5. Centrosome
6. Nuclear membrane
7. Nucleus
8. Nucleolus
9. Rough endoplasmic reticulum
10. Smooth endoplasmic reticulum
11. Mitochondrion
12. Lysosome
13. Lipid granule
14. Vacuole
15. Microbody
16. Desmosome
17. Pit
18. Microvilli
19. Packaged granule
20. Canaliculus
21. Capillary endothelial cell
22. Erythrocyte

FIGURE 1-2   *Structures of the cell.*

apparatus and smooth endoplasmic reticulum. The apparatus may serve as a synthesizing factory to produce endoplasmic reticulum and to produce "packaged" materials for export from the cell: materials such as mucus, lipids, enzymes, proenzymes, or hormones.

6. The endoplasmic reticulum is considered to be the skeleton of the cell. It may be either rough or smooth. Rough endoplasmic reticulum has dark granules of RNA (ribonucleic acid) closely applied to the reticular membrane. The RNA granules function in protein synthesis and replication and are called ribosomes.
7. Vacuoles are clear areas in the cytoplasm. They appear to be storage depots for nutrients or water, or they may be relicts of lipoid deposits.
8. Pinocytic vesicles are pocket or flask-shaped depressions in the cell membrane which function in the collection of fluid.
9. Pits, lined with hairlike bristles, are thought to be involved in protein metabolism and uptake of molecules from the bloodstream.
10. Nissl bodies are fine cytoplasmic granules of ribonucleo protein that appear to function in protein synthesis.

The Nuclear Membrane

The nuclear membrane forms the inner boundary of the cytoplasm, separating it from the nucleoplasm. It appears under the electron microscope to be a double membrane with openings (annuli) in the outer layer. The exact function of these holes is not known. Probably they transmit nutrients and fluid to the nucleoplasm. There are some cell nuclei in primitive life forms which do not have a nuclear membrane, but these will not be covered in this text.

The Nucleus

The principal component of the nucleus is nucleoplasm, or nuclear sap. This is a proteinaceous ground substance that supports the various nuclear structures. These structures are the chromatin (linin) filaments and the nucleoli.

The chromatin filaments contain virtually all of the cell's complement of DNA (deoxyribonucleic acid). In the "resting state" the chromatin is distributed diffusely throughout the nucleus giving the DNA greater opportunity to contact nuclear material and RNA from which it replicates itself. Prior to cell division, during the mitotic phase of the cell's existence, the chromatin condenses and coils into tight spiral masses called chromosomes. There is a fixed number of chromosomes for the cells of each species of animal.

The nucleoli are spherical bodies, one or more of which may be found inside the nucleus. Under the electron microscope these appear to be aggregations of ribosomes. Nucleoli contain considerable amounts of RNA and are probably involved in protein and RNA synthesis.

Structures of the Mitochondrion

It is presently believed that the mitochondria form the "power packs" of a cell and that their numbers are indicative of cell activity. Mitochondria apparently convert the energy contained in aliphatic molecules (straight chain molecules) into energy packed in ATP (adenosine triphosphate) which can be utilized by cells to power biochemical reactions. Enzymes which function in oxidative processes appear to be attached to the folds of the inner limiting membrane. Some investigators believe that the mitochondria can reproduce themselves, since they contain a molecule of self-replicating DNA and the equipment for producing protein. This could make the mitochondrion a kind of a cell within a cell and would support the hypothesis that the mitochondria were at one time independent (but symbiotic) organoids.

Mitochondria are the workers in the cellular vineyard. They are among the earliest recognized cellular organelles, yet little more is known about them than that they look somewhat like the accompanying sketch (Figure 1-3). There is still a great deal which must be learned if mankind is ever going to understand the mystery of life.

Undoubtedly, if mankind manages to survive the discoveries in nuclear physics, we will eventually be able to create life from basic chemicals. However, the occasional furor in the media about creating life is still a misnomer. The people who are claiming to be creating life are taking cellular "soups" made by macerating living cells and combining them with nuclei of other living cells, and, thus, from life, are making other forms of life. They are not creating in the sense that they are utilizing nonliving (or not previously living) matter. They are merely adapting what has already been made. This is not to discount the enormous achievements of these researchers; it is merely to put things into their proper perspective.

Some day, when we understand more about the functioning of DNA and RNA, the functional composition of such structures as the mitochondrion, and the functional purpose of the score or more cellular organoids, then we shall be able to create life, and in doing so we shall

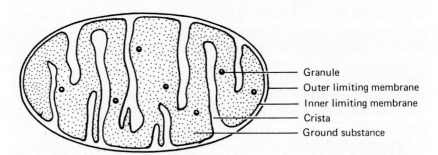

FIGURE 1-3   A mitochondrion.

possess one of the fundamental attributes of Deity. The others will be harder to attain.

PROTOPLASM

Living material, regardless of type of species, is composed of protoplasm. Again, we cannot define what this material is except in the most general terms. Protoplasm is organic matter in which the vital phenomena are manifested. It is the essential component of the cytoplasm and nucleus of cells and is a complex mixture of colloids and solutions that contain proteins, carbohydrates, fats, salts, and water. The exact composition of protoplasm varies between species.

Protoplasm is essentially colloidal. Colloidal systems, by definition, consist of small particles of solute dispersed in a solvent and occur in two types: suspensoids and emulsoids. Suspensoids have no affinity between the dispersed particles and the solvent in which they are suspended. Emulsoids have a high affinity between particles and solvent. Emulsoids can and do undergo distinct changes in their physical state through rearrangement of the colloidal particles; these changes are referred to as sols and gels. A sol is a colloid in the liquid state; a gel is a jelly-like mass similar to hardened gelatin (which is, incidentally, a gel).

PHYSICO-CHEMICAL FACTORS AFFECTING CELLS

The individual cells of the body act, interact, and react with each other in a variety of ways. They are not, with the possible exception of germ cells, independent. The normal functioning of an animal's body involves a continuous exchange of information which is usually chemically or electrochemically controlled. In some instances the control is exercised by chemical compounds (hormones or enzymes) in the interstitial fluid (p. 281) or by a direct transference of chemicals between adjacent cells.

The transport of substances involves a number of physical and chemical phenomena such as filtration, diffusion, osmosis, Donnan equilibrium, and dialysis. These affect protoplasm as well as nonliving chemical mixtures. Filtration is the passage of a liquid through a membrane due to a difference in hydrostatic pressure. Diffusion is the homogenous mixing of two or more miscible liquids due to Brownian movement of molecules. Osmosis is the passage of a solvent through a membrane from an area of low to an area of high molecular concentration. Donnan equilibrium involves the passage of ions through a semipermeable membrane to achieve ionic balance on both sides. Dialysis involves the passage of small molecules through a semipermeable membrane to achieve a molecular balance on each side.

All of these reactions, it will be noted, tend to harmonize the cell and its environment. This appears to be one of the principal functions of life aside from the vital criteria. A cell in harmony with its environment can be said to be at ease. A cell which is not in harmony is "dis-eased."

ORGANIZATION OF CELLS

One-celled animals which are surrounded by their environments readily adjust themselves to it and have a relatively easy time achieving harmony (or equilibrium) with their surroundings. Complex animals, however, must modify the environment to fit the requirements of their cells. A highly organized animal such as a horse or a dog is composed of billions of cells, aggregated into similar groups known as tissues. These tissues in turn are formed into groups working together that are known as organs. Organs in turn are aggregated to form systems and the various systems are integrated to form an animal body. It can be seen that such an organization requires a great deal of cooperation, specialization, and mutual interdependence. Cells in the complex body, therefore, must sacrifice some of their individual potentialities in the interests of the whole; for instance, only germ cells are normally capable of independent reproductive functions. Some somatic (body) cells can reproduce in a limited manner. Others cannot. But none, except germ cells, has the natural ability to produce a new organism.

Recently, a modification of a technique called "cloning" has brought about the theoretical possibility of producing a complete organism from a single body cell, which need not be a germ cell. The technique indicates that most (and possibly all) somatic cells have an inherent but masked capability for reproducing a complete organism. Under natural conditions this situation does not occur. Uncontrolled growth, however, is a relatively frequent occurrence and results in neoplasia (tumor, cancer). Yet in this quasi-independent reproduction, the neoplastic cells are dependent upon bodily organization for survival and will ultimately die when their uncontrolled growth causes interference with the vital functions or structures of the host. In no case yet known has a cancer survived the death of its host without artificial protection.

For the mechanism of cell reproduction see p. 418ff. The exact reasons why a cell reproduces itself are unknown. Undoubtedly a number of factors are involved which can be either normal or pathologic. These, moreover, can probably act individually or collectively. Drugs such as phytohemagglutinin can stimulate mitosis and in all probability mechanical factors such as wear and tear; hormonal factors such as growth hormone, erythropoietin, thyroxin, prolactin, and progesterone; homeostatic factors such as stress or increased bodily demand; and dietary factors and other stimuli that are either causative or influential in triggering the reproductive process in cells. Some kinds of cells are more capable of reproduction than others, and some are incapable of reproducing at all. The reasons for these responses are being studied, but considerable work yet remains to be done before the answers are known.

BIOLOGICAL RHYTHMS

It should be obvious that order and organization are of great importance in the functioning of a complex life form. Yet there is another factor which is not so apparent but which also exercises profound effects and should always be taken into consideration when dealing with living things. This factor is often called the biological clock or "temporal rhythm." Persons unacquainted with the importance of this aspect of life have a tendency to disregard the system of cycles that order the lives of most living organisms. Experience, however, indicates that noninterference with cyclic responses and stimuli is necessary to get the most efficiency from an organism. Those who do not take biological rhythms into account are seldom successful in their efforts to understand in depth the functional aspects of life.

We know of biological rhythm from our own lives, although mankind is a supremely adaptable species and can fit into numerous environments and situations. Nevertheless, there are some rhythms that remain. The human respiration rate averages about 16 breaths per minute. Cardiac rhythm in a resting state is about 70 beats per minute. The human life pattern is normally diurnal with the "day" divided roughly into 16 waking and eight sleeping hours that come in fairly regular succession. Interruption of this pattern can create problems if rapid or sudden dislocations occur, such as airplane flights into different time zones. In addition, there are seasonal rhythms. Shakespeare was quite right when he wrote about the winter of our discontent, and Keats was merely stating a biological rhythm when he wrote

> *In the spring a lovelier iris lightly glints the burnished dove:*
> *In the spring a young man's fancy lightly turns to thoughts of love.*

But cycles are even more important among other life forms. Plants are rigidly regulated according to seasons and length of days. Many life forms are strictly cycled according to the presence or absence of sunlight (diurnal or nocturnal). Seasons of the year affect migratory patterns, sexual activity, reproduction, coat color or plumage, hibernation or estivation patterns, communal activity, and a horde of other functions. Everywhere one looks in the world of living things, cyclic rhythm is present. The real wonder is why it is so often ignored.

Understanding of biological rhythms and utilizing them is of prime importance in managing domestic and wildlife populations, treating parasitic diseases, handling stress, and understanding stressful situations and responses. The list is long, and the penalties for ignorance or unconcern are severe. Ignoring or disturbing biological cycles can produce some profound and unexpected effects, most of which are still unknown and many of which are fertile areas for research and investigation. The principles of biological rhythms and their ramifications need a great deal more attention than has presently been given to them.

AGING

In the due course of time (with the possible exception of one-celled life forms which theoretically live forever) cells grow old and die, and the tissues, organs, systems, and organisms of which they are a part also age and die. The process of aging and death has occupied the attention of man for thousands of years, since man alone of all the animals on this world has foreknowledge of physical decline and death.

Until recently, thousands of investigations into the phenomena of aging were conducted with little or no success, and the mechanism of the aging process was as much a mystery as in prehistoric times. It is a tribute to the explosive nature of modern research that this state of affairs is at last being remedied, and new discoveries are apparently uncovering useful information that will not only prolong active lifespan, but may also add to the total number of years of life. Although the predictions of several decades ago that aging would be slowed or stopped and that lifespan would be doubled or trebled have not yet come to pass, the outlook is not as bleak as it was.

Although medicine has managed to bring longer lives to a greater percentage of people and pet animals, and has increased their activity, the biblical optimum of three score and ten is still the general limit for humans. Some spectacular general improvement in lifespan of laboratory mammals has been recorded, and this has stimulated the hope that life can be prolonged.

Of course, there have been stories of imminent breakthroughs before, and the latest ones may prove to be mirages, but the prospects for success seem good.

The major roadblock to understanding the phenomenon of aging is that we have never known too much about how the process works. We know what aging is, but we do not know the mechanisms which produce it. Theoretically, mammals should live much longer than they do. The discipline of gerontology, or the study of aging, was established to discover why they do not. The study, however, is still in its infancy, and even today gerontologists cannot agree upon a commonly acceptable definition of aging, let alone what causes it. But that may change in a few years.

Some things are known:

1. Aged tissues are more susceptible to stress and to damage by pathogenic organisms, metabolic disease, and cancer.
2. There are observable variations in the structure and function of body organs and systems which can be traced to impaired function of cells.
3. Radiation can, through ionization and alteration of RNA, DNA, and the genetic structure of cells, cause definitive changes in cell

structure, and these changes may be passed on to subsequent generations of cells.
4. Rapidly dividing cells tend to throw off or eliminate changes by radiation and other mutagens.
5. Cell death from mutagens probably will not occur if nutrition is not affected and the DNA molecule is not too severely damaged.
6. Damaged DNA can produce cell changes that can deplete vital stores of nutrients, RNA, and protein.
7. Cells that rarely or never divide retain mutagenic effects, playing a role in aging as "detriments," which exacerbate or trigger other damage and may further the aging process.

This isn't very much knowledge, but it is about all the solid information that exists. It has given rise to two theories of aging that are of interest mainly because of their historical qualities. It should be remembered that theories of aging are not necessarily basic to therapy. They are an attempt to explain why the phenomenon occurs, not to cure or relieve it. Yet these theories are important because they underlie methods of research that may someday produce answers to the problem of old age.

The oldest of the two theories has been called the "wear and tear," or physiological aging theory. It was developed from the obvious fact that all things wear out in the due course of time. Extended to living things, the theory states that each stress to which cells or tissues are subjected exacts a toll, and the cumulative toll is reflected in a loss of function. A variation of this theory states that cells are endowed with a finite capacity for forming and/or utilizing enzymes. When this capacity is exhausted, the cell dies, and when enough cells are dead or damaged, the organs or systems fail.

According to its proponents, each disease or biological insult exacts its price in shortened lifespan. This is fine for the purpose of raising contributions for "war" on disease, tobacco, alcohol, drugs, and other metabolic poisons and pollutants. However, experimental evidence so far has not supported the idea too well. Surviving mice subjected to LD^{50} chemical stresses live as long as healthy controls. However, mice subjected to LD^{50} doses of radiation have significantly shorter lifespans than the controls. If we except radiation, the cumulative toll part of the theory is not valid. Nevertheless, complex organisms do decline and die in a reasonably predictable time span.

Another variation of the "wear and tear" theory states that although there are few stresses from which the body cannot recover completely, if an organ or system is severely damaged it apparently triggers damage to other organs or systems which add to the damage, and the cumulative effects of these various detriments can produce age. This may well be, but the detriments are so varied in their natures and effects, that a generally applicable rationale of damage that proves this premise has not yet been reported.

The most popular current theory of aging is the "mutation" or pathological aging theory. This states that body cells gradually, through spontaneous mutation and exposure to mutagens such as radiation and chemicals, accumulate harmful or lethal genes which cause death or impaired function of cells, and eventually result in deterioration and death of the organism. All cells, even ones which do not multiply, i.e., nerve cells, can be affected, and in time the entire organism becomes riddled with cellular disorder and inefficiency.

Experiments could be designed to test this theory. This would entail a tremendous expense, and no prospect of such financial support exists.

Experiments on a less expensive scale indicate that there may be some validity to this concept, although they do not explain the long-lived exceptions other than on a basis of chance or possibly hereditary resistance. They do not explain why most long-lived enclaves or individuals are found in mountainous areas, which are considerably more exposed to cosmic radiation and thermal stresses.

Although most of today's students of aging have stopped searching for the elixir of life and have turned their efforts to more pragmatic channels, such as finding means to prolong vigor or increase meaningful existence up to the brink of death, a few workers still look for the ultimate solution to the problem of indefinitely prolonging lifespan.

A breakthrough may have been made at the National Institute of Aging Gerontology Research Center in Baltimore, Maryland. Electrochemical controls in the brain and compounds such as dopa, dopamine, norepinephrine, and serotonin seem to hold some promise in retarding age and increasing lifespan. This work is still in its infancy.

The action of dopamine in prolonging life is remarkable. In aged mice it has been noted that there is a depletion of dopamine in the hypothalamus and the corpus striatum. The reason why this occurs is unknown. However, massive doses of dopamine (160 mg/day) add 10 percent to the lifespan of mice. The drug, unfortunately, has unpleasant and dangerous side effects in some humans. Much work yet remains to be done in this area, but a new route of exploration of the old age syndrome through control of general metabolism via the pituitary, and control of muscle tone and metabolism via the corpus striatum has been established.

The immunologic system (p. 252ff) has become another area of gerontologic research. The lymphocytes are proving to be one of the most fruitful research fields in animal biology, and their possibilities seem far from exhausted. The function of the thymus and bone marrow in aging has been indicated by successful transplants of young marrow and thymus into old mice. Improvements of the immunological system of the old mice was spectacular and some of the mice lived 30 percent beyond their normal lifespan.

A third line of investigation is based upon the theoretical presence of a death hormone. This idea developed from studies of anadromous salmonids which pass from vigor to senility and death in a matter of weeks. In fish it is an excess of ACTH, a pituitary hormone, that causes death. Although this spectacular end of the mature Pacific salmon is not paralleled in mammals, the fish deaths have stimulated research into the possibility of a similar lethal hormone existing in mammals. Results so far indicate that a substance produced by the pituitary decreases tissue response to thyroxin, a general metabolic hormone.

None of this work is definitive, but it indicates promising avenues of research that may break the roadblock in the study of old age that has persisted for the past few decades.

Goethe once remarked "What man can conceive man can achieve," and perhaps man's dreams of immortality may some day come true.

chapter 2
Bone and Cartilage

BONE

The basic material of the skeleton is bone and cartilage. In an adult mammalian skeleton, bone is by far the largest component. Bone is a dense substance composed principally of two materials, tricalcium phosphate ($Ca_3(PO_4)_2$) and ossein, a gelatin-like protein. The ossein imparts resiliency and flexibility, reduces the danger of breakage, and forms the medium through which the bone derives most of its nourishment. The tricalcium phosphate gives strength and rigidity.

    Bones are not static masses of calcium compounds and proteins. They are living, constantly changing structures that grow, shrink, and adapt during the entire life of the animal. If bone is severely injured it can die, and, like any other dead tissue, it is subject to degenerative processes. While it is in the living state, bone must be properly supplied with nutrients in order to remain alive. Each bone, therefore, is supplied with one or more blood vessels, lymph vessels, and nerve trunks. These penetrate the bone through holes called nutrient foramina, ramify through the central portion of marrow cavity, and finally penetrate into the hard structure of the bone itself.

    In addition to being the framework of the body, bones perform certain other functions. They act as blood forming organs, protective coverings, and aids to locomotion and respiration.

CARTILAGE

In mammals, cartilage is a gristly flexible tissue usually found attached to, or associated with, the joint surfaces of bones or forming attachments between certain bones. Occasionally special forms of cartilage may be inserted between the joint surfaces of bones or form a supporting framework for softer structures, such as the trachea and bronchi, or furnish the skeletal base for the external ear. It is the substance from which the majority of the bones of the skeleton develop.

In the body, cartilage occurs in three forms: hyaline cartilage, elastic cartilage, and fibrocartilage. All of these forms consist of specialized cells called chondrocytes (Gr.*chondro*—cartilage; *cyte*—cell) embedded in a dense mucoprotein matrix. Cartilage is softer than bone, more flexible and resilient, and less capable of regeneration once it has been damaged.

Hyaline (Gr. *hyalos*—glass) cartilage is the commonest and most prevalent form. It is a translucent, bluish-white substance usually found on the joint surfaces of movable bones. It is composed of chondrocytes and matrix only.

Elastic and fibrocartilage are found in special locations in the body. These forms contain yellow elastic tissue or white fibrous tissue in addition to the chondrocytes and matrix. A thin membrane called the perichondrium (Gr. *peri*—surrounding; *chondro*—cartilage) covers all cartilage.

BONE FORMATION AND STRUCTURE

Essentially, bone is derived from two sources, cartilage and membrane, resulting in the names of "cartilage bone" or "membrane bone" which are sometimes used to describe adult structures. Cartilage bone forms the majority of the skeleton, while membrane bone is restricted to certain bones of the skull.

In the embryo, cartilage bars are formed early in development in the positions which will ultimately be occupied by true bone. Within these bars of cartilage, one or more centers of ossification develop, where special cells known as chondroclasts (Gr. *chondro*—cartilage; *clast*—break), osteoblasts, (Gr. *osteo*—bone; *blast*—forming), and osteoclasts work to transform the cartilage to true bone. Chondroclasts break down the cartilage, leaving gaps or spaces which are filled with primary bone by the osteoblasts. Later, this primary bone is removed by osteoclasts, and true or secondary bone is laid down by a secondary series of osteoblasts. At birth, most of the cartilage bars have been partly replaced by true bone, and from birth on the major changes are growth and secondary bone formation.

Membrane bone is formed on the outside of the fibrous tissue membranes covering the developing brain and skull. The process is one of simple bone formation and does not involve the more complex procedures of destruction and replacement of cartilage.

Bone Classification

Bones may be classed according to shape as long, short, irregular, or flat. The terms are descriptive enough not to require further elaboration except in the case of flat bones. In this book, flat bones will be considered as the membrane-derived bones of the skull only. Cartilage bones may occasionally be flattened, but due to their origin and manner of development the term "flat" is not precisely applicable and tends to cause confusion.

Structure of a Long Bone

Essentially a long bone consists of three primary parts: a shaft and two ends. The shaft is the diaphysis while the ends are epiphyses (singular: epiphysis). Between the shaft and ends of immature bone is a narrow band of cartilage called the epiphyseal plate. In young animals (particularly sheep) the distal epiphyseal plates of the metacarpal and metatarsal bones are referred to as "break joints" by packers and meat cutters. Epiphyseal plates are the locations where growth in length of bones takes place. In mature animals the epiphyseal plates are replaced by bone.

The ends of long bones are covered with articular (hyaline) cartilage at their joint surfaces. The remainder of the bone is covered with a closely adherent fibrous membrane called the periosteum (Gr. *peri*—surrounding; *osteo*—bone). The periosteum is lined on its inner surface with osteoblasts which are generally inactive in the adult, and are called osteocytes (*osteo*—bone; *cyte*—cell). In young animals, however, the osteoblasts are functional and are responsible for increasing the diameter of the shafts and ends of the bone. One or more holes, or nutrient foramina, penetrate the shafts and ends of a long bone. These holes contain blood and lymph vessels and nerves.

The shaft of a long bone is hollow in the center. This hollow is called the marrow cavity. The marrow cavity is lined with endosteum (Gr. *endo*—within; *osteo*—bone), a fibrous membrane similar in structure and function to the periosteum, and is filled with a fatty substance called marrow.

Marrow

Marrow is of four types: red, yellow, white, and gelatinous. These types are generally dependent upon the age of the animal, red bone marrow being found in young animals while gelatinous marrow is found in the aged. Some species, because of their short lifespan, never have any form of marrow other than red. Rats, mice, hamsters, and other small rodents are the commonest species in which the red bone marrow persists in the long bones throughout life.

Red marrow is a blood forming organ and is soft and well filled

BONE AND CARTILAGE

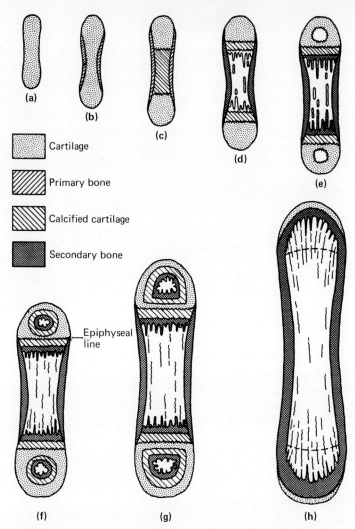

(a) Cartilage bar, (b) Periosteal bone collar, (c) Primary center of ossification, (d) Ossification of diaphysis, (e) Appearance of ossification centers in epiphyses, (f) Ossification of epiphyses (note epiphyseal line), (g) Growth of bone and further ossification of epiphyses, (h) Adult bone

FIGURE 2-1 *Development of a long bone. (Adapted from Maximow and Bloom,* A Textbook of Histology.*)*

with vessels and cellular elements. Yellow marrow is more fatty in nature, is harder, and does not perform blood forming functions, although it is capable of reverting to red marrow in times of extreme stress. White marrow is principally hard fatty tissue; it is nonfunctional as a blood forming organ and is incapable of reversion. Gelatinous marrow is a degraded semi-fluid form found in aged and senile animals. The age distinction of the various types of bone marrow is more characteristic of the human being than any other species, although animals which have an average lifespan in excess of 10 years will usually possess the age-related changes mentioned above.

Spicules and Trabeculae

The marrow cavity disappears near the ends of the shaft and its place is taken by tiny spines (spicules) and plates (trabeculae) of spongy bone. This loosely organized meshwork of bone fills the ends of diaphyses and the centers of epiphyses. It functions as a shock absorbing mechanism, and its arrangement gives great strength accompanied by lightness. Cross or longitudinal sections through the ends of long bones will reveal, even to persons unacquainted with engineering principles, that the spicules and trabeculae are arranged along definite lines of thrust and contribute tremendously to the strength of the bone.

It is interesting to note the development of spicules and trabeculae in improperly set fractures. The marked alteration in lines of thrust and directions of stress and tension produces a readily observable deviation in the arrangement of spicules and trabeculae, and provides contributory evidence that the spicules and trabeculae have a load bearing function in long bones.

Cancellous Bone

In the ribs, sternum (breastbone), and vertebrae of mammals, spicules and trabeculae fill the central portion of the bone, enclosing small pockets of red bone marrow which do not regress with age. The red marrow of the ribs and sternum is the principal source of blood cells to replace those destroyed by normal daily wear and tear. These bones, and some others with similar medullary (middle) portions, are sometimes called spongy marrow, or cancellous marrow bones. The term, "cancellous bone," which has been applied to these structures as a description of the entire bone, is incorrect unless only the marrow cavity is being considered. The term, "cancellous," is more properly confined to the formation of loosely organized bony outgrowths and processes which are the result of disease.

The Haversian System

The compact bone of animals larger than a rabbit, when observed under the microscope, reveals lamellated tubular structures which form the Haversian system. This type of bone organization is not found in spicules and trabeculae, nor is it found in birds or small mammals such as mice and rats. The various components of the Haversian system such as lacunae, canaliculi, and Volkmann's canals are more properly a study of microscopic anatomy or histology. For that reason it will only be mentioned here to indicate that the system of ducts and canals furnishes nutrition to the osteocytes buried in the compact bone structure.

chapter 3
The Skeleton

The skeletal system includes the bones, cartilage, teeth, and joints. The specific studies are osteology (Gr. *osteo*—bone; *logy*—study of), chondrology (Gr. *chondro*—cartilage), odontology (Gr. *odonto*—tooth), and arthrology (Gr. *arthro*—joint). Logically, it is the first system to be studied, for the skeleton forms the foundation upon which all the other systems are laid or built.

The skeletal system is divided into three parts: the appendicular skeleton, the axial skeleton, and the visceral skeleton. The appendicular skeleton consists of the bones of the limbs. The axial skeleton is composed of the skull, the vertebral column, the ribs, and the sternum. The visceral skeleton consists of such bones as may be developed in the soft tissue of certain organs or parts, such as the os cordis (Lat. *os*—bone; *cordis*—of the heart) in the ox, the os penis in the dog, mink, beaver, and other animals, and the scleral rings in the eyes of birds.

THE APPENDICULAR SKELETON

The appendicular skeleton is attached to the axial skeleton by means of two bony structures, the pectoral (shoulder) girdle and pelvic (hip) girdle.

The pectoral girdle attaches the forelimbs to the body and is incomplete in domestic mammals. A complete pectoral girdle consists of six bones, the right and left scapulas, coracoids, and clavicles. One or more of these pairs is usually missing in mammals. Climbing and burrowing mammals usually possess a scapula and clavicle, coursing and grazing mammals usually possess a scapula only, or, if a clavicle or coracoid is present, it is either small and/or incomplete or is a process attached to a larger bone. Examples of these two general exceptions are the clavicular remnants found in dogs and cats and the coracoid process on the scapulas of various members of the equine family. All three pairs of bones of the pectoral girdle are seen in birds, reptiles, and some of the lower mammalian forms. In animals which possess only a scapula, the attachment of the forelimb (pectoral limb) to the body is by means of fleshy and tendinous structures. A bony attachment of the forelimb to the body is found only in animals which possess two or more pairs of functional bones in the pectoral girdle.

The pelvic girdle usually consists of three pairs of bones, the right and left ilium, ischium, and pubis, and is present and complete in all domestic mammals. It connects the hind (pelvic) limbs to the body. The bones of the pelvic girdle are usually fused into the right and left innominate bone or os coxa. The two innominate bones may be fused along the mid-ventral line to form the pelvic ring (pelvic girdle, "H" bone, hip bone). In some aquatic mammals and in birds the pelvic girdle does not form a complete ring.

The hindlimb (pelvic limb) is attached to the pelvis by a ball and socket joint. The pelvis in turn is firmly attached to the spinal column either by direct bony junction to the sacrum, as in the adult horse, or by strong ligament connections, as in the cow, sheep, dog, and cat.

The pelvic limb, the main propulsive limb of most terrestrial animals, requires this firm attachment to allow the powerful posterior muscles to act with maximum efficiency. The pectoral limb functions more to support the body than propel it; hence a strong bony attachment is not necessary. In burrowing and climbing animals where the pectoral limbs have more functions than support and propulsion, a bony attachment to the axial skeleton (the clavicle) is usually present.

A comparison of forelimbs and hindlimbs reveals that the bones of one limb have a corresponding or analogous bone in the other, with the exception of the kneecap or patella.

A number of sesamoid bones are present in both limbs but are not listed in this table because their location and number vary between (and within) species.

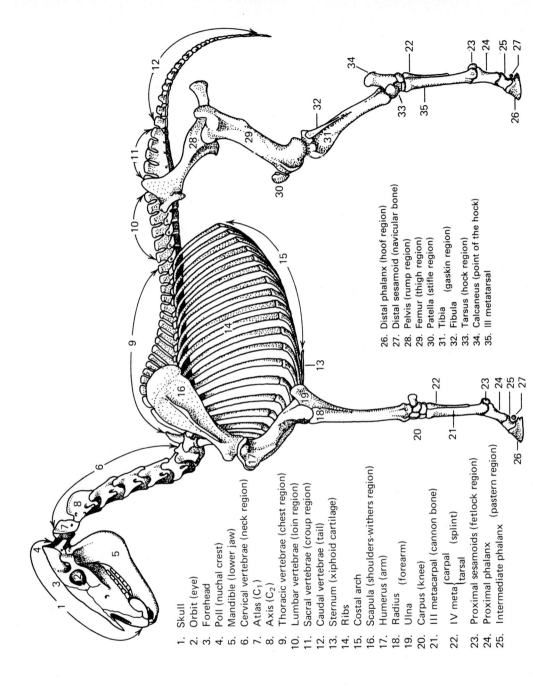

1. Skull
2. Orbit (eye)
3. Forehead
4. Poll (nuchal crest)
5. Mandible (lower jaw)
6. Cervical vertebrae (neck region)
7. Atlas (C_1)
8. Axis (C_2)
9. Thoracic vertebrae (chest region)
10. Lumbar vertebrae (loin region)
11. Sacral vertebrae (croup region)
12. Caudal vertebrae (tail)
13. Sternum (xiphoid cartilage)
14. Ribs
15. Costal arch
16. Scapula (shoulders-withers region)
17. Humerus (arm)
18. Radius (forearm)
19. Ulna
20. Carpus (knee)
21. III metacarpal (cannon bone)
22. IV meta {carpal (splint) / tarsal
23. Proximal sesamoids (fetlock region)
24. Proximal phalanx (pastern region)
25. Intermediate phalanx
26. Distal phalanx (hoof region)
27. Distal sesamoid (navicular bone)
28. Pelvis (rump region)
29. Femur (thigh region)
30. Patella (stifle region)
31. Tibia (gaskin region)
32. Fibula
33. Tarsus (hock region)
34. Calcaneus (point of the hock)
35. III metatarsal

FIGURE 3-1   Skeleton of the horse.

35

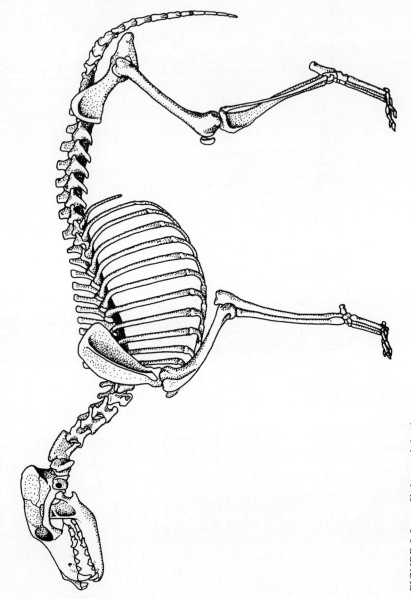

FIGURE 3-2 Skeleton of the dog.

Pectoral Limb (Forelimb)		Pelvic Limb (Hindlimb)	
1. Shoulder	Scapula Clavicle† Coracoid†	1. Rump or Hip	Ilium Ischium Pubis
2. Arm	Humerus	2. Thigh	Femur
3.		3. Kneecap	Patella
4. Forearm	Radius Ulna	4. Leg or Shank	Tibia Fibula†
5. Wrist ("Knee")	Carpals*	5. Ankle (Hock)	Tarsals*
6. Manus (Forefoot)	Metacarpals* Phalanges*	6. Pes (Hindfoot)	Metatarsals* Phalanges*

*One or more may be absent or fused.
†May be absent.

The Scapula

Classification: Modified long bone.

Location: The scapula covers the dorsal portion of the first four or five ribs. The long axis extends obliquely from the fourth thoracic spine to the sternal end of the first rib.

Description: The scapula of the horse has two surfaces, three borders, and three angles.

The surfaces are lateral and medial. The lateral surface is divided by the spine of the scapula into two unequal parts and possesses the following structures: the tuber spinae, which is located on and above the middle of the spine of the scapula; the supraspinous fossa, a depression cranial to the spine; the infraspinous fossa, a depression caudal to the spine; a nutrient foramen caudal to the spine and near the neck; and a vascular groove below the foramen and on the neck. The medial or costal surface is concave along its long axis. This concavity is called the subscapular fossa. Its borders converge to a point dorsally and separate two triangular areas. The lower third of the medial and lateral surfaces are marked by vascular grooves.

The borders are cranial, caudal, and dorsal. The cranial border is convex and rough in its dorsal portion and concave and smooth ventrally. The caudal border is slightly concave, thick in the upper third, rough in the middle, and smooth ventrally. On the dorsal (vertebral) border is the suprascapular cartilage which becomes ossified or partly ossified in older animals.

The angles are cranial, caudal, and glenoid. The cranial angle is formed by the junction of the dorsal and cranial borders. The caudal angle is formed by the junction of the dorsal and caudal borders. The glenoid angle is formed by the caudal and cranial borders. It is attached

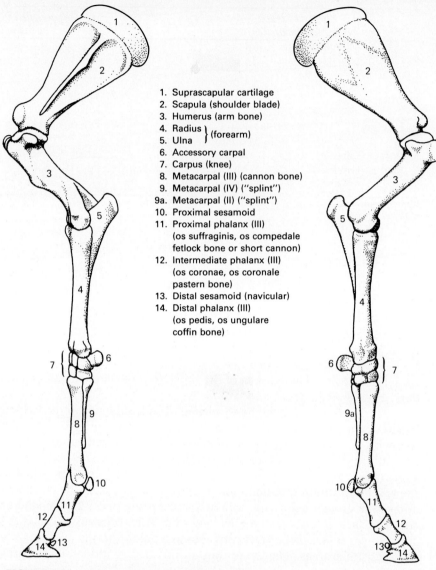

FIGURE 3-3   Bones of the pectoral limb—horse.

to the body by the neck and bears a glenoid cavity for articulation with the head of the humerus.

In the horse scapula there are several structures which are either unique in equines or differ appreciably from those in other domestic animals, i.e., the glenoid notch which cuts the rim of the glenoid cavity cranially; the tuber scapulae (supraglenoid tubercle), a prominence cranial to the glenoid cavity which is relatively large in horses; and the coracoid process, a small projection medial to the tuber scapulae.

The acromion, a projecting mass of bone located on the distal end of the spine of the scapula, is not found in the horse but is present in the cow, sheep, pig, dog, man, and other animals. It is particularly well developed in those species which possess a clavicle or collar bone.

COMPARATIVE ANATOMY OF THE SCAPULA

Cattle. Compared to the horse the scapula of cattle is a more perfect triangle. It is wider at the proximal end and narrower at the distal end. The spine is more prominent and set farther cranially which makes the supraspinous fossa narrow and tapering to a point ventrally. The spine bears an acromion on its distal end. The glenoid cavity does not possess a notch, and the tuber scapulae and coracoid process are much smaller than in the horse.

Sheep. The scapula of the sheep differs from the cow in size but only in minor structural details. The coracoid process is absent. It is virtually impossible to tell the scapula of a sheep from those of deer and goats by morphology alone.

Swine. The scapula of the pig is very wide in proportion to its length and possesses a somewhat rhomboid shape due to the convexity of the cranial border. The spine is triangular, and curves backward. It bears a large tuber in its middle portion and a small acromion distally. The suprascapular cartilage is not as extensive as in the horse and cow. The neck is well marked, the tuber scapulae small, and the coracoid process absent. The glenoid cavity is circular and does not possess a notch.

Dogs. In the dog the scapula is relatively narrow in proportion to its length. The cranial border is markedly convex, although not as much so as in the pig. The spine is thick, centrally located, and curves backward. It increases in size ventrally and has an acromion at its distal end. The tuber spinae is insignificant. The neck is short and thick with a well-developed tuber scapulae. There is no coracoid process or glenoid notch. The suprascapular cartilage is reduced to a thin band, and the caudal border is thickened in the proximal (dorsal) part. On the medial surface the subscapular fossa is shallow.

Cats. The scapula of the cat is more similar in outline to that of the pig than the dog. The cranial border is strongly convex, and the caudal border is relatively straight. The spine is straight or slightly concave cranially and is located centrally. It increases in size ventrally and bears on its ventral portion a tuber spinae and acromion. The tuber scapulae curves medially to form a coracoid process. The glenoid cavity is roughly circular and does not possess a glenoid notch.

THE SKELETON

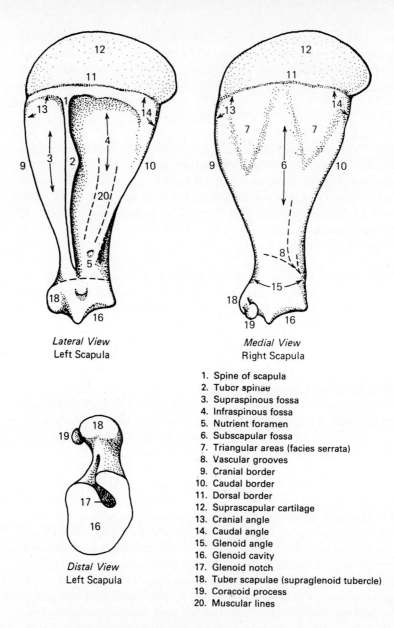

FIGURE 3-4 Scapula of the horse.

1. Spine of scapula
2. Tuber spinae
3. Supraspinous fossa
4. Infraspinous fossa
5. Nutrient foramen
6. Subscapular fossa
7. Triangular areas (facies serrata)
8. Vascular grooves
9. Cranial border
10. Caudal border
11. Dorsal border
12. Suprascapular cartilage
13. Cranial angle
14. Caudal angle
15. Glenoid angle
16. Glenoid cavity
17. Glenoid notch
18. Tuber scapulae (supraglenoid tubercle)
19. Coracoid process
20. Muscular lines

PHYSIOLOGY OF THE SCAPULA

The scapula serves as the bony support and attachment for a sling of muscles and connective tissue that in turn supports the cranial parts of the body. It also serves to transmit motion of the forelimbs to the body. Hence there is a certain uniformity of shape in this structure throughout the Class Mammalia. Regardless of whether it is or is not directly connected to the axial skeleton, the scapula is located between layers of soft tissues which help determine its basic shape and surfaces.

There are sound physiological reasons for its shape. Situated as it is between layers of muscle, the bone must be flattened to slide easily back and forth along the muscular layers without causing distortion of the muscles or changes in direction of angular pull and consequent reduced mechanical efficiency of the limb. The medial surface must be smooth, since it slides for some distance across a surface composed of muscle and connective tissue. The lateral side is rough to provide for attachment of the muscles of the upper leg and shoulder and of the suspensory apparatus, which holds it to the body and permits it to function in support and locomotion.

At this time it should be pointed out that it is a fairly constant rule that the superficial (closet to the surface) or lateral portions of bones will be rough and the deep (farthest from the surface) or medial portions smooth. This is particularly true of those bones which are close to or part of the axial skeleton. In the limbs, the lateral sides of the bones usually have the roughest surfaces and largest prominences.

Humerus

Classification: Long bone.

Location: It lies between the scapula and the radius and ulna.

Description: The humerus of the horse consists of a shaft and two extremities (proximal and distal). The shaft has four surfaces: lateral, medial, cranial, and caudal. The lateral surface is smooth and spirally curved to form the musculospiral groove. The medial surface is straight in length and rounded from side to side. Near the middle of the medial side of the shaft is the teres tuberosity, and below, in the distal third, is a nutrient foramen. The cranial surface is flattened and is wide and smooth above and narrow and rough below. It is separated from the lateral surface by a crest which supports the deltoid tuberosity. The caudal surface is smooth and rounded.

The proximal extremity consists of a head, neck, and two tuberosities (lateral and medial), separated by the bicipital groove (intertuberal groove). The head is a smooth, convex, articular surface in front of which is a fossa containing a number of small foramina. It is connected to the shaft by the neck. The lateral tuberosity is larger than the medial and is located cranio-laterally. It consists of cranial and caudal parts. The cranial part forms the lateral boundary of the bicipital groove (intertuberal groove). The bicipital groove is cranial to the head and is bounded by the cranial parts of the lateral and medial tuberosities. Cranially it is divided by an intermediate ridge.

The distal extremity is an oblique articular surface consisting of

THE SKELETON

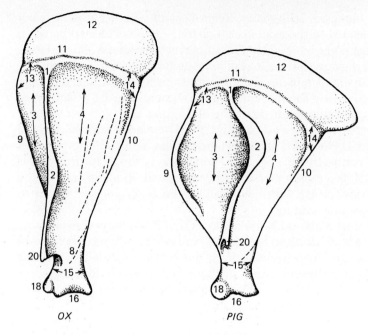

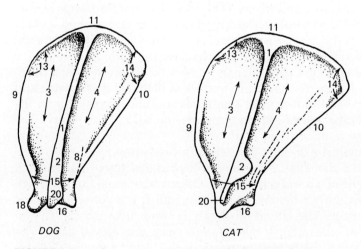

1. Spine of scapula
2. Tuber spinae
3. Supraspinous fossa
4. Infraspinous fossa
8. Vascular grooves
9. Cranial border
10. Caudal border
11. Dorsal border
12. Suprascapular cartilage
13. Cranial angle
14. Caudal angle
15. Glenoid angle
16. Glenoid cavity
18. Supraglenoid tubercle
20. Muscular lines
21. Acromion

FIGURE 3-5 *Comparative anatomy—scapula—lateral views.*

two roughly cylindrical articular surfaces* (trochlea and capitulum) separated by a groove, a lateral and medial epicondyle, a coronoid fossa, and an olecranon fossa. The capitulum is smaller and projects farther ventrally than the trochlea and contains a shallow groove. The coronoid (radial, supracondyloid, supratrochlear) fossa is a depression located above the articular surfaces and is bounded laterally and medially by the epicondyles.

COMPARATIVE ANATOMY OF THE HUMERUS

Cattle. The humerus of cattle is smoother than that of equines. The musculospiral groove is shallower and the deltoid tuberosity less prominent. The teres tuberosity is smaller. The lateral tuberosity of the head is quite large and curves medially over the intertuberal groove, which is undivided. The distal articular surface is set somewhat obliquely to the shaft. The articular surfaces, coronoid fossa and olecranon fossa are large and well defined.

Sheep. The humerus of sheep is smaller but otherwise similar to that of bovines. The deltoid tuberosity is closer to the proximal end and not as prominent. The lateral tuberosity of the head is smaller and does not extend as far medially over the intertuberal groove.

Swine. The humerus of swine is strongly curved and short. The shaft is laterally compressed, being flattest on the medial side. The musculospiral groove is shallow, and the deltoid tuberosity is relatively small. The teres tuberosity is absent. The head is relatively large, and the lateral tuberosity is extremely prominent, and divided into cranial and caudal parts. The cranial part curves sharply medially over the intertuberal groove. The distal articular surface is thrust outward and downward from the shaft. The distal articular surfaces (trachlea and capitulum) are of nearly equal size. The coronoid fossa and groove are prominent, and the olecranon fossa is narrow and deep. Occasionally the plate of bone which closes the bottom of the olecranon fossa and separates it from the coronoid fossa will be perforated, forming a supracondyloid foramen.

Dogs. The humerus of the dog is long and slender, has a slight spiral twist, and is somewhat S-shaped along its long axis. The deltoid

---

*The capitulum is called the capitellum or radial head in humans. The medial articular surface is called a trochlea in recent veterinary texts, although the term is not accurate for equines or ruminants. It is, however, a reasonably accurate description for cats and dogs. The obsolete term, "condyle" has thus been expanded to include the various fossae, which were separately named in older texts. The terms lateral and medial epicondyles, however, have been retained, which makes the terminology somewhat confusing since there are now no condyles to which the bony prominences are attached. It might be better to think of these structures as the "epicapitulum" and the "epitrochlea."

THE SKELETON

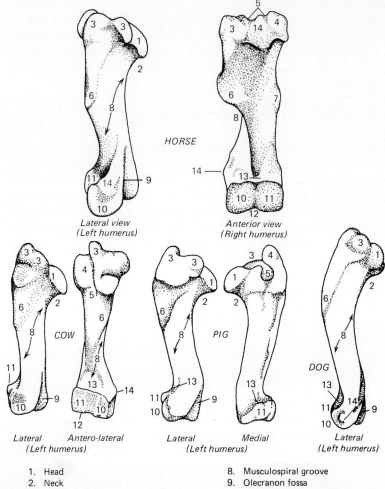

1. Head
2. Neck
3. Lateral tuberosity (greater tubercle)
4. Medial tuberosity (lesser tubercle)
5. Intertuberal (bicipital) groove
6. Deltoid tuberosity
7. Teres tuberosity
8. Musculospiral groove
9. Olecranon fossa
10. Lateral condyle (capitulum)
11. Medial condyle (trochlea)
12. Intercondyloid groove
13. Coronoid (supracondyloid, radial, supratrochlear) fossa
14. Intertuberal eminence (intermediate tubercle)

FIGURE 3-6 *Comparative anatomy—humerus.*

tuberosity is small. The teres tuberosity is missing. The head is long and curved backward. The intertuberal groove is shallow. The capitulum on the distal surface is prominent. The bony partition between the coronoid and olecranon fossae is usually absent, forming a large supratrochlear foramen.

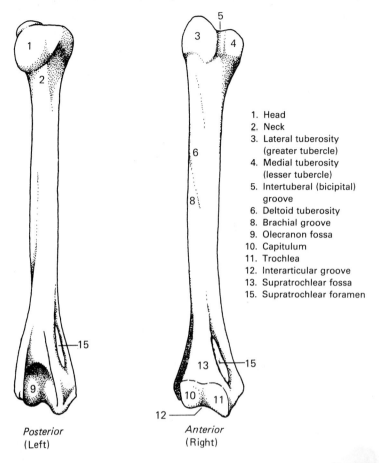

FIGURE 3-7   Humerus of cat.

Cats. The humerus of the cat is similar to that of the dog. The distal articular surface has a more prominent interarticular groove than does that of the dog. A supratrochlear (epicondyloid) foramen is present above the medial condyle. It is not similar to the structure of similar name in the dog and is found in most members of the cat family.

PHYSIOLOGY OF THE HUMERUS

The function of the humerus is threefold. It serves as a connecting link between the scapula and the distal portions of the limb. It furnishes attachments for the origin and insertion of the major muscles that control the motions of the upper and lower parts of the forelimb. Its angular arrangement in relation to the scapula and the distal parts of the forelimb allows it to act as a shock absorbing device to cushion the impact of the forefeet against the ground. In man and the primates, this third function

The Radius

Classification: Long bone.

Location: It lies between the humerus and the carpus. It is fused to the ulna along the posterolateral part of the shaft.

Description: The radius of the horse has a shaft and two extremities. The shaft has two surfaces and two borders. The surfaces are cranial and caudal. The cranial surface is smooth, convex lengthwise and rounded

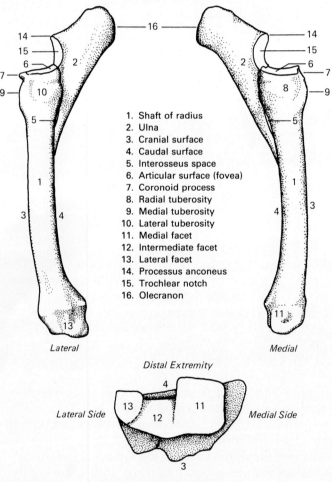

1. Shaft of radius
2. Ulna
3. Cranial surface
4. Caudal surface
5. Interosseus space
6. Articular surface (fovea)
7. Coronoid process
8. Radial tuberosity
9. Medial tuberosity
10. Lateral tuberosity
11. Medial facet
12. Intermediate facet
13. Lateral facet
14. Processus anconeus
15. Trochlear notch
16. Olecranon

FIGURE 3-8 Radius and ulna of the horse.

from side to side. The caudal surface is flattened. Proximally there is a groove which combines with the shaft of the ulna to form the interosseous space. The borders are medial and lateral. Both borders are concave lengthwise.

The extremities are proximal and distal. The proximal extremity has an articular surface, two facets, and three tuberosities (radial, medial, and lateral). The proximal articular surface is crossed by a sagittal ridge which divides the surface into medial and lateral facets. There is a synovial fossa caudally and a prominent lip cranially. The lip is the coronoid process. There are two facets for articulation with the ulna. They are separated by a depression and are located below the caudal border of the articular surface. The radial tuberosity is located medially. The medial tuberosity is continuous medially with the radial and extends around the craniomedial aspect. The lateral tuberosity is on the lateral side of the cranial surface.

The distal extremity articulates with the carpus and consists of three facets and two surfaces. The facets are medial, intermediate, and lateral. The medial facet is quadrilateral and largest in area. It is S-shaped when viewed from the side. It articulates with the radial carpal bone. The intermediate facet is similar to the medial but smaller. It articulates with the intermediate carpal bone. The lateral facet is smallest. It articulates with the ulnar carpal bone and the accessory carpal.

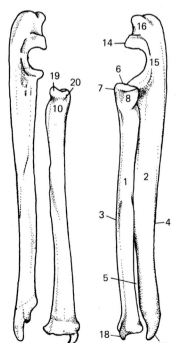

1. Shaft of radius
2. Ulna
3. Cranial surface
4. Caudal surface
5. Interosseus space
6. Articular surface (fovea)
7. Coronoid process
8. Radial tuberosity
10. Lateral tuberosity
11. Distal end of radius
12. Distal interosseus space
13. Distal end of ulna
14. Processus anconeus
15. Trochlear notch
16. Olecranon
17. Styloid process of ulna
18. Styloid process of radius

FIGURE 3-9   *Radius and ulna of cat.*

COMPARATIVE ANATOMY OF THE RADIUS

Cattle. The radius of cattle is shorter and wider than that of horses. The coronoid process is small. The proximal articular surface consists of two glenoid cavities separated by a groove. The distal extremity is enlarged and thick medially, giving the shaft an inward curving appear-

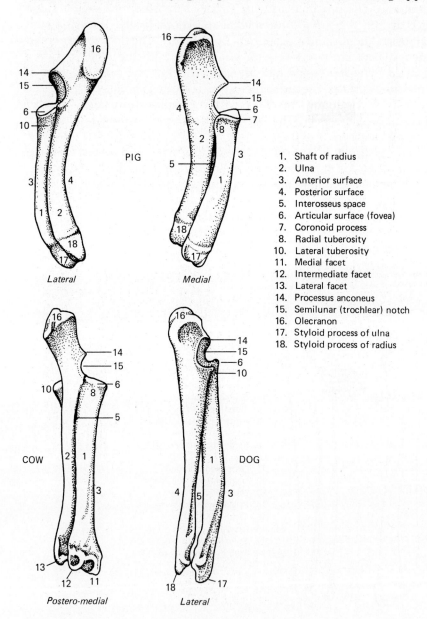

1. Shaft of radius
2. Ulna
3. Anterior surface
4. Posterior surface
5. Interosseus space
6. Articular surface (fovea)
7. Coronoid process
8. Radial tuberosity
10. Lateral tuberosity
11. Medial facet
12. Intermediate facet
13. Lateral facet
14. Processus anconeus
15. Semilunar (trochlear) notch
16. Olecranon
17. Styloid process of ulna
18. Styloid process of radius

FIGURE 3-10 *Radius and ulna—comparative anatomy.*

ance. The ulna is normally fused to the caudolateral aspect of the radius in its middle and proximal portion, leaving proximal and distal interosseus spaces.

Sheep. The radius of the sheep is similar to that of the cow except for smaller size and more pronounced curvature.

Swine. The radius of swine is short and thick and curved caudally. It increases in size distally and unites loosely with the shaft of the ulna.

Dogs. The radius of the dog is flattened craniocaudally and increases in size distally. The shaft curves backward and inward and is separate from the ulna. The proximal end is small and consists of a head connected to the shaft by a neck. The radial tuberosity is small, but the lateral is large. The distal extremity has a projection along its medial border (the styloid process of the radius).

Cats. The radius of the cat is less curved but otherwise similar to that of the dog.

The Ulna

Classification: Long bone.

Location: It is attached to the caudolateral surface of the radius.*

Description: The ulna of the horse consists of a shaft and two extremities. The shaft is incomplete and tapers to a point distally and has three surfaces and three borders. The surfaces are cranial, medial, and lateral. The cranial surface is in contact with the caudal surface of the radius, except at the interosseus space. The area that goes into the formation of the interosseus space includes a nutrient foramen. The medial surface is smooth and slightly concave. The lateral surface is flattened. The borders are medial, lateral, and caudal. The medial and lateral borders are thin and sharp, except at the interosseus space, where they are rounded. The caudal border is straight in length and rounded from side to side.

The extremities are proximal and distal. The proximal extremity (the olecranon) is a tuberosity that consists of a medial and a lateral surface, two borders, two facets, and a free extremity. The medial surface is concave and smooth, while the lateral surface is convex and rough. The borders are cranial and caudal. The cranial border bears on its middle the processus anconeus which overhangs the (trochlear) notch. The tro-

---

*This description is not adequate for cats and dogs and brachiating animals with unattached ulnas.

chlear notch is a curved articular surface which is convex transversely and articulates with the humerus. The caudal border blends into the caudal border of the shaft. Two facets are situated below the trochlear notch. These articulate with facets on the caudal part of the proximal end of the radius.

The distal extremity is fused with the radius.

COMPARATIVE ANATOMY OF THE ULNA

Cattle. The ulna of cattle is complete. The shaft is three sided and fused with the radius except at the proximal and distal interosseus spaces. The olecranon is large and bears a rounded tuberosity. The distal end forms the styloid process, which projects beyond the distal end of the radius.

Sheep. The ulna of the sheep is similar to that of cattle.

Swine. In swine, the ulna is considerably larger and heavier than the radius. The caudal surface is smoothly concave, and the cranial surface lies in close approximation to the radius to which it is attached by the interosseus ligament. In the proximal third a smooth concave area concurs with a similar area on the radius, forming the interosseus space. The proximal extremity is large, and the olecranon is prominent. The distal extremity is small and tapers to a blunt point.

Dogs. In dogs the ulna is complete, straight, and larger than the radius. It diminishes in size distally, is completely separate from the radius, and possesses an extensive interosseus space. The olecranon is grooved and has three bony prominences. The trochlear notch unites with the radius to form a compound joint. The distal end is small and bluntly pointed (styloid process of ulna).

Cats. The ulna of the cat is larger than the radius. It has an olecranon which is short and blunt and a shaft that decreases in size distally and is marked on its lateral surface by a number of prominent lines. The distal extremity of the shaft has an elongated styloid process and an articular facet.

PHYSIOLOGY OF THE RADIUS AND ULNA

The radius and ulna have markedly different functions, depending upon the species and habits of the animal. The two bones may be partially or completely fused as in the horse, cow, pig, and sheep, or they may be separate and movable as in the dog, cat, and man. In the first instance, the fusion of the bones prohibits rotation of the lower portion of the limb. Motion is restricted to a back and forth hinge-like action which is adapted for propulsion, but very little else. In animals with a separate radius and ulna, the lower limb is capable of rotation as well as a hinge-like motion. This is of considerable advantage to carnivores, and climbing and burrowing animals, whose existence depends upon grasping and holding objects. The development of forearm rotation in terrestrial mammals is highest in the human.

The Carpus

Classification: A group of short bones.

Location: The carpus lies between the radius and ulna and the metacarpals.

Description: The carpus consists of a group of six to eight bones, depending on the species of animal. It is commonly and incorrectly called the "knee" in animals. In man it is called the wrist. The knee is found only in the caudal limbs. The bones are arranged in two rows, proximal and distal, and are named and numbered as follows:

	PROXIMAL ROW			
	Radial	Intermediate	Ulnar	Accessory
(MEDIAL SIDE)		DISTAL ROW		(LATERAL SIDE)
	First	Second	Third	Fourth

The first carpal bone is small and may be missing in about half of the horses examined. In those where it is present, it may be fused to the caudal surface of the second carpal bone.

The carpus, while it is seldom studied intensively in courses of this nature, should not be dismissed as being of minor importance. Like the hock (tarsus) in the hind leg, it is an important structure in locomotion of cursorial animals since it functions both as a hinge joint and as a shock absorbing device through the series of inclined planes that form its joint surfaces. The combined movement of its joints forms a major flexion area which transmits forward movement to the body and contributes to the length of the stride. So-called knee injuries in the carpi of race horses and saddle horses are problems of great difficulty to trainers and veterinarians and can result in permanent lameness.

In brachiating animals, the wrist joint is equally important but for different reasons. In these animals the flexibility of the wrist joint is more important than shock absorption or locomotion functions, but, again, the complexity and delicacy of the joint make it easily subject to injury and hard to treat.

Since the bones are short, they have no true marrow cavity. The interior is filled with a cancellous structure composed of spicules and

52
THE SKELETON

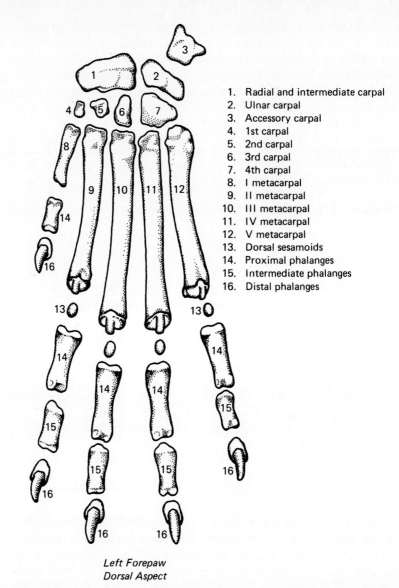

FIGURE 3-11 *Carpals, metacarpals, and digit of the dog. (Adapted from Miller, Guide to the Dissection of the Dog.)*

trabeculae. Physically, these bones are not as resistant to compression stresses as long bones and tend to develop bony growths or hairline fractures which are painful, restrict the movement of the joint, and often will not heal. The very complexity of the joint area restricts surgery and therapeutics to a great degree. It is one of the areas of the body that is being intensively studied by orthopedic surgeons.

THE APPENDICULAR SKELETON

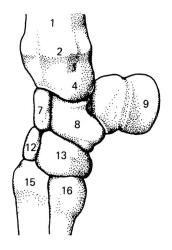
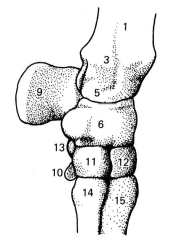

Medial

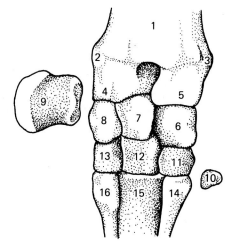

1. Radius
2. Lateral malleolus
3. Medial malleolus
4. Lateral facet
5. Medial facet
6. Radial carpal
7. Intermediate carpal
8. Ulnar carpal
9. Accessory carpal
10. 1st carpal
11. 2nd carpal
12. 3rd carpal
13. 4th carpal
14. II metacarpal
15. III metacarpal
16. IV metacarpal

FIGURE 3-12   Carpus of the horse.

Comparative Anatomy of the Carpals

Species	Number	Radial	Intermediate	Ulnar	Accessory	1st	2nd	3rd	4th
Horse	7 or 8	x	x	x	x	x/0	x	x	x
Cow & Sheep	6	x	x	x	x	0	fused—fused		x
Pig	8	x	x	x	x	x	x	x	x
Dog	7	fused—fused		x	x	x	x	x	x

The Metacarpals

Classification: Long bones.

Location: The metacarpals are located between the distal row of the carpals and the proximal phalanx.

Description: Three bones are present in the horse: the second, third, and fourth (II, III, and IV). The first and fifth are absent. Of the three, the third is fully developed, articulating with the carpus and digit. It is commonly called the cannon bone. The second and fourth are greatly modified, being much reduced in size and are commonly called splints. Both the second and fourth metacarpals are enlarged at their proximal ends and are firmly attached to the caudal, medial, and lateral surfaces of the third metacarpal for about halfway down its shaft. The splints become more slender distally, enlarge at their distal extremities, and terminate about two-thirds of the way down the shaft of the third metacarpal.

The third metacarpal is composed of a shaft and two extremities. The shaft has two surfaces, cranial and caudal. The cranial surface is convex while the caudal surface is somewhat flattened, giving the bone an oval cross section. The shaft is slightly curved cranially, giving it a bowed appearance.

The extremities are proximal and distal. The proximal extremity of the third metacarpal articulates with the distal row of carpals. The distal extremity of the third metacarpal is formed into a condyle which possesses a raised central area or sagittal ridge.

COMPARATIVE ANATOMY OF THE METACARPALS

The metacarpals vary in number with the species of animals. Their number conforms to the number of functional or visible digits. In cloven-hoofed ruminants the III and IV metacarpals are functional but are fused together to form a single bone.

The Digit

This is a group of bones analogous to the fingers and toes of man. With the exception of slight modifications in size and shape, The digits of the front and hindlimbs are the same in the large domestic animals.

In the horse each digit is composed of three phalanges, two proximal sesamoid bones, and one distal sesamoid bone.

The digits have great variation in number and composition among the various species of domestic mammals, i.e., cow—two per foot; pig—four per foot; dog—five front and four or five rear; cat—five front

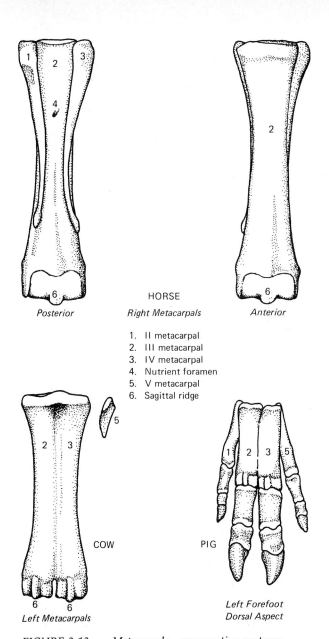

FIGURE 3-13   Metacarpals—comparative anatomy.

and four rear; man—five hand and five foot. However, they all have some things in common:

Each digit except the first is composed of three phalanges.
The first digit is composed of two phalanges.
Each digit articulates proximally with a metacarpal or a metatarsal bone.

The digit ends distally with a horny claw, nail, or hoof.

Digital sesamoid bones are usually present in variable numbers.

Some animals (notably the cow goat and water buffalo among the domestic animals) possess false digits or dewclaws which do not articulate with a metacarpal or metatarsal bone and are remnants of digits which have nearly disappeared with the evolution of the animal to its modern form (for ergot and chestnut in the horse, see p. 471).

PROXIMAL (FIRST) PHALANX
(SHORT CANNON—OS SUFFRAGINIS, OS COMPEDALE)

Classification: Long bone.

Location: Between the III metacarpal (metatarsal) bone and the intermediate phalanx.

Description: The proximal phalanx consists of a shaft and two extremities. The shaft has cranial and caudal surfaces. The cranial surface is convex from side to side, while the caudal surface is flattened. The proximal extremity consists of two facets separated by a groove and articulates with the third metacarpal (metatarsal) and the distal extremity articulates with the intermediate phalanx.

INTERMEDIATE (MIDDLE, SECOND) PHALANX
(PASTERN BONE—OS CORONAE, OS CORONALE)

Classification: Long bone.

Location: Between the proximal phalanx and the distal phalanx.

Description: The intermediate phalanx is somewhat wider than it is long. It is composed of four surfaces: proximal, distal, cranial, and caudal. The proximal surface articulates with the proximal phalanx, and the distal surface articulates with the distal phalanx and the navicular bone. The cranial surface is convex from side to side, and the caudal surface is flattened.

DISTAL (THIRD) PHALANX
(COFFIN BONE, OS PEDIS, OS UNGULARE)

Classification: Modified short bone.

Location: The distal phalanx articulates with the distal extremity of the intermediate phalanx and the navicular and is enclosed in the horny structures of the hoof.

Description: In the horse the distal phalanx has three surfaces, three borders, and two angles. The surfaces are articular, cranial (dorsal), and caudal (volar or plantar). The articular surface faces upward and backward. On its caudal border it has a flattened area for articulation with the distal sesamoid (navicular) bone. The cranial surface slopes downward and forward and is convex from side to side. The surface is rough and porous and contains many foramina. On either side a dorsal groove passes forward and ends at a large foramen. The caudal surface is arched and is divided into two unequal parts (sole* surface and flexor surface) by a curved semilunar crest. The sole surface is the large crescent shaped area anterior to the crest. It is comparatively smooth. The flexor surface is smaller and semilunar in shape. It has a foramen on either side which leads into the semilunar canal.

The borders are proximal, distal, and caudal (volar* or plantar). The proximal or coronary border separates the dorsal and articular surfaces and has an extensor process in its middle. The distal border is thin and irregular and separates the dorsal and volar surfaces. The caudal border separates the articular and caudal surfaces.

The angles or wings are masses of bone which project backward on the medial and lateral sides. The medial wing is usually shorter than the lateral. The proximal border of each wing carries cartilages (lateral cartilages).

Digital Sesamoid Bones

Classification: Short bones.

Description: The digital sesamoids of the horse consist of three bones (two proximal and one distal). The proximal sesamoids are located behind the distal articular surfaces of the third metacarpal. The bases of the bones are distal and the apices proximal. The distal sesamoid (navicular) is located behind the junction of the intermediate and distal phalanges. It is shuttle or boat shaped, with its long axis lying transverse to the long axis of the limb. It enters into the formation of the coronary joint. Its flat surface aids the free movement of the deep digital flexor tendon. As has been previously mentioned in the discussion of the carpus, short bones are not as resistant to compression injury as long bones. The navicular, sandwiched between the much larger and more dense masses of the intermediate and distal phalanges, is very subject to damage from the type of crushing forces that occur in race horses, steeplechasers, and jumpers. The result of such damage may be a condition known as navicular disease, a permanent, progressive, and ultimately disabling injury. Navicular disease may also arise from other causes, including contaminated puncture wounds.

---

*The volar surfaces and borders are also called palmar. The term solar is also used in place of sole.

THE SKELETON

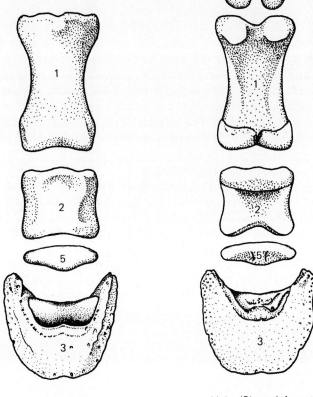

Dorsal Aspect *Volar (Plantar) Aspect*

1. Proximal phalanx
2. Intermediate (middle) phalanx
3. Distal phalanx
4. Proximal sesamoids
5. Distal sesamoid

FIGURE 3-14   *Digit of the horse.*

COMPARATIVE ANATOMY OF THE DIGITAL SESAMOIDS

Cattle and Sheep. In cattle and sheep, six sesamoids are present on each foot, three for each digit, and arranged as in horses. They are relatively smaller than equine sesamoids. The distal sesamoids are not flattened as in horses. In the sheep the abaxial sesamoids in the proximal row are compressed laterally. In cattle, accessory digits or dewclaws usually do not possess sesamoids or phalanges.

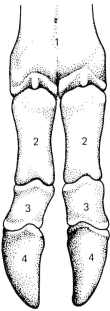
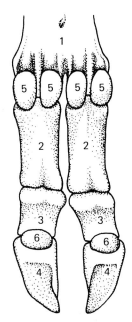

Dorsal Aspect   Volar Aspect

1. III and IV metacarpal
2. Proximal phalanx
3. Intermediate (middle) phalanx
4. Distal phalanx
5. Proximal sesamoids
6. Distal sesamoid

FIGURE 3-15   *Digit of the cow.*

Swine. In swine, three sesamoids are present for each chief digit and two for each accessory digit, making a total of ten sesamoids in each foot. The distal sesamoids are absent in the accessory digits.

Dog. In the dog each paw possesses a maximum of 13 sesamoids, four dorsal and nine volar (plantar). In the forefoot, the volar sesamoids are located in pairs at the metacarpophalangeal joint, except for the first digit (thumb), which (when present) possesses a single sesamoid. The dorsal sesamoids are found embedded in the joint capsule of the II to IV metacarpophalangeal joints. The I metacarpophalangeal joint does not possess a dorsal sesamoid. The distal sesamoids are small cartilaginous masses. The sesamoids of hind and forepaws are alike in size and location. However the word, "metatarsophalangeal," is used to identify the sesamoids of the hind paw.

Cat. There are nine digital sesamoids in the forepaw of the cat and eight in the hind. They are located in pairs on the volar or plantar surface of the metacarpo(tarso)phalangeal joints. The hind paw possesses only a rudimentary I metatarsal and no sesamoid is present. The I metacarpal usually has one sesamoid on its distal extremity as in the dog.

Man. Human digital sesamoids are irregular and may vary in number. Generally, nine sesamoids may be found in each hand and seven on each foot on the palm (volar) or sole (plantar) surface. Three sesamoids, two proximal and one distal, are found in the thumb, one proximal and one distal in the index finger, one proximal in the middle finger and ring finger, and two proximal in the little finger. Three sesamoids, two proximal and one distal, are found in the great toe; two (one proximal and one distal) in the first toe; and two proximal in the little toe.

PHYSIOLOGY OF THE SESAMOIDS

Functional sesamoid bones allow muscles to exert greater pull on the movable bones distal to the sesamoids. They give the tendons, which ride over the sesamoids, a greater amount of angular pull and, consequently, greater leverage. In some locations, the sesamoids (dorsal sesamoids and fabellae) have no apparent purpose. In running animals, the proximal volar sesamoids of the digit, and the patella (kneecap) are very important to give force and power to the stride and to improve muscle efficiency and increase speed.

The Pelvic Girdle

The pelvic girdle consists of right and left os coxae or innominate bones which are united to each other ventrally at the symphysis pelvis. The two os coxae are each formed by the fusion of their respective ilium, ischium, and pubis at the acetabulum. The pelvis articulates with the sacrum and the femur. The bony pelvis plus the soft tissues of the rectum, anus, and reproductive system form the pelvic cavity. This structure is relatively smaller and narrower in males than in females.

THE ILIUM

Classification: Modified long bone.

Description: The ilium is irregularly triangular and has two surfaces, three borders, and three angles. The surfaces are gluteal and pelvic. The gluteal or lateral surface faces dorsally and backward. It is wide and concave in front, narrow and convex behind. The gluteal

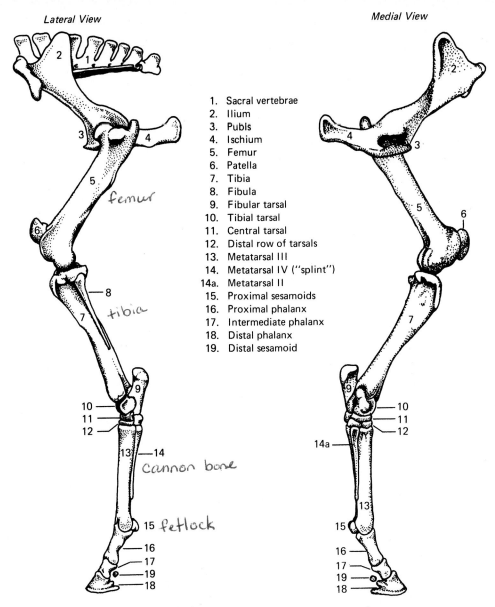

FIGURE 3-16 Pelvic limb of the horse.

surface bears the gluteal line, which crosses the wide part of the bone from the middle of the medial border toward the tuber coxae. The pelvic surface faces downward and forward, is convex, and consists of two parts: a medial triangular part and a lateral quadrilateral part. The medial (triangular) part is rough and bears an irregular facet (the auricular area) for articulation with the sacrum. The lateral (quadrilateral) part is smooth

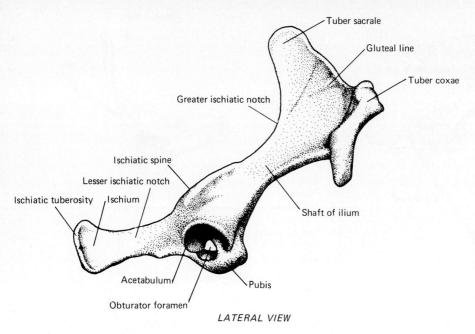
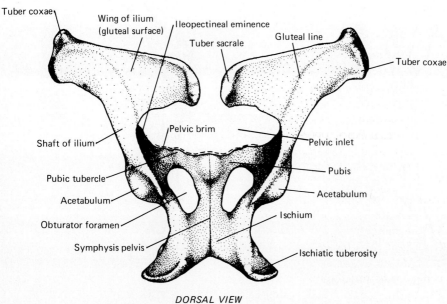

FIGURE 3-17   Pelvis of the horse.

and is crossed by the iliopectineal line which begins below the auricular area and continues on the shaft of the ilium to the anterior border of the pubis. The line is interrupted by vascular grooves. Below these grooves is the psoas tubercle.

The borders are cranial, medial, and lateral. The cranial border is concave, thick, and rough. The medial border is concave and smooth. The middle of the medial border forms the greater sciatic notch and continues as the ischiatic spine. The lateral border is concave and rough. The cranial part is crossed by vascular grooves and the caudal part contains a nutrient foramen.

The angles are medial, lateral, and acetabular. The medial angle (tuber sacrale) curves upward and backward. The lateral angle (tuber coxae) is a large rectangular mass, narrow in the middle and enlarged at both ends. The acetabular angle meets the ischium and pubis at the acetabulum and has a dorsal border and a shaft. The dorsal border forms part of the ischiatic spine which is rough laterally and smooth medially. The shaft is constricted and has three surfaces: the lateral surface is convex and rough, the pelvic surface is smooth and grooved by vessels and nerves, and the ventral surface is crossed by vascular grooves.

THE ISCHIUM

Classification: Modified long bone.

Location: The ischium forms the caudal part of the ventral floor of the pelvis, and joins the ilium and pubis at the acetabulum.

Description: The ischium is irregularly quadrilateral and has two surfaces; four borders, and four angles.

The surfaces are pelvic and ventral. The pelvic surface is smooth and concave from side to side. The ventral surface is flattened and rough.

The borders are cranial, caudal, medial, and lateral. The cranial border forms the caudal margin of the obturator foramen and is smooth and concave. The caudal border is thick and rough and forms half of the ischial arch. The medial border meets the opposite bone at the ischial portion of the symphysis pelvis. The lateral border forms the lesser sciatic notch.

The angles are craniomedial, craniolateral, caudomedial, and caudolateral. The craniomedial angle meets with the pubis to form the medial boundary of the obturator foramen. The craniolateral angle joins the ilium and pubis at the acetabulum and bears the ischiatic spine on its dorsal part. It is grooved medially by the obturator vessels. The caudomedial angle joins its fellow at the symphysis ischii. The caudolateral angle forms the (ischiatic tuberosity) and its lower border forms the ventral ischiatic spine.

THE PUBIS

Classification: Modified long bone.

Location: The pubis is joined caudally to the ischium. Cranially it meets the ilium and ischium at the acetabulum.

Description: The pubis has two surfaces, three borders, and three angles.

The surfaces are pelvic and ventral. The pelvic surface is convex cranially in young horses and in stallions but is concave and smooth in mares and geldings. The ventral surface is convex and rough and is crossed by the pubic groove on the cranial border.

The borders are cranial, medial, and caudal. The cranial border is thin medially except in young animals and in stallions, and bears the psoas tubercle near the symphysis pubis. The medial border meets the opposite bone at the symphysis pubis. The caudal border is concave and forms the cranial margin of the obturator foramen.

The angles are medial, acetabular, and caudal. The medial angle meets its opposite number at the symphysis pubis. The acetabular angle joins the ilium and ischium at the acetabulum. The caudal angle joins the ischium and forms part of the medial boundary of the obturator foramen.

THE ACETABULUM

The acetabulum is a cotyloid cavity which encloses the head of the femur. It is formed by the junction of the ilium, ischium, and pubis, and faces ventrolaterally. It consists of two parts, an articular part, and a nonarticular part. The articular part is crescent shaped and may be faceted. It is bounded internally by the acetabular fossa, which forms the nonarticular part, and externally by the rim of the cavity. The medial part of the rim is cut by the acetabular notch.

Embryologically, the acetabulum of the horse includes a separate bone, the os acetabuli or cotyloid bone, which arises from a separate ossification center. The os acetabuli is most distinct in the three-month-old equine embryo and usually fuses to the acetabulum prior to birth.

In the cow, sheep, and pig a similar situation exists, but in the dog the bone exists after birth and fuses to the pubis about the third month of life. In the cat the os acetabuli persists into adult life, and although it ultimately fuses with the pubis, retains its identity. In man, the bone exists as a Y-shaped bar that ossifies and fuses to the ilium, ischium, and pubis at about 18 years of age.

COMPARATIVE ANATOMY OF THE PELVIS

Cattle. In cattle, a ventral spinous process below the pelvic floor is present. The ischium is large with its long axis directed backward and upward, in contrast to the relatively horizontal axis in the horse. The transverse axes form a deep curve. The ischiatic spine is high and sharp edged and bears a series of rough lines on its lateral faces. The ischiatic tuberosity is massive and triangular with dorsal, lateral, and medial prominences.

The pubis is similar to that of the horse. The acetabulum is smaller than that of the horse and is marked by two notches (caudomedial and craniomedial) and is usually faceted on the articular surface.

Sheep. The pelvis differs considerably from the bovine. The long axis of the ilium and ischium forms a relatively straight line. The tuber coxae is only slightly thickened, and the tuber sacrale is pointed. The floor of the pelvis is flatter than that of the bovine. The ischiatic tuberosity is flattened rather than triangular. The pelvic canal axis slants backward and downward.

Swine. The axis of the ilium and ischium forms almost a straight line. The wing of the ilium is directed outward and downward, forming a moderately acute angle with the medial plane. The ischia diverge laterally and posteriorly. The ischiatic tuberosity has three prominences. The acetabulum is set caudally, its rim is thick, and the acetabular notch, though narrow, is relatively deep.

Dog. The wing of the ilium is nearly parallel with the median plane. The ischium has a twisted appearance because the acetabular portion is nearly vertical, while the posterior parts are almost horizontal. The ischiatic tuberosity faces laterally. The floor of the pelvis is flattened.

The acetabulum has a broad and deep acetabular fossa whose open end is ventral. A thin plate of bone separates the medial surface of the acetabulum from the pelvic canal. The obturator foramen is roughly triangular. An os acetabuli is present in young animals but fuses to the larger bones early in life.

Cat. In the cat, the os coxae is composed of four definitive bones: the ilium, ischium, pubis, and the os acetabuli (cotyloid bone), which lies with its apex in the acetabulum and its angles and base articulating with the ilium, ischium, and pubis. There is a slight angle between the ilium and ischium, which gives the pelvic canal a curved appearance. The wings of the ilium are nearly parallel with the median plane and the ischiatic tuberosity faces laterally. The pelvis is set obliquely to the spinal axis and forms an acute angle with the frontal plane. The obturator foramina are large and oval. The acetabulum is set in the caudal third of the os coxae and possesses a deep acetabular notch. The acetabular fossa is shallow.

PHYSIOLOGY OF THE PELVIS

In the terrestrial mammals, the hind legs are chiefly adapted for propulsion. This requires a direct attachment between the hindlimb and the axial skeleton. The attachment is made through the pelvic girdle at the sacroiliac joint. The pelvis, in order to accommodate the forces involved, is usually fused into a solid bony ring that gives more efficient transmission of propulsive force to the body. Modifications in size and shape of the pelvis can be correlated with the species, habitat, method of locomotion, and type of locomotion.

Since the central opening of the pelvis forms the birth canal in the female, the factor of reproduction enters into its structure. Further modifications depend upon the size of the offspring carried by the female. In females, the pelvic inlet is usually relatively larger in diameter and tends to be more circular than it is in males. This difference is so noticeable in cattle and sheep that one can determine the sex of carcasses from which all other identifying organs or structures have been removed.

The fusion of the pelvic bones and the subsequent distortions of pelvic shape to conform to the physical and physiological requirements of a given species produce a structure that is readily recognizable. It can be used to identify sex and species in the absence of other identifying remains.

The Femur

Classification: Long bone.

Location: It lies between the pelvis and the patella, tibia, and fibula.

Description: The femur has a shaft and two extremities.

The shaft is cylindrical, flattened on the caudal aspect and larger proximally than distally. It has four surfaces and two borders. The surfaces are cranial, medial, lateral, and caudal. The cranial, lateral, and medial surfaces are convex and smooth. The caudal surface is flattened and irregular and has a number of prominences and depressions. The borders are medial and lateral. The medial border bears the trochanter minor (lesser trochanter) on the proximal part, a nutrient foramen below a rough area in the middle portion, and a medial supracondyloid crest in the distal third. The trochanter tertius occurs on the proximal part of the lateral border. A supracondyloid fossa is present on the caudolateral aspect of the distal third of the shaft.

The extremities are proximal and distal. The proximal extremity is large and consists of a head, a neck, and a trochanter major (greater trochanter). The head is on the medial side and faces inward, upward, and forward. The head is cut by a notch, the fovea capitis. The neck is distinct. The trochanter major is lateral and is divided into two parts, cranial and caudal. The cranial part is convex and rises a little above the head of the femur. The caudal part is higher and is separated from the cranial part by a notch. The caudal border of the trochanter major forms the lateral wall of the trochanteric fossa. The trochanteric fossa is a deep depression immediately below the trochanter major.

The distal extremity consists of a trochlea, two condyles and two epicondyles. The trochlea is composed of two articular ridges separated by a groove. It articulates with the patella. The medial ridge is wider and more prominent. The medial and lateral condyles are bulbous and are separated by an intercondyloid fossa. A ridge connects each condyle to the trochlea. The medial and lateral epicondyles are rounded prominences located beside the condyles.

COMPARATIVE ANATOMY OF THE FEMUR

Cattle. The femur of cattle is roughly similar to that of the horse, except that the shaft is smaller and smoother. The trochanter major is not divided into cranial and caudal parts. The trochanter minor is less distinct and oriented more caudally. The trochanter tertius is absent.

Sheep. The femur of sheep is slightly curved cranially. The trochanter major is small. The trochanter minor is reduced, and the trochanter tertius is absent.

Swine. The femur of swine is similar to that of the bovine. The trochanter minor, however, is less distinct, and the shaft is relatively heavier and more massive.

Dog. The femur of the dog is relatively longer and smoother than that of the cow or sheep. The shaft is more nearly cylindrical. The trochanter major is small, the trochanter minor is very small, and the trochanter tertius is missing. The supracondyloid fossa is absent. Two facets are present on the caudal aspect of the condyles for articulation with the fabellae.

Cat. The femur of the cat is similar to that of the dog.

PHYSIOLOGY OF THE FEMUR

The function of the femur is fourfold. It serves to protect the pelvic viscera and support the birth canal. It is a connecting link between the pelvis and the distal portions of the hindlimb. It furnishes attachments for the origin and insertion of the major muscles that control its motion and the movements of the distal part of the limb. It gives a positive connection between the hindlimb and the axial skeleton resulting in a more efficient transfer of propulsive force to the body.

The Patella

Classification: Short bone.

Location: The patella is found on the cranial distal surface of the femur and articulates with the trochlea of the femur.

Description: The patella is the largest of the sesamoid bones. It is quite specialized and has two surfaces, two borders, a base, and an apex. The surfaces are cranial and articular. The cranial, or free surface, is quadrilateral in form, and is convex and rough. The articular surface is also quadrilateral in form, but is concave and smooth. It bears a low rounded ridge which separates it into medial and lateral areas. The medial area is the larger. The borders are medial and lateral and converge below at the apex. The medial border is concave, the lateral border is convex, and is less prominent. The base faces upward and backward. It is convex transversely and concave when viewed from the side. The apex is distal and is formed into a blunt point.

PHYSIOLOGY OF THE PATELLA

The structure of the patella and the femorotibial joint is designed for the maximum application of muscular force to extend the upper portions of the hindlimb. This results in a greater "kick" or drive of the hindlegs and a consequent increase in speed of movement. A corresponding structure does not exist in the foreleg, probably because of the greater adaptation of that limb to weight supporting (or manipulative) functions.

The Fabellae

Classification: Short bones.

Location: Fabellae are found on the caudal distal surface of the femur above the femoral condyles.

Description: The fabellae are specialized sesamoids which develop in the origin of the gastrocnemius muscle (calf muscle). In cats there are four fabellae; in most other animals with fabellae there are two. They are present in dogs, cats, and other carnivores, but are not found in horses, cattle, sheep, deer, or swine. If only two fabellae are present, they articulate with the dorsal portion of the femoral condyles. If there are four, the additional pair are located along the caudal part of the proximal end of the tibia.

THE APPENDICULAR SKELETON

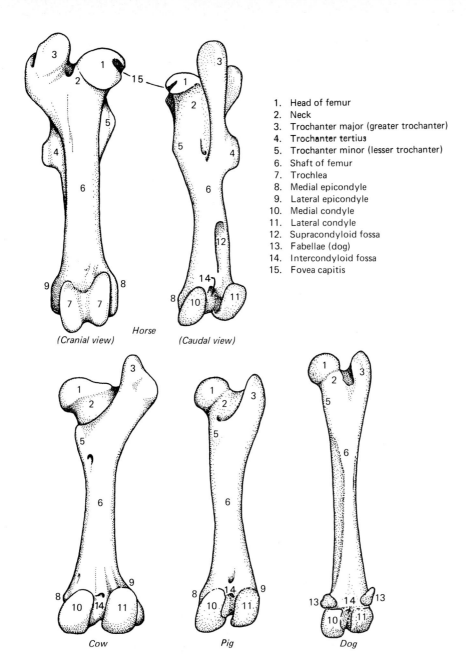

1. Head of femur
2. Neck
3. Trochanter major (greater trochanter)
4. Trochanter tertius
5. Trochanter minor (lesser trochanter)
6. Shaft of femur
7. Trochlea
8. Medial epicondyle
9. Lateral epicondyle
10. Medial condyle
11. Lateral condyle
12. Supracondyloid fossa
13. Fabellae (dog)
14. Intercondyloid fossa
15. Fovea capitis

FIGURE 3-18 *Right femur—comparative anatomy.*

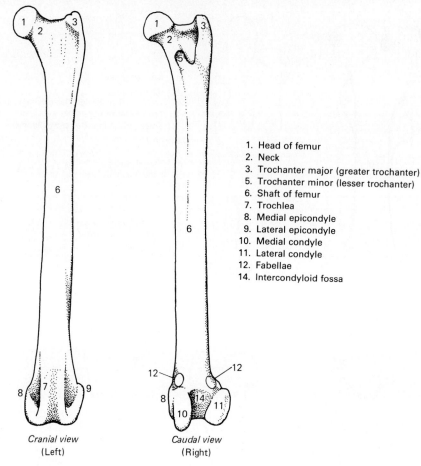

FIGURE 3-19   Femur of cat.

The Tibia

Classification: Long bone.

Location: The tibia extends obliquely downward and backward from the femur to the tarsus.

Description: The tibia has a shaft and two extremities. The shaft is large and three-sided proximally, becoming smaller and flatter in the middle portion, and widened and rectangular at the distal extremity. There are three surfaces, medial, lateral, and caudal, and three borders, cranial, medial, and lateral. The medial surface is broad and rough proximally; distally it narrows. The lateral surface is a smooth spiral curve. It is wide and concave in the proximal portion, narrow and convex centrally, wider and flatter distally. The caudal surface is flattened and divided into two parts by the popliteal line. Near the popliteal line is a nutrient foramen. The cranial border bears a crest in the proximal end.

The extremities are proximal and distal. The proximal extremity is large and three sided and bears two condyles (medial and lateral) and a tuberosity. The medial condyle is the larger and is slightly saddle-shaped. It is separated from the lateral condyle by the popliteal notch and the tibial spine. The lateral condyle has an overhanging margin below which is a facet for articulation of the fibula. The tuberosity is located on the proximal end of the crest and is separated from the lateral condyle by the sulcus muscularis.

The distal extremity is small and quadrilateral. It has an articular surface and two malleoli. The articular surface consists of two grooves, separated by a ridge. The two grooves face obliquely forward and laterally. The medial groove is narrower and deeper than the lateral groove. The malleoli are bony prominences on the outer margins of the articular grooves. The medial malleolus is smaller than the lateral, which is the distal extremity of the fibula.

The Fibula

Classification: Modified long bone.

Location: Along the lateral side of the tibia.

Description: The fibula has a shaft and two extremities. In the horse, the shaft is incomplete and is usually a slender rod that tapers to a point. It forms the lateral boundary of the interosseus space and generally terminates in the middle third of the tibia.

The extremities are proximal and distal. The proximal extremity is flattened transversely and has two surfaces (medial and lateral) and two borders (cranial and caudal). The medial surface bears a narrow articular area on the upper border. The lateral surface is rough, and the cranial and caudal borders are rounded.

The distal extremity is fused with the distal extremity of the tibia and forms the lateral malleolus.

COMPARATIVE ANATOMY OF THE FIBULA

Horse. The shaft is vestigial and incomplete. Both ends are fused with the tibia.

Cattle. No shaft is present. The ends are fused with the tibia. The proximal end is bluntly pointed.

Sheep. No shaft is present. The ends are fused with the tibia. The proximal end is incorporated into the tibial condyles.

Pig. The bone is complete and smaller than the tibia. The ends are free at birth and become united to the tibia at two to three years of age.

Dog. The bone is complete and smaller than the tibia. The ends are free and articulate with the tibia.

Cat. The cat is similar to the dog.

PHYSIOLOGY OF THE FIBULA

In all domestic mammals the fibula is gradually becoming vestigial. The degree of disappearance varies from virtually complete absence of this structure in cattle and sheep, to the physical (but not functional) persistence in the dog, cat, and human. In horses, the fibula is a common site for fractures and separation which produce lameness. It is often an incomplete structure with the shaft failing to develop in its middle portion. This results in the bone becoming three distinct parts—a head, middle piece, and lateral malleolus.

The Tarsus

Essentially, the tarsus performs the same functions for the hindlimb as the carpus does for the forelimb. The tarsus, however, flexes anteriorly while the carpus bends posteriorly. The tarsus of the horse consists of six or seven short bones arranged roughly in two rows:

PROXIMAL ROW

(Calcaneal tuber)   (Trochlea)
Calcaneus   Talus

Central
Tarsal

(LATERAL   DISTAL ROW   (MEDIAL
SIDE)   SIDE)
Fourth   Third   Second   First
Tarsal   Tarsal   Tarsal   Tarsal

The first and second tarsal bones are usually fused in the horse. The largest bone of the tarsal group is the calcaneus (fibular tarsal) which is located laterally and caudally in the proximal row. This bone has a long bony process known as the calcaneal tuber (tuber calcis) to which is attached the Achilles tendon of the gastrocnemius (calf) muscle. The calcaneus articulates with the tibial tarsal, the central tarsal, and the fourth tarsal bones.

The talus (tibial tarsal) is the second largest bone of the tarsal group and forms a fulcrum around which the hock joint is flexed and extended.

73
THE APPENDICULAR SKELETON

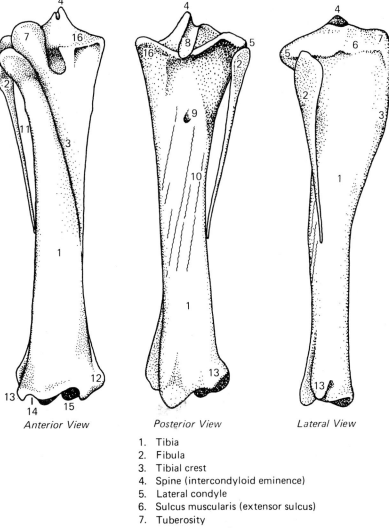

Anterior View   Posterior View   Lateral View

1. Tibia
2. Fibula
3. Tibial crest
4. Spine (intercondyloid eminence)
5. Lateral condyle
6. Sulcus muscularis (extensor sulcus)
7. Tuberosity
8. Intercondyloid fossa
9. Nutrient foramen
10. Popliteal line
11. Interosseus space
12. Medial malleolus
13. Lateral malleolus
14. Lateral articular groove
15. Medial articular groove
16. Medial condyle

FIGURE 3-20   Tibia and fibula of the horse.

74
THE SKELETON

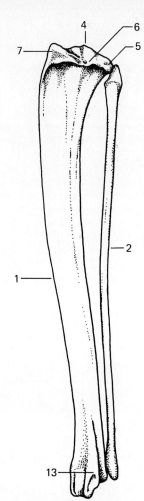

1. Tibia
2. Fibula
3. Tibial crest
4. Spine
5. Lateral condyle
6. Sulcus muscularis
7. Tuberosity (lateral tuberosity)
13. Lateral malleolus

FIGURE 3-21   Tibia and fibula of cat—lateral view.

It is characterized by a large pulley-like mass (the trochlea), which is located on its cranial surface and articulates with the distal end of the tibia. The two prominent oblique ridges of the trochlea curve forward, downward, and outward (laterally), and form an angle of about 15 degrees with the vertical. The trochlea is single in the horse, dog, and cat, but two trochleae (one on each end of the talus) are found in the cow, elk, deer, sheep, and swine. The talus articulates with the tibia dorsally, the calcaneus caudally, and the central tarsal ventrally.

The central tarsal is a flattened four-sided bone that lies between the talus and calcaneus and the four distal tarsal bones. It articulates with the talus and calcaneus dorsally and the four distal tarsal bones ventrally.

Comparative Anatomy of the Tarsus

Species	No.	Talus	Calcaneus	Central Tarsal	4th	3rd	2nd	1st
Horse	6	Massive one trochlea	+	+	+	+	fused—fused	
Cow & Sheep	5	Long & narrow 2 trochleae	+	fused—fused		fused—fused		+
Pig	7	Long & narrow 2 trochleae	+	+	+	+	+	+
Dog & Cat	7	Body, neck & head, 1 trochlea	+	+	+	+	+	+

The first and second tarsal bones are normally fused in the horse. The third tarsal has a flattened oval shape. The fourth tarsal is an irregular prism-shaped bone. The first three distal tarsals articulate with each other and with the central tarsal dorsally and the metatarsals ventrally. The fourth tarsal articulates with the calcaneus in addition to its articulation with the metatarsals ventrally.

As has been previously mentioned for the carpus, the tarsus is a locomotion and a shock absorption device. It is an important structure. Adapted as it is in cursorial animals to impart a high degree of leverage to the hind foot, the shape of the tarsus plays an important part in determining the speed and agility of an animal. Like the carpus, the tarsus is subject to injury and, although hock injuries are seldom totally disabling, they often impair an animal's efficiency of movement.

In most animals, as long as the trochlear function is unimpaired, the tarsal joint will function well, even though all the remaining bones of the group are fused into a solid mass.

The Metatarsals

Classification: Long bones.

Location: The metatarsals occur between the distal row of tarsals and the proximal phalanx.

Description: Same as metacarpals with the following exceptions: In the horse both cranial and caudal surfaces of the third metatarsal are curved, giving the bone a circular cross section. The shaft is straight rather than bowed cranially. The fourth metatarsal is attached intimately to the lateral caudal surface of the third metatarsal for a distance of one-quarter the length of the shaft. In general, in all ungulates, and in most plantigrade and digitigrade animals, the metatarsals are longer and less massive than the metacarpals.

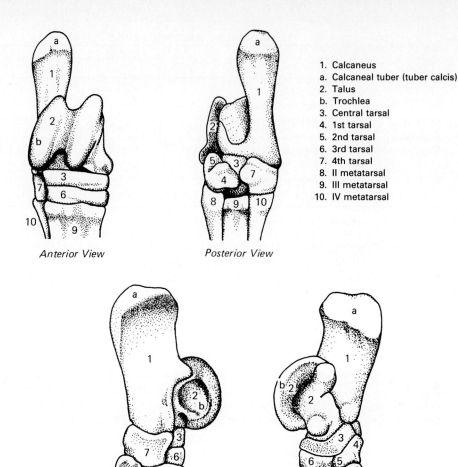

FIGURE 3-22   Tarsus of the horse.

The Digits

The hind digits are essentially similar in composition and structure to those of the forelimbs. The digits of multidigited animals also have considerable similarity. The dog, cat, and certain other digitigrade mammals usually have five digits on the front feet and four on the hind. Relics of a fifth digit (I phalanx) on the hind feet are much more frequent in dogs than in cats. When the digit is present, the bony attachment to the hind leg is usually vestigial or missing. The extra digit is generally called a "dewclaw" and is usually snipped off shortly after birth except in certain breeds when show rules dictate that it remain. It is a useless

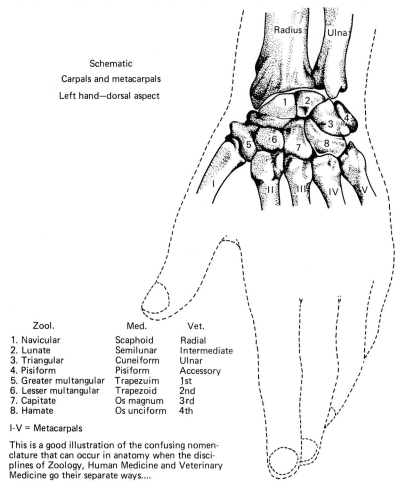

FIGURE 3-23 *Schematic of carpals and metacarpals, left hand, dorsal aspect.*

excresence and a show defect in some breeds and almost invariably becomes either irritated or injured in working dogs where it may be allowed to remain. Cats rarely have this digit visible externally although vestigial bony remnants are often present and incorporated into the tarsometatarsal joint.

THE AXIAL SKELETON

The axial skeleton consists of the vertebrae, the ribs, the sternum, and the skull.

Classification: Irregular bones.

Location: Along the midline dorsal axis of the body.

Description: The vertebrae form a chain or column of dorsal, median, and unpaired bones extending from the skull to the tail. The column is divided into five regions: cervical (C) (neck); thoracic (T) (chest); lumbar (L) (waist); sacral (S) (rump); and coccygeal or caudal (Cy) (tail). The number of vertebrae in each region for a given species is quite constant except for the coccygeal. However, except for the cervical, there is a variation in number of vertebrae between species.

The vertebrae in the sacral region are fused together more or less completely into a solid mass which articulates with the pelvis. These are called fixed (false) vertebrae in contrast to the movable (true) vertebrae composing the remainder of the vertebral column.

The vertebrae in each region have certain characteristics which serve to distinguish them from those of other regions, and individual vertebrae within a given region may further be distinguished from each other. However, all have a common plan of structure consisting of three basic parts: a body, an arch, and processes.

The body is a cylindrical mass, oriented craniocaudally, upon which the other parts are built. The body connects cranially and caudally with the bodies of adjacent vertebrae through intervertebral discs or cartilages (menisci). In most mammals, the cranial faces of the bodies of the cervical, thoracic, and lumbar vertebrae are convex, while the caudal surfaces are concave. The bodies of the sacral vertebrae are fused and those of the caudal vertebrae are convex on both ends. The ventral portion of the body is rounded. The dorsal portion is flattened, and forms the floor of the vertebral canal. In the thoracic region the body has lateral facets on its cranial and caudal borders which articulate with part of the heads of two successive pairs of ribs.

The arch is composed of two lateral halves and is constructed dorsally to the body. The halves meet medially to form a bony ring called the vertebral foramen. Successive foramina, together with the soft tissues which unite them form the vertebral canal which contains the spinal cord. The arches are notched before and behind, and the unions of adjacent vertebral notches form the intervertebral foramina through which the spinal nerves and vessels pass. There is considerable variation in the formation and location of the intervertebral foramina, particularly in the caudal thoracic vertebrae.

The processes can be divided into articular processes, of which there are usually two cranial and two caudal for each true vertebra; spinous processes, which are located dorsally and ventrally along the median plane; transverse processes, which project laterally from the arch or from the junction of arch and body; mammillary processes, which are found on caudal thoracic and cranial lumbar vertebrae between the transverse processes and cranial articular processes; and accessory processes, which are located between the transverse and the caudal articular processes. Caudal vertebrae of some species possess a ventral or hemal arch for passage of blood vessels.

The Cervical Vertebrae

The cervical vertebrae, or neck bones, are seven in number for all terrestrial and arboreal mammals except for a species of tree sloth. Aquatic mammals such as whales and dolphins (cetaceae) and the manatee and dugong do not abide by this rule and may have one or more cervical vertebrae missing. The first two are greatly modified and are frequently designated by name (e.g., C_1, atlas, C_2 axis). The sixth and seventh cervical vertebrae are also modified but do not differ greatly from the remainder. With the exception of the atlas, the cervical vertebrae are roughly quadrangular and are ordinarily longer through the body than vertebrae of other regions. The processes and structures of the cervical vertebrae are massive, but tend to become less so with each successive vertebrae. The length of the bodies decreases from C_2 to C_7.

Lateral to the body of most cervical vertebra are two openings, one on each side, running parallel with the vertebral foramen. These are called the foramina transversaria. Successive foramina transversaria form two lateral canals, or canalis transversaria, through which pass the vertebral vessels and nerves. The foramina begin with the sixth cervical vertebrae and end in the atlas.

The atlas, or first cervical vertebra, is formed as a strong flattened ring with markedly enlarged transverse processes or "wings." The ring encloses a large vertebral foramen and is composed of two lateral masses connected by dorsal and ventral arches. The atlas articulates cranially with the occipital condyles of the skull and caudally with the dens or odontoid process of the axis. The atlas is an exception to the general rule that vertebral bodies of cervical vertebrae are convex cranially and concave caudally; both ends are concave.

The axis, or second cervical vertebra, has the longest body of the cervical vertebrae and is characterized by the dens or odontoid process on its cranial face which fits into the caudal part of the atlas. It has a heavy and strong dorsal spine which has an articular process on its flattened caudal end.

To a large extent, the atlas and axis control movement of the head. The skull articulates in a hinge (ginglymus) fashion with the atlas, permitting an up and down movement. The atlas articulates on a pivot (trochoid) joint with the axis, permitting rotation of the head.

The third, fourth, and fifth cervical vertebrae are very similar and progressively shorter in length. These vertebrae have well developed cranial and caudal articular processes: short, thick dorsal spines; a rudimentary ventral spine; and large, plate-like transverse processes which extend laterally and backward.

The sixth and seventh cervical vertebrae are somewhat different

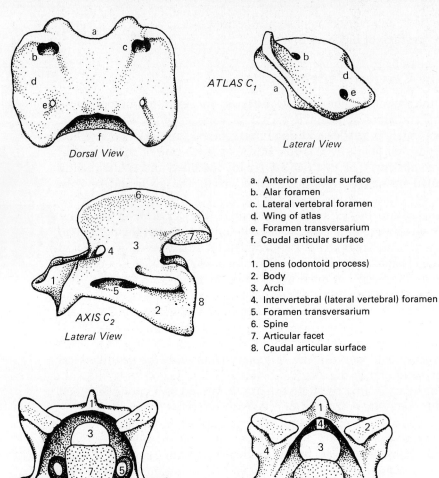
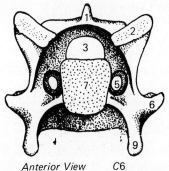
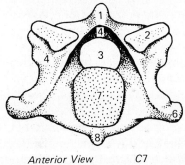

FIGURE 3-24   *Atlas, axis and cervical vertebrae of the horse.*

THE AXIAL SKELETON

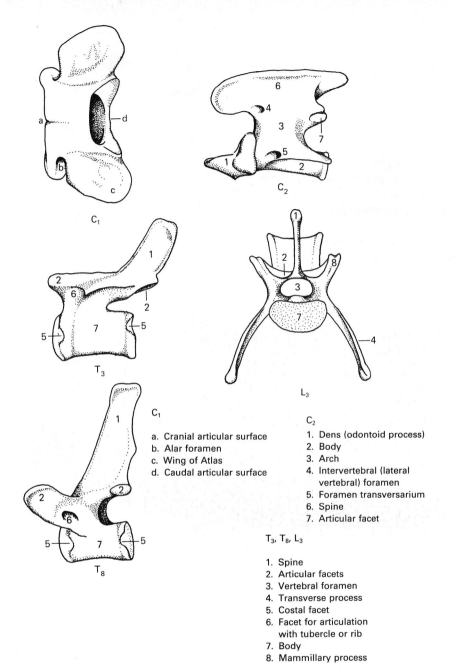

C₁
a. Cranial articular surface
b. Alar foramen
c. Wing of Atlas
d. Caudal articular surface

C₂
1. Dens (odontoid process)
2. Body
3. Arch
4. Intervertebral (lateral vertebral) foramen
5. Foramen transversarium
6. Spine
7. Articular facet

T₃, T₈, L₃
1. Spine
2. Articular facets
3. Vertebral foramen
4. Transverse process
5. Costal facet
6. Facet for articulation with tubercle or rib
7. Body
8. Mammillary process

FIGURE 3-25   Vertebrae of cat.

from the preceding three with larger vertebral foramina and more apparent spinous processes. The seventh, in addition, possesses facets on each side of the caudal articular surface of the body for articulation with the heads of the first pair of ribs. It does not possess foramina transversaria.

The Thoracic Vertebrae

There is considerable variation in the number of thoracic vertebrae between species of mammals, and occasionally there is variation within a species.

The bodies are short. Cranial and caudal paired lateral facets (costal

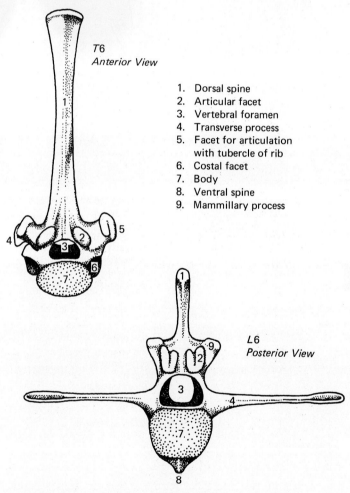

1. Dorsal spine
2. Articular facet
3. Vertebral foramen
4. Transverse process
5. Facet for articulation with tubercle of rib
6. Costal facet
7. Body
8. Ventral spine
9. Mammillary process

FIGURE 3-26 *Thoracic and lumbar vertebrae of the horse.*

facets) for articulation with the heads of the ribs are present. The arches and articular processes are small. The transverse processes are short and thick and possess facets for articulation with the tubercle of the rib which possess the same number as the vertebra, e.g., the tubercles of the first pair of ribs articulate with the body of T_1. In the horse the ventral spine is absent. The dorsal spine is prominent and is slanted backward. The length of the spines increases from the first to the fourth thoracic vertebrae and decreases from the fourth to the last. The summits of the spinous processes are expanded and rough.

The last thoracic vertebra usually lacks a caudal pair of costal (rib) facets. Exceptions to this are found in horses with floating ribs and in swine and sheep with supernumerary ribs. Other species may also be an exception to this statement when there are more pairs of ribs than there are thoracic vertebrae.

The Lumbar Vertebrae

These vary in number between species and are characterized by dorsal spines which slant forward, large and well-formed transverse processes which are flattened dorsoventrally, a distinct ventral spine, and relative equality in size and shape.

In animals other than horses and cattle, the transverse processes are more likely to be oval in cross section, less prominent, and tend to slant downward and forward.

The lumbar vertebrae of the so-called rigid-spined mammals (e.g., horses, cattle, sheep) have an interlocking arrangement of the cranial and caudal articular facets that bind the lumbar vertebrae into a relatively inflexible bony mass. In horses the last one or two vertebrae often become fused to the sacral vertebrae through exostosis (bony outgrowth) and ankylosis (fusion) of their transverse processes.

The Sacral Vertebrae

These vary in number and shape among the domestic animals, but consist essentially of a fused or partially fused mass of two or more bones that have rather well-marked dorsal spines that slant caudally. Numerous dorsal and lateral foramina are present, and a well-developed, roughened, articular surface on the cranial lateral face exists for articulation with the pelvis.

The Caudal (Coccygeal) Vertebrae

These vary considerably in number, even within a species, and decrease progressively in size and complexity from the first to last.

THE VERTEBRAL FORMULA

A vertebral formula is a shorthand method of listing the number of vertebrae in a given species of animal. A typical example would appear as follows:

$$\text{Horse:} \quad C_7 \quad T_{18} \quad L_6 \quad S_5 \quad Cy_{15\text{-}21}$$

This means that a horse has seven cervical vertebrae, 18 thoracic, six lumbar, five sacral, and from 15 to 21 caudal vertebrae.

PHYSIOLOGY OF THE VERTEBRAL COLUMN

The vertebral column, while morphologically similar in the various species, exhibits some functional differences between species. Flexibility is perhaps the outstanding of these. In animals that depend upon running in order to eat or to avoid being eaten, the vertebral column is one of two functional types: a relatively rigid structure (e.g., involving the thoracic, lumbar and sacral segments) such as is found in horses, cattle, and sheep, or the highly flexible spines of dogs and cats. In herbivores, the column not only serves to transmit the forces of the legs to the body, but it also serves to support a large digestive tract and a pregnant uterus in range or wild environments where the pregnant female carries one or more large fetuses for considerable distances during gestation.

Carnivores, however, do not have a large digestive tract, nor do they need to travel such great distances to acquire food. Their habits tend to be more sessile, their ranges more restricted, and during the latter stages of pregnancy the females generally move only when absolutely necessary. The spine of the carnivore, thus freed from constant major weight bearing functions, tends to be more flexible and a greater aid in running. Where herbivores must depend principally upon their legs, the carnivore can, by flexing its spine, add considerable distance and force to its stride. The flexible spine also aids acceleration. The enormous acceleration developed by a charging lion (0–45 mph in five or six strides) appears to be aided by the flexibility of the spine as much as any other factor.

Comparative Vertebral Formulae

	C	T	L	S	Cy
Horse	7	18	6	5	15-21
Cow	7	13	6	5	18-20
Sheep	7	13	6-7	4	16-18
Pig	7	14-15	6-7	4	20-23
Dog	7	13	7	3	20-23
Cat	7	13	7	3	5-23
Human	7	12	5	5	4

The Ribs (Costae)

Classification: Modified long bones.

Location: Between the thoracic vertebrae and the sternum. Each rib articulates dorsally with the bodies of two vertebrae and is continued ventrally by a costal cartilage which may or may not attach directly to the sternum.

Description: The number and shape of ribs varies with the species of mammal. A typical rib is composed of a shaft and two extremities. In horses and other large mammals the shaft is flattened and strongly curved. In small mammals the shaft is usually oval or round in cross section.

The proximal extremity consists of a head, neck, and tubercle. The head articulates with the bodies of two adjacent thoracic vertebrae. The neck is the bony junction of head and shaft. The tubercle projects backward at the junction of neck and shaft and bears a small facet on its dorsomedial surface for articulation with the transverse process of its thoracic vertebra.

The distal or costal extremity is roughened for connection with the costal cartilages. Those ribs directly connected to the sternum (breastbone) by costal cartilages are known as sternal ribs (true ribs). The remainder are asternal ribs (false ribs). Ribs at the caudal end of the thorax which have their ventral ends free are known as floating ribs. Spaces between ribs are called intercostal spaces. The ribs contain spongy red marrow and are important throughout the life of the animal as blood forming organs. The ribs also function to protect the thoracic viscera and, together with their muscles, act as respiratory aids.

The Costal Cartilages

These are bars of hyaline cartilage which serve to connect the ribs directly or indirectly to the sternum. The cartilages of asternal ribs overlap to form the costal arch. Cartilages of floating ribs are not attached

Comparative Anatomy of Sternal and Asternal Ribs

	Total No.	Sternal	Asternal
Horse	18-19 pair	8 pair	10-11 pair (1 pr. floating when 11 pr. present)
Cow	13 pair	8 pair	5 pair
Sheep	13-14 pair	8 pair	5-6 pair
Pig	14-15 pair	7 pair	7-8 pair
Dog	13 pair	9 pair	4 pair (last pr. usually floating)
Cat	13 pair	9 pair	4 pair
Human	12 pair	7 pair	5 pair (2 pr. floating)

to the costal arch. In mature animals, and most often in dogs, the costal cartilages may become partially ossified.

The Sternum

The sternum is formed by a group of medial, ventral, segmental, unpaired bones, which complete the thoracic skeleton ventrally. It is composed of three parts: the presternum (manubrium sterni) and cariniform cartilage; the mesosternum (body); and the metasternum and xiphoid (ensiform) cartilage.

The presternum has many shapes. These are determined to some extent by the presence or absence of a clavicle (collar bone). Animals with clavicles usually have a broad and strong presternum, while animals without clavicles normally possess a narrow (cow), laterally compressed (horse), or rudimentary (dog) presternum. The presternum has on its cranial surface a cariniform cartilage which is extensive in the horse, absent in the cow, sheep, and human, and is rudimentary or absent in most other animals. The body (mesosternum) of the sternum of the horse consists of seven bony segments called sternabrae which are roughly cuboidal in shape. The metasternum is the caudal bony segment of the sternum and bears the xiphoid cartilage on its caudal end. The xiphoid cartilage is present in all animals though greatly variable in size and shape. It is called the ensiform cartilage in man.

The Thoracic Cavity

The thorax is a closed cavity. It is bounded by the thoracic vertebrae dorsally, the ribs and costal cartilages laterally, and the sternum ventrally. The cranial aperture (the thoracic inlet) is formed by the first thoracic vertebra, the first pair of ribs and costal cartilages, and the presternum. The caudal (diaphragmatic) aperture is composed of the last thoracic vertebra, the last pair of ribs, the costal arch, and the metasternum. The diaphragm forms a partition between the thoracic and abdominal cavities. The thorax contains the heart, lungs, diaphragm, thymus, portions of the esophagus and trachea, the pleural and peri-

The Comparative Anatomy of the Sternum

Sternum	No. of Bones	Fused or Separated	Cartilages	
			Xiphoid	Cariniform
Horse	7	F	+	+
Cow	7	F	+	−
Sheep	6-7	F	+	−
Pig	6	S	+	±
Dog	8	S	+	±
Cat	8	S	+	±
Human	6	F	+	−

cardial membranes, the beginning and end of the blood vascular system, the termination of the lymphatic system, and other less important structures.

The Skull

Classification: A group of flat and irregular bones.

Location: At the cranial extremity of the vertebral column.

Description: The skull is divided into two parts, the cranium and the face. The cranium is composed of the bones of the skull which immediately surround the brain and which are incorporated into the floor and vault of the brain cavity.

The face consists of the bones of the skull which are not a part of the cranium.

THE BONES OF THE CRANIUM

The bones of the cranium consist of the following single and paired bones:

Single Bones	*Paired Bones*
1. Occipital	1. Interparietal
2. Sphenoid	2. Parietal
3. Ethmoid	3. Frontal
	4. Temporal

THE SINGLE BONES OF THE CRANIUM

The single bones of the cranium are highly modified cranial extensions of the vertebral column. The occipital bone is situated at the caudal part of the cranium, and forms the rear walls of the vault and floor of the cranial cavity. The ventral portion of the bone is perforated by a large foramen, the foramen magnum, through which the spinal cord passes. The foramen magnum marks the junction of the cranial cavity and the vertebral canal and is bounded dorsally by the squamous part, laterally by the lateral parts, and ventrally by the basilar part of the occipital bone. The lateral parts bear the occipital condyles, which articulate with the atlas (C_1). Lateral to the condyles are the paramastoid processes, blunt bony prominences which project ventrally. The basilar part extends rostrally* from the floor of the foramen magnum and attaches to the caudal aspect of the sphenoid.

---

*In the new terminology, the front (or nasal) end of the skull is called "rostral" rather than "anterior," and the rear (or vertebral) end is called "caudal" rather than "posterior."

THE SKELETON

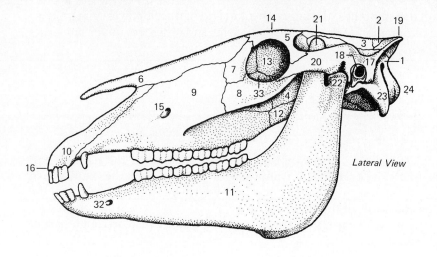

FIGURE 3-27a   Skull of the horse.

1. Occipital
2. Interparietal
3. Parietal
4. Squamous temporal
5. Frontal
6. Nasal
7. Lacrimal
8. Zygomatic (malar)
9. Maxilla
10. Premaxilla (incisive bone—os incisivum)
11. Mandible
12. Palatine
13. Orbit
14. Supraorbital foramen
15. Infraorbital foramen
16. Foramen incisivum (rostral end of skull)
17. Petrous temporal
18. External acoustic meatus
19. Nuchal crest (caudal end of skull)
20. Zygomatic arch (zygomatic process of temporal)
21. Coronoid process of mandible
22. Condyle of mandible
23. Jugular process (paramastoid process)
24. Occipital condyle
32. Mental foramen
33. Lacrimal foramen

The squamous part of the occipital bone is a roughly semicircular mass that lies above the foramen magnum. The external surface has a prominent ridge, the nuchal crest. The rostral borders unite with the parietal, interparietal, petrous temporal, and squamous temporal bones.

The sphenoid bone lies along the floor of the cranium rostral to the basilar part of the occipital. It consists of three parts: the body, the wings, and the pterygoid processes.

The body is medial, cylindrical, and flattened dorsoventrally. The rostral part of the body is concealed to some extent by the vomer and pterygoid bones. The caudal end articulates with the basioccipital. The body is hollowed rostrally by the sphenopalatine sinus and contains the optic canals and the pituitary fossa on its dorsal aspect.

The wings are composed of two parts, the orbital wings and the temporal wings. The orbital wings are larger and curve dorsolaterally from the sides of the presphenoid to unite dorsally with the frontal bone, rostrally with the ethmoid bone, and caudally with the squamous temporal, and an overlapping portion of the temporal wings. The temporal wings extend laterodorsally from the postsphenoid to unite dorsally with the squamous temporal, rostrally with the orbital wings, and caudally with the parietal bone. The pterygoid processes underlie the pterygoid bones.

The ethmoid bone is situated in front of the body and orbital wings of the sphenoid. It, too, consists of three parts: the cribriform plate, the lateral masses, and the perpendicular plate. The cribriform plate, a sieve-like partition, is located along a transverse plane between the cranial and nasal cavity. The lateral masses (also called the ethmoid turbinates) are cone-shaped structures composed of an extremely complex arrangement of coiled and folded sheets of thin bone covered with mucous membrane. The bases of the lateral masses are attached to the cribriform plate and are separated medially by the perpendicular plate. The entire complex that forms the lateral masses is called the ethmoid turbinates. Some 27 subsidiary parts have been recognized. Basically these consist of a primary group of 6 endoturbinates and a secondary group of 21 (or more) ectoturbinates. The perpendicular plate of the ethmoid is situated medially and runs along the median plane, and is continued rostrally by the nasal septum.

THE PAIRED BONES OF THE CRANIUM

The interparietal bones are situated centrally between the squamous part of the occipital bone and the parietal bones. Generally there are two bones in the young that fuse into a single bone in the adult. They may be so completely fused to the parietal bones in older animals that they cannot be identified.

The parietal bones compose the greatest part of the vault of the cranium. They are quadrilateral in outline and have two surfaces (outer

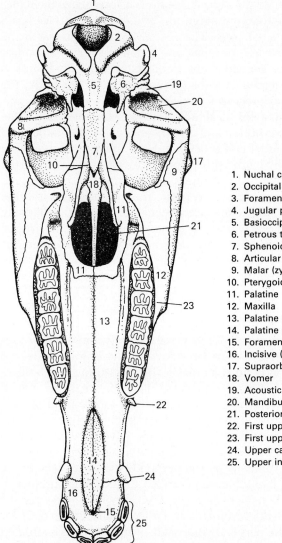

FIGURE 3-27b   Skull of the horse—ventral view.

1. Nuchal crest
2. Occipital condyle
3. Foramen magnum
4. Jugular process (paramastoid process)
5. Basioccipital
6. Petrous temporal
7. Sphenoid
8. Articular tubercle
9. Malar (zygomatic)
10. Pterygoid
11. Palatine
12. Maxilla
13. Palatine process of maxilla
14. Palatine process of premaxilla
15. Foramen incisivum
16. Incisive (premaxilla, os incisivum)
17. Supraorbital process
18. Vomer
19. Acoustic
20. Mandibular recess
21. Posterior nares
22. First upper premolar
23. First upper molar
24. Upper canine tooth
25. Upper incisor teeth

and inner), and four borders (medial, lateral, rostral, and caudal). The medial borders unite with each other along the midline of the skull. The lateral borders unite with the squamous temporal bones. The rostral borders unite with the frontal bones. The caudal borders unite with the interparietals and the occipital bone.

The frontal bones are situated between the parietal, and the nasal bones at the border of the cranium and face. They enclose the frontal sinuses and consist of three parts: nasofrontal, orbital, and temporal. The nasofrontal part forms the forehead. At its junction with the orbital

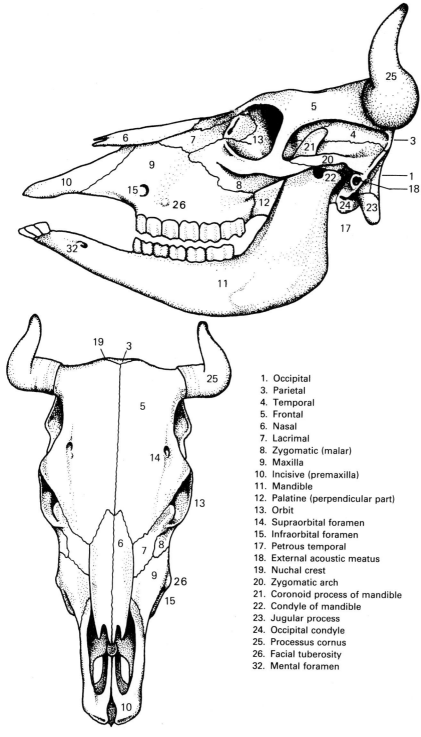

FIGURE 3-28   *Skull of the cow.*

THE SKELETON

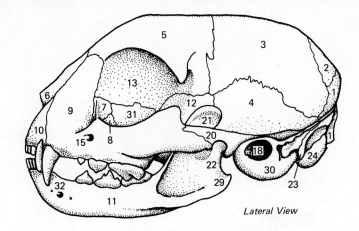

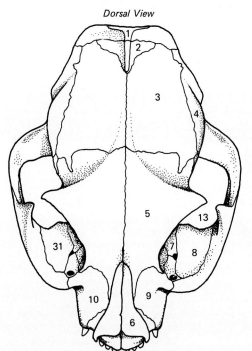

1. Occipital
2. Interparietal
3. Parietal
4. Temporal
5. Frontal
6. Nasal
7. Lacrimal
8. Malar zygomatic (malar)
9. Maxilla
10. Incisive (premaxilla)
11. Mandible
12. Palatine
13. Orbit
15. Infraorbital foramen
18. External acoustic meatus
20. Zygomatic arch
21. Coronoid process of mandible
22. Condyle of mandible
23. Jugular process
24. Occipital condyle
29. Angular process of mandible
30. Tympanic bulla
31. Orbitosphenoid
32. Mental foramen

FIGURE 3-29   Skull of the cat.

part it gives off the supraorbital process. It unites with the zygomatic process of the temporal bones. The orbital part forms most of the medial wall of the orbit of the eye. It is separated from the nasofrontal part by a ridge that forms the dorsal margin of the optic cavity. The temporal part is separated from the orbital part by a groove that is covered by the orbital wing of the sphenoid.

The temporal bones form the greater part of the lateral wall of the cranium. They are bounded dorsally by the parietal bones, ventrally by the temporal wings of the sphenoid bone, rostrally by the frontal and malar (zygomatic) bones, and caudally by the occipital bone.

In the horse the temporal bones consist of two parts, the squamous temporal and petrous temporal. The squamous temporal bone is the larger of the two. From the ventrolateral surface of this bone the zygomatic process extends to join the posterior end of the malar bone and form the zygomatic arch. The zygomatic process bears a condyle and a glenoid cavity for articulation with the mandible. The petrous temporal bone is characterized by the presence of the external acoustic process and the opening in it, which is called the external acoustic meatus. The petrous temporal bone lies between the occipital and squamous temporal bones and covers the mastoid processes and the structures of the middle ear. It is extremely dense and hard.

THE BONES OF THE FACE

The bones of the face consist of the following single and paired bones:

Single Bones	Paired Bones
1. Vomer	1. Maxilla
2. Mandible	2. Incisive (premaxilla)
3. Hyoid	3. Palatine
	4. Pterygoid
	5. Nasal
	6. Lacrimal
	7. Zygomatic (malar)
	8. Dorsal turbinates (conchae)
	9. Ventral turbinates (conchae)

THE SINGLE BONES OF THE FACE

The vomer is a medial bone which enters into the formation of the ventral part of the septum nasi. It is grooved to receive the perpendicular plate of the ethmoid bone and the septal cartilage. The rostral extremity lies above the palatine process of the premaxilla.

The mandible is the largest bone of the face and forms the lower jaw. It is composed of two halves, which fuse anteriorly at the symphysis mandibulae and is considered as a single bone thenceforth. It contains the lower teeth and articulates with the squamous temporal at the condyloid fossa. It consists of a body and two rami and possesses two pair of major foramina: the mental foramina, which are located laterally on the body, midway between the first premolars and the canine teeth, and the mandibular foramina, which are located medially about halfway up the rami.

In the horse the hyoid bone is composed of 12 distinct parts, the tympanohyoid (2), the stylohyoid (2), the epihyoid (2), the ceratohyoid (2), the basihyoid (1), the thyrohyoid (2) and the lingual process (1).* The temporal cartilages attach the hyoid to the styloid processes of the right and left petrous temporal bones. The various components of the hyoid are united by cartilage. Although technically the hyoid is a single bone, in actuality it is not. It supports the base of the tongue, pharynx, and larynx and through its cartilagenous "joints" aids the swallowing and breathing processes. Hyoids are similar in all domestic mammals, although the cat and dog do not possess a lingual process.

THE PAIRED BONES OF THE FACE

The maxillae form the basal part of the upper jaw and contain the sockets of the upper cheek teeth and canines. They articulate with almost all the facial bones and the frontal and temporal bones of the cranium. They each consist of a body, zygomatic process, and palatine process, and enclose the maxillary sinuses. They possess two major foramina (the infraorbital foramina) which open on the lateral surfaces about three inches above the third premolars.

The incisive bones form the rostral part of the upper jaw and sockets for the incisor teeth. They articulate with the maxillae and nasal bones and unite with each other rostromedially at the premaxillary symphysis. They each consist of a body, palatine process, and nasal process, and unite to form one major foramen, the foramen incisivum, which opens rostrally on the line of union and lies dorsal to the central incisors.

The palatine bones are located on the sides of the caudal nares. They consist of two parts: a horizontal part (hard palate) and a perpendicular part which forms the majority of the lateral walls of the caudal nares.

The pterygoids are two very small flattened bones situated on either side of the caudal nares.

The nasal bone forms the major portion of the roof of the nasal cavity. It is located anterior to the frontal bone and is triangular in outline, wide posteriorly, and narrow anteriorly. It forms the dorsal part of the nasal cavity.

The lacrimal bones are found in the anterior part of the orbit of the eye. They possess three surfaces: the facial surface, orbital surface, and nasal surface. The lacrimal bones are perforated by the lacrimal foramina which penetrate the orbital surfaces and form the posterior portions of the lacrimal ducts, which extend from the eyes to the anterior nares.

The zygomatic bones (malar bones) are ventral to the lacrimals.

---

*Older veterinary terminology calls these divisions of the hyoid bone the styloid cartilage, the great (or greater) cornu, the middle cornu, the small (or lesser) cornu, the body, the thyroid cornu, and the lingual process.

They possess three surfaces: the facial surface, orbital surface, and nasal surface. The caudal portion of the zygomatic bone unites with the zygomatic processes of the squamous temporals to form the right and left zygomatic arches.

There are four turbinate (conchal) bones arranged in two pairs, dorsal and ventral. These pairs are attached to the middle of the lateral walls of the nasal cavity and project into it, filling most of the open space. The dorsal conchae are larger than the ventral. The bones are very thin and are rolled into a scroll-like formation, which is covered with mucuous membrane. By their arrangement with each other and with other bones, they form four channels in the nasal cavity: the dorsal nasal meatus, the middle nasal meatus, the ventral nasal meatus, and the common nasal meatus. The dorsal meatus is the space between the dorsal concha and the nasal bone. The middle meatus separates the ventral concha and the palatine processes of the incisive and maxilla. The common meatus separates each pair of turbinates from the septum nasi and vomer bone. The ventral meatus lies between the ventral conchae and the palatine processes of the premaxilla and maxilla.

THE VISCERAL SKELETON

The visceral skeleton is extremely variable in animals and consists of bones that are developed in the soft tissues of the body. Among the domestic animals, only a few species possess a visceral skeleton, and one of these is open to question. The bones generally conceded to belong to the visceral skeleton are the os cordis and the os penis and the scleral ring (when ossified) in birds. The os rostri of the pig is open to question because it apparently derived from the basal septum. Remnants of other bones, such as the clavicle in dogs and sesamoid bones in various species, also fit this definition but are not included among the visceral skeletal structures.

Bone	Location	Species
1. Os rostri	in the soft tissues of the external nares	Swine
2. Os cordis	around the origin of the aorta and pulmonary artery at the base of the heart	All members of the bovine family *(bovidae)* and sheep *(ovidae)*
3. Os penis	within the penis	All members of the dog family *(canidae)* Members of the mink family *(mustellidae)* Some rodent species *(rodentia)* Seals, walruses, and raccoons
4. Scleral ring	in the sclera of the eye, surrounding the ciliary body and iris	Most birds (all domestic fowl)

chapter 4
Arthrology (Syndesmology)

There is a certain confusion of terms insofar as this chapter is concerned. At the moment "syndesmology" is the accepted term in one of the newer veterinary texts. "Arthrology" is the preferred term in human anatomy. Nomenclature committees have been waffling over these terms since 1893 when "syndesmology" was proposed as the name for the study of joints. Although the term is semantically incorrect, "syndesmology" persists in the literature because an initial name always carries some weight.

Arthrology literally means "the study of joints." Joints are divided into three major classes: synarthroses (immovable joints), diarthroses (movable joints), and amphiarthroses (partially movable joints).

Joints are formed when two or more bones come together and are united by fibrous, elastic, or cartilagenous tissue (or by a combination of two or more of these uniting media).

Here, I think, I had better warn you that if you read beyond this text, you are going to have problems with terminology. Veterinary anatomy is passing through a period of change where—in an effort to standardize nomenclature—an international naming organization (the Nomina Anatomica Veterinaria) has settled upon Latin as the language of choice for anatomy. There is a certain deadly parallel in this since Latin is static, dead and cumbersome, yet it is preferable to books whose authors believe they have carte blanche to alter terminology. An example of Latin terminology is given in Figures 5-2, 5-3, 5-4, 5-5 on pages 106–109.

CLASSIFICATION OF JOINTS

Synarthroses

Synarthroses are those joints which are united by fibrous tissue and/or cartilage in such a manner as to practically prevent all movement. There is no joint cavity. Many of these joints become firmly united to each other by bony growth. Immovable joints include sutures, syndesmoses, synchondroses, symphyses, and gomphoses.

Sutures are joints where bones are united by fibrous connective tissue (ligaments) along lines which can be extremely irregular; such joints are found in the skull. Syndesmoses are immovable joints in which white fibrous and/or yellow elastic tissues unite the bones; examples of these joints are the union of the shafts of the metacarpals in the horse. A synchondrosis is a joint in which the bones are united by cartilage; the best example of this joint is found in the union of the basioccipital and sphenoid in young animals. Symphyses consist of, and are limited to, certain median joints which connect symmetrical bones; an example is the symphysis pelvis. The uniting medium is cartilage and fibrous tissue. The gomphosis is the joint between a tooth and its socket.

Diarthroses

Diarthroses are those joints that possess a cavity, a capsule, and mobility. They may be simple, possessing only two articulating surfaces, or complex, with more than two articulating surfaces. All true diarthrodial joints must have the following structures: articular surfaces, articular cartilages, a joint capsule and a joint cavity.

Certain other structures may enter into the formation of a diarthrodial joint. These include ligaments, which are bands or sheets of inelastic but flexible white fibrous tissue which help bind the joint together. Other structures that may be found are articular discs (menisci), which are plates of fibrocartilage inserted between articular surfaces of a joint, and marginal cartilages, which are rings of fibrocartilage that circle the rim of joint cavities.

The movements of diarthrodial articulations are determined by the shape and extent of the joint surfaces and the arrangement of the ligaments. These movements are gliding, angular, circumduction, rotation, adduction, and abduction. Gliding movement is illustrated by the articular processes of the cervical vertebrae, which slide upon one another. Angular movement involves movement around one or more axes and is well illustrated by the elbow joint. Angular movement is composed of two parts, flexion and extension. In flexion the angle diminishes,

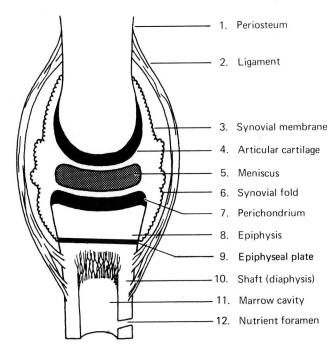

FIGURE 4-1   Composite schematic diagram of a movable joint.

while in extension the angle increases. Circumduction is the movement whereby the distal end of a limb describes a circle or a segment of a circle. Rotation is the movement of one segment around the long axis of another; the atlantoaxial articulation in the neck is a good example of this. Adduction is movement of a limb toward the median plane, and abduction is movement of a limb away from the median plane.

Diarthrodial joints are classified as to their specialized forms and movements, as arthrodia (gliding joint), ginglymus (hinge joint), trochoid (pivot joint), and enarthrosis (ball and socket joint).

Amphiarthroses

Amphiarthroses are those joints that share some characteristics of both synarthrodial and diarthrodial joints. These joints unite the bodies of the vertebrae. Except for the atlas (C_1) and the axis (C_2) they do not possess a true joint capsule. The vertebral bodies are united by ligaments which form a joint cavity and a kind of joint capsule. In some species, notably flexible-spined running animals, synovial capsules may be present throughout the vertebral column, which would exclude spinal joints in these species from the amphiarthrosis category.

In the cervical, thoracic, and lumbar regions menisci (intervertebral discs) are inserted between the bodies of adjacent vertebrae. The intervertebral discs are cushioning pads and function as shock absorbing

ARTHROLOGY (SYNDESMOLOGY)

devices. Each disc is composed of a tough fibrocartilage outer layer, the annulus fibrosus, which covers an inner gelatinous layer, the nucleus pulposus.

If the nucleus loses its elasticity or the annulus is damaged, the annulus can rupture and allow the nucleus to be displaced into the vertebral canal, putting pressure on the spinal cord. This condition is called a herniated nucleus pulposus, or a "slipped disc." It results in severe pain which may be accompanied by partial or total paralysis caudal to the point of injury.

Sarcoarthroses (Synsarcoses)

A joint such as the connection between the scapula and the thorax in the horse and other cursorial animals has no conjunction of bone to bone. Nevertheless the structure is a joint, has mobility, and contributes to the movement of the animal. I am inclined to refer to such structures as sarcoarthroses or "fleshy joints." They have also been called synsarcoses. However, there is a great deal of difficulty in making this particular structure conform to the classification scheme for joints. So, without attempting to alter the classification, this structure is included as a special case.

Synovia

Movable joints are usually lubricated. If they are not, they quickly become immovable. The lubricant is a proteinaceous fluid called synovia. It is ordinarily colorless to yellowish in color and is remarkably similar to the thin albumin found in a chicken's egg. The synovia is secreted from the albuminous fraction of the blood supply to the joint by synovial cells which are found in the synovial membrane inside the joint capsule. Synovia functions to prevent erosion of the joint cartilages.

chapter 5

The Muscular System

The muscular system is the most prominent and extensive anatomical feature of the body. However, this system is more of interest to students of veterinary medicine and surgery than to those whose interests lie mainly in agricultural and conservation fields. Therefore, this section will be covered rather briefly with major attention being paid to muscle physiology.

TYPES OF MUSCLE

In general, muscles are divided into three major groups depending upon their cellular structure. These are striated (striped or skeletal) muscle, smooth (unstriped or visceral) muscle, and cardiac (heart) muscle. All muscle tissue, regardless of its structure or location, possesses one common quality—the ability to contract or shorten its normal length. Muscles do not stretch except under unusual circumstances. Apart from this, there are considerable differences between the various types of muscles.

Striated Muscle

Striated (striped) muscle is composed of long, unbranched, multinucleated fibers. It derives its name from the fact that under the microscope each muscle fiber is crossed transversely by dark bands or "striations." Striated muscle is of two types: red and white (or pink) muscle. The red muscle is found in areas which perform sustained and continuous work, and white muscle is found in areas which perform quick, but intermittent movement. The difference in color appears to be related to the amount of myoglobin and cytochrome, and the number of mitochondria in the muscle cytoplasm. Red muscle apparently generates ATP (adenosine triphosphate) faster than white muscle and is more resistant to fatigue. Other features of striated muscle are the peripheral location of the nuclei in the muscle fibers and the fact that nerve endings apparently enter every muscle fibril.

These nerve endings are called motor end plates or neuromuscular junctions. Just prior to its termination, the axon of a motor nerve fiber (see Nervous System p. 311) loses its myelin sheath and divides into a number of fine terminal fibers which appear to end at individual muscle fibers. Electron microscope studies reveal that there is a concentration of mitochondria in the terminal nerve fibers which would indicate biochemical activity. The motor end plates contact the muscle fibril along an area of folded sarcoplasmic reticulum called the junctional folds. The end plate is separated from the terminal part of the nerve fiber by a cleft about 500Å wide. This is called the synaptic cleft and appears to function in the same manner as a cholinergic synapse (p. 317).

The muscle fibril (myofibril) is the basic unit of structure. Each fibril is composed of a number of longitudinal segments or sarcomeres crossed by a regular succession of cross bands labeled from smallest to largest in point of width M, Z, N, H, I, and A. The light colored I-band is called the isotropic band while the darker A-band is called anisotropic. The space from one Z-band to the next forms a sarcomere or functional unit of the fibril. Each fibril is surrounded by a network of reticular fibers, the sarcoplasmic reticulum. At the level of each Z-band a transverse membrane system (T-S) extends from the fibril to the sarcolemma, which surrounds the entire muscle fiber. Each muscle fiber is composed of a number of fibrils and is surrounded by a sheath of connective tissue called the sarcolemma, and another called the endomysium. Fibers combine to form a muscle bundle, or fasciculus, which is surrounded by a membrane called the perimysium. The size of these bundles varies between muscles and is responsible for the coarseness or "grain." A number of fasciculi are aggregated together into a muscle which is surrounded by still another membrane, the epimysium. Ordinarily a muscle consists of a body and two ends which terminate in cords or sheets of white fibrous connective tissue called tendons.

Striated muscle ordinarily possesses definite orgins and insertions, and is usually found attached by tendons at one or both ends to some part of the skeleton. It is voluntary, or under direct control of the will, and is capable of sudden and violent contraction. Each fiber in a striated

TYPES OF MUSCLE

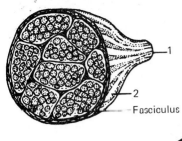

Muscle (gross)

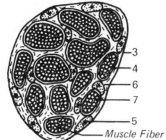

Fasciculus ("muscle bundle")

1. Tendon
2. Epimysium
3. Perimysium
4. Endomysium
5. Muscle cell nucleus
6. Connective tissue
7. Sarcolemma
8. Sarcoplasmic reticulum
T-S. Transverse membrane system

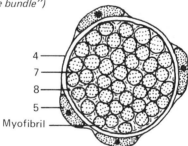

Muscle fiber

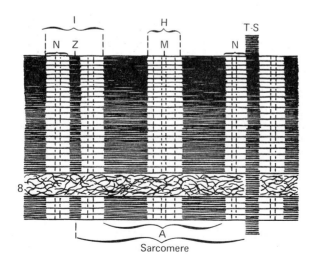

FIGURE 5-1   Gross and fine structure of muscle.

muscle is innervated, and extremely fine gradations of muscle activity can be obtained.

Since the ends of striated muscles are usually attached to some structure (usually the skeleton), the terms "origin" and "insertion" require definition. The classic statement that the origin of a striated muscle is the immovable part and the insertion the movable part is not entirely correct. It is better to think of the origin as the "anchoring" end and the insertion as the "working" end of a muscle. There are a few striated muscles in the body which do not possess origins and insertions. Examples of such muscle are the orbicularis oris muscle, which is the sphincter muscle of the mouth, the external anal sphincter, and the striated muscle sheathing the esophagus.

Muscles of the limbs are usually arranged in antagonistic pairs, that is, for every muscle that performs a given act, there is another muscle that performs an opposite action. For instance, the biceps brachii of the upper arm flexes the elbow joint while the triceps brachii extends the joint. Flexor and extensor, levator and depressor, adductor and abductor, and pronator and supinator muscles are arranged in this fashion.

The striated muscles of the trunk, however, are generally sheets or bands of muscle which are not arranged as are the muscles of the limbs. The counterbalancing muscle is usually found on the opposite side of the body in the same general area. However, assuming that the spinal column is a single bone rather than a chain, the principle of antagonistic pairs still holds reasonably well.

The real exceptions lie mainly in the subcutaneous muscles which are attached to either the skin or broad sheets of tendons called aponeuroses, and in the thoracic and abdominal muscles which support the viscera. These last function principally in respiration, defecation, and parturition. Although arranged in pairs they generally function as a unit.

Striated muscles are seldom completely relaxed even in the resting body. The general antagonism results in a phenomenon called tonus. This is a functional balance; the slight contraction of one muscle being counterbalanced by a concomitant contraction of its antagonist. The principle of muscle tone or tonus is characteristic of the normal healthy body, and the interplay of one muscle against the other serves to maintain the voluntary musculature in readiness for instant response to stimuli. Flaccidity or flabbiness of striated muscle is usually a response to nerve injury or debilitating disease. With loss of tone, a muscle quickly loses its power of response, and ultimately becomes atrophied and useless.

In describing striated muscles, the following classification is generally used

1. Name
2. Location
3. Origin
4. Insertion
5. Action
6. Structure (or shape)
7. Relationships
8. Blood supply
9. Nerve supply

The following description of the biceps brachii (quoted from Sisson and Grossman, *The Anatomy of the Domestic Animals* 3rd ed.) will show how the system works and why it is not used here.

Biceps brachii. This is a strong fusiform muscle, which lies on the anterior surface of the humerus.

Origin. The tuber scapulae.

Insertion. (1) The radial tuberosity; (2) the medial ligament of the elbow joint; (3) the fascia of the forearm and the tendon of the extensor carpi radialis.

Action. To flex the elbow joint; to fix the shoulder and elbow in standing; to assist the extensor carpi radialis; and to tense the fascia of the forearm.

Structure. The muscle is enclosed in a double sheath of fascia, which is attached to the tuberosities and the deltoid ridge of the humerus. The tendon of origin is molded on the intertubal or bicipital groove; it is very strong and dense and is partly cartilaginous. It is bound down here by a tendinous layer which furnishes attachment to part of the posterior deep pectoral muscle. Its play over the groove is facilitated by the large intertuberal or bicipital bursa (*bursa intertuberalis*); the synovial membrane extends somewhat around the edges to the superficial face of the tendon. A well-marked tendinous intersection runs through the muscle and divides distally into two portions. Of these, the short, thick one is inserted into the radial tuberosity and detaches fibers to the medial ligament of the elbow joint. The long tendon (*lacertus fibrosus*) is thinner, blends with the fascia of the forearm, and joins the tendon of the extensor carpi radialis; thus the action is continued to the metacarpus.

Relations. Laterally, the brachiocephalicus and brachialis muscles; medially, the posterior deep pectoral and the superficial pectoral muscles; in front, the anterior deep pectoral muscle; behind, the humerus, the coracobrachialis muscle, the anterior circumflex and the anterior radial vessels, and the musculocutaneous nerve.

Blood Supply. Branches of the brachial and anterior radial arteries.

Nerve Supply. Musculo-cutaneous nerve.

To carry this description through some 450 muscles which are present in an animal's body would result in a great deal of bulk that would not add material value to this text, especially since this book is designed primarily for nontechnically oriented students. For those who intend to pursue a medical or veterinary career, excellent descriptive works already exist that cover this field thoroughly.

However for the benefit of those who want to see what muscles are like in the animal body, I include four illustrations which show the complete dissection of the muscles of the cervical region of the horse. This may reinforce the reasons why I have not gone into detail on musculature in this text. For the basic student, such detail is unnecessary. For the advanced student, unless (s)he wishes to become a professor of anatomy, it is almost equally redundant. The muscular anatomy

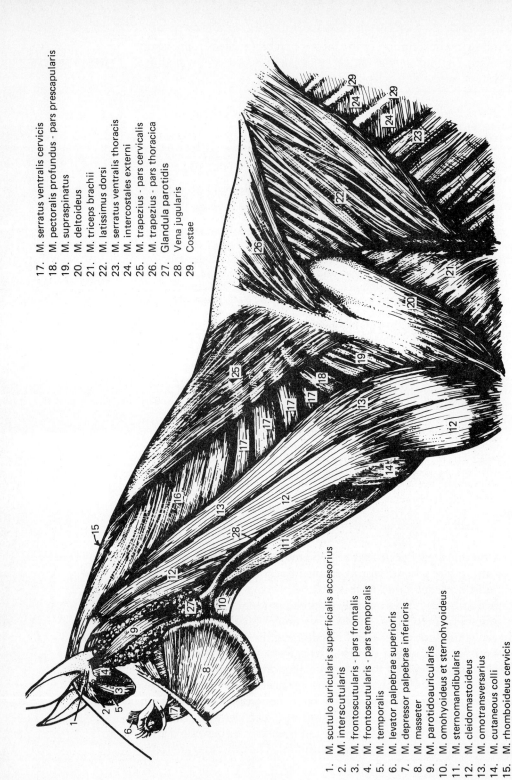

FIGURE 5-2 Horse superficial musculature cervical region.

1. M. scutulo auricularis superficialis accesorius
2. M. interscutularis
3. M. frontoscutularis - pars frontalis
4. M. frontoscutularis - pars temporalis
5. M. temporalis
6. M. levator palpebrae superioris
7. M. depressor palpebrae inferioris
8. M. masseter
9. M. parotidoauricularis
10. M. omohyoideus et sternohyoideus
11. M. sternomandibularis
12. M. cleidomastoideus
13. M. omotransversarius
14. M. cutaneous colli
15. M. rhomboideus cervicis
16. M. splenius
17. M. serratus ventralis cervicis
18. M. pectoralis profundus - pars prescapularis
19. M. supraspinatus
20. M. deltoideus
21. M. triceps brachii
22. M. latissimus dorsi
23. M. serratus ventralis thoracis
24. M. intercostales externi
25. M. trapezius - pars cervicalis
26. M. trapezius - pars thoracica
27. Glandula parotidis
28. Vena jugularis
29. Costae

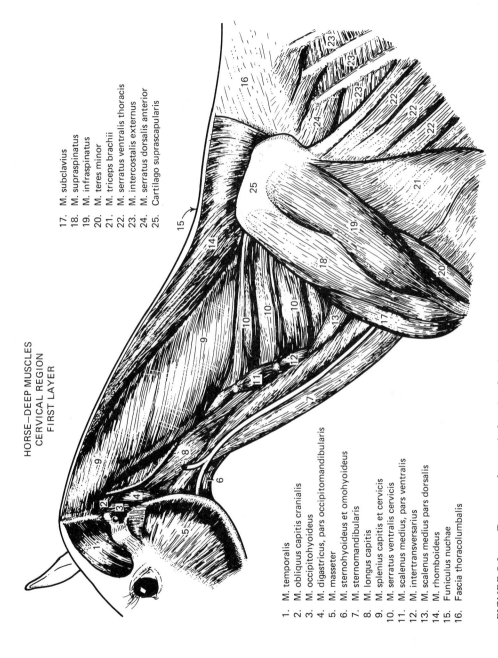

HORSE—DEEP MUSCLES
CERVICAL REGION
FIRST LAYER

1. M. temporalis
2. M. obliquus capitis cranialis
3. M. occipitohyoideus
4. M. digastricus, pars occipitomandibularis
5. M. masseter
6. M. sternohyoideus et omohyoideus
7. M. sternomandibularis
8. M. longus capitis
9. M. splenius capitis et cervicis
10. M. serratus ventralis cervicis
11. M. scalenus medius, pars ventralis
12. M. intertransversarius
13. M. scalenus medius pars dorsalis
14. M. rhomboideus
15. Funiculus nuchae
16. Fascia thoracolumbalis
17. M. subclavius
18. M. supraspinatus
19. M. infraspinatus
20. M. teres minor
21. M. triceps brachii
22. M. serratus ventralis thoracis
23. M. intercostalis externus
24. M. serratus dorsalis anterior
25. Cartilago suprascapularis

FIGURE 5-3   Deep muscles cervical region first layer.

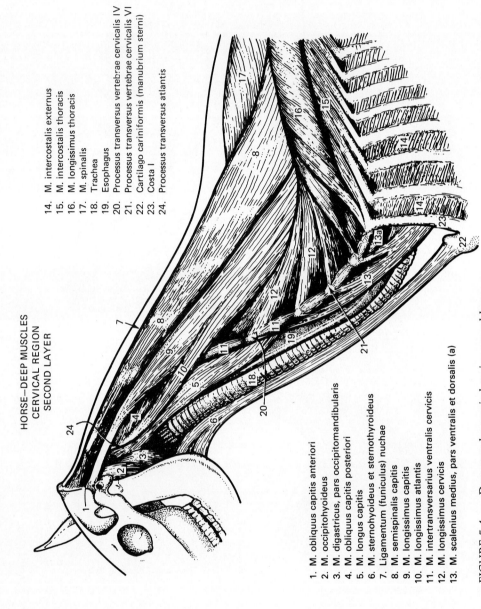

HORSE–DEEP MUSCLES
CERVICAL REGION
SECOND LAYER

1. M. obliquus capitis anteriori
2. M. occipitohyoideus
3. M. digastricus, pars occipitomandibularis
4. M. obliquus capitis posteriori
5. M. longus capitis
6. M. sternohyoideus et sternothyroideus
7. Ligamentum (funiculus) nuchae
8. M. semispinalis capitis
9. M. longissimus capitis
10. M. longissimus atlantis
11. M. intertransversarius ventralis cervicis
12. M. longissimus cervicis
13. M. scalenius medius, pars ventralis et dorsalis (a)
14. M. intercostalis externus
15. M. intercostalis thoracis
16. M. longissimus thoracis
17. M. spinalis
18. Trachea
19. Esophagus
20. Processus transversus vertebrae cervicalis IV
21. Processus transversus vertebrae cervicalis VI
22. Cartilago cariniformis (manubrium sterni)
23. Costa I
24. Processus transversus atlantis

FIGURE 5-4 *Deep muscles cervical region second layer.*

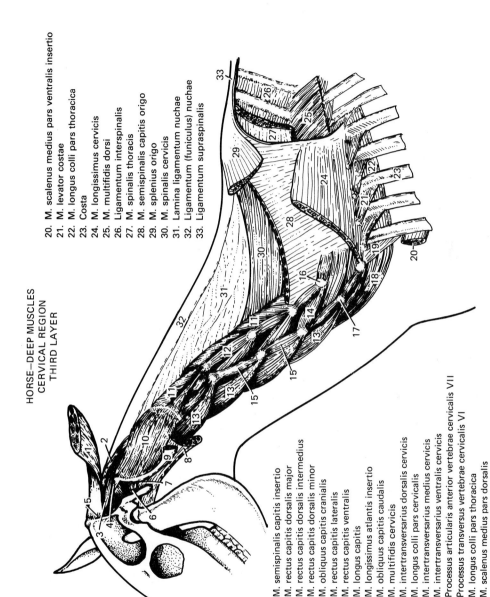

FIGURE 5-5 *Deep muscles, cervical region third layer.*

HORSE—DEEP MUSCLES
CERVICAL REGION
THIRD LAYER

1. M. semispinalis capitis insertio
2. M. rectus capitis dorsalis major
3. M. rectus capitis dorsalis intermedius
4. M. rectus capitis dorsalis minor
5. M. obliquus capitis cranialis
6. M. rectus capitis lateralis
7. M. rectus capitis ventralis
8. M. longus capitis
9. M. longissimus atlantis insertio
10. M. obliquus capitis caudalis
11. M. multifidis cervicis
12. M. intertransversarius dorsalis cervicis
13. M. longus colli pars cervicalis
14. M. intertransversarius medius cervicis
15. M. intertransversarius ventralis cervicis
16. Processus articularis anterior vertebrae cervicalis VII
17. Processus transversus vertebrae cervicalis VI
18. M. longus colli pars thoracica
19. M. scalenus medius pars dorsalis
20. M. scalenus medius pars ventralis insertio
21. M. levator costae
22. M. longus colli pars thoracica
23. Costa
24. M. longissimus cervicis
25. M. multifidis dorsi
26. Ligamentum interspinalis
27. M. spinalis thoracis
28. M. semispinalis capitis origo
29. M. splenius origo
30. M. spinalis cervicis
31. Lamina ligamentum nuchae
32. Ligamentum (funiculus) nuchae
33. Ligamentum supraspinalis

involved in certain surgical procedures, however, is extremely important, but this is a professional study and not applicable to basic books or procedures. Nevertheless, I suppose that a little detail does no harm, if only to show the extent of the subject that is so briefly covered in this chapter.

Smooth Muscle

There appear to be two types of smooth muscle, one of which is called visceral muscle and the other multi-unit muscle. Both use the myosin-actin mechanism of contraction which seems to be standard for all muscle tissue. The essential difference seems to be in the arrangement of the individual muscle fibers and the innervation. Visceral muscle is found in the muscular layer of the stomach and intestines; in the urogenital system; and in the arrectores pilorum (piloerector) muscles found at the base of certain hairs, particularly those in the neck and shoulder region and along the spine. Multi-unit muscle is found in areas where more precise control of muscle contraction is essential. Such areas include the sphincter and dilator iridis muscles of the eye, the smooth muscles in the walls of blood vessels, and the smooth muscle fibers around the alveoli and alveolar ducts of the lungs.

Smooth muscle, as the name implies, is smooth in structure (has no myofibrils) and does not possess the cross striations characteristic of skeletal and cardiac muscle. It is white to pinkish white in color, and usually does not possess origins or insertions, and does not attach to the skeleton. Physically, the individual cells are smaller in size than either skeletal or cardiac muscle cells. Under the microscope the individual muscle fibers appear to be spindle-shaped with a centrally located nucleus. There is no sarcolemma. The innervation is autonomic, and in visceral muscle is relatively poor since only a few cells of each group receive direct nerve supply. This is particularly true in the smooth muscle of the digestive tract.

The multi-unit smooth muscle cells receive total innervation, as do the piloerector (arrector pili) muscles attached to the bases of certain hairs which act to make the hairs "stand on end." However, in most smooth muscle tissue, the nervous stimuli are passed mechanically from cell to cell once the innervated cells are stimulated. Mechanical passage of the contraction impulse would cause one to suspect that the action of those muscle fibers where this occurs would not be as quick or responsive as skeletal muscle. This is particularly true for the musculature of the digestive tract where the contractions tend to be relatively slow and sustained. It is quite opposite of striped muscle which contracts rapidly and is more readily fatigued. Furthermore, since the smooth muscle nerve supply is not directly connected to the brain, the muscle contractions are involuntary and tend to keep occurring regardless of conscious desires.

Specific groups of smooth muscle cells are sensitive to certain drugs introduced into the circulating blood and will respond to either adrenalin or acetylcholine stimulation, but not to both. Intestinal smooth muscle, for example, responds to acetylcholine, and piloerector muscles respond to noradrenalin. The contraction mechanism seems to be the same as that for skeletal and cardiac muscle fibers.

Cardiac Muscle

Cardiac muscle is ordinarily found only in the heart. There are exceptions such as the "lungenherz" muscle fibers around the bronchi of mice and rats. Structurally, cardiac muscle has some of the qualities of both striated and smooth muscle cells. It possesses the cross striations of striped muscle, but the nuclei are located centrally rather than on the periphery of the muscle bundles. Unlike striated muscle, the cardiac muscle fibers under light microscope examination appear to branch and intermesh with each other, giving the impression that the heart muscle is a syncytium or union of cells that forms a single structure. A sarcolemma is present, but the muscle bundles or fasciculi are not as apparent as they are in skeletal muscle.

Two structures peculiar to cardiac muscles, i.e., intercalated discs and Purkinje fibers, serve to further differentiate it under light microscopy. Intercalated discs are dark bands which extend transversely across one or more muscle fibers, and are distinctly different from cross striations. Their number apparently increases with age. Purkinje fibers (p. 112) are modified muscle fibers which are confined to ventricles of mammalian hearts, although they are found in the atria of birds. These carry contraction impulses to the ventricular muscle and are a part of the contractile system. They apparently operate by contraction and do not carry electrochemical nervous impulses. This system can function independently and explains why a transplanted heart continues to function in another body. The force and rate of contraction is regulated to some extent by the autonomic nervous system and by certain reflexes and hormones.

Exogenous chemicals (cardiac glucosides) and the concentration of sodium, potassium and calcium ions in the blood plasma and interstitial fluid can also affect the contractions of the heart. In addition, cardiac muscle cells have an inherent capability to contract rhythmically. This has been demonstrated in experiments on cardiac cell function and is particularly true of the sino-atrial (S-A) node or "pacemaker" where the contraction impulses for a normal heartbeat originate. Probably the end-to-end fusions of the cardiac muscle cells at the intercalated discs contribute to the rhythmicity of normal heart contractions and assure that the heart follows the "all-or-none" law; i.e., all contractions are maximal in response to an effective stimulus and their force is maximal regardless of the strength of the stimulus.

Cardiac muscle fibers are of three kinds, i.e., atrial muscle, ventricular muscle, and Purkinje fibers. The first two of these have the property of changing in length during periods of intense activity. This appears to be a normal peculiarity of cardiac muscle and is somewhat similar to the laboratory effect of isotonic contraction (p. 114). The effect in the heart results in a physiological distension and a greater amplitude, although the force remains constant. The increased flow of arterial blood thus attained is an automatic adjustment of the heart muscle to an increased demand for blood by the body.

Cardiac muscle cells have numerous mitochondria that apparently contain oxidative phosphorylation enzymes needed to permit continuous rhythmic contraction of the heart without building up an oxygen debt.

Under electron microscopy cardiac muscle appears, insofar as the muscle fibers are concerned, to be much the same as striated muscle. The entire structure differs in the relative reduction of the sarcoplasmic (endoplasmic) reticulum and the large number and size of the mitochondria (sarcosomes). Electron microscopy has demolished the old idea that intercalated discs are scar tissue resulting from broken fibrils. The discs appear, beyond reasonable doubt, to be elaborate attachments, similar to the T-S bands in skeletal muscle, that link contractile fibers of cardiac muscle end to end. The discs always occur at the level of the Z lines, but may be more than one sarcomere apart, which indicates that the boundaries of all cells in a given group do not occur on the same plane. This fine distinction is lost in light microscopy.

Functionally the heart behaves like a syncytium, and the contraction impulses, which normally originate at the sinoatrial (S-A) node, pass outward from that structure as a contraction wave which spreads across the atria. The wave moves at the speed of about one meter per second and involves the right and left atria in quick succession. The wave contacts the atrio-ventricular (A-V) node at the junction of the septa separating the atria from the ventricles, and is delayed for about 0.08 second before it is conducted into the A-V bundle of His. The A-V bundle is the sole physiological connecting link between the atria and the ventricles. It is composed of Purkinje fibers and commences at the A-V node located at the base of the interatrial septum. It penetrates the interventricular septum and divides into two secondary bundle branches which supply the right and left ventricles. These branches quickly subdivide into arborizations which lie under the endocardium and form an extensive network that can be seen particularly well beneath the endocardium in the hearts of cattle and sheep. The fibers penetrate the myocardium and ramify through the ventricular muscle. The ventricular conducting apparatus is sufficiently elaborate to contact all parts of the ventricular musculature, and the Purkinje fibers which form it fuse with the ventricular muscle fibers, so that any impulse will be carried directly to the muscle. Since the Purkinje fibers by their contraction propagate a conduction wave at a speed of about three to seven meters per second and the conduction wave in atrial and ventricular muscle *per se* is only about one meter per second, the Purkinje network is essential to cause the virtually simultaneous contraction of the ventricles which is necessary for efficient pumping action.

It appears that, although the depolarization wave that produces contraction of heart muscle is the same or similar chemically and physically to the depolarization and "firing" of a nerve impulse, the impulse that causes contraction of the heart is not excited by the simultaneous neural volleys which motivate skeletal muscle but by the syncytial effect produced by the physical union of the muscle cells in the atria and the ventricles, i.e., the contraction impulse is essentially physical and is passed from one muscle cell to another. However, any stimulus (tactile, thermal, or electrical) can originate a contraction of heart muscle. This probably explains why the S-A node, which fires more rapidly than the A-V bundle or the muscle cells themselves, is the control or "pacemaker" of the heart. If the S-A node is damaged or inoperative, the pacemaker function will be taken over by the A-V node and the heart will beat more slowly.

Damage to the conducting apparatus of the heart may result in flutter or fibrillation, both of which are incoordinate movements of heart muscle. The explanation of these conditions is sufficiently complex that it will not be presented here. These and other impairments of heart rate and contractions are essentially pathological, and should not be included in a general discussion of muscle movements. For more information on heart action see Chapter 9.

Comparative Structure of Muscle Tissue

Type	Color	Fiber	Location of Nuclei	Cross Striations	Intercalated Discs	Sarcolemma	Innervation	Total Innervation	Purkinje Fibers
Striated	red	straight	peripheral	yes	no	yes	voluntary	yes	no
Smooth	white	spindle-shaped	central	no	no	no	involuntary	sometimes	no
Cardiac	red	branched	central	yes	yes	yes	involuntary	no	yes

MUSCLE PHYSIOLOGY AND JOINT MOVEMENTS

The skeletal muscles of an animal have the function of altering the position of the various parts of the body in relation to one another. This function, which is a response to nervous stimuli, affects not only posture, but motion. With the exception of the involuntary muscles (e.g., heart, digestive system) most of the muscles in the body are under direct control of the will. Therefore, the application of conscious direction will result in movements that enable the animal to better adapt itself to its environment and the pressures of that environment.

MUSCLE CONTRACTION

There are three periods involved in muscle contraction; a latent period, a contraction period and a recovery period. The latent period involves the time from stimulation to the beginning of muscle contraction or about 0.01 of a second. The contraction period is somewhat longer, taking about 0.04 second; and the recovery period is longer still, requiring about 0.05 second. The entire contraction cycle, therefore, involves about one-tenth of a second. Stimuli applied at about 0.10-second intervals will result in normal muscle contractions. Stimuli applied either continuously or at shorter time intervals than the normal contraction cycle will not give the muscle sufficient time to relax before the next stimulus starts. Such stimuli will produce a condition called tetany where the muscle is in a state of continuous contraction. This is seen occasionally in specific diseases (e.g., tetanus infection) or in disorders of the nervous system (convulsions, tetanic spasm), or in certain toxic states where the contraction cycle of the muscle is prolonged, and normal stimuli can trigger a tetanic response.

Striated Muscle Contraction

There are three types of striated muscle contraction which occur in the intact animal. These are concentric, isometric and eccentric. A fourth type, the isotonic contraction, occurs in laboratory preparations. Concentric is the usual form and is defined as the contraction which occurs in muscle operating against a movable resistance. Isometric contractions are those which occur in muscles operating against an immovable (stationary) resistance. Eccentric contractions are those which occur in muscle that has relaxed beyond normal tonus. An example of this is the raising of the head of a long necked animal, which has been lowered against tonus through the application of gravity, i.e., a gradual losing to a constantly applied (but not overpowering) force such as the weight of the animal's head. Isotonic contraction is a special form of eccentric contraction where the length of the muscle changes in response to a constantly applied force, but the force of contraction remains approximately the same as it was in its previous unstretched state.

The Contraction Mechanism

The anisotropic band (A-band) of a muscle fibril (Figure 5-1) is composed mainly of a substance called myosin. The isotropic band (I-band) of the fibril contains another substance called actin. These substances interdigitate at the ratio of one myosin to six actin elements. Muscle contraction involves the sliding of myosin and actin elements over each other, which may result from a folding of protein molecules in the muscle fibril. The exact mechanism is still unclear. However, considerable work has been done on the chemistry of muscle contraction and the following sequences seem to be reasonably precise and may survive without major change.

ATP (adenosine triphosphate), a high energy phosphate, reacts with myosin and calcium ions to produce ADP (adenosine diphosphate), phosphoric acid and energy:

$$ATP \xrightarrow{Myosin + Ca^{++}} ADP + H_3PO_4 + E\uparrow$$

The energy release apparently triggers the interdigitation of myosin and actin thus causing muscle contraction.

If ATP is present in the presence of myosin and calcium ions but fails to split, the result is muscular rigidity. An excellent example of this is rigor mortis, which was once thought to be the result of the protein-denaturing effects of lactic acid.

If ATP is present, but the splitting mechanism is not, the result is muscular relaxation. Muscle relaxation can also be obtained by AGP (alpha glycerophosphate) plus magnesium ions (Mg^{++}) which inhibit myosin activity.

The contraction of muscle is associated with a number of reactions. Under conditions of normal or minimal activity the muscles derive their energy from aerobic processes, e.g., energy derived from foods by oxidative metabolism (see Chapter 13). These oxidative reactions result in the production of ATP which furnishes the energy for ordinary muscle work.

Under conditions of strenuous activity, however, the oxidative mechanisms are inadequate to meet the energy demands placed upon them, and additional energy is derived from anaerobic processes which involve (a) the metabolism of ATP present in the muscle tissue; (b) the formation of lactic acid from glucose; (c) the degradation of one-fifth of the lactate via the pyruvic acid cycle (tricarboxylic cycle) to carbon dioxide, water, and energy; and (d) the conversion of creatinine phosphate to phosphoric acid and creatine.

After the activity has ceased, the normal oxidative mechanisms continue to operate for some time to convert the lactic acid back to glucose and to restore the muscle levels of ATP and creatinine phosphate. This process is called restoration of oxygen debt which was created by the anaerobic metabolic processes. The biochemical reactions involved are considerably more complicated than this brief summary indicates, but they essentially serve to release quantities of energy, the principal source of which is the high energy phosphate ATP. The energy

116
THE MUSCULAR SYSTEM

derived from glucose apparently serves in the resynthesis of the phosphoric acid-creatine complex that has been hydrolyzed by muscle activity. Four-fifths of the lactic acid produced by the anaerobic metabolism of glucose and glycogen is again converted to glycogen within the liver. It should be noted that the glycogen-lactic acid conversion is accomplished only under conditions of active work and that the cellular metabolic processes also produce carbon dioxide which is removed by the mechanisms of internal respiration. Some of the energy set free by muscle work is heat that is partly utilized to help maintain body temperature. The essential reactions in muscle activity are as follows:

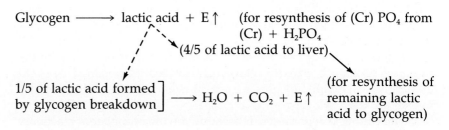

$$\text{ATP} \xrightarrow{\text{Myosin + Ca}^{++}} \text{ADP} + H_3PO_4 + E\uparrow$$
$$\text{Creatine PO}_4 \longrightarrow (Cr) + H_3PO_4 + E\uparrow \quad \text{(the resynthesis reaction which provides the energy phosphate for conversion of ATP from ADP)}$$

$$\text{Glycogen} \longrightarrow \text{lactic acid} + E\uparrow \quad \text{(for resynthesis of (Cr) PO}_4 \text{ from (Cr) + H}_2PO_4$$
(4/5 of lactic acid to liver)

1/5 of lactic acid formed by glycogen breakdown $\longrightarrow H_2O + CO_2 + E\uparrow$ (for resynthesis of remaining lactic acid to glycogen)

It is a well-known fact that exercise improves muscular efficiency. This is partly due to the fact that exercise develops muscle size and partly because muscles tend to function more efficiently with repeated usage. This is probably due to facilitation (p. 337).

ATTACHMENT OF STRIATED MUSCLE

Muscles are normally attached to other structures by cords or sheets of white fibrous connective tissues called respectively tendons or aponeuroses. In general, tendons serve to attach muscles to two or more separate bones. Muscles which are attached to skeletal structure function principally in locomotion, respiration and movements of the head, neck, and tail.

LOCOMOTION

In locomotion, the muscles of the limbs play the most important part, although those of the flexible spines of carnivores also enter into movement. In the forelimbs of cursorial animals, the scapulas are supported in a sling of muscles and connective tissue (a sarcoarthrosis) and are attached to the trunk by two large (left and right) serratus ventralis muscles that have distinct cervical and thoracic parts. The muscle fibers are attached to the last few cervical vertebrae, the anterior half of the rib cage, and to the medial surfaces of the scapulae. Contraction of the cranial portion draws the scapula forward. Contraction of the caudal portion pulls the scapula to the rear. The major muscle, however, that is involved in the forward motion of the shoulder is one that is not directly attached to that structure. This is the brachiocephalicus which attaches at the one end to the head and back of the neck and at the other to the middle cranial portion of the humerus and serves to rotate the scapula forward, to straighten the angle between the scapula and humerus.

Two large muscles, the triceps and biceps brachii, respectively, extend and flex the joint between the humerus and the radius and ulna. The upper part of the shoulder is connected to the trunk by two trapezius muscles, which form a sling extending over the posterior surface of the neck and withers. Movement of the lower portions of the limb is chiefly effected by the muscles of the forearm which act through long tendons connected to the bones of the digit. Distal to the carpus, the forelimb of the horse is devoid of effective muscles.

The hindlimb does not operate on exactly the same plan since the hindlimb is fastened by a direct bony connection to the axial skeleton. The femur is extended by a group of large gluteal muscles, the principal of which are the biceps femoris, the semitendinosus and the semimembranosus, and is flexed by the quadriceps femoris and the tensor fascia lata. The gluteal muscles, i.e., gluteus superficialis, gluteus medius, and gluteus profundus, serve to extend the femoroacetabular joint through their attachment on the greater trochanter of the femur. They act in conjunction with the semimembranosus and semitendinosus muscles, which pass over the ischiatic spine and attach to the proximal posterior surface of the tibia, and serve to flex the femorotibial and extend the femoroacetabular joints. The lower portions of the hindlimb function as do the lower portions of the forelimb except that the flexion of the hock is anterior.

Almost all the joints of the limbs function on the principle of the lever. Levers may be considered to be first order, second order, and third order. Levers of the first order are those where the fulcrum (F) lies between the power (P) and the resistance (R), for example:

$$\frac{\downarrow P \qquad R \downarrow}{F}$$

In the limbs, this is the lever of extension.

Levers of the second order are those where the fulcrum is at one end and nearer the resistance than the power, for example:

$$\frac{\uparrow P \qquad R \downarrow}{F}$$

This lever is rarely found in the body.

Levers of the third order are those where the fulcrum is at one end, but nearer the power than the resistance, for example:

$$\frac{\qquad \uparrow P \qquad R \downarrow}{F}$$

In the limbs, this is the lever of flexion. The nearer the power is to the fulcrum, the greater the degree of flexion gained by a given amount of contraction. This type of leverage produces speed but sacrifices power. This is not necessarily a disadvantage, since the principal function of the limbs is to propel the body a given distance in the shortest time. Levers of the third order occur in the elbow joint, the stifle, the hock, the carpus, and the joints of the lower leg and digits. Ideally, the power should be applied at right angles to the arm of the lever, but in practice this seldom occurs.

The types of lever and the effective application of power explain why the limbs of a mammal become fixed in extension during a convulsive seizure. The extensor muscles, although ordinarily smaller and weaker than the flexors, have a much more effective lever arrangement and can therefore apply more effective force to the movement of the limbs.

In locomotion of the horse it was once thought that the muscular attachment of the forelimbs to the trunk inhibited the forelimbs from performing any major function other than support and that the hind legs did virtually all the work of propulsion. This view is not correct. Although the hindlimbs do perform a greater portion of the propulsive work, the forelimbs aid considerably in the forward motion of the body. High speed photography of galloping horses has shown that the forelegs propel the animal 10 or more feet forward and, in doing so, raise the body four inches vertically from the ground. The forelimbs, however, are secondary to the hind.

In regard to the motion of the limbs, the joints are, to a large extent, determiners of limb movements and are discussed in Chapter 4. In the foreleg, the shoulder, because of its muscular attachment to the body, is capable of considerable forward and backward movement but has very little lateral or medial movement. The elbow joint is a pure hinge joint that restricts limb motion to a fore and aft movement. The carpus, which consists of a series of inclined planes, has two types of movement: one is illustrated by the joint between the distal end of the radius and the proximal row of carpal bones and the other by the joint between the proximal row and distal row of carpals. The intracarpal joint has a hinge-like action, while the joint between the distal carpals and the proximal metacarpals has a gliding movement. The summation of these movements results in flexion and extension of the carpus. The inclined planes formed by the joint surfaces of the carpal bones serve as shock absorbing devices. In both forelimbs and hindlimbs the fetlock, pastern and coffin joints form yielding articulations which absorb shock and flex and extend the lower portions of the limb, thus contributing an additional impetus to forward motion.

In the hindlimb, the hip joint is of the ball and socket type, but its lateral motion is restricted by the heavy gluteal muscles. It is further restricted by the accessory ligaments and the ligamentum teres of the head of the femur. These ligaments prevent any great amount of lateral movement. The stifle joint is restricted in its movement to flexion and extension. It consists of two joints, the femoropatellar and femorotibial. Its motion is modified by the outward angulation of the tarsal trochlea and the articular grooves in the distal end of the tibia. These produce a mild rotating motion of the femorotibial joint, which moves the knee clear of the abdomen during flexion of the leg. The hock joint is a multiple joint similar in some respects to the carpus, but its action is modified from the pure flexion and extension movement of the forelegs by the angular arrangement of the trochlea. The movement between the rows of tarsal bones is extremely restricted. In horses, fusion of these bones may occur, but this does not appreciably interfere with the movement of the hock. The motions of the metatarsals and hindfeet are similar to those of the comparable bones in the foreleg.

Gait

The summation of motion of the limbs and body results in gait. The horse has three basic gaits:

Slow gait
Medium gait
Fast gait

Each of these three major categories is subdivided into a number of secondary categories. The problem of classification arises because "slow gait" as used in shows can mean several gaits, i.e., running walk, fox trot, or slow pace which may be involved in the schooling of a "five-gaited horse."

The walk is a slow, flat-footed, four-beat, diagonal gait. A four-beat gait is one where each foot touches the ground at a distinctly separate time interval. A diagonal gait is where the forefoot on one side of the body and the hindfoot on the other are intimately related to the action of the gait. If the feet of a horse were numbered as follows:

1. Left front
2. Right front
3. Left rear
4. Right rear
   (near)
   (off)

the movement of the feet at the walk would be 1, 4, 2, 3.

The amble is also a slow, flat-footed four-beat gait. It differs from the walk in that it is parallel (or lateral) rather than diagonal, i.e., 1, 3, 2, 4.

The medium gait has six subdivisions. The term "medium gait" is something of a misnomer since two of the gaits within its compass (i.e., trot and pace) are quite fast. Perhaps "intermediate gait" would be a better term. From slowest to fastest, the subdivisions are:

Running walk
Fox trot
Slow pace
Rack
Trot
Pace

The running walk has a hereditary as well as a training background and could be classed as a partially learned gait. It is a fast, extended walk that has been described as being on the edge of breaking from a walk to a trot. It has also been called the "singlefoot." It is faster than a walk, with higher action and a more pointed step, but slower than the trot. Animals with this gait are highly prized by stockmen and people who have to be on horseback for long periods of time. The foot movement is as for the walk but with a much shorter time interval between steps, i.e., 1, 4, 2, 3.

The fox trot and slow pace are variations of parallel gait where the forefeet and hindfeet on the same side move together, i.e., (1, 3), (2, 4). The fox trot is a short, broken, nodding gait in which the legs on the same side move more or less together. It is an extended amble, faster than the amble and slower than the slow pace. The slow pace (stepping pace) is another variant of the parallel two-beat gait and is characterized by slower leg movement than the pace and a slightly broken cadence of the action of the lateral pairs of legs. It is essentially an extended fox trot without the excessive nodding of the head. It is somewhat more comfortable to the rider than the pace since the essential two-beat character of the gait is broken. Foot action is similar to the fox trot being (1, 3), (2, 4), with a short but appreciable interval between each footfall on the same side and a longer interval between opposite sides.

The rack is a wholly learned gait and is essentially a four-beat gait where the two forelegs are moved as in the canter and the two hind legs are moved as in the trot. It is very comfortable for the rider but quite uncomfortable for the horse. Generally a horse has to be held in the gait by "heel and hand," i.e., pressure on the reins (hand) to slow the hind legs to a trot, and touching with the spurs (heel) to speed the forelegs to a canter. It is a very "flashy" gait and one highly favored in horse shows. The foot movement is 4, 1, 3, 2, with the feet coming down in quick succession. The gait is about as fast as a medium trot.

The trot is a natural two-beat diagonal gait where the near front and off hindfoot strike the ground at the same instant followed by the off front and near hind. For a brief period in the trot the horse is entirely off the ground. The foot movement is (1, 4), (2, 3). It is comfortable to the horse but somewhat uncomfortable to the rider.

The pace is a two-beat parallel gait that is similar technically to the trot but has a foot movement of (1, 3), (2, 4). The pace is considered by some to be slightly faster than the trot. It is comfortable for the horse but deceptively uncomfortable for the rider, the twisting motion of the body of the rider produced by the parallel gait will often feel easy at first but in the end will be quite painful.

The fast gait has three divisions:

Canter

Gallop

Run

All three involve the same foot movement and differ only in the speed of movement and degree of extension of the limbs. A canter is a contained (or shortened) gallop. A run is an extended gallop. Since the gallop is the basic fast gait, it is the only one which needs to be described.

The gallop is a three-beat gait in which, normally, one pair of diagonal feet strike the ground simultaneously and the remaining two feet strike separately. The foot movement is 3, (1, 4), 2, or 4, (2, 3), 1. The difference depends upon the lead (leed), i.e., which singly moving forefoot strikes the ground. The inherent characteristics of this gait put excess wear on the simultaneous feet; so the lead should be changed frequently to balance the wear. At the gallop, canter or run, the horse is in flight for a considerable portion of the stride.

Some writers describe the canter as a different gait, since the weight of the horse tends to be supported more by the hindquarters and the stride is more compact, thus giving a bounding up-and-down movement.

Occasionally a horse will "cross canter" or parallel canter where the simultaneous footfall is executed by the front and rear foot on the same side of the body, i.e., 4, (1, 3), 2, or 3, (2, 4), 1. Like the pace, this gait is uncomfortable for the rider.

THE MUSCULAR SYSTEM

The Mechanics of Locomotion

While other aspects of the skeleton and musculature, i.e., aids to respiration, digestion, vocalization, reproduction, and elimination, are important, they do not have the dramatic impact of locomotion. It is here that the bones and muscles play their greatest roles in bodily function.

Mammals have forms of locomotion as diverse as their physiques, but the large domestic animals with the possible exception of swine are all cursorial. In the wild state, cursors depend upon fleetness of foot and ability to travel long distances for survival. These abilities have some important advantages. They give cursors (running animals) a certain amount of freedom from environmental pressures because of their ability to range over broad areas to find food and water. They also give cursors a degree of independence from seasonal changes and an improved ability to escape enemies or secure prey.

The specialization of an animal for running has resulted in a number of bodily changes and adaptations for getting the most speed out of a basic set of bones and structures. The ultimate developments to date have produced horses that can run at 40 miles per hour, antelope that can do 60, and cheetahs that can reach 70 for short distances.

These speeds involve a number of factors. In order to run fast, an animal must be able to move its limbs with force and rapidity. It must be able to support itself against the force of gravity and utilize every possible action in its stride to achieve forward motion. In addition, it must be able to stride rapidly with a minimum of muscular effort.

In adaptation for running, mammals have developed two distinct types of locomotion: one wherein the spine is flexible and by its flexion and extension contributes to forward progress and the other wherein the spine (the body segments, i.e., thoracic, lumbar, and sacral) is relatively rigid and forward motion is almost entirely a function of the legs.

Regardless of which type is employed, the legs of running animals have become relatively long in proportion to the length of their bodies. This is an obvious development, since longer legs take longer strides, but it has limitations. There is an effective length beyond which the efficiency of the musculature is impaired and results in a slower stride. Since speed is the product of the length of the stride times its rate, leg length is held to rather definite limits for each species.

Increase in size does not necessarily result in increased speed, since in addition to the duty of propulsion, the locomotor system must also act as a means of support. In this connection, the classical square-cube rule should be remembered. Body weight is proportional to the cube of its linear dimensions, but muscle power is only proportional to the square of its cross section. Therefore, doubling the cross-sectional diameter of a muscle would produce a fourfold increase in strength, and an eightfold increase in weight. Furthermore, if one considers that long muscles cannot contract as fast as short ones since the rate of contraction varies inversely with any linear dimension, it can be seen that beyond a certain size of animal, muscles simply cannot accommodate the combined demands of weight, rapidity of movement, and length of stride. This explains why the largest terrestrial animals can neither gallop nor jump and why moderately large cursors must be highly specialized in order to move rapidly enough to survive.

The increase in length of leg has taken place in the distal portion of the limbs (Figure 5-6) and particularly in the metatarsal and metacarpal bones. The increase in length has been accompanied by a rise of the bearing surface from the palm (plantigrade) to the toe (digitigrade) to the toenail (unguligrade). The highest development for running is found in the feet of equines and ruminant ungulates where the animals literally walk on their fingernails and toenails. At the same time there has been

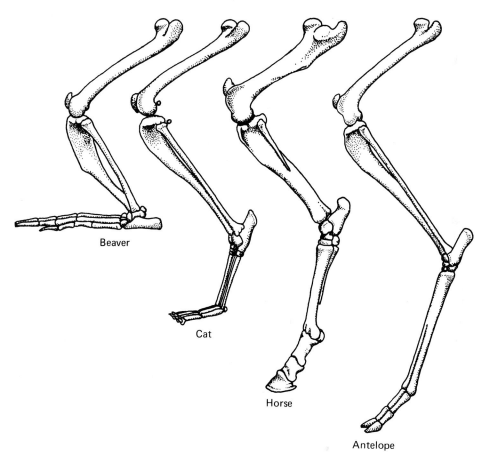

FIGURE 5-6   *The evolution of the cursorial leg. Note the lengthening of the metatarsals, and the modification of the digits. (The legs are all drawn to the same relative scale, the length of the femur being used as the base measurement.)*

a loss in the number of digits. Equines have gone farthest in this respect. Their hooves are highly developed structures which serve not only as bearing and propulsive mechanism but also as shock absorbing devices and blood pumps.

The effective length of the forelimb of cursors is increased by the relocation of the scapula and the reduction in the number of bones of the pectoral girdle. Of the original three bones (scapula, coracoid, and clavicle) only the scapula remains, and it has been reoriented until it lies with its long axis parallel to the median plane of the body. There is no direct bony connection between the scapula and the thorax. Held in a sling of muscles and connective tissue, the scapula pivots around a point near the proximal third of the bone and swings back and forth in rhythm with the movement of the leg. This fleshy attachment permits some movement of the entire bone in the four quadrants of the sagittal plane (dorsal, ventral, cranial, caudal), which results in lengthening or shortening of the forelimb in relation to the body.

Fast runners must stride rapidly. A racehorse takes about 2.5 strides per second at the dead run and each stride covers between 20 and 25 feet. A cheetah, with essentially the same length of stride takes better than 3.5 strides per second. The speed advantage is obviously the cat's, but how does the cat with its shorter limbs and shorter muscles (which give the faster rate of stride) manage to get the same length of stride as the horse? The answer lies in the flexible spine of the cat. By flexing and extending its spinal column as its legs swing under its body, the cat is capable of moving its hind feet farther forward under the body as the spine is flexed and reaching farther forward with the front feet when the spine is extended. This adds a considerable distance to the length of the stride.

However, the spinal movement also creates a problem. The body tends to rise as it passes over a planted foot and fall when the feet are not underneath the body or are off the ground. Ideally, the body should move along a line parallel with the surface of the ground for maximum efficiency of propulsion. Flexible-spine runners compensate by extending the carpal and tarsal joints, flexing the knee and allowing the scapula to move dorsally in its sling of muscles and connective tissue. Rigid-spine runners extend the fetlock and pastern joints in addition to the joint movements listed above. Appropriate leg movements will damp out most of the up and down "bobbing" movement of the body.

The attachment of muscles to the bones also has an effect upon the length and the rapidity of the stride. Cursorial animals have their muscles of locomotion attached closer to the joints than do nonrunners. This results in a shorter muscle and quicker muscle contraction. It also results in a longer arc of movement of the distal end of the bone in which the muscle is inserted. This arrangement sacrifices power, but this is not particularly disadvantageous, since running animals do not need the power developed by climbers, burrowers and swimmers. Air resistance and body weight offer the principal impediments to motion, and these are compensated partly by body conformation, partly by the concentration of muscle masses near the axis of the body, and partly (in ungulates) by suspensory ligaments. These last structures are best developed in equines and extend from carpus or tarsus to the distal phalanges. As the foot strikes the ground, the fetlock and pastern joints are extended. This puts tension on the suspensory ligament, which snaps back to its normal tension as weight is removed from the foot, aiding flexion of the fetlock and pastern joints and giving the distal end of the limb an upward push. This relieves some of the burden on the leg muscles and adds force to the stride.

Since the speed of contraction of muscles is limited, it follows that the speed of the limb or body movement must be correspondingly limited, even though that speed is increased by the closer attachment of the muscle to the pivot of movement in the joint. The total speed of movement of the distal end of a limb (or of a chain of vertebrae), however, can be markedly increased if different muscles move different bones in the same direction at the same time. Therefore, the more joints in a given limb, the more rapidly the distal extremity of the limb will move.

Extra joints have been acquired through the evolution of the distal portions of the leg, particularly in ungulates. These animals have, at least in effect, acquired an extra joint at the fetlock and also have acquired still another joint through the altered location and movement of the scapula. In flexible-spined runners the spinal column adds yet another angular movement to the limbs, since both the anterior and posterior ends of the spine contribute to the effective movement, respectively, of the front and hind legs (Figure 5-7). This action contributes to the rate of stride by speeding the motion of the body as well as the legs.

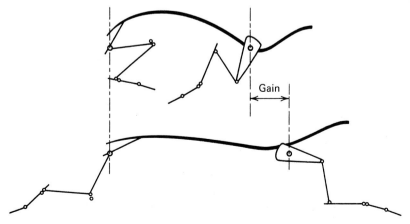

FIGURE 5-7   *Effect of spinal flexion on increasing length of stride.*

126
THE MUSCULAR SYSTEM

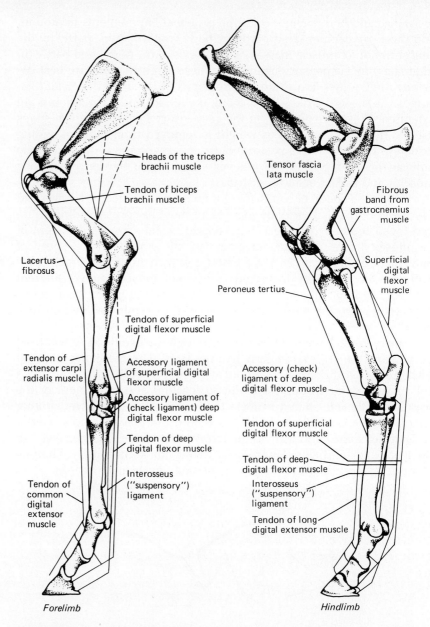

FIGURE 5-8 *Stay apparatus. Only a slight contraction of the muscles will lock the bones of the front and hind legs into a rigid unit.*

In the rigid-spined cursors, where the legs are primarily involved in locomotion, the evolutionary adaptations are more extensive. Essentially these function to reduce weight in the distal extremities of the limbs and economize muscular effort. The suspensory ligaments have already been mentioned as have been the reduction in the number of bones in the distal ends of the legs. There is also a concomitant loss of soft tissue with the distal tissues reduced to bones, tendons, ligaments, nerves, vessels, and skin. A fourth adaptation is the placement of muscles and the arrangement of joints so that virtually all movement of the limbs is confined to motion along a sagittal plane. Adductor and abductor muscles are reduced or modified to produce fore and aft motion. The ulna and fibula have lost their ability to pronate and supinate, and the bones either fuse to the radius or tibia, or disappear. In flexible-spined runners, the ulna and fibula persist, but rotation has disappeared in the hind legs and has been reduced in the forelegs.

A fifth modification involves the concentration of the muscle masses closer to the central axis of the body. This results in greater efficiency. The load on the muscles is not only reduced, but the shortening of the muscle masses increases the speed of their contraction, and consequently increases the rate of the stride.

The efficient design embodied in cursorial animals represents one of nature's finest efforts to develop a highly efficient organism within the limits of the materials involved.

THE "STAY APPARATUS"

It is a reasonably well-known fact that healthy horses have no real need to lie down and can actually sleep while standing. The explanation for this phenomenon lies in the so-called "stay apparatus," an assortment of tendons, muscles, and ligaments in the front and hind legs that can stiffen these structures into a firm supporting mechanism that requires a minimum amount of muscular energy to keep the animal erect. All equine species possess this mechanism, but it is not present in other domestic animals. The drawings in Figure 5-8 show how the stay apparatus is constructed. How it works precisely is a matter of functional anatomy. For that information you will have to read more widely than this text.

chapter 6
Teeth

TYPES AND ANATOMY OF TEETH

Teeth may be classified according to permanence, according to manner of growth, and according to function.

Teeth of mammals fall into two general classes according to permanence: temporary or milk teeth and adult or permanent teeth. The milk teeth are present at birth or shortly afterward and persist from a few weeks to several years depending upon the species of animal. They are usually less hard and fewer in number than the adult teeth which replace them. The adult teeth are permanent structures and usually persist throughout the life of the animal.

Teeth fall into three general classes according to manner of growth: true teeth, constantly growing teeth, and constantly erupting teeth. True teeth occur in carnivores, omnivores, and in the incisors of ruminants. They possess a crown, neck, and root, grow to adult size, and then wear away without further growth or eruption to compensate for wear. Examples of constantly growing teeth are the tusks of swine and the incisor teeth of rodents. These teeth continue to grow throughout the life of the animal and do not possess a definite root or neck. They can be considered to be rootless teeth growing from persistent active pulp. Constantly erupting teeth are exemplified by the entire dentition of

TEETH

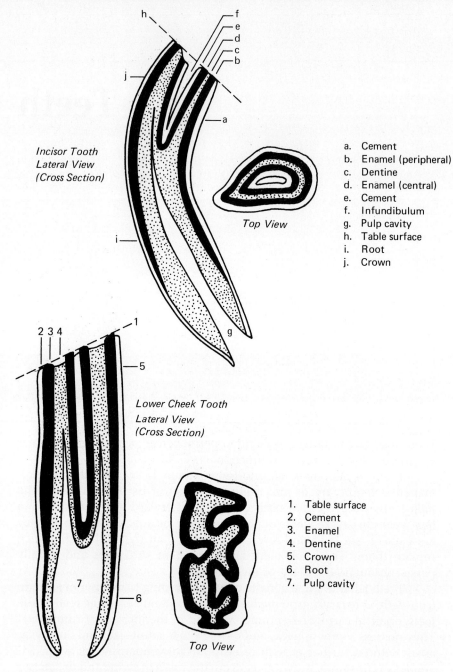

FIGURE 6-1   Teeth of the horse.

equines and the cheek teeth (premolars and molars) of ruminants. Teeth of this type are complexly layered masses which possess extremely long roots and no definite neck. As the animal ages the tooth sockets gradually fill from below with bone, which slowly pushes the teeth from the socket to compensate for wear. Very old animals with teeth of this type may have them pushed completely out of the jaw.

Teeth of domestic animals fall into several classes according to function and diet: cutting teeth (incisors), tearing or seizing teeth (canines), and grinding or shearing teeth (premolars and molars).

A tooth is composed of five parts and four tissues. The following list applies to true teeth, which have all the parts, as opposed to the other varieties. Constantly growing teeth have no neck or root, and constantly erupting teeth have no neck.

Parts	Tissues
Crown	Enamel
Neck	Cement
Root	Dentine
Pulp Cavity	Pulp
Table Surface	

The crown of a tooth is that part which protrudes above the gum line. The neck is that part which is found at the gum line. It appears as a constriction which separates the crown from the root. The root is that part which is embedded in the tooth socket (alveolus) in the jaw. Roots may be either single (incisors and canines), paired (premolars), or multiple (molars). The pulp cavity is a central cavity found in the tooth. It extends from the base of the root for a variable distance into the crown. It contains the dental pulp, vessels, and nerves. This cavity is large in young animals, but with age it is encroached upon by dentine and in old animals may become virtually obliterated. The table surface is the cutting (incisor), piercing (canine), or grinding (premolar and molar) surface of a tooth. Practically, only the premolars and molars have true table surfaces.

The tissues of a tooth, ranging from hardest to softest are enamel, cement, dentine, and pulp. Enamel, the hardest tissue in the body, is usually found as a layer covering the dentine of a tooth. It is generally translucent white to ivory in color. Cement, the outermost covering of a tooth, may range from yellow to black in color and is very hard, but not as hard as enamel. In true teeth, cement covers the roots only; in constantly erupting teeth, it may be more extensive. Dentine forms the bulk of most teeth, covers the pulp, and is hard and yellowish-white in color. Pulp is a soft gelatinous tissue in the center of the tooth that supports blood vessels and nerves.

A certain number of descriptive terms are necessary for study of the teeth. The dental arcade (or arch) is the horseshoe-shaped line along which the teeth grow. Its shape is dependent upon the shape of the skull and length of the jaw, being long and narrow in horses, cattle and other herbivores, and short and wide in humans and apes. The surfaces of the teeth are labial (vestibular), toward the lips; lingual, toward the tongue; buccal, toward the cheeks; masticatory (table or occlusal), toward the corresponding tooth surface in the opposite jaw; and interdental, between teeth in the same jaw.

THE AGE OF THE HORSE

One of the things most veterinarians, stockmen, livestock buyers, and judges learn rather early in their careers is how to estimate the age of animals. Teeth are most often examined as a basis for age estimation. Other physical signs are also used, since no single system is infallible, but if one knows what goes on within an animal's mouth one is well on the way to becoming proficient at age estimation.

Lack of knowledge about age can be expensive since horse traders and livestock dealers are not above teaching costly lessons to inexperienced purchasers. If one is to deal with animals, at least a rudimentary knowledge of how to determine age can be of considerable value professionally and—if nothing else—will improve one's status in the environment of sales barns, show rings, riding academies, kennels, and veterinary clinics.

To properly tell the age of a horse, one should follow a fairly standard routine. First, look at the entire horse from a distance, and decide whether it is a colt, mature, or old. This is not too hard to do. Certain signs such as size, length, and quality of mane and tail, knobbiness of the joints, presence or absence of pits above the eyes, muscling of the head and neck, the roughness of the wall of the hooves, and the quality and color of the hair coat can usually give enough information to classify the horse into a general age category.

Then handle the horse. Feel the rear edge of the ribs, the lower border of the jaw, the premolars and molars beneath the cheeks, and the quality and flexibility of the skin. Young horses (under 5 years old) will have rounded rear edges on the ribs, the lower border of the jaw will be rounded and convex (i.e., bulge downward). There will be less than six cheek teeth (e.g., premolars and molars) in each side of the upper and lower dental arcade, and the skin will be pliable and elastic.

In older horses (5 years and older) the rear borders of the ribs get progressively sharper, the lower jaw line will flatten out, sharpen along its edges, and become concave when the animal is old. There will be six cheek teeth in each side of the upper and lower dental arcade, and the skin will be less pliable.

Now inspect the teeth. This is not as easy as it sounds because horses usually do not like to have their mouths examined. The horse should be haltered in such a manner that its head is at a convenient height and it cannot toss its head or rear back. If the animal is a quiet type, have a helper restrain it with a hand-held halter. In my opinion, hand holding is better than tying, since it usually is less disturbing to the horse. If the horse looks young, feel the cheek teeth in the upper or lower arcade by pressing the fingers against the outside of the cheek and identifying the teeth. If there are three, the horse is under 1 year old, and if there are four, the horse is between 10 months and 2 years old. If there are five, the horse is between 2 and 3½ years of age, and if there are six, the animal is probably over 4 years old.

Face the horse from the front. If you are right handed, use the left corner of the horse's mouth. Part the lips at the left corner. Reach in through the interdental space and grasp the horse's tongue. Your thumb and little finger should be under the tongue, and your first, second, and third fingers should be on top of it. Clamp down firmly, since the tongue is slippery, and draw it out through the left side of the mouth and into the left corner. The horse will open his mouth. A light cotton glove will help you hold onto the tongue. Heavy work gloves will not do. If you are left handed, go in through the right side of the mouth.

While holding the tongue in the corner of the mouth, raise or lower the upper and lower lips with the free hand. Observe the table surfaces of the incisor teeth. Look well and quickly, since horses do not enjoy having their lips handled or their tongues held for long periods while you try to make up your mind.

Special attention is paid to the lower incisors, although it is good practice to look at both rows. There are 12 incisors in the horse's mouth, six in the upper jaw and six in the lower. These are arranged in three pairs in each jaw. The central pair is called the centrals, or nippers; the middle pair, the dividers; and the outer pair are the corners. Upper and lower and left and right are used to designate specific teeth, such as the right upper divider or left lower nipper.

There are differences between the temporary and the permanent incisors. The milk teeth are shorter, narrower at their bases, smoother on their surfaces, and lighter in color than the permanent teeth. They have a definite constriction or "neck" at the gum line, may be chalkier in texture, and will tend to be slightly convex on their labial surfaces. Adult teeth are harder, larger, and longer. They have no neck and tend to be flat on their labial surfaces.

Note the following:

1. Are canine teeth present? How many incisors are present?
2. Do the incisor teeth bulge at their bases? Do they have a distinct neck?

3. Is the angle of the bite straight or angled?
4. Is there a groove in the outer surface of the corner incisors?
5. Are there any angles (notches) on the table surface of the upper corner incisors?
6. Do the teeth diverge from one another or are they crowded together?
7. Is the canine-incisor arcade round, squarish, or compressed laterally?
8. What is the shape of the table surfaces of the incisor teeth?

1. Canine teeth, when present (usually absent in mares and sometimes absent in geldings), indicate that the horse is over four and a half years old. Ordinarily the canines break through the gums at four and a half years of age, are almost fully up at five, and show a small amount of wear at six years of age. A mature horse will have six incisors in each arcade. A colt will have two incisors from six days to six weeks of age, four from six weeks to six months of age, and six over six months. This is the so-called "rule of six."
2. If the incisor teeth bulge at their bases, they are milk teeth. Adult incisor teeth have no neck, and their outer surfaces do not bulge. A mouth filled with milk teeth indicates an animal younger than two and a half years old, and a mouth filled with adult teeth indicates an animal over four and a half years old.
3. The angle of the bite increases with age. The incisor teeth of a young animal usually meet at a straight angle (180°) or close to it. The angle becomes increasingly acute as the horse grows older.
4. If there is a groove (Galvayne's groove) on the outer surface of the upper corner incisors, the horse is over 10 years old. Galvayne's groove appears at 10 years, is halfway down the tooth at 15, and is gone by 30 years of age. However, it is not an accurate sign.
5. If notches are present on the upper corner incisors, the horse is either seven years old or over 11 years old. The seven year notch is fairly accurate. The 11 year notch may persist for five or six years.
6. If the teeth tend to grow straight, or to diverge, the horse is under 10 years old. In horses over 10 the teeth tend to crowd together and form a straight line across the front of the mouth. In aged horses, i.e., 20 years or older, the incisors are crowded together.
7. The shape of the dental arcade changes with age (Figure 6-2). If the arch of the incisors is round, the horse is probably under eight years old. If it is flattened, the animal is over 11; the ages from eight to 11 are indeterminate.
8. The shapes of the table surfaces of the incisors are fairly reliable indicators of age. From years one through seven, the transverse dimensions of the table surfaces are greater than the distance from the labial to the lingual borders. By nine years of age, the dimen-

sions are about equal. Above 10 years of age, the labial-lingual distance is greater than the interdental distance. At about 14 years of age, the lower central incisors are noticeably triangular, with the apex of the triangle toward the tongue. The dividers are triangular by 15, and the corners by 17 years of age. The centrals become biangular at 18, the dividers at 19, and the corners at 21 years of age.

In addition to the above eight items, there are a few other things about the teeth which should be remembered. These include the infundibuli (dental cups), dental stars, prevalence of canine teeth, and the function of cheek teeth in age determination.

Either a black cavity in the center of the incisors (the infundibulum) or a yellowish line or circular mass (the dental star) should be visible on the table surface. The infundibuli persist in the lower incisors for eight to 12 years, disappearing first in the centrals and last in the corners. The dental star is the dentine filled apex of the pulp cavity (Figure 6-2). As the teeth wear down, it changes from a thin transverse line to a round mass that fills the center of the table surface.

Canine teeth occur in two or three percent of mares. About 60 to 70 percent of males have canines in both upper and lower jaws; seven to 10 percent have them only in the upper jaw, and 26 to 30 percent have them only in the lower jaw. The canines occur in geldings as well as stallions but they are not as well developed.

The cheek teeth (molars and premolars) are not included in precise age determination, although they may be of some value in indicating a horse's approximate age. Difficulty of access is why they are not used.

All of a horse's permanent teeth have erupted by the time the animal is five years old. A five-year-old is said to have a "full mouth." Between the ages of four and five a horse erupts four incisors, four canines and eight molars. This explains why a three-year-old can sometimes outwork a four-year-old of comparable size and musculature. The four-year-old is too bothered by teething.

Now let's put it all together:

The life of the horse can be divided into five periods corresponding to the changes which occur in the mouth.

The Eruption of the Milk Teeth

At birth to one week the upper and lower pairs of nippers appear. At four to six weeks the upper and lower dividers appear. At six to nine months the upper and lower corners appear. At 12 months, both upper and lower nippers are worn level, and the upper and lower corners are beginning to wear at their rostral parts. The foal has then passed through the first period of its life and enters the second period.

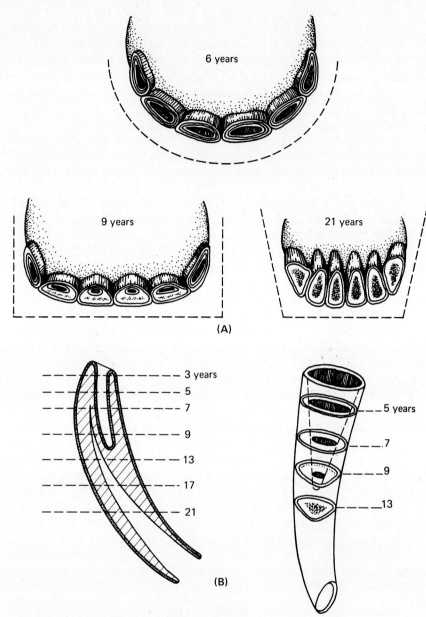

(A) Changes in shape of arcade with advancing age
(B) Wear patterns of incisor teeth at various ages

FIGURE 6-2   *Lower incisor teeth of the horse.*

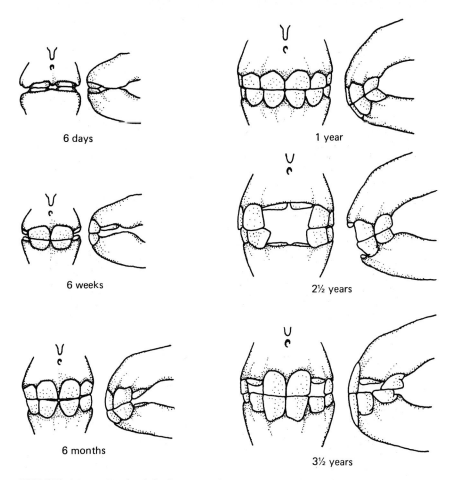

FIGURE 6-3 Teeth of the horse—6 days to 3½ years. (Adapted from Huidekoper, Age of the Domestic Animals.)

Period of Wear of the Temporary Teeth

At 16 months the corner teeth are beginning to wear at their caudal borders, and the nippers in both jaws are beginning to bulge just a little at their bases. The animal now enters the third period of its life.

Period of Eruption of the Permanent Teeth

At two and a half years the lower nippers are shed and replaced by permanent teeth. At three years the permanent nippers have met, but are not greatly in wear. At three and a half years the lower temporary dividers are shed and the permanent dividers appear. At four years the permanent dividers have met but are not worn. At four and a half years the temporary corners have disappeared, and the permanent corners

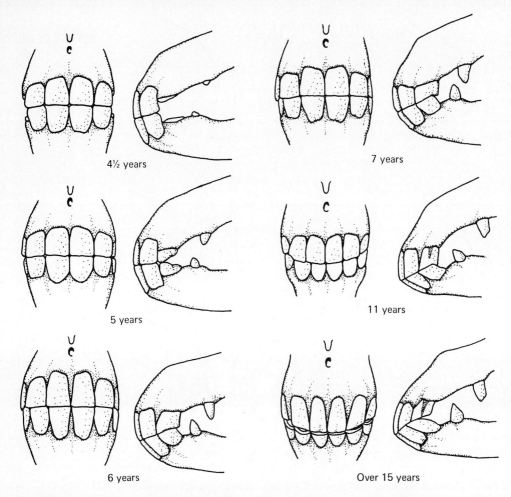

FIGURE 6-4　Teeth of the horse—4½ years through over 15 years. (Adapted from Huidekoper, Age of the Domestic Animals.)

appear. Usually at about this age the bridle teeth (canines) appear in stallions and geldings. In some animals these teeth appear at a younger age. At five years the rostral part of the permanent corners have met but are not worn, and the caudal parts are buried in the gum. The animal is now said to have a full mouth, and it enters the fourth period of its life.

The Leveling of the Permanent Incisors

At six years the cups in the lower nippers have worn level, and the enamel button is prominent. The lower corner teeth are wearing at their anterior border, and their caudal border is just beginning to wear. At seven years the lower dividers have worn level. The lower corners are wearing considerably at their caudal borders. The upper corners usually have a well-defined notch in their caudal parts. At eight years the cups are about level in the lower corners. The cups in the lower nippers have become narrower from side to side and somewhat wider from labial to lingual sides. Frequently, the white line or dental star has appeared halfway between the cup and the rostral border of the lower nippers. This is a white line when it first appears, but it gradually yellows and becomes shorter and broader until it is almost circular in shape. The animal now enters into the fifth period.

The Wearing Away of the Crowns

At nine years the lower nippers have their dental stars more distinct than they were at eight, and the enamel button may disappear. The cups in these teeth have become triangular in shape and are near the caudal borders of the teeth. The upper nippers are frequently worn level. The notches in the upper corners have disappeared. At 10 years the cups in the lower nippers have become small and triangular. The dental stars are plain in the lower nippers and dividers and frequently can be seen in the lower corners. The upper dividers are often worn level at this age. At 11 years the cups in the lower nippers are small and round. The cups in the lower dividers are triangular in shape and are located near the lingual border of the table surface. A notch in the upper corner teeth may again appear at 11 years. At 12 to 14 years the cups are worn away in the lower nippers and (usually) in the dividers. This results in what is called a "smooth mouth." The white line, or dental star, has become roughly circular in all the lower teeth and is in the center of the table surfaces, which have become longer from lips to tongue, and narrower from side to side. The incisor portion of the arcade is laterally compressed. The upper corner teeth are long at their rostral borders but are short caudally. The lower dividers are broadly triangular in shape. At 15 to 18 years of age the cups are usually gone in all the lower incisors but may still be present in the upper corners. At 20 and older, the cups are gone from all the incisor teeth. The table surfaces are very long from labial to lingual margins, but so narrow from side to side that they are often called biangular. The angle of bite has become so acute that a greater length of the crown is exposed above the gum line than in young animals. This gives rise to the term "long in the tooth," which has at times been applied to humans as well as horses.

Attempts to make an old horse look younger are occasionally seen. One trick known as "gypping" is to dye graying hair its original color. "Puffing the glen" is an attempt to fill the depressions above the eyes which tend to become deeper with age. A third, and somewhat more common trick, is known as "bishoping." This involves drilling holes in the table surfaces of the incisors of an aged horse and staining the cavities black. This will indicate to the novice horse buyer that the doctored

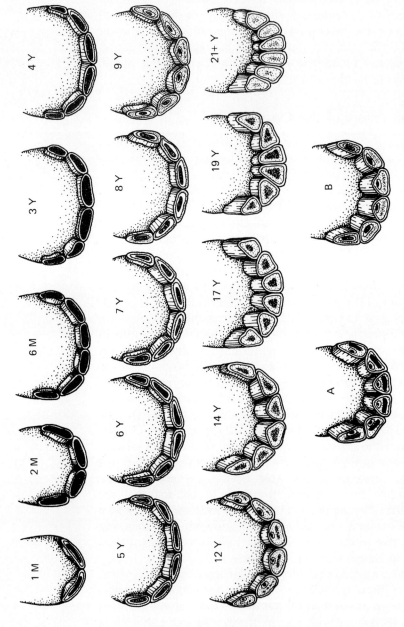

FIGURE 6-5   Table surfaces—lower incisor teeth.

(A) Bishopping—the horse is 13. Hopefully, it looks eight or nine. Table surface shape is giveaway, also remaining enamel buttons.

(B) Excessively hard teeth or long infundibuli give an appearance of youth. Horse is 17 and looks about 12. Long teeth and narrow jaw are the giveaways.

animal is under 10 years of age. The lack of an enamel rim around the edge of the artificial cup and the shape of the table surfaces of the teeth should be the giveaway.

Tooth Eruptions (Lower Incisors)	I_1	I_2	I_3
Milk teeth ("rule of 6") erupt	6 days	6 weeks	6 months
Permanent teeth erupt	2½ years	3½ years	4½ years
Cup gone in lower incisors	6 years	7 years	8 years
Dental star appears in lower incisors	8 years	9 years	10 years
Table surface becomes round	9 years	10 years	11 years
Dental star appears in center of tooth	12 years	12 years	13 years
Table surface becomes triangular	14 years	15 years	17 years
Dental star appears round	15 years	15 years	16 years
Table surface becomes biangular	18 years	19 years	21 years
Galvayne's groove			10-30 years
Seven-year notch			7 years
Eleven-year notch			11-15 years

THE AGE OF THE DOG

A dog's age may be hard to determine due to habits and discoloration from dietary deficiencies or distemper. However, certain basic periods can be listed that will form a rule of thumb for estimating age. At birth there are no teeth. By the time the puppy is three weeks old the temporary incisors and canines erupt. At three months the temporary incisors and canines are well worn. At six months the permanent incisors erupt, and by one year the permanent teeth are all in. These teeth are fresh and white, and the incisors show a tricuspid pattern on their cutting edges.

At 15 months the lower central incisors show wear, and by one and a half to two years the cusps on the lower central incisors disappear. At two and a half to three years the cusps on the intermediate lower incisors disappear, and the upper centrals show wear. At four years the cusps on the upper central incisors disappear, and the upper intermediates are worn. At about this time the teeth become yellowish and may have tartar deposits at the base of the canines.

By five years all incisors are markedly worn, and above five years, estimating a dog's age is a matter of inspired guesswork. A few guides to aid the guesses are given herewith, but there is no assurance that these will either be present or accurate. At about six and a half years a brownish or greenish discoloration of the canines may occur. Usually at this time there is also a graying of the hairs on the muzzle and around the eyes. The muzzle becomes enlarged and boxy. The digits are enlarged and claws become elongated. These conditions become progressively more apparent until the dog becomes senile. Senility is at varying ages in dogs; some of the smaller breeds maintain physical activity and health for 15–18 years, while members of large breeds may be dead of old age at 10 years.

THE AGE OF THE BOVINE

Ruminants have no upper incisor teeth. Their place is taken by a dental pad. A calf usually has the central incisors cut at birth. By two weeks all of the milk teeth have erupted. In the incisors the process begins with the centrals and extends to the corners. By one year, all milk teeth are markedly worn. At one and a half years the centrals are either worn to gum line or have fallen out, and at one and a half to two years the permanent centrals erupt. These teeth are large and grayish-white in color. At two years the permanent centrals are usually up and coming into wear. By two to two and a half years the permanent second incisors are up. At two and a half years they are in wear. Four large teeth are apparent at this period. At three years, the permanent third incisors appear and are fully up and are in wear at three and a half years. At four years the corner incisors appear and are fully up and in wear at four and a half years. At five years all incisors are up and the corners show slight wear.

A rule of thumb for estimating age of cattle up to five years old is:

2 permanent teeth 2 years old
4 permanent teeth 3 years old
6 permanent teeth 4 years old
8 permanent teeth 5 years old

After five years, age determination is principally shown by wear. At seven years the corner incisors show wear. At 10 years the centrals are halfway down to the gum line. At 15 to 20 years the central and second incisors are at gum line or are missing. The wear of teeth depends greatly on pasture conditions, type of diet, and presence of minerals, e.g., fluorine is soil. Wear in cattle is an inaccurate means of telling exact age, and even under the best conditions is merely an approximation.

THE AGE OF THE SHEEP

As in cattle, only the incisors are used in estimating the age of sheep. A chart which is fairly correct follows.

Age of the Sheep (also goats)	
Milk teeth: 1st incisor (I 1)	Birth-1 week
2nd incisor (I 2)	1-2 weeks
3rd incisor (I 3)	2-3 weeks
4th incisor (I 4)	3-4 weeks

Permanent Incisors Erupt at	Rule of Thumb
I 1—1 to 1½ years	1 year
I 2—1½ to 2 years	2 years
I 3—2½ to 3 years	3 years
I 4—3½ to 4 years	4 years

The age beyond four years is estimated by wear as in the bovine. "Broken-mouthed" sheep have one or two teeth worn out or missing. "Smooth-mouthed" sheep have over two teeth missing. This is grounds for salvage of the animal.

THE AGE OF THE PIG

Eruption of Temporary Teeth	Eruption of Permanent Teeth
I 1—2-4 weeks	I 1—1 year
I 2—2 months	I 2—1½ years
I 3—Birth	I 3—10 months
	C —10 months

DENTAL FORMULA

A dental formula is a convenient shorthand method of listing the type and number of teeth in mammals. It is based on the following symbols: I-incisors; C-canines; P-premolars; M-molars; and is a graphic picture of one-half the dental arcade. For instance, a horse has 40–42 teeth arranged in similar pairs around the upper and lower dental arcade. By taking one-half of the upper and lower arcades, location and number of teeth in a normal skull can be readily recorded.

TEETH

	Incisors	Canines	Premolars	Molars
(upper arcade)	3	1	3 or 4	3
Gum line				
(lower arcade)	3	1	3	3

$\times 2$

Horse*   ICPM $\dfrac{313 \text{ or } 4^* \; 3}{313 \quad\quad 3} \times 2 = 40 = 42$

Ox (Sheep, Deer and Goat)   ICPM $\dfrac{0033}{4033} \times 2 = 32$

Dog   ICPM $\dfrac{3142}{3143} \times 2 = 42$

Cat   ICPM $\dfrac{3131}{3121} \times 2 = 30$

Pig   ICPM $\dfrac{3143}{3143} \times 2 = 44$

Human   ICPM $\dfrac{2123}{2123} \times 2 = 32$

*First premolars will also rarely occur in the lower jaw of a horse and thus give an occasional animal a maximum of 44 teeth. I have not seen lower first premolars without the presence of uppers. The dental formula listed here is the accepted norm.

In addition to this variations, dental abnormalities such as ectopic (out of place) teeth, and supernumerary (too many) teeth also occur, but these deviations from normal are subjects more properly covered in dental pathology than dental anatomy and physiology.

Age of Eruption of Temporary Teeth

Teeth	I_1	I_2	I_3	I_4	C	P_1	P_2	P_3	P_4
Horse	B-7D	4-6W	6-9M	0	0	B-2W	B-2W	B-2W	0
Ox	B-3W	B-3W	B-3W	B-3W	0	B-3W	B-5D	B-5D	0
Sheep	B-1W	1-2W	2-3W	3-4W	0	2-6W	2-6W	2-6W	0
Swine	2-4	U.2-3M	B	0	B	5M	5-7M	U.4-8D	4-8D
		L.1½-2M						L.2-4W	2-4W
Dog	4-5W	4-5W	4W	0	3-4W	4-5W	4-5W	3-4W	3-4W

I—incisor; C—canine; U—upper; L—lower; B—birth; D—days; W—weeks; M—months; P—premolar

Age of Eruption of Permanent Teeth

Teeth	I_1	I_2	I_3	I_4	C	P_1	P_2	P_3	P_4	M_1	M_2	M_3
Horse	2½	3½	4½	0	4-5	5-6M	2½	3	4	10-12M	2	3½-4
Ox	1½-2	2-2½	3	3½-4	0	2-2½	1½-2½	2½-3	0	5-6M	1½	2-2½
Sheep	1-1½	1½-2	2½-3	3½-4	0	1½-2	1½-2	1½-2	0	L.3M U.5M	9-12M	1½-2
Swine	12M	16-20M	8-10M	0	9-10M	5M	12-15M	12-15M	12-15M	4-6M	8-12M	18-20M
Dog	4-5M	4-5M	4-5M	0	4-5M	5-6M	5-6M	5-6M	5-6M	4M	U.5-6M L.4½-5M	6-7M

Figures are years unless otherwise indicated. M = months.

PHYSIOLOGY OF THE TEETH

Since teeth are a part of the skeletal system structurally, and the digestive system functionally, it is logical to presume that the diet of animals will be reflected in their tooth structure. Like any logical presumption, this is subject to exceptions, but on the whole it holds true. Teeth are constructed according to the work they must perform and are adapted to diet. The adaptive changes are most evident in the molars and premolars although the canines (elephant, walrus, pig) can vary greatly between species, and the incisors (rodents) may also be highly modified.

Dietary habits differentiate animals into three classes: herbivores, carnivores, and omnivores. A fourth group, the edentates, are of no importance in this discussion, since they have no teeth. Edentates are anteaters, armadillos, aardvarks, pangolins, and some species of sloths. All are insectivores. Small insectivorous mammals, such as moles and shrews, have teeth of the carnivore type. Each of these groups has different dental requirements. The herbivores, subsisting principally on plant material which is relatively indigestible, must prepare this material throughly so that the digestive juices and enzymes make maximum contact with the surface of the food. The food material must be ground or pulverized into small particles with relatively large surface areas for their total weight or bulk. In mammals the teeth are the structures responsible for accomplishing this. The molars, and premolars with their flat and complexly folded components and rough table surfaces are excellent grinders. As a result of the angle of the surface and rotary movement of the lower jaw, they do an excellent job of reducing herbage to a finely ground pulp that has many times the surface area (and consequently many times the digestibility) of the original material.

Carnivores, on the other hand, deal digestively with food that is less resistant to the action of digestive juices and enzymes and is consequently more reducible and absorbable. As a result, careful preparation of ingesta is not so important, for what the teeth miss will be readily attacked and broken down by the digestive juices. The cheek teeth of carnivores are adapted to shearing movements to cut through muscles and connective tissues and reduce them to a size which can be conveniently swallowed. The teeth, therefore, have uniformly hard sharp surfaces and are not complexly folded. The jaw motion is confined to a chopping or scissors-like movement which allows the sharp cheek teeth to shear rather than to grind.

Omnivores, or animals which normally subsist on a mixed diet of animal and plant matter, have teeth which are a mixture of herbivore and carnivore dentition. The surfaces of the premolars have a shearing function, while the molars are grinders. The incisors and canines appear to be more closely related to carnivorous dentition. The molars and premolars do not have the complex folding characteristic of herbivore cheek teeth. The jaw movements of omnivorous animals are a combination of the hinge motion of carnivores and the rotary movements of herbivores.

Among omnivorous animals, the pig is an exception. Swine teeth do not fit any convenient classification. From a structural viewpoint they appear to have both shearing and grinding functions. The functional separation found in the premolars and molars of other omnivores is not apparent.

An interesting adaptation is found in the mouths of aquatic carnivores (seals, dolphins, porpoises) whose main diet is fish. In these animals the teeth are sharp and conical. These are seizing teeth rather than cutting, grinding, or tearing teeth, and are adapted for securing slippery prey, which is usually swallowed whole. Still another adaptation is found in sharks and certain bony fishes whose diet consists mainly of clams or coral polyps. The cheek teeth have become rounded masses primarily used for crushing, while the front teeth bear a strong resemblance to human incisors.

Another unusual form of dentition is found in elephants. The usual dental formulas of the Indian and African elephants differs slightly:

$$\text{Indian } \frac{0(0\text{-}1)01}{0001} \times 2 \qquad \text{African } \frac{0101}{0001} \times 2$$

This is because Indian cow elephants do not ordinarily have tusks, while African cow elephants do. However, the most peculiar specialization exists in the cheek teeth. The tusks are simply constantly growing upper canines that develop from an active pulp which produces layers of ivory that increase the length and diameter of the tusks as the animal grows older. Ivory is a form of dentine and is soft enough to be worked with tools, which accounts for its value in the jewelry trade. The cheek teeth, however, are complexly layered masses with an extensive root system and contain the four tissues (cement, enamel, dentine and pulp) found in other mammalian teeth. There are a total of 24 of these teeth which grow successively from 12 tooth buds in each jaw (six on each side in both upper and lower jaws). New teeth grow virtually simultaneously and displace the previous cheek teeth rostrally; so for a short time an elephant may have eight cheek teeth present in its mouth before the rostral molars are shed. The first three teeth of the series are apparently immature (i.e., "milk teeth") since they are smaller, less complex, and do not have the developed layering of the last three sets. These are displaced at one, five and ten years of age. The last three sets persist for a much longer time. A reasonably accurate method of determining the age of dead elephants from their teeth was worked out by Dr. Richard Laws in the 1960s. Unfortunately it cannot be applied to live animals, but it has shown that elephants can live 60 to 70 years.

chapter 7
The Digestive System

The digestive system is essentially a tube which extends from mouth to anus. This tube undergoes considerable modification in its various parts to accomplish such specialized functions as chewing, swallowing, gastric and intestinal digestion, absorption, and excretion. Certain exocrine glands (pancreas, liver, salivary) empty their products into the tube via long ducts. Other exocrine glands (goblet cells, chief cells, Brunner's glands, etc.) are incorporated into the wall of the digestive tube. Endocrine glands are also associated with the pancreas or as a part of the crypts of Lieberkühn. Specialized muscular layers form part of the wall of the tube, and function to propel ingested material through the system. An autonomous (independent) nervous system regulates its activity.

PARTS OF THE ALIMENTARY CANAL

The alimentary canal (digestive tube) is divided into nine major parts, some of which may be subdivided differently depending upon whether the animal examined has a simple or compound stomach or colon variations.

1. Mouth
2. Pharynx

3. Esophagus
4. Stomach
   a. Simple Stomach
      (1) Cardiac region
      (2) Esophageal region
      (3) Fundic region
      (4) Pyloric region
   b. Compound Stomach
      (1) Rumen
      (2) Reticulum
      (3) Omasum
      (4) Abomasum
         (a) fundic region
         (b) pyloric region
5. Small intestine
   a. Duodenum
   b. Jejunum
   c. Ileum
6. Cecum
7. Large intestine
   a. Large colon
   b. Small colon
8. Rectum
9. Anus

The Mouth

The mouth is the beginning of the digestive system. It includes the lips, tongue, teeth, and salivary glands and is responsible for the initial breakdown of food material by mechanical action, so that the smaller particles may be more readily acted upon by the digestive juices in the stomach and intestines. Saliva, a mixture of serous and mucous fluids, is produced principally by three sets of glands: two parotids, two submaxillary (mandibular), and four sublingual salivary glands. The oral mucosa also contains scattered mucous and serous buccal glands, but these are of minimal importance.

The two parotid salivary glands are located below the ears and behind the angle of the jaw. They produce a serous fluid that is emptied into the mouth via the right and left parotid ducts which open in the vicinity of the last upper premolars. The submaxillary (mandibular) salivary glands lie medial to the parotids and in some species may be covered by the rami of the mandible. They secrete mixed serous and mucous fluid which flows through ducts that run forward along the medial surface of the body of the mandible, and empty through two raised areas (caruncula sublingualis) located close to the base of the lower canine teeth. In rodents the submaxillary glands may be entirely mucous secreting.

The four sublingual salivary glands are located two on each side of the mouth beneath the mucous membrane along each side of the base of the tongue. They produce a mixed serous and mucous fluid that empties via a number of sublingual ducts (about 30 in the horse) that open along the sublingual folds of the tongue. In man and rodents the sublingual salivary glands are entirely mucous and in dogs and cats they are primarily mucous. In dogs and cats there is a fourth pair of salivary glands (the zygomatic), which are found at the rostral end of the zy-

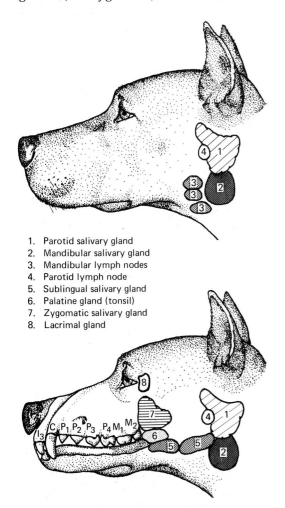

1. Parotid salivary gland
2. Mandibular salivary gland
3. Mandibular lymph nodes
4. Parotid lymph node
5. Sublingual salivary gland
6. Palatine gland (tonsil)
7. Zygomatic salivary gland
8. Lacrimal gland

FIGURE 7-1 *Salivary glands and lymphatics of the face.*

gomatic arch below the orbits of the eyes. These are sometimes called the orbital or suborbital salivary glands and produce serous fluid that empties into the right and left caudal buccal areas via a major and three or four minor ducts.

The mouth is closed rostrally by the lips, dorsally by the hard and soft palates that separate the oral and nasal cavities, laterally by the cheeks, and ventrally by the tongue. At its caudal end, at the base of the tongue, the mouth joins the pharynx.

The Pharynx

The pharynx is a funnel-shaped musculomembranous organ belonging to both the digestive and respiratory systems. The enlarged opening of the funnel faces rostrally and opens into the mouth and nasal cavity. The caudal opening is continued by the esophagus and the trachea. Dorsally, the pharynx is bounded by the base of the cranium, ventrally it is bounded by the larynx and esophageal opening. The pharynx has seven openings: caudal nares (two), eustachian tubes (two), oral opening, laryngeal opening (open except when swallowing), and esophageal opening (closed except when swallowing).

The Esophagus and Stomach

The esophagus is a musculomembranous tube extending from the pharynx to the stomach. It is divided into three parts according to location, i.e., cervical, thoracic, and abdominal. It is covered along most of its length by striated muscle. In the horse the esophagus is 50 to 60 inches long and one inch in diameter. The cervical part is six inches longer than the thoracic part, and the abdominal part is about one inch long. The last six to nine inches of the esophagus of the horse are covered with smooth muscle. In the bovine, the esophagus is shorter, larger in diameter, and more dilatable (three to three and one-half feet long, and two inches in diameter). The esophageal smooth muscle in cattle begins only an inch or so cranial to the junction of the esophagus with the stomach. The esophagus of sheep and goats resembles that of bovines except that it is about 1 inch in diameter.

The lumen of the esophagus is closed by deep longitudinal folds when empty and tends to remain closed unless something is being swallowed. The mucous membrane lining the esophagus is composed of stratified squamous epithelium. The submucosa consists of fibrous and elastic tissue and contains many mucous glands which act to lubricate the esophageal lining.

It should be mentioned that there are at least two major kinds of stomachs in mammals and a third kind in birds. These are the simple stomach, the compound (ruminant and pseudoruminant) stomach, and the gastric complex in birds, usually composed of a crop, proventriculus and ventriculus (p. 523).

The Simple Stomach

The simple stomach is a large specialized dilation of the alimentary tract, situated between the esophagus and the small intestine. It is a complete structure in all monogastric animals. In ruminants and pseudoruminants it is incomplete. In the horse, it is a J-shaped sac located slightly to the left of the median plane. It has two surfaces, two borders, and two extremities. The surfaces are parietal and visceral. Both are convex. The parietal surface lies against the diaphragm. The borders or

PARTS OF THE ALIMENTARY CANAL

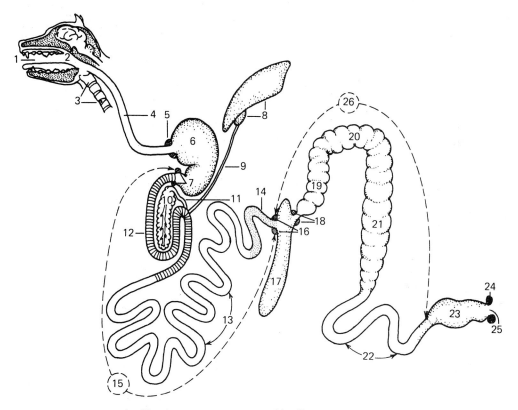

1. Mouth
2. Pharynx
3. Larynx and trachea
4. Esophagus
5. Cardiac sphincter
6. Stomach
7. Pyloric sphincter
8. Liver and gall bladder
9. Common bile duct
10. Pancreas
11. Pancreatic duct
12. Duodenum
13. Jejunum
14. Ileum
15. Small intestine
16. Ileocecal orifice and valve
17. Cecum
18. Cecocolic orifice and valve
19. Ascending colon ⎫
20. Transverse colon ⎬ Large colon
21. Descending colon ⎭
22. Small colon
23. Rectum
24. Anal sphincter
25. Anal orifice
26. Large intestine

FIGURE 7-2   *Schematic (nonrepresentational) drawing of the digestive system.*

curvatures are the greater and the lesser. The left (dorsal) extremity is a blind sac. The cardia and the esophageal orifice are located somewhat caudal to the dorsal extremity along the lesser curvature. A muscular ring, the cardiac sphincter, separates the stomach from the esophagus. The right extremity is smaller than the left. It is composed of the pyloric sphincter and orifice, and is continued by the duodenum. At the junction of stomach and duodenum the strong annular pyloric sphincter muscle separates these two parts of the alimentary tract.

The simple stomach is divided into four regions: cardiac, esophageal, fundic, and pyloric. These vary considerably in size and shape between species. The fundic and the pyloric regions are the principal centers of glandular activity.

STRUCTURE OF THE SIMPLE STOMACH

The wall of the stomach is composed of four layers of tissue, the serosa, the muscularis, the submucosa, and the mucosa. The serosa (visceral peritoneum) covers the outer surface of the stomach and is attached to the underlying muscular layers. It is continuous with the parietal peritoneum, the serous membrane that lines the abdominal cavity.

The muscular portion consists of three layers of smooth or unstriped muscle, an outer longitudinal layer, a middle circular layer, and an internal oblique layer.

The submucosa consists of connective tissue which contains vessels, glands and nerves. The mucosa is divided into two parts; an esophageal part and a glandular part. The esophageal part lines the esophageal area of the stomach. It is white in color, has no glands, and is composed of stratified squamous epithelium. In the horse it forms a saclike structure, the saccus cecus, that joins the glandular part at the margo plicatus, a raised area which marks the cardiac region. The margo plicatus is a structure peculiar to equines and is not present in other domestic animals. Caudal to the esophageal region is an area that is soft and velvety to the touch. This contains the gastric glands and is called the glandular part. The glandular part is divided into three regions: the cardiac region, fundic region, and pyloric region.

The cardiac region is closest to the esophageal and contains the cardiac glands. They are mucous glands, and do not produce enzymes. They are thought by some investigators to be regressive structures left over from an ancestral species which had a larger fundus.

The body of the stomach is called the fundic region and contains the fundic glands. These are the true gastric glands and are composed of three types of cells; body chief (zymogenic) cells, neck chief cells, and parietal cells. Body chief cells are found in the body and deeper parts of the gastric glands. They are enzyme producers and contain so-called zymogen granules (substances from which gastric enzymes are derived).

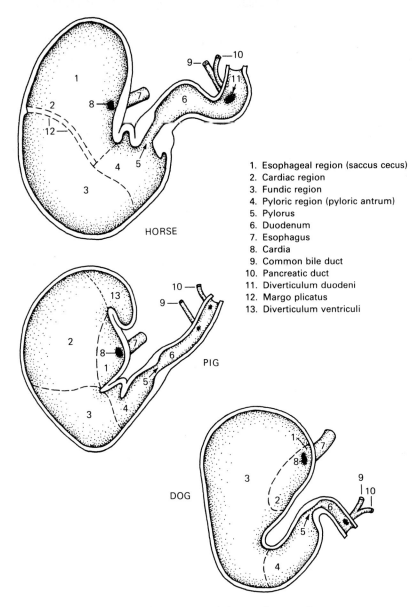

FIGURE 7-3   Comparative anatomy of the simple stomach.

Neck chief cells line the gastric glands near their openings and are mucus secreting cells. Parietal or border cells produce hydrochloric acid and "intrinsic factor." A fourth cell type, the argentaffin (surface) cell, is also found in both fundic and pyloric glands. These cells secrete mucus. They apparently differ from neck chief cells only in their staining reaction.

The caudal part of the stomach is called the pyloric region and contains the pyloric glands. The products of their secretion are mucus and small amounts of proteolytic enzymes. No parietal cells are present.

The Compound Stomach

The compound stomach is characteristic of ruminant animals and pseudoruminants and is very large. In ruminants it fills three-quarters of the abdominal cavity, and is located generally on the left side. It consists of four parts: rumen (paunch), reticulum (honeycomb), omasum (manyplies), and the abomasum, or true stomach, which contains the glandular structures. The capacity of the compound stomach is 30-40 gallons in cattle.

The rumen is by far the largest part of the adult compound stomach. It comprises approximately 80 percent of the total stomach volume. In cattle its long axis extends from the seventh or eighth rib to the pelvis. It is compressed laterally. It possesses two surfaces and two borders. The two surfaces are the parietal, or left surface, which is convex and contacts the diaphragm, abdominal wall, and spleen; and the visceral; or right surface, which is irregular and contacts the omasum, abomasum, intestines, and liver. The right and left surfaces are marked by grooves which separate the rumen into dorsal and ventral sacs.

The two borders are the dorsal curvature which conforms to the diaphragm and the sublumbar muscles, and the ventral curvature which conforms to the abdominal floor. The cranial or reticular extremity of the rumen is divided by a groove into two sacs, dorsal and ventral. The ventral sac is blind, and the dorsal sac leads to the reticulum. The line of demarcation between the rumen and reticulum is known as the rumeno-reticular fold. The rumen and reticulum form a domed vestibule (atrium ventriculi), on which the esophagus terminates. The esophageal opening is called the cardia, which opens on the inner surface above a fold of mucous membrane called the esophageal groove. The esophageal groove extends from the cardia to the opening between the reticulum and the omasum. In young animals it can be stimulated to form a tube which allows food to bypass the rumen. Caudally the rumen ends in a dorsal and ventral sac.

The rumen functions as a fermentation vat, an organ of maceration, a site for bacterial digestion, and an organ of absorption. Sixty to 70 percent of the cellulose ingested by ruminants is digested here by bacterial action. The bacteria, protozoa, and their metabolic products are then digested or absorbed by the ruminant.

The reticulum is the smallest section of the compound stomach comprising approximately five percent of the total volume. It is the most anterior of the compartments and is usually located underneath the left sixth to eighth rib. It lies entirely to the left of the median plane. It is lined with mucous membrane folded into hexagonal patterns. It receives

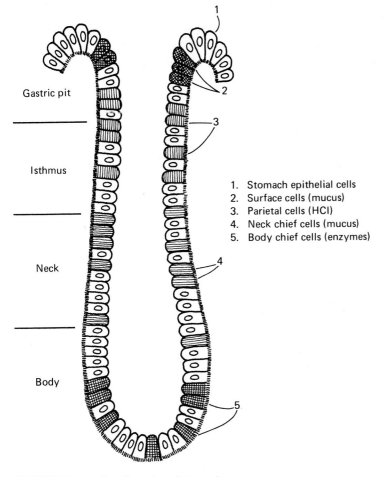

FIGURE 7-4   Fundic region of stomach mucosa.

heavy matter in food and acts as a liquid reservoir to soften these materials. It is not an essential structure as it can be removed and an animal will still live and ruminate.

The omasum contains approximately seven percent of the volume of the compound stomach. It lies to the right of the median plane and is a ball-shaped mass which contains broad, thin folds of mucous membrane. It supposedly grinds food more finely than the rumen and reticulum, and absorbs excess moisture. It is not concerned in rumination and is not an essential structure.

The abomasum is the "true" or glandular stomach. It secretes gastric enzymes and hydrochloric acid. In ruminants it appears to be an essential structure. It comprises about eight percent of the total stomach volume and lies for the most part on the abdominal floor along the median plane, to the right of the reticulum and ventral sac of the rumen.

It consists of three parts: the fundic portion, the body, and the pyloric portion. The fundic portion forms the cranial end. It connects with the omasum via the omaso-abomasal orifice. The body is the caudal extension of the fundus. The small terminal pyloric portion adjoins the small intestine, and is separated from it by the pyloric sphincter. The mucous membrane of the abomasum is glandular. Cardiac glands are found in a small area around the omaso-abomasal orifice. The fundic glands occur in the cranial part which contains large folds of mucous membrane. Pyloric glands are found in the body and pyloric portions of the abomasum.

Variations in Stomach Structure

One should never forget that there are few, if any, absolutes in animal biology and that any classification is a relative scheme that must deal with a number of qualifications, gradations, and exceptions. This is true of the stomach as well as other body structures. It is convenient to classify stomachs as simple and compound, but it should be remembered that the stomachs of dogs are not really characteristic of all simple stomachs, nor are the compound stomachs of cattle entirely characteristic of ruminant stomachs. Basically, a simple stomach performs chemical aspects of digestion without previous fermentation, and a compound stomach performs chemical digestion after a preliminary fermentative process and rumination. However, that does not necessarily mean that all simple or compound stomachs will follow a similar structural plan.

In the so-called "simple stomached" herbivores there are some truly remarkable variations which range from the simple diverticulum ventriculi of the pig and the more complexly sacculated stomach of the peccary, to the highly complex stomach of the hippopotamus which contains three compartments, a diverticulum and an esophageal groove. Structurally the hippopotamus stomach is closer to compound than simple, yet it is apparently not a ruminant stomach, since hippopotami do not ruminate.

Many rodents (notably the golden hamster among those used by man) have sacculated stomachs. The pack rat (*Neotoma* spp.) and a number of others (*Dipodomys, Notomys, Perognathus, Peromyscus*) have sacculated stomachs with a nonglandular portion where fermentation can take place.

Members of highly specialized herbivore genera such as the three-toed tree sloth (*Semnopithecus* sp.) have remarkably complex stomachs which have many similarities to those of ruminants although the sloths do not ruminate.

Among animals that ruminate, the Camelidae (which includes dromedary and bactrian camels, vicuñas, alpacas, guanacos and llamas) are classed as pseudoruminants. Although they possess compound stomachs, the differences in microscopic and gross anatomy tend to place them in a different category than true ruminants. The omasum, for instance, is small and tubular and is structurally a part of the abomasum, and the reticulum contains glandular cells.

The Small Intestine

The small intestine is a tube connecting the stomach to the cecum and large intestine. It is suspended from the dorsal part of the abdominal cavity by a fold of peritoneum called the great mesentery. It consists of three parts: duodenum, jejunum, and ileum. In the horse, the total length is about 70 feet, with a normal capacity of about two gallons. The small intestine is where most of the absorption from gastric and intestinal digestion occurs. The duodenum (about five percent of the total length) is the fixed part of the small intestine, and is closely attached to the stomach. It is formed in an S-shaped curve which contains the pancreas. The pancreatic and bile ducts of horses enter the duodenum five or six inches caudal to the pyloric valve.

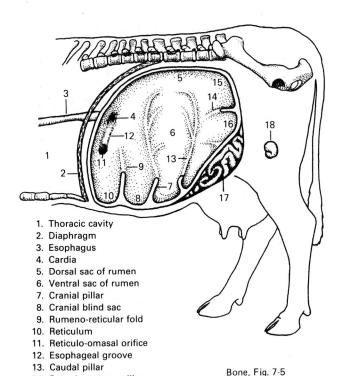

1. Thoracic cavity
2. Diaphragm
3. Esophagus
4. Cardia
5. Dorsal sac of rumen
6. Ventral sac of rumen
7. Cranial pillar
8. Cranial blind sac
9. Rumeno-reticular fold
10. Reticulum
11. Reticulo-omasal orifice
12. Esophageal groove
13. Caudal pillar
14. Dorsal coronary pillar
15. Caudal dorsal sac
16. Caudal ventral sac
17. Small intestine
18. Supramammary lymph node

Bone, Fig. 7-5

FIGURE 7-5   *Stomach of ruminant.*

160
THE DIGESTIVE SYSTEM

The jejunum is about 90 percent of the total length of the small intestines and has no distinct demarcation from either the duodenum or the ileum. The jejunum and ileum form the mesenteric part of the intestines and are found generally in the left dorsal position of the abdominal cavity. In dead animals the jejunum is usually empty. The last four or five percent of the length of the small intestine is usually contracted, and this part is called the ileum.

The small intestine, like the stomach, consists of four layers: serosa, muscularis, submucosa, and mucosa. The serous layer is complete except at the mesenteric edge, where the vessels and nerves reach the intestine from the mesentery. The muscular layer consists of an external longitudinal and an internal circular layer of smooth muscle. The submucosa is composed principally of connective tissue and contains the duodenal (Brunner's) glands, the bases of lymph nodules, the crypts of Lieberkühn, and isolated smooth muscle cells. Mucous membrane is present

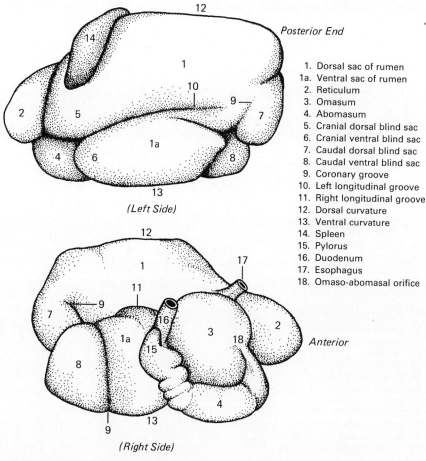

1. Dorsal sac of rumen
1a. Ventral sac of rumen
2. Reticulum
3. Omasum
4. Abomasum
5. Cranial dorsal blind sac
6. Cranial ventral blind sac
7. Caudal dorsal blind sac
8. Caudal ventral blind sac
9. Coronary groove
10. Left longitudinal groove
11. Right longitudinal groove
12. Dorsal curvature
13. Ventral curvature
14. Spleen
15. Pylorus
16. Duodenum
17. Esophagus
18. Omaso-abomasal orifice

FIGURE 7-6   Ruminant stomach (cow).

and extensive and is composed of columnar epithelium. A short distance caudally from the pyloric valve the diverticulum duodeni is found, into which the bile and pancreatic ducts open.

The mucosa of the relaxed small intestine contains longitudinal folds (rugae), which are covered with small nipple-like folds (villi) interspersed with basal pockets (crypts of Lieberkühn). The villi are covered with columnar epithelial cells interspersed with goblet (mucus) cells. Underlying the epithelium is a basement membrane that contains scattered smooth muscle cells (muscularis mucosae) and a loose connective tissue framework that supports blood and lymph capillaries.

The lymph capillaries (lacteals) are one of the notable features of the villi. They derive their name—lacteals—from the milky appearance of their contents during the absorptive phase of intestinal activity. Lacteals are principally concerned with the absorption of fats and fatty acids and to a lesser extent with absorption of carbohydrates and proteins.

The blood capillaries are principally involved in the transport of carbohydrates. Protein absorption appears to be shared about equally between lacteals and capillaries. Lacteals ultimately connect with the vascular system at the cranial vena cava (Figure 9-6). Blood capillaries unite to form the hepatic portal vein, which ends in the liver (Figures 9-4, 9-5).

There are three kinds of glands in the small intestines: intestinal glands, duodenal glands, and Peyer's patches. Intestinal glands (crypts of Lieberkühn) are found throughout the small intestine and are simple tubular structures which contain secreting (Paneth) cells and epithelioid cells. Duodenal (Brunner's) glands are found in the first part of the small intestine and are branched tubuloalveolar glands situated in the submucosa. Aggregated nodules of lymphoid tissue called Peyer's patches occur throughout the small intestine.

THE INTESTINAL EPITHELIAL CELLS

The columnar cells that form most of the epithelium of the intestine have long been known to have a specialized structure which would agree with the specialized function, but much of the nature of this specialization has been obscure until the advent of electron microscopy. Under the light microscope a "striated border" can be seen on that part of the cell which is next to the lumen of the gut. The electron microscope resolves this border into a closely packed arrangement of microvilli whose free surfaces are covered by a protective mucopolysaccharide coat which apparently contains the enzyme adenosine triphosphatase (ATP-ase). Both the coat and the enzyme are supposed to be involved in the early stages of biochemical degradation of food and the absorptive process. Beneath the microvilli is a somewhat amorphous fibrillar zone called the terminal web, which does not contain the usual cytoplasmic organelles. Beneath the web, the cytoplasm is filled with normal cytoplasmic organelles including a large number of mitochondria which indicate an active function.

Laterally the epithelial cells are connected to each other by interdigitating membranes that adhere to each other along specialized regions called desmosomes, tight junctions, and terminal bars. These seal off the lumen of the gut from the epithelial intercellular spaces. Ultramicroscopic studies of fat metabolism indicate that lipid droplets first collect in spaces between the microvilli and pass intact into the cytoplasm by a process similar to pinocytosis. Once inside the cell, the droplets are assembled in the lumens of the smooth endoplasmic reticulum and are transported to the region of the basement membrane where they are extruded from the cell to the intercellular spaces surrounding the lacteals. No significant differences in structure exist between light microscopic and electron microscopic appearance of goblet cells.

COMPARATIVE ANATOMY OF THE SMALL INTESTINE

The small intestine of cattle is about 130 feet long. The intestines are located largely in the right abdominal region. The duodenum is three to four feet long and is formed into an S-shaped curve which contains the pancreas. The bile duct opens at the beginning of the S-shaped curve. The pancreatic duct opens about a foot behind the bile duct. Intestinal glands are found throughout its length. Duodenal glands are found in the first 12 to 15 feet of the small intestine. Peyer's patches are larger and more distinct than those in horses.

The small intestine of sheep is about 80 feet long, and generally similar in structure to that of cattle.

The small intestine of swine is about 50 to 65 feet long and is similar to that of the horse, except that the bile and pancreatic ducts open separately into the duodenum.

The small intestine of the dog is approximately five times the body length of the animal. Owing to the great variation in size between breeds, no specific lengths can be given. In 80 percent of all dogs, the pancreas empties into the duodenum via two ducts, one of which is associated with the terminal portion of the common bile duct at a structure called Vater's ampulla. Those dogs that have one pancreatic duct possess separate openings for bile and pancreatic secretions. Structurally the various parts of the small intestine are similar to those of other mammals.

The small intestine of the cat is similar to that of the dog.

The Cecum

The cecum is a blind sac situated between the ileum and the large colon. In a medium-sized horse it is about four feet long with a capacity of about 12 gallons. It is somewhat comma-shaped, and lies to the right

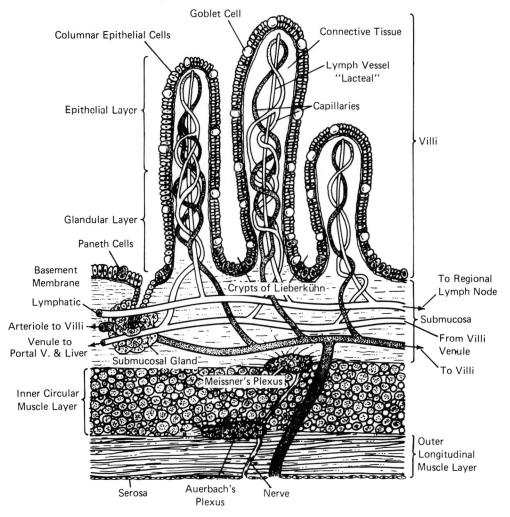

The submucosa consists of connective tissue and contains muscle fibers, glands, fat and the larger blood and lymph vessels which supply the mucosa and muscle layers

FIGURE 7-7   *Long section of small intestine (schematic).*

of the median plane. It consists of a body, base, and apex. The cecum is characterized by four longitudinal bands, dividing its surface into four rows. The rows are sacculated by successive rings of circular muscle. It has two openings at the base: the ileocecal opening, which possesses a thick muscular layer (sphincter ilei), and the cecocolic orifice, which has a valvular fold (sphincter ceci). These two openings lie next to each other, separated by a fold of mucous membrane.

The Large Intestine

The large intestine extends from the termination of the ileum to the anus. It does not include the cecum. In a medium sized horse it is about 25 feet long. The large intestine differs from the small intestine in that it is larger, sacculated, has longitudinal bands (incomplete musculature), and a more fixed position. It is divided into the large colon and small colon.

The large colon of the horse can vary from 10 to 21 feet in length, depending upon the size of the animal, and has a maximum capacity of about 24 gallons. It is divided into four parts and three flexures, which are identified by anatomic location, the number of longitudinal bands (which vary from one to four), and the spatial position of the part within the abdominal cavity.

No. of Bands	Parts	Flexures
4	Right ventral colon	Sternal
4	Left ventral colon	Pelvic
1 band at beginning, 3 at end	Left dorsal colon	Diaphragmatic
3	Right dorsal colon	

In the horse, the small colon extends from the termination of the right dorsal colon to the rectum. It has a length of 10 to 12 feet in the average size animal and a diameter of approximately three inches. It possesses two muscular bands and a number of sacculations.

In horses the serous coat is not present in all parts of the large intestines. Two muscular coats (outer longitudinal and inner circular) are present. The bulk of the longitudinal muscle coat of the horse is found in longitudinal bands. In other animals (cattle, sheep, swine, and dogs) the outer muscular layer is complete. The rectal mucosa is so abundant that it hangs in folds when the rectum is empty. Connective tissue is seen in all parts. The mucous membrane does not contain any Brunner's glands. Intestinal glands are present. There are no villi (except in the ceca of equines) and no Peyer's patches.

COMPARATIVE ANATOMY OF THE LARGE INTESTINE

There is no essential difference except size between the large intestine of the donkey and the horse. The large intestine of cattle is much smaller than that of horses. It does not possess bands or sacculations since the musculature is complete. It is located in the right dorsal part of the abdomen. The cecum is 30 inches long and five inches in diameter. The colon (ansa spiralis) is 35 feet long, five inches in diameter in the first part, two inches in diameter in the last part, and is coiled in a double flat whorl. The rectum is shorter than in horses.

165
PARTS OF THE
ALIMENTARY
CANAL

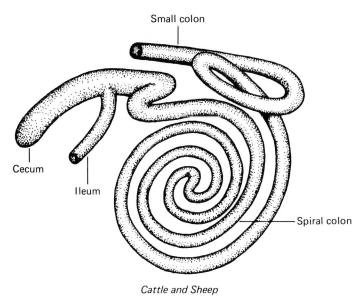

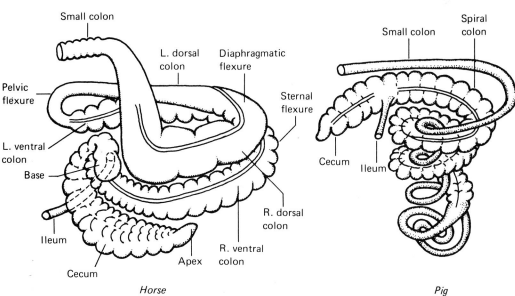

FIGURE 7-8   Comparative anatomy of the large colon.

Sheep have a 10-inch cecum and a 15-foot colon, which are similar in shape and position to those of cattle.

The cecum of the pig is a blind sac 15–25 inches in length and three inches in diameter. The great colon is quite long, and is a direct continuation of the open end of the cecum. It is large in diameter, sacculated throughout most of its length, and is folded into a double loop which is twisted into the form of a helical spiral. The rectum is short and straight, and is connected to the great colon by a short undifferentiated small colon.

The dog does not possess a small colon. The cecum is short and nonfunctional, and opens directly into the colon which is not sacculated. The colon is considered to be divided into ascending, transverse, and descending parts as in man. The descending colon empties into the rectum.

The colon of the cat is similar to that of the dog.

The Rectum and Anus

The rectum is the terminal part of the intestine and is found in the pelvic cavity. It is about 12 inches long in horses, slightly shorter in cattle, and proportionately smaller in sheep, swine, dogs, and cats. It is essentially an organ of storage.

The anus is the terminal part of the alimentary system, and is continuous with the skin. Three major muscles are present: the sphincter ani internus, a terminal thickening of the circular muscle; the sphincter ani externus, which is located outside the internal sphincter; and the retractor ani, which retracts the partial prolapse that the anus undergoes during defecation. At the rectoanal junction the epithelium changes abruptly from columnar to stratified squamous. Around the anal opening can be found circumanal glands (sebaceous glands). In the dog there are also two laterally located anal sacs that lie between the internal and external anal sphincter muscles. These sacs may become filled with a semi-solid secretion that has an extremely foul odor and can be very irritating to both the dog and its owner and may require veterinary attention.

EXOCRINE GLANDS

The Pancreas

The pancreas is a gland found in the first loop of the duodenum. It communicates with the duodenum by a duct which may or may not be united with the bile duct from the liver. In horses, the ducts terminate in a pouchlike structure called the diverticulum duodeni (hepatopancreatic ampulla), which extends into the lumen of the intestine. In sheep, the bile duct and pancreatic duct are united at their termination. In cattle and swine, they are separate. In cats and dogs, there are usually two pancreatic ducts. The caudal duct is single and is the larger of the two. The cranial duct often unites with the bile duct. The two ducts are usually two or three centimeters apart. The pancreas possesses two functions and a double structure: an acinar portion, which is exocrine and secretes enzymes, and the islets of Langerhans, which are endocrine structures and secrete hormones.

The exocrine portion of the pancreas is a compound tubuloalveolar gland. It does not possess a distinct capsule but is covered by fibroelastic connective tissue that passes into the gland and divides it into primary lobules. Groups of primary lobules are enclosed by connective tissue to form secondary lobules. The pancreatic duct extends almost the entire length of the pancreas and gives off branches to all primary lobules. The branches to the primary lobules undergo further branching and give off branches to each secondary lobule. The branches to the secondary lobules give off long narrow intercalated ducts, which in turn branch before entering the terminal secreting acini. The excretory ducts are lined with simple high columnar epithelium. As the ducts decrease in size the epithelium becomes lower. In the intercalated ducts the epithelium is cuboidal.

The pancreatic acinar cells are notable for their large amounts of rough endoplasmic reticulum and an extensive Golgi apparatus which functions to enclose enzymes in membranous packets that are released into the acinar ducts when needed, and are there freed of their protective coverings.

Islets of Langerhans are cellular aggregations of endocrine substance interspersed irregularly among the acini. They have no duct system and their products are secreted directly into the blood stream. Two cell types, alpha and beta, appear to be important. Alpha cells contain packaged secretion droplets of a glycogenolytic hormone called glucagon. Beta cells contain no droplets, but secrete insulin.

The Liver

The liver is the largest gland in the body and is essential to life. It performs a number of functions among which are storage and formation of animal starch (glycogen), secretion of bile, detoxification of poisons, breakdown of uric acid, formation of urea, and desaturation of fatty acids. The liver is situated caudal to the diaphragm and its cranial surface conforms closely to the shape of that structure. Generally, the liver lies to the right of the median plane, although this location varies depending upon the species of animal.

The liver is usually reddish brown in color but may range from tan to purple and is covered with a connective tissue capsule. It has two surfaces and four borders. The surfaces are parietal and visceral, and the borders are dorsal, ventral, right, and left. Two to seven lobes may be present depending on the species of animal.

The parietal surface is convex and faces cranially and dorsally and conforms closely to the diaphragm against which it lies. It possesses a medial groove, the fossa venae cavae, in which the caudal vena cavae and the terminal part of the hepatic vein lie. Surrounding the fossa are the hepatic attachments of the coronary, falciform, and round ligaments of the liver.

The visceral surface is concave, faces caudally and ventrally, and possesses a number of ridges and depressions which conform to the visceral organs that lie against it.

The dorsal (caudal) border lies upward and backward and is generally quite thick. The ventral border lies downward and forward. It is usually thin and bears on its surface the round ligament of the liver which is the adult remnant of the umbilical vein of the fetus. The right border is thin and long. It extends along the inner surface of the costal arch to a point near the last rib, and bears the right lateral ligament on its dorsal margin. The left border is thin and conforms to the shape of the viscera which lie against it.

Blood enters the liver via the hepatic artery and the portal vein. The hepatic artery is a branch of the coeliac artery, and the portal vein is the main trunk of the portal system that extends from the intestines to the liver. Blood leaves the liver via the hepatic vein which enters the caudal vena cava at or near the cranial parietal surface in the region of the fossa venae cavae.

The nerve supply is automatic and visceral sensory and is derived chiefly from the vagus nerve and the hepatic plexus of the sympathetic nervous system.

The liver possesses a system of excretory ducts which serve to convey bile to the duodenum. The ducts begin as small canaliculi in the lobules of the liver that empty into bile capillaries in the center of each lobule. These capillaries are aggregated into interlobular ducts which unite to form interlobar ducts. The interlobar ducts combine near the portal fissure to form the hepatic duct. Intercalated in the course of the hepatic ducts in most animals is a sac-like structure called the gall bladder, which is connected to the hepatic duct by a short cystic duct. The gall bladder functions as a storage place for bile. Most mammals possess a gall bladder but some do not, notably the equidae, deer, elk, and moose (also elephants, giraffe, rhinoceri, camels, tapirs, and rats). In some species—notably dogs, cats, mice, and humans—the gall bladder has the ability to concentrate bile. Domestic ruminants and swine apparently lack this ability. In mammals that do possess a gall bladder, the continuation of the duct system beyond the junction of the cystic and hepatic ducts is called the common bile duct (ductus choledochus communis). In those animals without gall bladders or cystic ducts, the word "common" is usually omitted and the structure is simply called the bile duct (ductus choledochus).

COMPARATIVE ANATOMY OF THE LIVER

The liver of horses lies with its long axis in a somewhat oblique direction. It is quite large, and weighs from 12 to 20 pounds depending upon the size of the animal. It is held in position by six ligaments and the pressure of surrounding organs. The ligaments are coronary, falciform, round, right, left, and caudate.

Three lobes and one process are present. There is no gall bladder. The lobes are left, middle, and right. The process is the caudate process, which is a part of the right lobe and surrounds the caudal vena cava and portal vein, and forms the major portion of the portal fissure. The caudate process bears the caudate ligament on its caudal surface.

The right lobe is the largest of the three and lies to the right of the median plane. It bears upon its dorsal (caudal) border the right lateral ligament of the liver.

The middle lobe is the smallest and bears on its cranial surface a vascular notch (fossa venae cavae) which contains the caudal vena cava and hepatic veins, together with the attachments of coronary ligaments of the liver. The falciform ligament is a ventral continuation of the coronary and extends ventrally from the vascular notch. It is continued as the round ligament of the liver.

The left lobe lies to the left of the median plane. Its ventral (cranial) border may extend as far cranioventrally as the sternal portion of the diaphragm. On the caudal (dorsal) border of the left lobe is the left lateral ligament of the liver. The visceral surface of the left lobe encloses the terminal portion of the portal vein and the hepatic ducts that unite to form the bile duct which empties together with the pancreatic duct into a small outpouching of the duodenum called the diverticulum duodeni.

The liver of cattle lies almost entirely to the right of the median plane. It is smaller in surface area, but thicker than the liver of the horse. Its weight ranges from 10 to 14 pounds in mature animals. It consists of a body that forms the major part of the organ and two small lobes (caudate and papillary) which are located caudomedially. The papillary lobe is usually present only in calves, and consists of a tonguelike mass which overlaps the portal tissue.

The ligaments are coronary, falciform, round, right lateral, and caudate. The left lateral ligament is absent.

A gall bladder is present and empties into the duodenum by a single duct, the common bile duct, which enters the duodenum separately about 10–12 centimeters (five inches) cranial to the opening of the pancreatic duct.

The liver of sheep lies entirely to the right of the median plane, and weighs about one and one-half pounds. It is divided laterally into two chief lobes (dorsal and ventral) by a transverse umbilical fissure, and possesses a small caudomedial caudate lobe. The liver is relatively shorter and more compressed than that of the cow. A gall bladder is present and the common bile duct unites with the pancreatic duct before entering the duodenum to form Vater's ampulla.

The liver of swine is of moderate size and weighs about four pounds in adult animals and lies somewhat to the right of the median plane. It is divided by three deep fissures into four principal lobes, which are right lateral, right central, left central, and left lateral. On the caudal portion of the right lateral lobe is found a fifth lobe, the caudate lobe, which is clearly demarcated by a fissure. A gall bladder is present, and the common bile duct opens into the duodenum separately from the pancreatic duct. The liver of the swine can be distinguished from that of other domestic animals by the dense and relatively thick connective tissue stroma that can be readily seen in gross specimens and gives a reticulated appearance to the organ.

The liver of dogs is relatively large, comprising about three percent of the body weight. It lies along the midline of the cranial abdominal cavity and extends approximately equal distance on both sides of the median plane. It is composed of five major and two minor lobes, separated by fissures. There is some argument about the number of lobes. Miller (*Anatomy of the Dog*) states that there are six, with the papillary forming a process of the caudate lobe. Getty (Sisson and Grossman, *Anatomy of the Domestic Animals*) states that there are five major lobes, with the quadrate and papillary lobes being processes of the right central lobe and the caudate lobe, respectively. Other (older) authors state that there are seven lobes. The major lobes are right lateral, right central, left central, left lateral, and caudate. The sixth lobe, the papillary lobe, is a branch of the caudate. The seventh lobe, the quadrate, is marked off from the right central lobe by a deep fissure that contains the gall bladder.* The bile duct may or may not unite with the pancreatic duct to form Vater's ampulla prior to entering the duodenum.

Comparative Anatomy of the Liver (General)

Species	Location with Regard to Median Plane	Number of Lobes	Gall Bladder	Bile Duct Termination
Horse	Oblique, mainly right	3	Absent	Single
Cow	Almost entirely right	2-3	Present	Single
Sheep	Entirely right	2-3	Present	Common
Pig	Majority right	5	Present	Single
Dog	Central	5-7	Present	Single or Common
Cat	Central	5-6	Present	Common

The liver of cats is similar to the liver of dogs. It lies along the median plane of the body and is composed of five or six lobes: right and left lateral, right and left central, and caudate. The right lateral lobe is divided into cranial and caudal parts by a transverse fissure and is

---

*This number varies with the various authors of texts and ranges from five through seven.

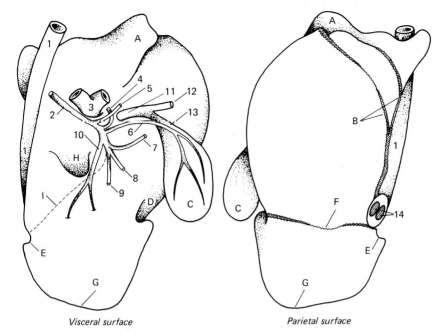

Visceral surface *Parietal surface*

1. Caudal vena cava
2. Hepatic artery
3. Portal vein
4. R. hepatic branch of 2
5. Pancreatic branch of 2
6. Cystic artery
7. Cranial pancreaticoduodenal artery
8. Gastroepiploic artery
9. Gastric artery
10. Left hepatic branch of 2
11. Hepatic duct
12. Cystic duct
13. Bile duct
14. Hepatic veins

A. Caudate lobe
B. Coronary ligament (cut)
C. Gall bladder
D. Notch for round ligament
E. Esophageal notch
F. Falciform ligament
G. Cranioventral margin
H. Papillary process
I. Line of lesser omentum

FIGURE 7-9   *Bovine liver.*

considered to be two lobes by some anatomists. The caudate lobe is small and is composed of two tongue-shaped masses that are united cranially and lie to the left of the caudal vena cava. The gall bladder lies in the fissure separating the central lobes and receives three heptic ducts. It gives off a single cystic duct, which is continued as the common bile duct. The common bile duct unites with the pancreatic duct to form Vater's ampulla just prior to entering the duodenum.

ACCESSORY STRUCTURES OF THE DIGESTIVE SYSTEM

The Peritoneum

The peritoneum is a serous membrane which covers the abdominal viscera and lines the inner surface of the abdominal cavity and a portion of the pelvic cavity. It appears smooth and glistening. It is covered with a flattened layer of mesothelial cells and is kept moist by the peritoneal fluid that these cells secrete. The peritoneum forms lubricating surfaces which cover the viscera and slide upon one another, allowing the enclosed viscera to move with a minimum of friction.

One should remember that the peritoneal cavity *per se* exists only potentially and should not be confused with the abdominal cavity. The visceral organs all lie outside the peritoneum in the abdominal cavity and the peritoneal cavity itself only contains a small amount of serous fluid.

The peritoneum is composed of two major and several minor parts. The major portions are called the parietal and visceral peritoneum. The minor parts are the omentum, mesenteries, and ligaments.

The parietal peritoneum lines the inner surface of the abdominal cavity and covers one surface of the retroperitoneal structures which include the kidneys, ureters, and urinary bladder.

The visceral peritoneum is that portion which almost completely encloses the visceral organs.

The Mesenteries

The parietal peritoneum meets dorsally along the midline of the body. Then it is reflected ventrally as a double layer which encloses the vessels and nerves to the digestive tract and forms a hammocklike support for the intestines. These reflected folds of peritoneum are called mesenteries.

There are two major mesenteries in the abdominal cavity: the great (cranial) mesentery, which supports the small intestine, pancreas, and cecum; and the lesser (caudal) mesentery, which supports the large intestines and the cranial portion of the rectum. Certain specialized folds of peritoneum that support the ovaries and fallopian tubes, and the duct system of the testicles are called the mesovarium and mesorchium, respectively.

The Ligaments

The suspensory ligaments of the abdominal cavity are folds of peritoneum enclosing sheets or cords of connective tissue which function to support organs other than the gut. Such structures include the ligaments of the liver, the round ligament of the urinary bladder, the broad ligament of the uterus in the female, and the genital folds in the male. A double fold of peritoneum that extends into the scrotum and covers the testicles is called the tunica vaginalis.

The Omentum

The omentum is a fold of peritoneum that passes from the stomach to the adjacent viscera. This structure is particularly well developed in mammals and usually consists of three parts: the gastro-splenic omentum, the lesser omentum, and the greater omentum. Structurally, the omentum differs from the remainder of the peritoneum. It is very thin, almost lacelike in appearance, and is interspersed with cords and masses of fatty tissue. The gastro-splenic omentum extends from the greater curvature of the stomach to the spleen. The lesser omentum passes from the lesser curvature of the stomach to the liver. The greater omentum passes from the greater curvature of the stomach and the spleen to the origin of the small colon. It does not pass directly from one organ to the other but instead forms an extensive sac-like structure that lies loosely on the abdominal floor, enclosed by the peritoneum.

The omentum has the ability to migrate to areas in the abdominal cavity that are damaged and serves to plug perforations or reinforce weakened or inflamed portions of the viscera. Indeed, this property, together with its ability to confine and reduce infection, is so marked that the omentum has been called "nature's surgical dressing." The omentum is frequently utilized by surgeons who use a fold or piece of it to help close operation sites in the intestine or other viscera.

The Abdominal Cavity

The abdominal cavity is the largest of the body cavities. It contains the abdominal viscera and the peritoneum cavity and is separated from the thoracic cavity by the diaphragm. It is continuous caudally with the pelvic cavity from which it is separated by the brim of the pelvis.

The cavity is roughly egg-shaped, with the small end directed caudally. Depending upon the species, it may be laterally compressed. The dorsal wall is formed by the caudal portion of the diaphragm and the lumbar vertebrae and muscles. The lateral walls are formed by the abdominal muscles, a part of the diaphragm, and that part of the rib cage which lies caudal to the diaphragmatic border. The ventral wall is formed by the rectus abdominus muscles and the xiphoid cartilage of the sternum. The cranial wall is formed by the diaphragm.

The Pelvic Cavity

The pelvic cavity is enclosed within the pelvis and is a direct continuation of the abdominal cavity. It contains the rectum and parts of the urogenital apparatus. It is partly lined with peritoneum and partly with pelvic fascia, a form of fibrous connective tissue. The walls are formed by the sacral vertebrae, the pelvic girdle, and a variable number of coccygeal vertebrae. The caudal part of the pelvic cavity is closed by the perineal fascia, the anus and its muscles, the vulva (female), or the root of the penis (male).

PHYSIOLOGY OF DIGESTION

General Considerations

The term, "digestion," includes all processes by which foods in the alimentary canal are prepared for absorption.

While the definition of foods can be made broad enough to include water, minerals, and vitamins, these latter materials do not furnish energy, and hence should not be classified as food. Food materials serve to yield energy necessary for the activity of the body. They are also necessary to build and repair tissues and to regulate body processes. All foods are organic in nature and are originally derived directly or indirectly from plants which have stored energy from the sun.

An animal body may be thought of as a hollow cylinder, one end of which is the mouth and the other the anus. Food passed into the anterior end of this cylinder is propelled by a variety of muscular contractions through the tube and in the process is broken down to its basic components and absorbed. After absorption, the basic components are reformed into material that can become a part of the wall of the cylinder, and the waste products that remain in the digestive tube are excreted. Since most foods are ingested in complex and nonabsorbable forms, a number of processes must take place before they are degraded to their basic components which can be utilized.

Essentially foods are composed of four materials: carbohydrates, fats, proteins, and vitamins. The first three must be reduced respectively to simple sugars (monosaccharides), fatty acids and glycerol, and amino or nucleic acids, since these are the only food substances that may pass across the mucous membrane of the absorptive portions of the gut. Other substances of larger molecular size can be passed across the intestinal epithelium if they are lipid or water soluble (e.g., vitamins and drugs), if they interact with specific areas of the gut (e.g., hemopoietic factor in the ileum), or if they can combine with carriers (such as choline) capable of passing substrates across the epithelial barrier.

Carbohydrates

Carbohydrates, as the name implies, are hydrated-carbon atoms collected together in chains called simple sugars. These in turn are polymerized into larger molecules which may contain hundreds of simple sugars joined into a single unit. Carbohydrates may contain carbon chains of 3, 4, 5, or 6 carbon atoms in the simple sugars which form their basic structures. In the animal body the 3, 5, and 6 carbon monosaccharides are the only ones that appear to be readily utilized, and of these the hexose (6-carbon) sugars are the most common form the body assimilates and uses. The types of carbohydrates result from the number of sugar molecules composing them. Monosaccharides are simple sugars and contain but one chain of carbon atoms.

The hexose (6-carbon) sugars are the commonest ones involved in nutrition. These have the empirical formula $C_6H_{12}O_6$ and are classed as a single group, although their structural configurations differ. They are soluble in water and can be readily absorbed across the intestinal mucous membrane. They are the units from which all other hexose carbohydrates are formed. The ordinary hexose monosaccharides—glucose (dextrose), galactose, and fructose—all have the same number and kind of atoms, but differ in their spatial arrangement. Disaccharides or double sugars consist of two molecules of simple sugar combined into a larger molecule, with the loss of one molecule of water. Double sugars have the empirical formula $C_{12}H_{22}O_{11}$. Depending upon the specific monosaccharides that go into their formation, the structural characteristics and sweetness of the disaccharides vary. The commonest of these double sugars—lactose (milk sugar), sucrose (table sugar), and maltose (corn sugar)—are respectively derived from one molecule of glucose and one of galactose, one molecule of glucose and one of fructose, and two molecules of glucose. Polysaccharides are compound sugars which include starches, gum, cellulose and lignin. They are formed from a great number of simple sugars (principally glucose) combined into one large molecule with the empirical formula of $(C_6H_{10}O_5)^n$ where "n" is the number of molecules in the structure. These substances (except starch) are generally resistant to digestion by mammals.

Gums, cellulose, and lignins cannot be digested by the enzymes in the animal gut. However, bacteria and protozoa in the rumen and cecum of herbivores can and do supply enzymes necessary to degrade some of these complex molecules.

Newborn animals have considerable lactase activity in their gut, but may not have any appreciable amounts of other disaccharide-reducing enzymes. Calves, in particular, lack maltase and have virtually no sucrase. In consequence, they have difficulty digesting starch and maltose and virtually no capability of digesting table sugar. Furthermore, mammals may lose lactase activity in the gut with the onset of puberty and can develop purging diarrhea from drinking milk during adolescence or adulthood.

Maltase and sucrase activity levels appear to increase with age, although in some genera (notably cattle) sucrase levels are never high.

Upon hydrolysis and absorption of simple sugars into the body, the metabolized carbohydrates may go in one of two directions. The simple sugar may remain in the blood stream, to be directly utilized by cells, or it may be formed into a substance called glycogen (animal starch) which is stored in the liver or muscles. Glycogen is a reserve supply of carbohydrate which is readily available to the animal during periods when dietary carbohydrate is not available. It usually consists of a polymer of about 14 molecules of glucose but the exact number of molecules varies.

Triose (3-carbon), tetrose (4-carbon), and pentose (5-carbon) sugars also exist in the diet and to a certain extent can be utilized by the animal. Of these three, triose and pentose sugars are more commonly used. Pentose sugars enter into certain metabolic schemes without change. Triose compounds, principally glycerol (which is extracted from fat digestion), are formed into ordinary 6-carbon sugars by synthesis in the liver, or are broken down into carbon dioxide and water.

A certain amount of glucose is necessary in the internal environment for normal cell functioning. Since glucose is the most readily available source of energy, some is always present in the circulating blood and in tissues. Glucose also forms a small part of protoplasm in combination with other compounds and forms the principal energy source for metabolic reactions in the body.

Fats or Lipids

Fats are triglycerides, i.e., combinations of three molecules of fatty acid and one molecule of glycerol. Fatty acids contain less oxygen and more carbon and hydrogen than do the carbohydrates. Therefore, they are a more concentrated form of energy, yet are less available because they must be changed to glucose before they can be used. Since fats are relatively stable, they can be held for longer periods of time as reserve energy stores and form a more permanent reserve than glycogen. Fats are stored within the body in specialized (fat) cells which occur in more or less specific locations or deposits throughout the body. Since fat is a poor conductor of heat, it also serves as insulating material. Fat enters into the structure of all cells (particularly of the cell membranes as phospholipid), and is found in relatively large quantities in the myelin sheaths of nerve fibers.

The quality and nature of ingested and absorbed fat has an influence on the physical nature of fat deposits in the body. Monogastric animals that feed upon unsaturated fats will have unsaturated (semifluid) fat in their bodies. The situation is different in ruminants where body fats tend to be more completely saturated regardless of the degree of unsaturation of ingested fat. This is due, at least in part, to the reducing activity in the rumen that hydrogenates dietary fat, and reduces the amounts of unsaturated fatty acids that are absorbed.

Triglycerides are usually hydrolyzed before they are split apart by lipolytic enzymes into their fatty acid and glycerol moieties. After the fatty acids and glycerol enter the bloodstream, the fatty acids are carried by the plasma as an albumin-bound complex mediated by choline (p. 185). The glycerol molecule is not directly utilized by body tissues, but is carried to the liver where it is either broken down into carbon dioxide and water or is synthesized into glucose. Its fate depends upon the nutritional state of the animal and the requirement for carbohydrate.

Metabolism of fatty acids is more complex, and may involve anabolic as well as catabolic schemes. While the precise reactions are important for an in-depth knowledge of bodily functions, the more complex mechanisms are not suitable subjects for this basic text. The brief discussions in Chapter 13 are about as far as one should go. It is sufficient at this stage to state that anabolic reactions and desaturation reactions exist primarily to provide types and qualities of fat needed by the body. Catabolic reactions that involve fat exist to provide energy substance in times of glucose deficiency in the diet. The rate of catabolism of fat and its conversion to energy is dependent on the nutritional state of the animal and glucose demand. If carbohydrate intake is low, stored fats are mobilized and their fatty acids converted into glucose by the liver and (minimally) by other tissues. If the fat conversion process occurs at too rapid a rate, two undesirable side effects can occur. Lipoid can accumulate in liver cells and impair their function, and metabolic by-products known as "ketone bodies" (acetoacetic acid, beta hydroxybutyric acid, and acetone) can be produced too rapidly to be detoxified and excreted. These substances are toxic and can damage liver and kidney function and result in a serious metabolic disease called acetonemia or ketosis.

Proteins

These compounds are far more complex in structure than either carbohydrates or fats and consist of polymers of amino acids which contain ammonia and may also contain sulfur, phosphorus, iron, and other elements besides carbon, hydrogen and oxygen. There are some 23 amino acids which have been isolated and identified by hydrolysis of proteins in the laboratory. The great variety of proteins in the body are compounded from these 23 amino acids in much the same manner that the letters of the alphabet are formed into different words.

Proteins, in order to be utilized, must be broken down by the digestive system into their component amino or nucleic acids which are then resynthesized or utilized in the body. Protein forms the bulk of the organic compounds in the protoplasm of cells and is used to a far greater extent than either carbohydrates or fats to form new protoplasm or to repair or restore that which already exists. Protein is not stored in the body. Amounts over the dietary requirements and above those needed to manufacture and maintain protoplasm are deaminized. The amino groups (NH$_2$ groups) are removed; the CHO moiety is formed into glucose or glycogen or utilized for energy, and the ammonia moiety becomes nitrogenous waste. Deaminization is normally performed in the liver, kidneys, and muscle tissue. Since proteins are not stored and since no single protein contains all 23 amino acids, it is necessary that the diet contain several different proteins. Of the amino acids, 13 can be synthesized in the body and are therefore not essential in the diet. The remaining 10 are termed "essential amino acids" and must be supplied in the ration.

An essential amino acid is one that cannot be synthesized by the body or cannot be synthesized fast enough to meet bodily requirements.

The essential amino acids vary for different kinds of animals but for most mammals, they consist of tryptophane, threonine, histidine, arginine, lysine, leucine, isoleucine, methionine, valine, and phenylalanine, resulting in the mnemonic T.T. Hallim V.P. Two others sometimes needed are serine and glycine.

It should be noted that protein can be converted into fat and that both protein and fat may be converted to carbohydrates and subsequently converted into energy. It should also be noted that neither carbohydrates nor fats can be made into proteins although there is a relatively common convertibility between carbohydrate and fat. The utilization of tissue protein as an energy source in a starving animal can be done for a short period but will rapidly result in permanent damage or death of the animal. In this particular case the body is consuming itself rather than its reserves.

Steroid Metabolism

Cholesterol is the principal steroid absorbed by the normal animal from a normal diet. The substance is also synthesized by the body from acetate and the amino acid leucine. The liver is the principal organ of synthesis although most body tissues, except nervous, may be involved. Of the body tissues the kidney, adrenal, and gut mucosa are paramount in cholesterol synthesis (p. 388ff).

The subject of cholesterol has received great publicity because of its apparent involvement in cardiovascular disease, yet it should be remembered that steroid compounds are substances of great physiological importance since they form corticosteroids, sex hormones, and vitamin D. The extent to which cholesterol is involved in hormone and corticosteroid synthesis is incompletely known, but it is a precursor of vitamin D-3, which is vital for the metabolism of calcium and phosphorus.

Cholesterol is ordinarily excreted in the feces, and to a minimal extent in danders. Ordinarily fecal excretion is preceded by conversion of cholesterol into bile acids by the liver. Some cholesterol is also converted into cholestanol by the liver. This is a neutral excretory product that has no apparent metabolic function.

Water

Water forms the major portion of the total body weight of mammals. Loss of over 10 percent of normal body water will cause disturbance of body functions; loss of over 20 percent will result in death. Between 6–10 percent water loss will produce sensations of thirst, headache, muscular incoordination, labored breathing, and thickened slightly bluish blood. Water loss between 10–20 percent results in shrivelled skin, delirium, and coma. Above 20 percent loss there is a failure of metabolic heat removal mechanisms that result in a fatal rise in body temperature. Water may be derived from a number of sources: from drinking water and water-containing liquids, from water in foods, and from metabolic water formed within the body as a result of biochemical reactions, principally the hydrolysis of fats and sugars ($CHO \rightarrow CO_2 + H_2O$ + energy).

Water yields no energy but is one of the essential substances in an animal's body. It is a necessary part of protoplasm. It serves as a transfer medium for dissolved nutrients and waste products to and from tissue cells. It is the principal vehicle for the transport and formation of glandular products and is concerned in the broad aspects of secretion and excretion. It is directly concerned with temperature regulation and acts as a lubricant for body surfaces. It is interesting to note that the lung-vapor transfer of body water to the outside is such an efficient cooling mechanism that certain animals (e.g., dogs, cats, and other carnivores) have virtually no sweat glands.

Intake and elimination of water by the body require expenditures of energy. In the gut where water enters the body and in the principal exit structures, i.e., kidney, sweat, sebaceous, salivary, and milk glands, cell activity moves water across barrier membranes. During life, these glands and organs continuously adjust amounts of body water and the substances it transports. In addition to these structures there are others, such as, skin and lungs, which function passively in water elimination, and hormones which control kidney function. Excessive water intake and/or excessive body loss via the gut can result in diarrhea; excessive body retention and/or reduced body water (dehydration) results in constipation. However, both diarrhea and constipation may result from a number of other causes. Other controls of body water involve hormones such as aldosterone (p. 354) and intermedin (p. 348) and parathormone (p. 352) and the juxtaglomerular apparatus (p. 308).

Under stable "normal" conditions mammals' water content is approximately 65 percent (range 55–75 percent) of total body weight. The percentage varies with age. An embryo may consist of 95 percent water. This reduces to 85 percent in the second trimester of pregnancy and to

75 percent in the newborn. At puberty body water is about 65 percent of normal body weight. Through adult life this percentage slowly drops until at old age or senility the amount stabilizes at about 50 percent. This gradual loss of body water appears to be associated with skin changes, loss of adnexal fat and altered fat distribution characteristic of old age, although body fat *per se* neither appreciably increases nor decreases body water content. Concomitant with the lowering of water percentage in the body there is an increasing percentage of fat, calcium salts in the bones, and body protein. Through the maturation period prior to old age the percentage of body fat continues to increase in well nourished animals. Old age (p. 21ff) appears to be associated with a physiological desiccation of the body cells and the intercellular spaces.

The greatest amount of water movement takes place between the gut and the blood-vascular system. An amount of water equivalent to the total body weight of a mammal will ordinarily be moved between these two structures in a 24-hour period. Fecal consistency is due almost entirely to the percent of retention of ingested and metabolic water.

By volume, a 100-kilogram mature normal mammal will contain about 40 liters of water. This is total body water and its amount can be determined directly or indirectly with reasonable accuracy. Direct determination involves weighing the animal, then grinding or homogenizing it in a blender and then removing the water by desiccating at 90–100°C or by freeze-drying (lyophilization). The dry matter is then weighed and the difference in weight represents the total body water. Indirect methods involve techniques using specific gravity or heavy water (tritium) dilution. The specific gravity method has problems because it cannot accurately account for lung volume and gases in the gut or elsewhere. At present, the better method of indirect measurement is through analysis of the dilution of heavy water by body fluids. The animal is weighed and then kept from food and water. A measured dose of tritium is given by intramuscular injection and the animal is held until equilibration occurs. This may take considerable time depending upon the age, dehydration, and species of animal being tested. A dehydrated camel, for instance, can take up to 20 hours. After equilibration is attained—which usually takes about six hours—the animal is weighed, a blood sample is withdrawn, and the blood analyzed for tritium. From the data the total body fluid can be calculated with acceptable accuracy. Moreover, other data such as water turnover and half-life of a water molecule can be calculated by returning the animal to a normal food and water regimen and withdrawing blood samples at predetermined intervals for testing.

Total body water varies according to the condition of the individual, the species, the percentage of body fat (the higher the fat the lower the water) and habitat. Regardless of the amount of water intake, over a period of time the amount of body water will stabilize and remain relatively constant. This is done through removal of excess water via the kidneys, alimentary canal, lungs, skin, sweat and mammary glands (Figure 7–10). In times of water deprivation, all of these mechanisms except the lungs can be inhibited, which considerably enhances water conservation.

The rate at which water passes through an animal's body is highly variable in both individuals and species. Interspecies differences appear to be associated with metabolic rate. Under similar conditions, cattle and water buffalo have a high rate of water turnover, sheep and goats have about half the turnover rate of cattle and buffalo, and camels and oryx have still lower rates. All of these animals are ruminants, but each pair evolved in different environments, which affected their turnover rate. Water turnover is also high in nursing animals.

Total body water can be divided into two major parts, intracellular fluid (ca. 70 percent) and extracellular fluid (ca. 30 percent). Extracellular fluid can be further subdivided into transcellular fluid, interstitial fluid, digestive fluid, and blood plasma.

Transcellular fluid is found in special locations in the body, e.g., pericardial fluid, peritoneal and pleural fluid, intraocular fluid, synovia

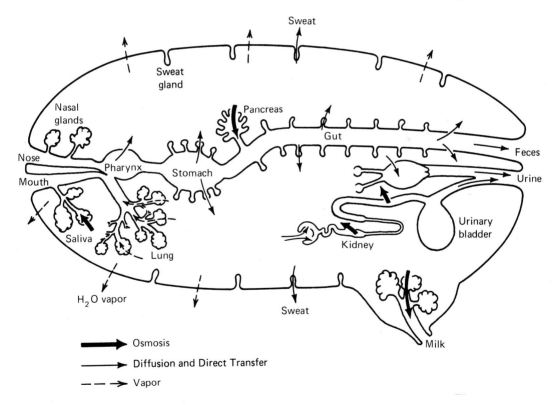

FIGURE 7-10   Means by which water enters and leaves the body. (Adapted from Phillis, Veterinary Physiology.)

and cerebrospinal fluid. It is not directly related to transudation through the capillary endothelium but is produced by secretory cells. It is ordinarily less than 1 percent of total body fluid.

Interstitial fluid forms the bulk of the extracellular fluids, or about 20 percent of total body fluid. It contains about two percent protein and is found in the tissue spaces surrounding cells. It is derived from blood plasma and cell exudates. It returns to the bloodstream either by osmosis through the capillary endothelium or via the lymphatic system (p. 281). In the process of moving between the bloodstream and the cells, exchanges take place in which metabolites, oxygen and electrolytes are given to the body cells and waste products are received from them.

Digestive fluid (gut water) accounts for about six to eight percent of the total body water in simple stomached animals. In ruminants and monogastric herbivores with large ceca the percentage is higher. This water is usually derived from drinking but it can be derived from body fluids if necessary.

Blood plasma constitutes about five percent of total body water and contains about seven percent protein giving a positive osmotic gradient in the capillary bed, which draws interstitial fluid back into the bloodstream. Details of plasma are discussed elsewhere (p. 234ff).

Mineral Salts

Mineral salts also yield no energy to the body, yet they play necessary roles in metabolism and nutrition. There is no point in detailing the long list of common and trace minerals required by the body. The list would include far too many substances for convenient presentation. It should be sufficient to indicate a few of the most important of these and if a more detailed study is desired, readers are referred to biochemistry and nutrition texts and courses which adequately cover this subject.

Among the more common minerals required by the body are calcium, phosphorus, iron, copper, cobalt, iodine, sodium, potassium, and sulfur. Calcium is a constituent of all protoplasm and body fluids and is present in large amounts in bone and teeth. It is a necessary part of the internal environment of all cells and is related to the irritability of muscle and nerve tissues. In its ionic form calcium is concerned with the regulation of blood and milk formation. Recent investigations indicate that calcium forms an essential part of cell membranes.

Phosphorus is closely related to calcium in both metabolism and nutrition, and is usually found in association with calcium in most structures described for that mineral.

Iron is found in the hemoglobin of red blood cells and also in all body tissues. Although iron is carefully saved and reused by the body, its storage is quite limited. The feed of an animal should contain a small amount of this element. Milk contains very little iron. Young animals, whose basic nourishment is milk, compensate for this by having a considerable store in their livers at birth. Baby pigs, which have the least amount, are subject to iron deficiency anemia or "thumps." This can be treated effectively by supplying iron to the animal either by injection or by addition to the diet.

Copper is also present in all cells and is a necessary supplement to iron in hemoglobin formation. Iodine is essential to the proper functioning of the thyroid gland which controls the general metabolism of the body. This mineral is present in the hormone thyroxin. A deficiency of iodine results in goiter and a lowered metabolism. Sodium and potassium salts occur mainly in the blood and tissue fluids, with sodium salts being found in the fluid portion and potassium salts in the cells. Their principal function is the transport of waste materials and the regulation of body pH. Sulfur enters into the composition of hair, horn and other keratinized structures.

Vitamins

Vitamins are organic food substances necessary for growth and metabolism. They are essentially exogenous catalysts, i.e., substances similar to enzymes but are not produced by the body and must be present in the diet. There are about 40 of these (including their analogs) presently known and there are probably more which await discovery. Basically, however, there are relatively few groups vital to health. These are the fat soluble vitamins (A,D,E,K), and the water soluble vitamins (B-complex and C). These are justifiably called "essential metabolites." Their absence from the diet causes what are known as deficiency diseases. Several vitamins (e.g., B_1, B_2, and C) are factors in tissue respiration and it may be possible that all vitamins serve as respiratory catalysts. In practice, adequate supplies of vitamins are usually synthesized from or occur in natural diets. The only ones that may be deficient in normal farm operations are vitamins A, B_1 and D. Care should be taken in the usage of some vitamins, particularly vitamin D, which can be toxic in excess. Long term damage can also be caused (at least in rabbits) through the administration of high, but not immediately toxic, levels of vitamin D. The lesions of vitamin D excess occur principally in the heart and aorta, and consist of calcification in the aorta wall and the myocardium and fibroelastosis in the endocardium. It has been shown that vitamin B_{12} is the "extrinsic factor" component of hemopoietic factor which is necessary for blood formation. Cobalt is a part of this molecule and it is probably here that this element enters the blood-forming picture. Although it is often said that animals "synthesize vitamins," this statement is not precisely true. It is the bacteria in the digestive tract (the ruminal and intestinal flora) which perform the synthesis.

Vitamins are primarily absorbed in the anterior part of the small intestine. Both fat soluble (A, D, E, K) and water soluble (B-complex, C) vitamins diffuse readily through the intestinal mucosa and into the capillaries and lacteals, except for vitamin B_{12}, which has a specific transport system in the ileum. Vitamin B_{12} exercises its antianemic effect through combination with intrinsic factor in the liver to form hemopoietic factor (p. 239ff). Similarly, fat soluble vitamins are conjugated with dietary fats and their absorption is directly related to fat absorption. Water soluble vitamins do not have this limiting requirement.

In many of the reactions involved in intermediate metabolism, vitamins of the B-complex play an important part.

Thiamin functions as a coenzyme in a number of metabolic reactions and as a cocarboxylase in the decarboxylation of pyruvic acid to yield acetyl Co-A and in the decarboxylation of alphaketoglutaric acid to yield succinyl Co-A. It also functions in the transketolase reaction involving pentose sugars and in several reactions involving the synthesis of fatty acids from carbohydrate.

Riboflavin functions in hydrogen transport, particularly in transfer of hydrogen atoms removed in the catabolism of amino acids, beta hydroxy butyric acid, aldehydes and purines, and in oxidative transfers involving adenosine triphosphate (ATP) formation.

Niacin forms coenzymes for dehydrogenases and transports hydrogen atoms into oxidative reactions which yield ATP.

Pantothenic acid is a component of coenzyme A.

Pyridoxine is a component of coenzymes which function in amino acid metabolism. As pyridoxine phosphate the vitamin enters into transamination reactions, the interconversion of serine and glycine, and the decarboxylation of 3–4 dihydroxyphenylalanine (DOPA) and glutamic acid. It is also involved in the transport of amino acids and ions across cell membranes.

Folic acid functions in the interconversion of serine and glycine, in the catabolism of histidine, in nucleic acid synthesis and in CO_2 utilization.

Vitamin B_{12} (Cyanocobalamin) is the extrinsic factor portion of the antipernicious anemia factor. It is also a coenzyme for the metabolism of labile methyl groups, aids in the synthesis of nucleic acid, helps maintain sulfhydryl groups and functions in protein formation from amino acids.

In 1972 vitamin B_{12} was synthesized by a research team from Harvard University and the University of Zurich. It was an 11-year effort. The synthesis is extremely complex and, since the natural vitamin is cheap and plentiful, will have no commercial application. The usefulness of the synthesis is in the "spinoff." The techniques developed the Law of Conservation of Orbital Symmetry, which tells what compounds can be constructed by synthesis. There also may be some medical use for analogs of B_{12}.

The classification of vitamins is an interesting example of how biological science progresses. In 1911, Hopkins *et al.* made the original observations on "accessory food factors." In 1912, Funk discovered a nitrogenous factor in yeast and rice polishings that cured beri-beri. He called this "vitamine" (life-amine) and the name, with the terminal "e" deleted, became a generic title for all exogenous, nonsynthesizable, essential metabolites regardless of whether or not they contained amine groups. Shortly thereafter the vitamins were categorized as fat soluble or water soluble and at the same time were given alphabetic classification symbols. Some time later it was found that certain of these vitamins, notably B and D, were family groups with more or less similar characteristics. This resulted in subclassifications such as B_1, B_2, B_6, etc. Eventually all the known vitamins were categorized chemically and in 1972 with the synthesis of B_{12} the old classifications became unnecessary—but they are still kept because it is easier for nonprofessionals to remember letters and numbers than chemical names.

Other Metabolites

Certain substances such as choline and methionine have vitamin-like functions. Choline is perhaps the best example of these, since it has from time to time been called a vitamin, and its position even today is ambiguous. The substance can be found in normal bile. In fact, its original isolation in 1862 was from pig bile. Bayer, in 1867, established its structure—which indicates that knowledge of the substance has been around for a long time. Seventy years after its isolation, its importance in nutrition was recognized. Choline functions in fat transport, exerts a sparing action on methionine (an essential amino acid), and furnishes labile methyl groups which can be used in many body reactions, and particularly in liver detoxification reactions. Choline is also a structural part of phospholipids and acetylcholine, and functions in energy-source reactions and in nerve impulse transmission. It can be synthesized by animals, which is one way in which it differs fom vitamins, and it is not an organic catalyst which is another of its differences. Nevertheless it is essential to growth and metabolism, and can with perfect justice be called an "essential metabolite."

Enzymes

Enzymes can be defined as endogenous organic catalysts. A catalyst is any substance that promotes a chemical reaction without itself entering into the reaction or undergoing any change. Enzymes are present in every living cell and catalyze the chemical reactions which occur within the cell. Although cellular components are relatively inert, enzymes enable the cells to become active chemical laboratories in which reactions proceed rapidly at low temperatures, under controlled conditions, without strong reagents. Enzymes may be produced and retained within cells or they may be liberated to function elsewhere. They are remarkably specific in their reaction and probably every chemical reaction in the body has its corresponding enzyme system. The study of enzyme chemistry and reactions is a fertile field. Although it has been known to the human race as long as the making of cheese, it has never been thoroughly explored, nor have determined efforts been made to find commercial usages for enzymes until recently.

Enzymes have a number of properties which serve to identify them. They are produced by the body. They are colloidal in nature. Their structure is generally protein. They need to be present only in small amounts since a tiny amount of enzyme is capable of catalyzing reactions involving enormous amounts of substrate. The chemical reactions into which they enter may be the hydrolytic breakdown of complex substances by the addition of water (this reaction is typical of the digestive tract), or they may be reductive, oxidative, fermentative, synthetic, or transportative. Reduction reactions involve either removal of oxygen or addition of hydrogen. Oxidative reactions involve either removal of hydrogen or addition of oxygen. Fermentation reactions involve anaerobic degradation of molecules with the production of alcohol and volatile acids. Synthetic reactions involve the forming or combining of substrates into new molecules.

Enzymatic reactions are usually reversible and each enzyme functions best at a specific hydrogen ion concentration and a specific temperature. Most enzymes are destroyed by heat or inactivated by cold. All enzymes possess specificity regarding the substrate which they affect and this forms one of the common bases for their classification and nomenclature.

Several systems of nomenclature have been used for enzymes. In the older systems the suffix "-in" was added to the base word and resulted in names like pepsin, trypsin, and ptyalin. These are confusing since the "in" suffix is also used for hormones. A newer system using the suffix "ase" is now in general use and serves to more readily identify these structures. Names such as salivary amylase and peptidase are replacing older words like ptyalin and erepsin. As knowledge of enzyme chemistry increases, the older terms, many of which covered a complex group of enzymes, are being discarded.

Saliva

Salivary secretion is primarily under control of the autonomic nervous system (p. 330ff). Parasympathetic stimulation results in active secretion of saliva, and inhibition of the parasympathetic can cause cessation of salivary flow. Sympathetic stimulation produces variable results, depending on the genera or species of animals studied. The scent or sight of food, the presence of food in the mouth, or the thought of food can stimulate saliva secretion.

The amount of saliva produced in a day can be quite large. This is particularly true in herbivores. A cow, for instance, can produce up to 200 liters of saliva per day. Secretion rates are highest during feeding and lowest shortly after feeding. In domestic animals, the parotid salivary gland apparently secretes continuously.

In certain animals, notably young dogs, pigs, chickens, and humans, saliva contains the enzyme salivary amylase and has amylolytic (sugar reducing) activity, but in most animals saliva functions mainly as a wetting agent and a lubricant. Other functions include providing a fluid to help form food boluses, maceration, solvent for removing flavor molecules from taste buds, and possibly some antibacterial action.

Swallowing

The act of swallowing is a reflex resulting from the presence of food in the pharynx. Following mastication and ensalivation, food is formed into a bolus at the base of the tongue. The tongue acts as a plunger and moves the food into the pharynx. The pharyngeal wall contains sensory receptors of the glossopharyngeal nerve (p. 330) which trigger reflex arcs that produce complex, coordinated movements which introduce the food into the esophageal inlet (aditis esophagus). Once food has reached the pharynx, the act of swallowing is involuntary, although in some animals it may be stopped by a "gagging" mechanism that is identical with the initial act of the vomiting sequence.

The motor branches of the glossopharyngeal nerve connect the swallowing (deglutition) center with the pharynx and the upper esophagus and respiratory tract. Stimulation produces the following events: the tongue moves caudally, the hyoid bones and larynx move cranially, dilate the esophageal inlet, bring the pharynx into close proximity to the base of the tongue, move the soft palate over the caudal nares, close the glottis, and move the epiglottis over the tracheal opening. The food bolus is pushed into the dilated esophageal inlet by the combined backward movement of the tongue and forward movement of the pharynx. The pharynx then relaxes, the larynx and esophageal inlet move caudally and the caudal pillars of the oral cavity close to prevent the return of the bolus. Peristaltic waves commence at the aditis esophagus and move toward the stomach propelling the bolus down the esophagus and ejecting it through the cardiac sphincter into the stomach or the rumen.

PHYSIOLOGY OF THE STOMACH

The complex structure of the stomach is reflected in a relative complexity of function which appears to depend upon the habits and food consumed by the animal.

The stomach has two primary and several secondary functions. The primary functions are to serve as a storage space for swallowed food and to act as an organ of digestion. The secondary functions include absorption, endocrine secretion and production of "intrinsic factor" (p. 240).

The stomach is an organ that is not essential to life. This fact has been proven by total gastrectomy in experimental and diseased animals, and in man. Its absence results in considerable stress to the animal, producing excessive demand upon the digestive functions of the small intestine, a tendency to develop anemia, and a depletion of the calcium stores of the body (which are mobilized to reduce the "acid tide" that develops during digestion in a gastrectomized animal).

Stomach function is correlated with structure. The distensibility of the organ makes it a repository for undigested food. The digestive capacity in herbivorous animals is associated with two phases—a microfloral phase, and an intrinsic digestive phase. The microfloral phase, which occurs in the rumen of ruminants and the cecum of monogastric herbivores, involves enzymes produced by bacteria and protozoa that reside in the stomach and aid in the reduction of plant material. The intrinsic digestive phase depends upon the functioning of the gastric glands. In carnivores and omnivores only the intrinsic phase is present.

True (intrinsic) gastric digestion in ruminants is confined to the abomasum, and in monogastric animals to the fundic and pyloric portions of the stomach. In these regions, three gland groups occur which are called cardiac, pyloric and fundic glands. The cardiac glands occupy the more forward (closer to the esophagus) regions of the glandular stomach. In ruminants they are confined to a narrow band around the omaso-abomasal orifice. These glands do not secrete enzymes but produce mucus. One widely held theory states that they are regressive structures which probably were derived from a more extensive fundic area that existed early in evolutionary history. The mucus produced by the cardiac glands has two functions. It helps macerate food and provides protection for the mucous membranes of the stomach against the powerful proteolytic enzymes and hydrochloric acid secreted by cells of the fundic and pyloric glands.

The fundic glands are most important to the digestive action of the stomach. These consist of long columnar structures composed of three types of cells: neck chief cells, body chief cells, and parietal cells (Figure 7-4). The neck chief cells secrete mucus. The body chief cells produce gastric enzymes. The parietal cells, which lie along the lower border of the glands next to the gastric epithelium, produce hydrochloric acid and "intrinsic factor."

The pyloric glands, like the cardiac glands, are apparently regressive structures. In these glands the parietal cells are missing and the body chief cells are less numerous. Functionally, this region produces mucus and small amounts of a proteolytic enzyme that does not appear to be the same as the pepsinogen secreted by the fundic glands.

Gastric Juice

Gastric juice is principally a secretion of the fundic glands. It is a clear, watery to viscous, colorless fluid with an acid reaction and a bitter taste. It consists of water, organic substances, inorganic salts, and hydrochloric acid. Mucus is secreted by surface cells, neck chief cells, and esophageal and cardiac glands. It is the principal secretory product (in amount) of the stomach and functions in maceration. It also supplies a protective coating for the epithelium against hydrochloric acid and gastric enzymes.

Associated with the organic material of gastric juice is an enzyme complex consisting of three identifiable enzymes. The enzymes are pepsin, rennin, and gastric lipase. Pepsin is a proteolytic enzyme which itself is a protein. It is secreted by the body chief cells as the proenzyme pepsinogen. Pepsinogen is activated to pepsin by hydrochloric acid. Pepsin converts proteins into proteoses and peptones. These are intermediate products of protein degradation. Complete hydrolysis of protein to amino acid is not accomplished by gastric enzymes.

Rennin is a milk coagulating enzyme found in the stomach of calves and probably other young ruminants. It apparently does not occur in monogastric animals. It functions to change the casein fraction of milk to paracasein, which then reacts with free calcium ions in the stomach to produce a gel called calcium paracaseinate. Functionally, this retards the passage of milk through the stomach and gives pepsin a longer time to act upon the protein substrate in the milk. Rennin, like pepsin, is activated by hydrochloric acid.

The third system, gastric lipase, has been the subject of considerable controversy since it was first reported early in the twentieth century. It occurs in small amounts and is most abundant in the digestive juices of carnivores. Like other lipases found in the digestive tract, it reduces fats to fatty acids and glycerol. Since most of the fat in the stomach is not emulsified (broken down into small droplets), the action of gastric lipase is minimal. One theory held by a number of investigators states that gastric lipase is produced by the neck chief cells of the fundic mucosa. Another school of thought believes that gastric lipase is derived from cranial migration of the enzyme from the intestines during the opening of the pyloric sphincter.

Hydrochloric acid produced by the parietal cells is one of the most important constituents of gastric juices. The acidity of stomach contents varies among the domestic animals, being highest in carnivores (pH 1 or less in dogs), and lowest in monogastric herbivores (pH 1.1–6.8 in horses). Hydrochloric acid formation functions to macerate food, to activate pepsin and rennin, to hydrolyze small (insignificant) amounts of sugar, and to act as an antiseptic solution for the stomach.

Control of Gastric Secretion

Gastric secretion may or may not be controlled by a neurohormonal system depending upon the animal involved. In humans and horses, gastric secretion is apparently continuous. In dogs it is intermittent. However, the rate of gastric secretion increases in the presence of food. There are apparently three phases in the stimulation of gastric secretion: a cephalic phase, a gastric phase, and an intestinal phase. The cephalic phase occurs from thoughts of eating or from thoughts of the sensations of eating. Physical stimuli such as taste sensations in the mouth and the presence of food mass in the mouth and esophagus can result in stimulation of the gastric sensory center in the brain, which in turn sends impulses via the vagus nerve to stimulate the glandular elements of the stomach.

In the gastric phase, mechanical stimulation from the presence of food causes the glands to secrete more abundantly. In this phase secretion is not the result of local nervous reflexes or extrinsic stimuli. It is probably controlled by the hormone gastrin, which is produced by the epithelial cells of the pyloric region (p. 360).

The duodenal portion of the small intestine is involved in the intestinal phase of gastric secretion. It also operates through hormonal stimulation. The hormone enterogastrone inhibits gastric secretion. By slowing digestion and acid production it regulates passage of food from the stomach to the small intestine. Some reports indicate that there may be another hormone, intestinal gastrin, which is secreted by the duodenal epithelium and augments or supplements stomach gastrin.

The Filling and Emptying of the Stomach

The stomach possesses a remarkable ability to adapt itself to the amount of food ingested. Generally, in monogastric animals, the empty stomach is contracted into a small tubular structure that appears to be hardly more than a dilation of the caudal end of the esophagus. With the ingestion of food, however, the stomach can become greatly enlarged. In monogastric herbivores, ingested food tends to form layers within the stomach. The first food ingested passes to the fundic portion and subsequent ingesta lies in layers on top of the first food eaten. Drinking of water does not greatly disturb the layering effect; the water either passes rapidly through the stomach wall or into the intestinal tract, leaving the food relatively unwetted. The layering effect in monogastric herbivores is also present in carnivores but to a lesser degree and in these animals can be disturbed by heavier masses sinking downwards into the fundic region. In ruminants, layering does not occur since the rumen movements and the large quantities of ruminal water facilitate mixing of the lighter ingesta, while the heavier substances are separated quickly by specific gravity and are transferred to the glandular stomach.

Depending upon the species and diet, the emptying of the stomach may be a quick or slow process. In carnivores the stomach will virtually empty itself between each meal. In herbivores, ingesta may remain within the stomach for several days. Emptying is a progressive process which occurs at intervals throughout the course of gastric digestion. At irregular intervals, small amounts of digested food from the stomach are emptied into the intestine through the pyloric sphincter. At one time it was thought that the pyloric sphincter was of great importance in this emptying process. More recent investigations, however, have shown that the pyloric region can be removed without appreciably modifying emptying. In the normal animal the pylorus is apparently stimulated to relax by the presence of ingesta containing a high level of acidity. Peristaltic waves moving down the fundus of the stomach propel ingesta toward the pylorus and force food through the sphincter into the duodenal bulb, which contracts and propels the ingesta into the duodenum.

Morphologically and functionally the stomach can be separated into two parts: one that includes the esophageal and cardiac regions is a place of temporary storage and the other, which involves the fundic and pyloric regions, is concerned with enzymatic activity. In the horse the margo plicatus demarcates these regions clearly. The separation is not so apparent in other animals although the pinkish color of the storage area contrasts with the darker red of the digestive region. The pyloric region is usually more muscular than the fundus, and apparently conducts most of the mixing activity of the lower stomach. The end product of these various operations is a grayish, semi-fluid acidic substance called chyme.

Peristalsis in the stomach generally originates in the esophageal region and passes caudally over the cardiac, fundic, and pyloric regions and ends at the pyloric valve. As gastric digestion proceeds, the chyme becomes more acidic, its volume increases, and the peristaltic waves increase in force. A small amount of chyme is forced into the pyloric valve. This stimulates a vagus nerve reflex which causes the sphincter to dilate. A subsequent peristaltic wave sends a squirt of chyme into the duodenum. The decrease in volume and acidity in the stomach as a result of this loss reduces the level of vagal stimulation and allows the pyloric sphincter to contract and remain closed until the conditions in the stomach again become favorable for vagal stimulation. The result is that the duodenum receives injections of chyme at intervals.

The opening and closing of the pyloric valve may also be affected by hormonal activity. Enterogastrone and secretin are probably involved as agents which inhibit opening of the sphincter, while gastrin probably acts as a stimulant.

Hunger Contractions

Hunger contractions are strong peristaltic waves that travel down the empty stomach from the esophageal opening to the pylorus and are considerably more severe than ordinary stomach contractions. These contractions are stimulated by a decrease in blood sugar, which exerts its function on the stomach through the brain and the vagus nerve. With the exception of hunger contractions, there are no physiological pains in the stomach. Any other pain sensation produced in this organ is probably the result of disease. Manipulations of the stomach ordinarily do not produce pain.

Vomition

Vomition (vomiting) is the forcible evacuation of the stomach via the mouth and esophagus in response to stimuli originating in the wall of the stomach, the gut, the vestibular apparatus of the inner ear, or as the result of certain drugs. The activity, once started, is controlled by a vomiting center in the medulla oblongata of the brain.

The act of vomiting is preceded by nausea, salivation, and pumping movements of the abdominal muscles. The evacuation phase is initiated by a forced inspiration with the glottis closed. This increases negative pressure (p. 211) in the thorax, balloons the esophagus, and stimulates the cardiac sphincter to dilate. At the same time a spasmic contraction of the abdominal muscles forces the abdominal viscera against the stomach and causes contraction of the pyloric and fundic musculature in a reversed peristaltic wave. The stomach contents are forced into the relaxed cardiac sphincter and food material is sucked into the ballooned esophagus. The acid stomach contents stimulate reverse peristalsis and the stomach contents are rushed up the esophagus.

The soft palate in the pharynx rises to occlude the caudal nares, the epiglottis is closed over the larynx, and the ingesta is expelled through the mouth.

In some animals, notably carnivores and omnivores, vomiting is easy, but in others, i.e., rats, mice, horses, and ruminants, vomiting is virtually impossible. The stimuli may occur, but because of the anatomic arrangement of the digestive tract, the stomach cannot be emptied. In ruminants the cranial passage of food from the abomasum is interdicted by the omasum. In rats and mice the cardiac sphincter cannot be dilated by stimuli arising from the gastric side. In horses, not only is the cardiac sphincter refractory, but the soft palate hangs like a curtain across the posterior part of the oral cavity and causes the ingesta to be forced out of the nose. In these latter three animals the stomach will rupture if the stimuli are too strong.

This response has been used to eradicate mouse and rat populations. The rodenticide red squill is a powerful emetic. Since carnivores vomit easily, the drug causes dogs and cats little harm, but it kills rats and mice since the powerful contractions of the stomach muscles cause gastric rupture and death from peritonitis.

PHYSIOLOGY OF THE INTESTINAL TRACT

The intestinal tract and its accessory glands functions in the final phases of digestion and absorption of nutrients. The intestinal tract also propels nutrients and waste materials from stomach to anus.

In general, the digestive phase of intestinal activity is accomplished primarily in the duodenum and the first third of the jejunum. Beyond this point the principal intestinal activity is absorption. Some compound stomached herbivores, and all simple stomached herbivores with functional ceca, are exceptions to the previous statements. In these animals (e.g., horses, swine, rodents) the cecum and occasionally the anterior colon replaces the rumen as an organ of bacterial digestion, and the colon has nutrient absorption functions. These functions are almost entirely the result of the presence of bacteria and protozoa rather than enzyme and hormone secretory activity of the epithelium lining these parts of the gut. Some digestion and absorption occurs in the small intestine, but this is relatively minor when compared with the activities of the cecum and colon of these animals.

Duodenal Functions

Digestion in the stomach is an acid medium or "acid phase" activity, since gastric juices contain hydrochloric acid. The small intestine, however, functions in an alkaline medium or "alkaline phase." This alkalinity is produced by products of the pancreas (pancreatic juice), the liver (bile), and Brunner's glands in the duodenum. These sources produce sufficient base to neutralize the acidity of chyme and maintain the pH of intestinal contents on the alkaline side. After intestinal digestion, chyme becomes chyle. Chyle is a milky alkalized material that contains emulsified fats and other products of intestinal digestion. It passes through the epithelial barrier in the gut and is picked up by the lymphatics and capillaries in the villi, and during the intestinal digestion phase and shortly thereafter can be found in considerable quantity in the lacteals. It ultimately enters the venous system through the thoracic duct or is passed directly to the liver through the portal circulation.

The walls of the duodenum, together with associated glandular structures (pancreas, liver, and intestinal glands), are responsible for producing the enzymes and hormones necessary to convert the partially digested products of the stomach into materials that can be absorbed by the body. The pancreas is a dual purpose, loosely organized, lobulated gland which lies along the duodenum and empties into that organ by means of one or more ducts. The pancreas produces a hormone (insulin), an alkaline fluid (pancreatic juice), and a number of digestive enzymes (ferments) which aid in the chemical reduction of food. The secretion of the pancreas is controlled and regulated by two hormones, pancreozymin and secretin (p. 365), which are produced by the wall of the small intestine.

The liver empties bile into the duodenum through the bile duct. Bile is a product of red blood cell destruction, and possibly of muscle metabolism. It is somewhat slimy in consistency, greenish to greenish gold in color, alkaline in pH, and consists essentially of bile pigments (biliverden and bilirubin), bile salts (bile acids, urea, cholesterol and iron), and fats. In the intestine, bile functions to alkalize the intestinal contents, dissolve cholesterols, and emulsify fats. Biliary secretion is continuous, but in those animals that possess gall bladders, the rate of delivery to the small intestine is controlled by a hormone, cholecystokinin, secreted by the wall of the small intestine.

There are four general types of intestinal glands, e.g., Brunner's glands, crypts of Lieberkühn, specialized cells, and Peyer's patches. Brunner's glands are in the submucosa of the duodenum and secrete an alkaline mucoid fluid, which apparently has some amylolytic (sugar reducing) activity. The crypts of Lieberkühn (Figure 7-7) are deep folds in the mucous membrane of the gut and are not glands in a true sense, but their function is apparently glandular. They are lined with large numbers of goblet cells that produce an alkaline mucus and contain Paneth cells which secrete enzymes. Crypts of Lieberkühn are found throughout the gut. Those in the large intestine, however, have few Paneth cells and are filled almost entirely with goblet cells. In addition to these definitive structures, the wall of the small intestine produces a number of enzymes and hormones, and if taken in its entirety, can be considered to be a glandular structure as well as an organ of digestion, absorption, and excretion. Secretion of the glandular structures of the small intestine appears to be regulated by the amount of chyme present.

The specialized secreting cells in the wall of the intestine have been generally identified as Paneth cells and epithelioid cells in the crypts of Lieberkühn and the mucus-producing goblet cells found throughout the intestinal tract.

Peyer's patches are aggregations of lymphoid tissue embedded in the submucosa of the intestinal wall and extending into the gut lumen. At one time these structures were considered to be of marginal importance, but with expanding knowledge of lymphocyte function (p. 241ff) their importance has increased. Peyer's patches are not found in any part of the gut except the small intestine. The large intestine appears to be devoid of these structures. They function, at least in part, to control bacterial populations, in antibody production, and in filtration of tissue fluids in the submucosa.

Intestinal Movements

Intestinal movements are of several kinds and function to mix ingesta with digestive secretions; bring ingesta into contact with the mucous membrane for absorption; move ingesta through the gut; expel feces from the rectum; and assist the flow of lymph and blood through the vessels of the intestine wall. There are at least seven separate movements of the small intestine (i.e., rhythmic segmentations, pendular movements, peristaltic rushes, slow peristaltic waves, tonus waves, tonus rings, and antiperistaltic waves). For practical purposes, however, the movements can be broken down into three groups: propulsive movements, which include peristaltic rushes, slow peristaltic waves, and antiperistaltic waves; absorptive movements, which include rhythmic segmentation and pendular movements; and control movements, which include tonus waves and rings and seem to be concerned with starting or stopping peristalsis.

The intestine tends to be polarized, i.e., the movement of the muscle contractions and the direction of flow are from mouth to anus. However, reverse peristalsis (antiperistalsis) can occur and may be either physiological or pathological in origin. Excessively strong antiperistalsis can produce a pathological condition known as intussusception, or telescoping of the gut.

In the lower bowel the intestinal movements are more sluggish than in the small intestine, but can occasionally develop great power. Generally, the major movements of the large intestine are propulsive.

Digestive Enzymes

Enzymes are organic catalysts that tend to accelerate chemical reactions within the body. In the process of digestion, enzymes help reduce complex food materials to simple substances which can be absorbed. Essentially, digestive enzymes fall into three major groups which act with considerable specificity upon the major food materials (proteins, carbohydrates, and fats). A few other enzymes exist within the gut to potentiate certain of the digestive enzymes. All digestive enzymes that have been isolated to date have proved to be proteins. Specifically, the digestive enzymes are as follows:

A. Proteases (protein digesting enzymes, proteolytic enzymes)
 1. Pepsin
 2. Rennin
 3. Trypsin
 4. Chymotrypsin
 5. Carboxypeptidase

6. Erepsin
7. Aminopeptidase
8. Polynucleotidase
9. Nucleotidase
10. Nucleosidase

B. Amylases (carbohydrate digesting enzymes, amyblytic enzymes)
1. Salivary amylase
2. Pancreatic amylase
3. Sucrase
4. Maltase
5. Lactase

C. Lipases (fat digesting enzymes, lipolytic enzymes)
1. Gastric lipase
2. Pancreatic lipase
3. Intestinal lipase
4. Lecithinase (phospholipase)

D. Potentiating enzymes
1. Enterokinase
2. Trypsin

Origin	Enzyme	Substrate	End Product
Mouth salivary glands	Salivary amylase	Starch	Maltose, dextrins
Stomach body chief cells	Pepsin	Proteins	Proteoses, peptones
	Rennin	Casein	Paracasein
fundic neck chief cells	Gastric lipase	Fats	Fatty acids, glycerol
Pancreas acinar portion	Trypsin	Proteins, proteoses peptones, poly- peptides	Peptones, peptides, amino acids; also activates chymo- trypsinogen
	Chymotrypsin	Same as for trypsin	Same as for trypsin except for activation
	Pancreatic amylase	Starch and dextrins	Dextrins, maltose
	Pancreatic lipase	Fats	Fatty acids, glycerol
	Carboxypeptidase	Peptides with a free carboxyl	Amino acids
	Deoxyribonuclease	Nucleic acids	Polynucleotides
	Ribonuclease	Nucleic acids	Polynucleotides
	Lecithinase	Lecithin	Lysolecithin

Origin	Enzyme	Substrate	End Product
Small intestine crypts of Lieberkühn	Erepsin*	Peptides, proteoses, peptones	Amino acids
	Animopeptidases	Peptides	Amino acids
	Sucrase	Sucrose	Glucose, fructose
	Lactase	Lactose	Glucose, galactose
	Maltase	Maltose	Glucose
	Nucleosidase	Mononucleotides	Nucleosides, H_3PO_4
	Nucleosidase	Nucleosides	Purine & pyrimidine bases, pentoses
	Enterokinase**	Trypsinogen	Trypsin

*Erepsin contains a number of aminopeptidases, dipeptidases, and polypeptidases.
**Enterokinase is a potentiating enzyme with the sole function of activating trypsin.

The above table gives a list of the origin and mode of action of the digestive enzymes. The list is still incomplete since all the enzymes have not been identified.

Digestive Hormones

The following table shows the known digestive hormones, their location, and action:

Location	Hormone	Action
Stomach	Gastrin	Stimulates gastric secretion
Small intestine	Cholecystokinin	Stimulates contractions of gall bladder
	Secretin	Stimulates bile, duodenal, and pancreatic (fluid) secretion
	Enterogastrone	Inhibits gastric secretion
	Pancreozymin	Stimulates pancreatic enzyme production
	Enterocrinin	Stimulates enzyme secretion in the small intestine
	Intestinal gastrin*	Stimulates gastric secretion
Kidney	Urogastrone*	Inhibits gastric secretion

*The exact status of intestinal gastrin and urogastrone have not been determined.

The Cecum and Large Intestine

With the exception of equines, rabbits, rodents, and other simple stomached herbivores, the cecum and the large intestine function chiefly as organs of water absorption and concentration of the fluid and semifluid intestinal contents. In the excepted species and other nonruminating herbivores, the large intestine and cecum are usually important organs of digestion and nutrient absorption and do not confine their activity to water conservation. They tend to compensate for the lack of a rumen and the alimentation advantages which the rumen gives to its possessor. The large colon and cecum produce fermentation products and microbiota which can be digested and absorbed. This cecal and colonic activity does not occur to any appreciable degree in ruminants or pseudoruminants.

The Rectum and Anus

Essentially, the rectum is an organ of storage where fecal products are retained until such time as sufficient quantity has been accumulated to result in nervous stimulation and defecation.

The anus is the caudal termination of the digestive tract and consists of two sphincter muscles and a retractor muscle which are normally contracted except during defecation.

PHYSIOLOGY OF RUMINATION

The physical characteristics of rumination that separate a ruminant from other herbivorous animals are that herbaceous food materials are processed twice within the mouth and twice within the compartments of the rumen before they are passed to the glandular stomach. If one watches a ruminant eat, one will notice that—in contrast to a horse—the animal quickly crops and swallows large amounts of herbage with relatively little chewing. Having thus gotten a "fill" of plant material, the ruminant will cease eating, lie down, and proceed slowly to regurgitate one bolus or "cud" after another. Each bolus will be thoroughly and slowly chewed and again swallowed. The mechanism of rumination, however, includes a great deal more than these observable acts. To illustrate what is involved, it is advisable to trace in detail one mouthful of food through the ruminant digestive tract from mouth to small intestine.

As the food is first ingested it is chewed perfunctorily, ensalivated, and swallowed. In this condition it enters the rumen. The rumen contains considerable amounts of water and enormous numbers of bacteria and protozoa. The initially swallowed plant material possesses a low specific gravity and thus remains more or less on the surface of the liquid contained within the rumen. Rumen contractions churn this light material, mixing it thoroughly with the rumen bacteria and protozoa. These organisms break down the complex cellulose into sugars and various organic acids. In the process, the plant material is hydrolyzed and softened by the fluid and bacterial action. Some time later this material is brought to the cardia and is passed into the posterior esophagus as an oval mass or bolus which is formed partially by the cardia and partially by the rumen muscles surrounding it. This mass is regurgitated up the esophagus, received again in the mouth as the cud which is thoroughly and leisurely chewed, liberally ensalivated, and reswallowed. At the second passage down the esophagus, the food material has been thoroughly ground into small particles that have a relatively high specific gravity. This material falls into the reticulum where it is passed to the omasum. There the excess water is extracted, and the food is passed from the omasum to the abomasum where true gastric digestion takes place. In this process the food has been subjected to two mechanical preparations and a phase of bacterial digestion prior to gastric and intestinal digestion and absorption. As the food passes into the abomasum, it brings along with it a number of bacteria and protozoa which are also digested. It has been said facetiously that a ruminant is not truly an herbivore but rather a meat-eating alcoholic. This statement is at least partially true. If one substitutes acetate for alcohol, it is quite precise, since cattle do consume great quantities of protozoa—which are animals—and absorb a large amount of short chain organic acids directly from the rumen which have been formed by bacterial and protozoal digestion of the lower molecular weight cellulose molecules.

From the glandular stomach onward, the digestive tract of ruminants is not essentially different from other mammals.

Ruminant Metabolism

The question of whether the dog, cat, or horse is "man's best friend" is shocking to pet lovers if viewed from one aspect of semantics. If a friend is to be defined as someone who gives something beside the intangible and ephemeral things called loyalty and affection, neither dogs, cats, nor horses would rank in the same category as sheep or cattle. And of the latter two, as far as American communities are concerned, cattle would win in a walk. Since ruminants provide the bulk of animal proteins, leather, and animal fibers, and do it without competing with mankind for food, they can further be categorized as unselfish, which is the highest form of friendship.

The machinery that makes the domestic ruminant such an unselfish contributor to mankind's welfare is found in that remarkable organ—the compound stomach—and the equally remarkable metabolism which it supports.

The ruminant has four stomach compartments. The first two, the rumen and reticulum, serve as a fermenting vat, and the last two, the omasum and abomasum, serve respectively to prepare food for gastric digestion by removing excess water and to accomplish the normal acid-enzyme digestive phase of the glandular stomach.

In the adult ruminant, the first two compartments of the stomach are enormous structures, making, when full, almost one-sixth the weight and one-fifth the volume of the animal. Large cows will often have a rumen capacity from 40 to 50 gallons and the capability of supporting countless billions of microorganisms which share a symbiotic relationship with their host. In return for a constant environment and a steady supply of raw food materials, the ruminal microbiota not only break down cellulose into digestible polysaccharides but also synthesize substances that can be directly utilized by the ruminant for food. From the complex cellulose of plant material in the rumen, they form large amounts of short chain fatty acids, principally acetic (60 percent), propionic (20 percent), and butyric (20 percent) and some alcohol. These substances pass into the bloodstream through the rumen wall in much the same way that monogastric mammals take up alcohol and water.

The act of carbohydrate digestion is quite different in ruminants than in monogastric animals. Since the simple stomached animals digest carbohydrates to form simple sugars that are absorbed, their bloodstreams contain appreciable amounts of glucose but almost no short-chain fatty acids. Ruminant blood is different. It has been calculated that about 90 percent of a ruminant's energy requirements are supplied by short chain acids produced in the rumen. Apparently, the ruminant's body tissues are extraordinarily efficient in utilizing acetate, propionate, and butyrate. This ability, however, does not begin at birth. A young calf, feeding on its mother's milk, metabolizes food in much the same manner as do other mammals but when the calf reaches about 4 months of age, the rumen takes over the main nutritive function. The pattern of metabolism changes. The blood sugar falls and volatile fatty acid content increases. The animal responds to this change by producing enzymes that aid in the breakdown of volatile fatty acids to energy by the tissues.

In the utilization of the volatile fatty acids, acetate provides a nonspecific source of energy and can be synthesized into fatty acids or ketone bodies. It apparently cannot produce the metabolic derivatives of glucose, which are produced in the glycolysis scheme and the tricarboxylic acid cycle, nor can it readily be transferred into glucose.

Propionate is of great importance in ruminant nutrition. Its configuration makes it amenable to synthesis into glucose by the liver and it provides about half the total glucose which enters a ruminant's metabolism. It is, therefore, a specific energy source and shares with carbohydrates and amino acids a specific energy function not shared by acetate and butyrate.

Butyrate is not particularly useful in nutrition. It can give rise to acetate through normal metabolic channels and may be involved in the synthesis of glucose. Its significance in this latter operation is doubtful. Like acetate, butyrate is considered to be a ketone former.

An additional metabolic pathway in the cow results from the synthesis of proteins from fatty acids. These mechanisms are in marked contrast to the metabolism of nonruminant mammals, which does not utilize short chain fatty acids to such a degree and lacks the ability to synthesize protein from fatty acids.

Although the ruminant digestive system gives great advantages insofar as food is concerned, it has certain disadvantages. A principal weakness is that fermentation within the rumen generates large quantities of gas which must be eliminated by frequent belching. If something interfers with this mechanism, ruminants develop the spectacular syndrome known as bloat. The exact mechanism of this disease is not precisely known, although a number of factors have been shown to be involved. Under natural conditions bloat occurs in animals grazing on lush legume pastures and the common variety of gaseous bloat is apparently caused by foaming agents (saponins) which are present in these grasses. It has been shown that the addition of antifoaming compounds, fats, or vegetable oils to the diet will reduce the occurrence of bloat.

A second serious defect of ruminant metabolism is its predilection toward formation of excess ketones in the body. This appears, for practical purposes, to be a specific disease of ruminants. The biochemistry of ketone production indicates that the primary causes of this condition are probably defects in metabolism of fatty acids, the wrong type of rumen flora, or deficiencies in the diet which result in a reduced production of fatty acids in the rumen. High producing dairy cows and multiparous ewes in late stages of pregnancy are particularly susceptible to ketosis because cows must have large amounts of glucose for milk production and ewes need available glucose for the support of twin or triplet fetuses. The primary cause is thought to be a deficiency of the liver's ability to detoxify or metabolize the short chain fatty acids in the bloodstream due to increased stresses imposed by lactation and pregnancy. It is quite possible that any one or all of the above factors may enter into the ketosis syndrome.

Ruminants, while able to convert some synthetic nitrogen-containing compounds (urea) into protein, are not entirely successful in utilizing the higher carbohydrate polymers such as cellulose and lignin. Neither the ruminal flora nor the animal's own digestive system is capable of breaking down lignin to absorbable carbohydrates or short chain fatty acids. It would be of considerable advantage to both ruminants and their owners if there was some simple way of "cracking" the high molecular weight carbohydrates to smaller digestible fractions that could be utilized.

chapter 8
The Respiratory System

At all times, and in all living cells, energy liberation continues without interruption. This energy results essentially from the oxidation of complex carbon-containing molecules; therefore, a constant supply of oxygen (O_2) is necessary for existence. The oxidation of the simple sugar glucose is roughly characteristic of the cellular oxidation processes that go on in the body, although the bodily reactions are vastly more complex than the simple breakdown shown here:

$$6O_2 + C_6H_{12}O_6 \rightarrow 6CO_2 + 6H_2O + ENERGY$$

One of the principal end products of this reaction is carbon dioxide (CO_2). Since this waste product is toxic, there must be some means for eliminating it from the body. Respiration provides the major route of elimination of CO_2 as well as the principal source of the oxygen needed to continue further oxidation processes in the body's metabolic schemes.

Since respiration is essentially an exchange of gases between an organism and its environment, there are a few principles dealing with the properties of gas that should be remembered. A gas or a mixture of gases behaves as a mixture of discrete particles. These particles (gas molecules) are separated by relatively large distances and are in a state of continuous random (Brownian) movement. The particles exert no force on each other unless they collide, and even then the energy loss is inconsequential since the molecules behave as elastic objects.

There are four laws that pertain to gases: Boyle's Law, Charles' Law, Henry's Law, and Dalton's Law. Boyle's Law states that at a constant temperature the volume of a gas varies inversely with the pressure to which the gas is subjected. Charles' Law states that at a constant pressure the volume of a gas varies directly with the temperature, and proportionately to absolute temperature. Henry's Law states that under a constant temperature, the amount of gas dissolved in a liquid with no affinity for the gas varies directly with the pressure of the gas in the surrounding medium. Dalton's Law states that the pressure exerted by each component of a mixture of gases is independent of the other gases in the mixture. A corollary to Dalton's Law deals with partial pressure (tension) of a gas, and states that when the partial pressure of a gas leaving a solution is equal to the partial pressure of the same gas entering the solution, the system is in equilibrium for that gas.

The application of these principles will be indicated later in the text, but they are included here since they should be brought to attention at the outset. A few terms such as eupnea, dyspnea, hyperpnea, polypnea, and apnea should also be understood and remembered.

Respiration increases in complexity with the complexity of the animal involved. In simple unicellular forms of life, respiration is a relatively simple process since the organism is in direct contact with its environment, and gaseous exchange is directly between the surrounding medium and the cell. Such a form of respiration is called direct respiration and is possible only if all the cells of an organism are in contact with their environment. In the complex multicellular forms of life direct respiration is impossible since the cells of the body have become specialized and the large majority of them are not in contact with the external environment. Therefore, some means of supplying oxygen to these cells must be present. The higher animals have developed a system of respiratory organs that exist for the purpose of bringing oxygen to the cells of the body and removing carbon dioxide, water, and other waste products of a gaseous or vaporous nature. While the structures differ among classes of animals, they all possess the common feature of a thin, moist, semipermeable membrane separating an oxygen-containing medium on one side and circulating blood on the other.

It can be seen, since the term "respiration" refers to a gaseous exchange, that there are two types of respiration occurring within the body of higher animals; external respiration involving the respiratory organs and bloodstream; and internal respiration which is an exchange of gases between the bloodstream and the body cells. In each type an exchange of useful and waste products takes place.

GROSS STRUCTURES OF THE MAMMALIAN EXTERNAL RESPIRATORY SYSTEM

The external respiratory system in mammals is essentially a series of passages and tubes that commence at the exterior of the animal and end blindly in a multitude of tiny, thin-walled, closed sacs or alveoli, which are in intimate connection with the bloodstream. The structures involved in the system are: the nasal cavity, pharynx, larynx, trachea, bronchi, lungs, pleurae, and thoracic cavity.

The nasal cavity includes the following structures: external nares and nasal hairs, nasal septum, dorsal, ventral, and ethmoid turbinates, paranasal sinuses, olfactory (smell) region, and the nasal mucous membrane.

The pharynx has been previously described (p. 152).

The larynx is a movable framework of cartilages and muscles that connects the pharynx and trachea. It supports a valvular apparatus (the epiglottis) which regulates the passage of air and prevents the aspiration (breathing) of food or other foreign bodies. It also contains the vocal cords, which are the chief vocal organ in mammals.

The trachea is a heavily walled tube extending from the larynx to the stem bronchi. Its lumen is kept permanently open by closely spaced C-shaped rings of cartilage set in its wall. The open parts of the cartilage rings are dorsal. The trachea passes downward and backward along the ventral surface of the neck and enters the thoracic cavity at the thoracic inlet. At the region of the heart the trachea divides into primary bronchi which enter the lungs at the hilus. Successive subdivisions of the bronchi terminate ultimately in the alveoli. The general scheme of the duct system of the lungs is as follows:

1. Trachea
2. Primary or stem bronchi
3. Secondary bronchi
4. Bronchioles
   a. Lobular bronchioles
   b. Intralobular bronchioles
5. Terminal bronchioles
6. Respiratory bronchioles
7. Alveolar sacs
8. Alveolar ducts
9. Alveoli

The C-shaped tracheal cartilages become plates in the stem bronchi; and the cartilages disappear entirely in the secondary bronchi when these are reduced to one millimeter in diameter. The secondary bronchi become bronchioles that continue to branch and diminish in diameter until they become respiratory bronchioles. Respiratory bronchioles can be recognized microscopically by their thin walls and because they terminate in alveolar sacs. The alveolar sacs are central to clumps of alveoli that are connected to the sacs by delicate alveolar ducts. All structures from the terminal bronchioles onward are too small to be seen or recognized by the unaided eye.

The respiratory bronchioles, down to and including the alveoli, form a pulmonary unit or lung unit which is the basic functional structure of mammalian respiration.

The alveoli are the true respiratory structures. Here the exchange of gases between the bloodstream and the inspired air takes place. Entwined about each alveolus is a dense network of capillaries that brings the blood in contact with the gases in the alveolus. Fresh air taken into the alveolus on inspiration replaces certain of the waste products held in the bloodstream. These waste products (mainly CO_2 and water vapor) are expelled from the alveolus on expiration.

At one time it was believed that the changes in the size of the lung during inspiration and expiration were mainly through expansion and contraction of the alveoli, but it has been demonstrated that air movement into and out of the lungs has very little effect on alveolar size. Inspiration and expiration are effected by dilation and lengthening, and compression and contraction of the bronchial tree down to and including structures anterior to the alveoli. The principal breathing activity probably occurs in the terminal structures, i.e., from the terminal bronchioles through the alveolar ducts. Ventilation of the alveoli is caused by air currents generated by the expansion and contraction of the duct system rather than bellows-like movement of the alveolar walls. Diffusion gradients also help alveolar exchange as does a widening of the alveolar ducts during expiration.

The smooth muscle in the bronchial tree aids expiration by contracting and inspiration by relaxing. These muscles are innervated by sympathetic and parasympathetic (vagus) fibers.

The alveoli and terminal respiratory passages are coated on their inner surfaces with a fluid which is probably secreted by the large alveolar (Type II) cells. It is a complex mix of protein, polysaccharide, and phospholipids (lecithin and sphingomyelin) that functions as a surfactant. It reduces the surface tension of the pulmonary fluids and helps maintain the alveoli and alveolar ducts as open structures, and also helps maintain the elasticity of the lung.

Unlike water, which has a constant surface tension regardless of area, surfactant has low surface tension in small areas and high surface tension in large. This is essential for alveolar functioning since alveoli in mammalian lungs are a mix of sizes in any given area. Without surfactant, and only water as a moistening medium to keep the alveolar membrane functional, the smaller alveoli would collapse into the larger

GROSS STRUCTURES OF THE MAMMALIAN EXTERNAL RESPIRATORY SYSTEM

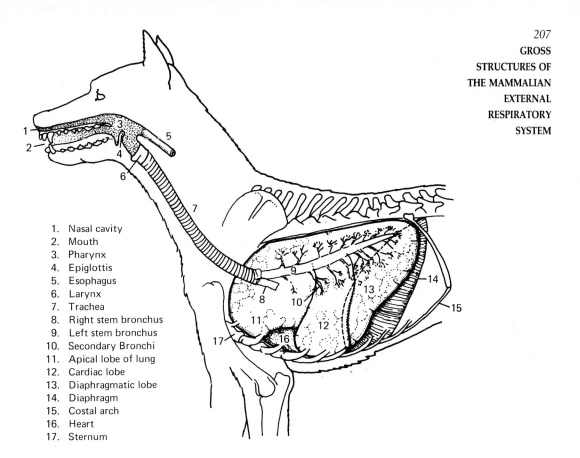

1. Nasal cavity
2. Mouth
3. Pharynx
4. Epiglottis
5. Esophagus
6. Larynx
7. Trachea
8. Right stem bronchus
9. Left stem bronchus
10. Secondary Bronchi
11. Apical lobe of lung
12. Cardiac lobe
13. Diaphragmatic lobe
14. Diaphragm
15. Costal arch
16. Heart
17. Sternum

FIGURE 8-1   Schematic drawing of the respiratory system.

because of the forces operating on dissimilar sized areas with equal surface tension. If the smaller alveoli did collapse, the area would become progressively atelectatic since all the residual air would be exhausted in the collapsed alveoli. Moreover, these collapsed alveoli would probably not reinflate, but would remain collapsed. However, with surfactant the smaller alveoli lose surface tension in proportion to their size and thus can coexist functionally with the larger alveoli.

Newborn animals without sufficient surfactant exhibit this collapse, with attendant labored breathing (dyspnea) and poor gaseous exchange. Often the affected animals die and on postmortem reveal lungs with severe patchy collapse or failure of inflation (atelectasia) of the lung tissue. Excessive watery secretion into the lungs in certain diseases and parasitism may impair surfactant production, cause increased surface tension and airway closure, with subsequent collapse and/or edema of the lungs.

Prior to electron microscopy there was considerable argument as to whether the walls of the alveoli were derived from capillary walls (the vascular system), or, conversely, whether the capillary walls at the pulmonary units were derived from the alveolar membranes (the respiratory system). The interalveolar septa, as these membranes are now called, are the locales through which gaseous interchange between the alveoli and bloodstream takes place. Since they are only in the neighborhood of 100 millimicrons thick, it appeared improbable prior to electron microscopy that more than one membrane could be involved. It has now been definitely shown that the alveolar-capillary wall is composed of elements of both the respiratory and vascular systems, and consists of a continuous, extremely thin, alveolar membrane, a basement membrane, and an equally thin capillary endothelium without apparent openings or breaks. These form a three-layered structure derived partly from the vascular system and partly from the respiratory system.

The Lungs

Two lungs (right and left) occupy the greater part of the thoracic cavity. They are divided into a varying number of lobes depending on the species of animal, and are essentially two elastic membranous sacs whose interior is in free communication with the outside air via a system of passages. The interior is highly modified, and its surface area greatly increased by the numerous alveoli. With the exception of the lung root

Comparative Gross Anatomy of the Lungs

	Tracheal Cartilage	*Bronchi*	*Lungs*
Horse	Incomplete dorsally	2 stem bronchi. Each gives off a cranial bronchus R stem bronchus gives off an accessory bronchus	R larger. No lobes except accessory lobe on right lung
Ox	Ends of rings in apposition	3 stem bronchi; 1 to R lung; 1 to L lung; 1 to R cranial lobe	Right much larger with 4 or 5 lobes, e.g., cranial (2) Cardiac, caudal, accessory, L lung bas 3 lobes; cranial cardiac caudal
Sheep	Similar to ox	Similar to ox	Similar to ox
Pig	Rings meet dorsally	Similar to ox	Right larger with 4 lobes Left similar to ox
Dog and Cat	Rings do not meet dorsally	2 stem bronchi; 1 branch to R cranial lobe, 1 branch to R lung; 1 branch to L cranial lobe, 1 branch to L lung	Right lung larger; lobes as in pig
Rat	Rings incomplete dorsally	2 stem bronchi; R bronchus gives off 2 branches, one to cranial lobe, the other to remainder of lung	R lung larger, with 4 lobes, e.g., cranial, cardiac, caudal, accessory. L lung has one lobe

(radix or hilus) and the pulmonary ligaments, each lung lies free in the thoracic cavity.

Pulmonary Vascular Systems

In addition to the air passages, the lungs have four separate vascular systems: a pulmonary blood vascular system, a bronchial blood vascular system, and two lymph systems.

The pulmonary vascular system begins with the pulmonary artery, which leaves the heart at the right ventricle. This artery carries unoxygenated ("venous") blood to the lungs. The pulmonary artery divides into right and left pulmonary arteries and then subdivides, ultimately forming the capillaries surrounding the alveoli, where the waste products carried by the blood are discharged and the blood cells and plasma receive a fresh charge of oxygen. The oxygenated ("arterial") blood is then carried back to the heart, and enters the left auricle through a series of pulmonary veins. From here the oxygenated blood is pumped through the body over the systemic circulation. (See pages 271 and 273.)

Since the blood surrounding the alveoli is derived from a circulatory system connected to the heart by short and direct major vessels, the blood flows and volumes tend to fluctuate with the heart rate and amplitude. This fluctuation, because of variable bodily demand, may vary from one area of the lung to another, between elements of each area, or even between segments of the meshwork surrounding the alveoli. So far, no overall control system for blood supply similar to the juxtaglomerular apparatus of the kidney (p. 308) has been demonstrated. Functionally, the lungs operate more or less according to the demands placed upon them, with the pulmonary circulation taking the most direct route to available alveoli and back to the heart. This throws the major burden of work onto the alveoli in the greatest cross-sectional area of the lungs. Lung tissue in the apical portion of the lungs and along the thin posterior margins is less used and is more subject to infection by microorganisms and parasites since it does not receive as much attention from the blood defense mechanisms.

The bronchial circulation of the lungs begins with the right and left bronchial arteries which invade the lung tissue and furnish it nourishment. The blood returns by the bronchial veins to the cranial vena cava via the vena azygos. The principal structures involved in the bronchial circulation are shown in the accompanying table.

The pulmonary lymphatic system, although dual, is not completely separate since the terminal radicles are united. The two lymph systems separately drain the pleural and subpleural tissues of the lungs and the pulmonary tissues located around the tubules down to, but not including, the air sacs and alveoli. This dual system has considerable significance in cases of pulmonary disease. A characteristic of the lymphatics of the lung is that they are arranged so that it is possible, under certain

Comparative Lung Morphology of Eight Mammalian Species

Species	Lobules	Pleura	Typical Distal Airway	Structures Supplied by the Bronchial Artery
Horses, Humans	Incompletely separated. Extensive interlobular connective tissue	Thick	Terminal bronchioles usually open into alveolar sacs. Respiratory bronchioles are rare and poorly developed	Bronchi, vasa vasorum of pulmonary artery and pulmonary vein. Pleura and interalveolar tissue. Some interalveolar septa
Cattle, Sheep, Swine	Complete separated. Extensive interlobular connective tissue	Thick	Similar to horse	Bronchi, vasa vasorum of pulmonary artery and pulmonary vein. Pleura and interlobular tissue
Dogs, Cats, Monkeys	Very poorly defined. Little interlobular connective tissue	Thin	Respiratory bronchioles open into alveolar sacs. Terminal bronchioles are short	Bronchi, vasa vasorum of pulmonary artery and pulmonary vein

Adapted from Phillis: *Veterinary Physiology.*

conditions, for lymph to flow toward the pleurae from the pulmonary tissue. This is particularly important in cases of bronchial pneumonia, when infection blocks the lymph drainage of the pulmonary tissues. Lymph backs up in the deeper tissues, causing a breakdown of normal drainage and a reversal of lymph flow. The pulmonary lymph system then drains over the pleural channels; and actually aids the spread of infection through the lungs, involving bronchial systems and lobules that would otherwise be unaffected.

The Pleurae

The pleurae are serous membranes which cover the inner wall of the thorax and the thoracic structures, the lungs, and the thoracic viscera. The pleurae are named for the structures which they cover; the parietal pleura covers the inner wall of the thorax; the visceral pleura covers the lungs; the pericardial pleura covers the heart; the diaphragmatic pleura covers the diaphragm; and the mediastinal pleura covers the midline area or mediastinum which contains the trachea, esophagus, aorta, vena cava, thoracic duct, and other structures and divides the thoracic cavity into more or less equal and separate halves. The anterior portion of the mediastinal pleura is fenestrated (lacelike) in some animals, resulting in a union of the two halves of the thoracic cavity. This is true of horses and dogs but not of humans, sheep, or cattle.

The Diaphragm

The mammalian diaphragm is a musculomembranous partition that completely separates the thoracic cavity from the abdomen. It is dome-shaped and bulges anteriorly. The peripheral portion of the diaphragm is composed of striated muscle, and the central portion is composed of a thin tendinous sheet. The diaphragm is pierced by four openings, the hiatus aorticus, the hiatus esophagus, the foramen vena cava and the crus, which are filled respectively with the aorta, the esophagus, the posterior vena cava, and the cysterna chyli of the lymphatic system. On the thoracic side, the diaphragm is covered by the diaphragmatic pleura. On the abdominal side it is covered by peritoneum.

The Thoracic Cavity

The thoracic cavity is a closed space with no external openings. It is bounded by the thoracic inlet cranially; the diaphragm caudally; the thoracic vertebrae and muscles dorsally; the ribs, costal cartilages, and costal muscles laterally; and the sternum, sternal muscles, and transverse thoracic muscles ventrally. It surrounds and includes the heart and its structures, the lungs, parts of the trachea and esophagus, mediastinal structures, and the great vessels entering and leaving the heart.

THE PHYSIOLOGY OF RESPIRATION

External Respiration

The external respiration of mammals consists of two mechanisms, breathing and transportation. Breathing is the act of bringing air and blood into intimate relationship within the lungs and consists of two phases, inspiration and expiration. Transportation is the carrying of respiratory gases between the alveoli and the bloodstream.

In mammals, the alveoli are the functional respiratory structures. These tiny sacs fill and empty with air through changes in size and shape of the lungs. Since the mammalian lungs are held within a closed cavity, changes in size and shape of this cavity are accompanied by corresponding changes in the lungs.

Negative Pressure

The lungs change shape to conform to the movements of the thoracic cavity through a phenomenon known as negative pressure. The pleural cavity is only a potential one since the space between the visceral and parietal pleurae contains a partial vacuum and a thin film of fluid that serves to moisten the two serous membranes. This partial vacuum, or negative pressure, holds the lungs in conformity with the inner walls of the chest. This is a critical feature of the mammalian breathing apparatus and without it the whole respiratory mechanism would fail.

During fetal life the lungs fill the thoracic cavity except for the heart, vascular, and digestive viscera. These structures contain no air, and the muscles of the chest wall are flaccid. At the first breath after birth the muscles tighten and the thorax of the newborn animal enlarges.

THE RESPIRATORY SYSTEM

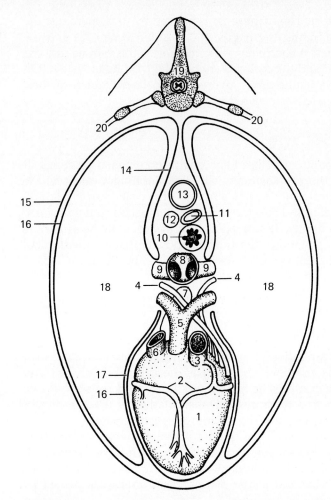

FIGURE 8-2 *Structures of the thoracic cavity.*

1. Heart
2. Coronary vessels
3. Aorta
4. Pulmonary veins
5. Pulmonary arteries
6. Cranial vena cava
7. Caudal vena cava
8. Trachea
9. Stem bronchi
10. Esophagus
11. Vena azygos
12. Thoracic duct
13. Thoracic aorta
14. Mediastinal pleura
15. Parietal pleura
16. Visceral pleura
17. Pericardial pleura
18. Lungs
19. Thoracic vertebra
20. Ribs

The partial vacuum already existing in the anaerobic thoracic cavity of the fetus causes the lungs to expand and air is drawn into them through the trachea. Since no free air normally exists within the fetus, the initial negative pressure between the lung and the inner wall of the chest was present from the beginning. Upon first expiration the lungs collapse but this collapse is not complete. The thoracic cavity never returns to its fetal dimensions. The chest muscles have acquired tonus during this first breath and the chest wall has assumed a new resting position which is much larger than it was.

Rupture of the chest cavity will cause a loss of negative pressure by the admittance of outside air. This will occasionally be severe enough to cause collapse of both lungs and death (by suffocation) of the animal. If the animal survives (e.g., incomplete collapse or only one lung collapsed), negative pressure is slowly reestablished by the absorption of gases through the pleurae by the circulatory system.

In certain animals, rupture can be produced surgically (thoracocentesis), without causing suffocation and death by introducing a filter-equipped hypodermic needle into the thoracic cavity and permitting air to leak into the space between the chest wall and the lung. The condition thus produced is called pneumothorax. The technique can be used in certain species with complete mediastina to allow diseased lungs to rest and heal without the mechanical irritation of expansion and contraction. It cannot be used in animals which have a fenestrated mediastinum. No restoration of negative pressure is required since the air introduced around the lung is absorbed into the bloodstream over a period of time; the negative pressure is reestablished, and within one or two weeks the lung again conforms to the chest wall. If a relatively insoluble gas, such as sulfur hexafluoride, is introduced instead of air, the gas absorption can be delayed about a month.

The Significance of Negative Pressure

The partial vacuum that exists between the lungs and the inner wall of the chest causes the lungs to accurately follow dimensional changes of the thorax during breathing. There is no pushing or pulling by the chest wall or by the diaphragm on the lungs as they inflate or deflate. When the thorax increases in size on inspiration, the lungs are stretched in an effort to fill the increased vacuum between them and the chest wall. On expiration, the reverse occurs.

Thin-walled, hollow organs in the thorax (esophagus, venae cavae, etc.) are affected by negative pressure. Thus, concomitantly with the stretching of the lungs by negative pressure, stretching or ballooning also occurs in these structures. In the veins this results in blood being aspirated toward the heart. More rapid breathing during work thus tends to speed the rate of venous return. Negative pressure also influences the act of vomiting (p. 192) which is commenced by a respiratory movement with a closed glottis. This act greatly increases the negative pressure within the chest, causing ballooning of the walls of the thoracic esophagus and aspiration of gastric contents toward the mouth.

The Mechanics of Breathing

On inspiration the thoracic cavity is enlarged by contraction of the diaphragm, by the lateral forward and outward movement of the ribs and chest muscles, and by the relaxation of the smooth muscles around the alveoli and bronchioles. This creates a partial vacuum inside the bronchial tree and air is drawn into the alveoli.

Contraction of the diaphragmatic muscles flattens the dome of the diaphragm and increases the intrathoracic space. Concomitantly with the contraction of the diaphragm, the abdominal muscles relax, allowing the abdominal viscera to move ventrally and caudally. Usually diaphragmatic respiration is called "abdominal" respiration because of this visceral movement. The contraction and relaxation of the diaphragmatic muscles are controlled by the phrenic nerves. If these are severed, paralysis of the diaphragm occurs. In dogs, paralysis of the diaphragm does not result in impairment of lung function since the expansion and contraction of the thoracic cavity can be accomplished by the movements of the ribs and thoracic muscles.

On expiration the thorax is decreased in size and air flows out as the elastic lungs contract. The muscles that affect expiration may be divided into three groups: muscles of the abdominal wall, chest muscles, and smooth muscles around the bronchioles. The muscles of the abdominal wall contract, and put pressure on the abdominal viscera. This pressure is transferred to the caudal face of the diaphragm and pushes it forward into its resting position. The muscles of the chest rotate the rib cage caudally and inward. The smooth muscles around the bronchioles contract, decrease the size of these structures and force air out of the lungs.

The intercostal muscles, transverse thoracic muscles, levatores costarum muscles, and the ribs form the principal group of structures that can supplant the diaphragm. Other thoracic and abdominal muscles can and do enter into the breathing process, but to a much lesser degree. The levatores costarum and the external intercostal muscles rotate the ribs around longitudinal axes passing through the heads of the ribs and their sockets in the vertebrae. This moves the ribs outward, upward, and forward, thus increasing the size of the chest cavity. The internal intercostal muscles rotate the ribs back into their resting positions and the transverse thoracic muscles compress the thoracic cavity laterally, completing the expiratory movement. The movements of the cranial four or five ribs are quite limited, but the movement of the caudal ribs can be extensive. Caudal rib movement in the normal animal is often employed to aid the diaphragm in accentuating lung filling and emptying to either increase oxygen content in the circulating blood or "blow off" excess carbon dioxide, or both.

In quiet breathing, the muscles of inspiration do not contract to their maximum extent and the expiratory muscles (other than the alveolar smooth muscle) do not contract at all. The recoil of the chest wall and collapse of the stretched lung tissue do the major part of the work. At the end of a normal expiration with all muscles at rest, the lung has a tendency to collapse still further but is prevented from doing so by the negative pressure between its surface and the parietal pleura.

Associated with inspiration and expiration is intrapulmonic pressure (air pressure inside the lungs). On inspiration the air within the lungs becomes rarefied and outside air rushes in to equalize the pressure.

On expiration the contracting lungs increase the intrapulmonic pressure until it is higher than the outside air. The air, therefore, moves outward. During the brief pause between inspiration and expiration the intrapulmonic pressure and atmospheric pressure are the same, since there is free communication through the air passages with the exterior environment. Since the pressures are equal at this time, there is no movement of air into or out of the lungs.

The changes in air pressure within the lung are ordinarily minimal. In quiet breathing the pressure in the air spaces of the alveoli of the lungs rises and falls only slightly during inspiration and expiration, resulting in small variations in pressure sufficient to move air in and out. When breathing is obstructed as in asthma and pneumonia, the changes in air pressure are greater.

Since the object of breathing is to get a certain volume of fresh air to the bloodstream in a given period of time, the question arises, which is more efficient: rapid shallow breathing or slow deep breathing? To settle this question, one must remember that the exchange of gases takes place only in the alveoli and not in the dead space of the trachea and bronchial tree. Hence, the more breaths one takes per minute the greater the amount of useless ventilation of dead space. From this point of view, it is more economical to take a few large breaths than many small ones. On the other hand, the work done for each breath increases as the square of the volume. From this viewpoint, it is cheaper, insofar as energy expenditure is concerned, to take many small breaths per minute. It is apparent that the greatest efficiency will lie somewhere between these two extremes. Calculations for man and domestic animals reveal that each species has a breathing frequency that will do the job with the least amount of work. In general, the rate of respiration varies inversely and the depth varies directly with the size of the animal.

Flow of air into and out of the lungs may be either laminar or turbulent. Laminar flow passes through the respiratory tree during the normal quiet breathing of an animal at rest. Turbulent flow can occur during peaks of respiratory activity or in cases of partial blocking of the airways. In laminar flow the air tends to travel smoothly and in straight lines, and the rate of flow increases or decreases in direct proportion to pressure changes. Turbulent flow is characterized by eddies and vortices, and the flow rates usually relate poorly to pressure changes.

Respiratory Rate

The rate of respiration varies between species and is expressed in number of breaths per minute for normal resting animals. The average number of respirations per minute for the domestic animals ranges from 8–16 in horses to 24–42 in cats. The physical factors affecting the rate of respiration apparently correlate with body size, e.g., females usually

Respiratory Rates per Minute of 8 Domestic Animals

Animal		Rate
Horse		8–16
Cow	Dairy	18–28
	Beef	12–20
Sheep		12–24
Goat		12–20
Pig		15–24
Dog		19–30
Cat		24–42
Human		12–30

breathe faster than males of the same species, smaller animals have faster respiratory rates than large ones, and immature animals have a faster respiration rate than adults. In pathological states (such as pneumonia or anemia) where the lung area or blood capacity are reduced, respiration will also be faster.

Lung Air Components

There are two terminology systems that apply to components of the respiration process. The older of the two deals with "air" and "capacity"; the newer deals with "volume", "level" and "capacity." The new terminology has been around since the 1950s and is now dominant. The "old," however, has not died out and a large number of relatively equivalent terms still exist. A comparison of the terminologies is shown below:

New Terminology	*Old Terminology*
Total lung capacity	Total lung capacity
Functional residual capacity	1/2 Total lung capacity
Vital capacity	Vital capacity
Inspiratory capacity	Tidal plus complemental air
Expiratory capacity	Tidal plus supplemental air
Tidal volume	Tidal air
Residual volume	Residual air
Minimal volume	Minimal air
Inspiration reserve volume	Complemental air
Expiration reserve volume	Supplemental air
Maximum inspiratory level	1/2 Total lung capacity
Maximum expiratory level	1/2 Total lung capacity minus residual air

Total lung capacity is all the air the lungs can hold, or about 42,000 cubic centimeters in the horse.

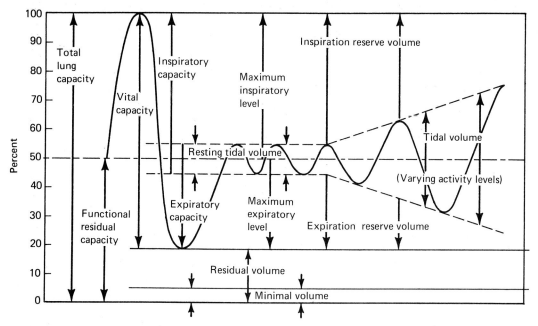

FIGURE 8-3   Lung air components. (Adapted from Dukes, Physiology of Domestic Animals.)

Vital capacity is the total functional capacity of the lungs, i.e., the volume of air that can be breathed out by the deepest possible expiration following the deepest possible inspiration. Using the horse as a base, the average vital capacity is in the neighborhood of 30,000 cubic centimeters. Normal capacity is the amount of air remaining in the lungs after a normal quiet expiration. In the horse this is about 24,000 cubic centimeters. Tidal volume is the volume of air inspired or expired in an ordinarily quiet breath. In a horse this is about 6,000 cubic centimeters. Inspiration reserve volume is the volume of air that can be inspired by the deepest possible inspiration after an ordinary inspiration. In the horse this is about 12,000 cubic centimeters. Expiration reserve volume is the volume that can be expired by the deepest possible expiration after an ordinary expiration. This also amounts to about 12,000 cubic centimeters in the horse. Thus vital capacity represents the sum of tidal, inspiration and expiration reserve volume.

A certain volume of air remains in the lungs even after exhaling supplemental air, for the functioning lungs in a live animal never collapse completely. This remaining air is known as residual volume and in the horse this amounts to about 12,000 cubic centimeters. A certain portion of residual air still remains in the lungs even after the chest cavity is opened. This is termed minimal air and represents the small volume trapped in the alveoli after the lungs have collapsed to their greatest possible physical extent.

Minute volume of air is tidal air times the number of respirations per minute.

All of these data have practical significance that may not be apparent to one who has not had experience with thoroughbred horses, but in the race-horse business the lungs of a newborn colt are, at times, of both vital and legal importance to both the breeder and the owner, particularly when the two occupations are filled by different people (or different businesses).

In horse breeding, service to expensive studs is frequently sold on the basis of a "foal that will stand and suck." The legal interpretation of this statement is that a foal is alive at birth. Many foals will die shortly after being born, and legal action has often been taken to recover stud fees. Presumptive evidence of life at birth is that the animal has breathed. A postmortem examination is performed on the dead foal and the lungs are removed and placed in water. If the lungs float this is evidence that breath has been taken and the foal was alive at birth. If the lungs sink, the animal is not considered to be alive and the stud fee can be recovered since the conditions of service have not been met. This is based on the fact that the lungs of the fetus do not contain air and the uninflated lung tissue is heavier than water. Any breath, no matter how small, will put enough air into the lungs to enable them to float (which is the reason why artificial respiration in nonbreathing newborn thoroughbred foals is discouraged).

Anatomical and Alveolar Dead Space

Anatomical dead space is the term used to describe the air passages between the external nares and the alveoli. In the dead space, there is no interchange between air and blood. The dead space consists of about one-third of the tidal air volume or approximately 2,000 cubic centimeters in the horse. The air contained in these passages mixes with the new air inhaled with each inspiration, warms and humidifies it and prevents sudden alterations in temperature and composition of the air reaching the alveoli. This is necessary to protect delicate terminal structures of the lungs from injury, since sudden changes in their environment may produce inflammation or other conditions which will lead to pneumonia or other lung disease.

A second form of dead space called alveolar dead space exists in alveoli that are not being used. This can be a fairly large percentage of the lungs in a resting animal, but the amount decreases with increased activity. It can reach zero in animals involved in heavy respiratory efforts to "blow off" excess carbon dioxide arising from severe oxygen debt.

Regulation of Breathing

Breathing is regulated by the complex action of the various muscles of respiration. The muscles of the chest, diaphragm, thorax, and those within the lungs function in so integrated a manner that it is obvious they are coordinated by a regulating mechanism. The existence of a respiratory center was postulated long before one was actually discovered. Its location was roughly determined by transections of the brain stem and spinal cord of experimental animals.

Transection of the spinal cord caudal to the origin of the phrenic nerves paralyzes the intercostal and abdominal muscles but does not affect the diaphragm. Transection of the cord at the level of the foramen magnum paralyzes the intercostal, abdominal, and diaphragmatic muscles. Transection of the brain stem rostral to the medulla oblongata causes little if any change in respiratory movements. From the results obtained from these transections it is obvious that the control center lies somewhere in the medulla oblongata of the brain.

Further experiments, mainly on cats, using electrodes inserted into the medulla, located two oval areas posterior to the acoustic tubercles in the rostral part of the medulla below the IV ventricle (Figures 11-4, 11-5, 11-6, p. 322ff) that had stimulating and inhibiting effects on respiratory movements. The areas lay one above the other and overlapped. The dorsal area was somewhat larger and slightly rostral and controlled expiratory movements. The ventral area regulated inspiratory movements. Both areas were bilateral and were connected by a large number of association fibers. Their functional portions were found to be scattered through this relatively extensive area, rather than confined to one or two discrete locations. The connections between the inspiratory and expiratory portions of the center were inhibitory, thus providing that expiratory movements would be inhibited while inspiratory movements were stimulated, and vice versa.

The respiratory center is influenced by both sensory nerves and chemical changes in the blood and controls the rate and depth of breathing. It is fundamentally automatic since it will continue to discharge rhythmic impulses down the motor nerves to the muscles even after all sensory nerves to the center are cut. The motor fibers leave the center and proceed down the spinal cord to emerge as peripheral motor nerves which innervate the various respiratory muscles in the thoracic and abdominal regions. The right and left phrenic nerves emerge from the cervical part of the spinal cord at the level of the seventh cervical vertebra and pass through the thoracic inlet and the mediastinum to activate the diaphragm. Other motor fibers emerge from the thoracic and abdominal levels of the spinal cord and go to various thoracic and abdominal muscles concerned with inspiration and expiration. The motor output of the respiratory center gains its automaticity from two origins: vagal stimulation through the inspiratory inhibition reflex and impulses from the pneumotaxic center in the pons which periodically inhibit inspiratory impulses from the center.

The respiratory center has a potential connection via sensory nerves to practically all parts of the body. Pain, no matter what origin, can cause the number and depth of respirations to increase and expiratory movements to become exaggerated. Blows to certain nerve plexuses such as the solar plexus may result in complete blocking of the center. A specific vagal respiratory reflex from the lungs and thoracic wall (the inspiratory inhibition reflex) results when sensory nerves in these two structures are stimulated by expansion of the lungs. Sensory nerves from the structures reach the center through the vagus nerve at the height of respiration and inhibit the center. This momentarily stops the discharge of motor impulses to the inspiratory muscles and expiration follows.

Since expiration in ordinary breathing is essentially a recoil mechanism, no specific expiratory impulses are sent from the respiratory center. These are only present in hyperpnea and polypnea. With excessive deflation of the lungs a group of deflation receptors are stimulated, which cause the respiratory center to initiate the next inspiration earlier and more forcefully than would ordinarily be the case. The combination of the inspiratory inhibition reflex and the deflation reflex is called the Hering-Breuer reflexes, which have been known since 1868. Transection of the vagus nerve abolishes these reflexes and results in slower, deeper respirations. With abolition of vagal control the pneumotaxic center takes over and gives a different character to the respiratory movements.

In addition to transporting the sensory impulses for the Hering-Breuer reflexes, the vagus nerve also carries sensory impulses from the aortic body and aortic baroceptors (pressure receptors) that can inhibit the inspiratory messages from the respiratory center. In general, vagal impulses inhibit the inspiratory portion of the respiratory center and are dominant over impulses from the pneumotaxic center. If both vagal and pneumotaxic stimulations are removed from the respiratory center, respiratory movements stop in full inspiration.

Chemical Regulation of External Respiration

The respiratory center responses can be modified by chemical stimulation via the bloodstream. This chemical stimulus depends upon the relative acidity of the center itself and not of the blood passing into it. An increase in the carbon dioxide tension of the blood decreases the rate of removal of hydrogen ions (by diffusion) from the respiratory center into the bloodstream and results in an accumulation of hydrogen ions (acidity) within the center. The rise in hydrogen ion concentration (acidity) produces a stimulation of the center and stronger and more frequent stimulation of the motor nerves which control respiration.

This explains why the breath cannot be held indefinitely. Voluntary control of respiration can stop breathing for a time. Ultimately one of two conditions occurs: either unconsciousness, where the respiratory center because of its increased acidity takes over, or the center will overcome the voluntary inhibition and force breathing. By the same token, an animal that is exercised vigorously has a considerable residue of carbon dioxide and lactic acid in its bloodstream as a result of muscle metabolism. The acidity of the blood prevents the loss of hydrogen ions from the respiratory center and an increase in the frequency and depth of respiration results. Newborn animals normally commence breathing through this mechanism, which becomes effective within moments after birth. Occasionally newborn animals will not respond to chemical stimulation. In such cases stimulation must be supplied from a different source. It is a common practice in human medicine to spank babies sharply on the rump, shortly after birth. It has been known for generations that this act stimulates the respiratory center. More recently, the mechanism affecting this reflex has been localized. Around the anal region are located numerous receptors that have a direct connection with the brain. The so-called "anal reflex" abruptly stimulates the respiratory center. Veterinarians have discovered that a finger inserted into the rectum of the newborn animal immediately after birth will result in precisely the same reflex as obtained in the human by a smart slap on the buttocks.

The respiratory center can be voluntarily inhibited by forced rapid conscious breathing. This tends to eliminate carbon dioxide and other acid products from the bloodstream; the blood acidity falls and the hydrogen ion removal from the respiratory center is so rapid that no acidity occurs in that organ. As a result, involuntary breathing can become temporarily suspended.

Voluntary Control of Breathing

Respiration ordinarily proceeds in an involuntary fashion but may be controlled voluntarily within limits. The breath can be held voluntarily, but only for a time. Respiration rate can be speeded, but only for a time. Phonation and other related acts are voluntary modifications of the respiratory tract involving breathing. They are secondary uses of respiration whereby air is passed over the vocal cords and sound results. The abdominal press or strain is another example of modifying respiratory movements. When an expiration is attempted with the glottis closed the tightening of the abdominal muscles causes a rise in pressure within the abdominal cavity because the contraction of these muscles is opposed by a fixed diaphragm. This is the abdominal press. It exerts pressure on the hollow abdominal organs and aids in emptying those that have a connection with the exterior. The mechanism aids defecation, urination, and parturition.

Principles of Ventilation

The ill effects of poor ventilation include headache, depression, drowsiness, and general physical and mental inefficiency. This is not necessarily due to deficient oxygen or to poisonous substances exhaled from the lungs but is more often caused by heat stagnation within the body. The principal means of losing body heat are by radiation and evaporation of water. A high environmental temperature inhibits radiation, high humidity limits evaporation, and lack of moving air limits both because an envelope of warm moist air tends to form around motionless bodies. In stables there is also an additional factor of pollution from decomposing urine, feces, and bedding. Decomposition produces chemical reactions that contribute to increased humidity and environmental temperatures.

Clearing of the Air Passages

The airways of the smaller bronchi and bronchioles are kept clear of obstruction by the cilia of the epithelial cells that line them. These fibrils beat in rhythmic waves to push fluid, mucus, or foreign matter up toward the glottis.

Other mechanisms for clearing obstructions from the airways include coughing and sneezing. Coughing involves a forcible expiratory effort made with the glottis closed. This raises the air pressure within the chest. The glottis then opens suddenly, reducing the pressure in the larger air passages to atmospheric level while high pressure remains in the deeper air spaces of the lung. The sudden drop of pressure in the trachea causes it to collapse by inward folding of the membranous portion of its structure, thus narrowing the lumen. Air forced out of the deeper air spaces of the lung passes through the narrowed trachea with considerable speed, thus expelling foreign matter. The sneeze is essentially an upper respiratory cough. An abnormal quantity of air is inspired and is expired with explosive force. The glottis in this case is wide open, and the air meets its chief resistance in the mouth or nasal passages.

Yawning and Hiccups

Yawning aids respiration by ventilating the lung more completely. Ordinarily, not all the alveoli of the lung receive equal ventilation. Some may actually be inactive. The blood that passes through these collapsed alveoli enters the arterial stream without being oxygenated and lowers the average oxygen content of the blood. This stimulates the respiratory center and may result in yawning. The yawn is essentially an exceedingly deep inspiration and involves most of the muscles of the body. It may also serve the purpose of squeezing stagnant blood out of peripheral vessels.

A hiccup is an abnormal response that apparently has no useful purpose.

AIR, GASEOUS EXCHANGE AND BLOOD TRANSPORT

Normal atmospheric air has a composition of roughly 79 percent nitrogen, 20 percent oxygen, and a variable amount of carbon dioxide, rare gases, and water vapor make up the remaining one percent. Air expired from the lungs will contain approximately 79 percent nitrogen, 16 percent oxygen, four percent carbon dioxide and will be saturated with water vapor. The essential difference between expired and atmospheric air is an exchange of four percent of oxygen for four percent carbon dioxide. The water vapor is derived principally from tissue water and is not a part of respiration except because it tends to moisten the mucous membranes of the respiratory tract.

The venous blood coming to the lungs is relatively high in carbon dioxide and low in oxygen content. Diffusion takes place across the alveolar membrane. Oxygen is absorbed by the red blood cells and carbon dioxide is released from them. The principal oxygen transporting medium is hemoglobin, an iron-containing pigment formed by the linkage of four iron porphyrin molecules to a polypeptide called globin. Alterations in this polypeptide are under genetic control, which accounts for interspecies differences in hemoglobin. Normal blood hemoglobin levels vary from about 10 grams/100 milliliters of blood to 17 grams/100 milliliters (see table p. 247). Hemoglobin has the capacity to combine loosely with oxygen and carbon dioxide. It is part of the blood buffer system, which maintains body pH at about 7.4, or slightly on the alkaline side of neutrality. When fully saturated, one gram of hemoglobin contains 1.34 milliliters of oxygen. Hemoglobin can combine readily with other chemical compounds, such as cyanide and carbon monoxide, to form methemoglobin, which is incapable of carrying oxygen (p. 237).

Anoxia

Anoxia is a condition of oxygen deprivation in the body. It is a serious state and can lead to death. Four kinds of anoxia exist:

1. Anoxic anoxia: This is caused by a failure of oxygen supply to normal blood, resulting in insufficient saturation of hemoglobin and low blood oxygen tension. The condition is associated with high altitudes, breathing of inert gases, pneumonia, parasitism, cancer, and impaired respiratory movements.
2. Anemic anoxia: This results from decreased hemoglobin in the

blood which produces low oxygen transport. Causes include anemia, hemorrhage, or methemoglobinemia.

3. Ischemic anoxia: This results from reduced blood flow rate which produces pooling or stagnation in certain tissues or in the entire circulation. Causes include congestive heart disease, right heart failure, and fibrillation.
4. Histotoxic anoxia: This occurs when body cells are so badly damaged or toxic that they cannot utilize available oxygen. Causes include cyanide poisoning, carbon monoxide poisoning, alcohol poisoning, anesthetic overdose, and histotoxic diseases.

Effects of anoxia range from tiredness and weakness to collapse and death. Signs include cyanosis, pallor, dyspnea, and hyperpnea. Fulminating anoxia has a rapid onset and ordinarily a fatal outcome. Acute anoxia has a total body effect but principally produces convulsions, tachycardia, and dyspnea, and may result in death. Chronic anoxia produces weakness, tachycardia and hyperpnea, and finally results in acclimatization or collapse.

Gaseous Exchange

In external respiration, the gaseous exchange between the alveoli and the bloodstream results in a change of "venous" to "arterial" blood. The oxygenated (arterial) blood then passes back to the heart and into the systemic circulation. At the tissues, the oxygenated blood within the capillaries encounters a region of low O_2 and high CO_2 tension. Diffusion then occurs across the capillary endothelium. (The arrows indicate the direction of diffusion of the two gases.)

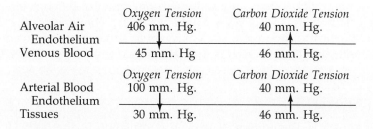

	Oxygen Tension	Carbon Dioxide Tension
Alveolar Air	406 mm. Hg.	40 mm. Hg.
Endothelium	↓	↑
Venous Blood	45 mm. Hg	46 mm. Hg.
	Oxygen Tension	Carbon Dioxide Tension
Arterial Blood	100 mm. Hg.	40 mm. Hg.
Endothelium	↓	↑
Tissues	30 mm. Hg.	46 mm. Hg.

Note that the carbon dioxide tension in the tissues is the same as that in the venous blood, and the carbon dioxide tension of the arterial blood is the same as that of the alveolar air. Thus, there always remains within the blood a residual carbon dioxide tension of approximately 40 millimeters of mercury. This constant carbon dioxide tension is important in regulating the respiratory and cardiac centers and the pH of the blood itself.

Internal Respiration

Three operations are involved in internal respiration: (a) transportation of oxygen and carbon dioxide to and from the cells of the body, (b) exchange of oxygen and waste products between the body cells and the transport mechanism, and (c) exchange of oxygen and waste products within the cells. Since the last operation is more metabolic than respiratory, and since it can be extraordinarily complex, it will not be discussed here.

The respiration process must reach every living cell in the body. However, the rate of respiration in cells varies widely. Nervous tissue, for instance, is an enormous consumer of oxygen and cannot normally exist for more than four minutes without it. Osteocytes, on the other hand, have little demand for oxygen and can live for hours in its absence. The problem of oxygen utilization is not one of concern since it is solved automatically by the cell's metabolic rate and by diffusion. As long as an ample supply of oxygen is present the cells will take care of their needs and balance the amount of oxygen taken in by the amount of waste (mainly CO_2) that is given off.

Oxygen and carbon dioxide concentrations in cells and the intercellular fluid depend upon the concentration of the gases in the capillary blood, the distance of the capillaries from the cells, and the rate of blood flow through the capillary bed. The greater the number of capillaries in a given area, the greater the area available for diffusion, and the greater the transfer of gases. This explains why hard working tissues (brain, liver, kidney, and cardiac and striated muscle) are liberally supplied with capillaries, and why relatively inactive tissues (bone, cartilage, tendon, and cornea) are not.

Hemoglobin changes its configuration as it binds or releases oxygen. In oxidation of hemoglobin there is a dissociation of hydrogen ions for each molecule of oxygen taken up; as a result oxyhemoglobin is more strongly acid than hemoglobin. The change in acidity from oxygen uptake is called the Bohr Effect. It accomplishes two purposes.

The increased hydrogen ion concentration aids release of oxygen to the body cells, and the increased oxygen partial pressure in the lungs aids hydrogen ion and carbon dioxide release. Hydrogen ion concentration from cell metabolism is relatively high in the capillaries, which results in a greater amount of oxygen released to the cells. The increased oxygen release causes a concomitant carbon dioxide release from the cells and its uptake by the bloodstream. The more carbon dioxide in the capillary blood, the higher the level of carbonic acid, and the lower the pH, which produces a greater release of oxygen. A cyclic pattern is thus established that allows active cells and tissues (which release more carbon dioxide) to obtain more oxygen for their metabolic activities.

The second major result of the Bohr Effect is a buffering action that keeps the blood pH at 7.4. When one mole of oxygen leaves the blood to go to the body cells, 0.7 mole of ionized hydrogen (produced from the hydration of carbon dioxide to carbonic acid and its immediate recombination into bicarbonate by reaction with a base) combines with the free hemoglobin from the bicarbonate reaction to form acid hemoglobin. The result is a buffered system (HHb-BHCO$_3$) with no change in pH. This process is called the isohydric shift.

Oxygen transport is principally through a loose chemical combination with hemoglobin in the red blood cells. One hundred milliliters of blood can absorb approximately 20 cubic centimeters of oxygen. This is considered to be 100 percent saturation. In the alveoli, however, blood only becomes 95 percent saturated. Thus, 100 milliliters of blood passing through the lung capillaries will absorb but 19 cubic centimeters of oxygen. Of this, about 18.75 cubic centimeters is chemically combined with hemoglobin and the remaining 0.25 cubic centimeters is in solution within the plasma. Dissociation (freeing) of oxygen from hemoglobin is accelerated by the presence of acids and higher temperature. Both of these conditions occur in working tissues and organs; hence, oxygen is freed rapidly in the areas where it is most needed. Conversely, if carbon dioxide and heat are liberated from venous blood within the lungs (as a result of exercise), the combining power of oxygen with hemoglobin is increased.

Oxygen Dissociation Curves

The amount of oxygen that can be taken up by a sample of blood, or by an aqueous solution of hemoglobin at varying oxygen pressures, can be plotted on standard graph paper with the percentage of hemoglobin saturation as the ordinate and the oxygen pressure in millimeters of mercury as the abscissa.

The curves are basically similar over a wide range of animal species. Curves of aqueous solutions of hemoglobin differ from those of whole blood and take the form of a rectangular hyperbola rather than a sinusoidal curve. This is because hemoglobin in blood releases its oxygen at a much higher pressure than it does in aqueous solution.

The movement of carbon dioxide into and out of the blood is of great importance to determine the combining capacity of oxygen with hemoglobin at a given pressure. However, a number of physical and chemical variables such as temperature, age and species of animal, hemoglobin concentration, and ionic concentration, can affect the shape of dissociation curves and affect the accuracy of information. Nevertheless, dissociation curves are useful in studies of oxygen and carbon dioxide uptake and transport in the blood.

Insofar as oxygen is concerned, dissociation curves help determine the degree of saturation of hemoglobin with oxygen, the total oxygen content of the blood as related to oxygen tension, the conditions affecting

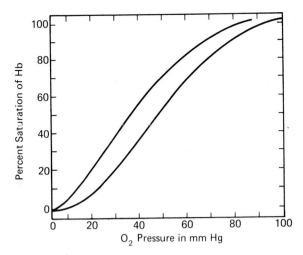

FIGURE 8-4   Oxygen dissociation curves. The curves indicate the range of variation in oxygen dissociation in blood obtained from nonpregnant cows. The CO_2 pressure in the atmosphere to which the bloods were exposed was 40–42 mm Hg. (Adapted from Dukes, Physiology of Domestic Animals.)

oxygen absorption and release by hemoglobin, and transitions from fetal (hyperbolic) to adult (sinusoidal) curves of oxygen dissociation.

It has been found that carbon dioxide dissociation curves of blood and of aqueous bicarbonate solutions differ greatly. This indicates that blood transport of carbon dioxide may not be through bicarbonate and dissolved carbon dioxide in the plasma, as is shown in the accompanying Figure 8-5 dealing with transport mechanism. Evidence indicates that the hemoglobin in the red cells also functions to carry carbon dioxide.

A large amount (about 40–60 cubic centimeters) of carbon dioxide can be extracted from a 100-milliliter sample of blood. A small amount of this (about 5 percent) is held in solution within the plasma. A larger amount (20 percent) is combined directly with the hemoglobin in the red blood cells. The remainder is carried as bicarbonate ions in the plasma. Figure 8-5 illustrates in simplified form the method of release and uptake of carbon dioxide by the red blood cells and the fluid elements of the bloodstream. The actual mechanism is considerably more complex and involves the formation of carbamino hemoglobin (HHb–NH_2 + CO_2 ⇌ HHb–NH–COOH) in the erythrocytes, which has a good affinity for CO_2 and poor affinity for oxygen, and the enzyme carbonic anhydrase, which makes possible the transportation of CO_2 as bicarbonate in the plasma.

A certain amount of nitrogen is present in the arterial and venous blood. The small amount (1–2.5 percent) indicates that it is present mainly in physical solution within the plasma. Nitrogen is not used by the tissues and is of no physiological significance.

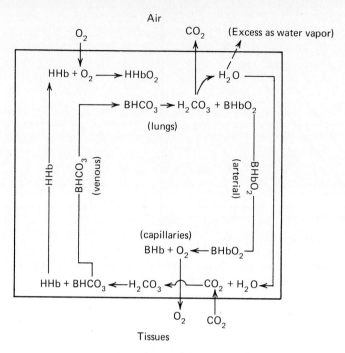

FIGURE 8-5　*Transport mechanism for oxygen and carbon dioxide.*

Pathologically, however, nitrogen transport is another story. Individuals and animals who work or are maintained under conditions of increased atmospheric pressures will have all atmospheric gases in greater amounts in both blood and tissues, concomitant with the increase in partial pressures of the gases in the alveolar air. When these individuals or animals return to areas of normal pressure the dissolved gases in the blood and tissues are at a higher pressure than the outside environment. If the return to normal environmental conditions is rapid, nitrogen bubbles may form in the blood and produce gaseous embolism in the tissues and capillary beds with resultant hemorrhage. This syndrome is known as "the bends" or decompression injury. It can be averted by replacing nitrogen with helium in the high pressure environment, or by slow decompression during the return to normal environmental conditions.

The Respiratory Quotient

Expired air contains the same gases as inspired air, but their proportions are different. Although the nitrogen volume in expired air is the same as for inspired air, which should be expected since nitrogen is a biologically inert gas and does not enter into body reactions, the proportions of oxygen and carbon dioxide are considerably different. Compared to inspired air, the oxygen content is lower and the carbon dioxide content is higher.

In humans based on a large number of studies of normal healthy individuals, the following averages were obtained:

	$O_2\%$	$CO_2\%$	$N_2\%$
Inspired air	20.93	0.03	79.04
Expired air	16.29	4.21	79.50
Difference	4.64	4.18	0.46

The percentage of oxygen intake is slightly greater than the carbon dioxide released, which accounts for the relative increase in nitrogen percentage. However, the absolute increase is zero; the same volume of nitrogen is expired as was inspired. The difference in the percentage of oxygen intake and carbon dioxide release represents the oxygen that is not utilized in the oxidation of carbon, but which is utilized to oxidize other elements such as hydrogen and sulfur. Some of this oxygen is also in the water vapor found in expired air; the remainder leaves the body by other routes, e.g., feces and urine.

The respiratory quotient (R.Q.), which is defined as the ratio of carbon dioxide output to oxygen intake, is of great value in developing information about the kind of food being oxidized in the body and is used in indirect calorimetry for measuring the relative rates of fat, carbohydrate, and protein metabolism. The respiratory quotient is usually written as follows:

$$\text{R.Q.} = \frac{\text{vol. } CO_2}{\text{vol. } O_2}$$

In the oxidation of the simple sugar glucose, the respiratory quotient is 1.00, e.g.,

$$C_6H_{12}O_6 + 6O_2 \rightarrow 6CO_2 + 6H_2O$$

or, in R.Q. terms

$$\frac{6 \text{ vol. } CO_2}{6 \text{ vol. } O_2} = \frac{6}{6} = 1$$

However, in the oxidation of triolein, the commonest body fat, the results are different:

$$C_{57}H_{104}O_6 + 80\, O_2 \rightarrow 57\, CO_2 + 52\, H_2O$$

or, in R.Q. terms

$$\frac{57 \text{ vol. } CO_2}{80 \text{ vol. } O_2} = \frac{57}{80} = 0.71$$

And in the oxidation of protein, using the amino acid alanine for a type reaction, different results are obtained:

$$2(C_3H_7O_2N) + 6\ O_2 \rightarrow (NH_2)_2CO + 5\ CO_2 + 5\ H_2O$$

or, in R.Q. terms

$$\frac{5\ \text{vol. } CO_2}{6\ \text{vol. } O_2} = \frac{5}{6} = 0.83$$

In dealing with mixtures of the basic foodstuffs, the R.Q. of a carbohydrate is always 1.00 because the quantity of oxygen in the carbohydrate molecule is exactly enough to oxidize the hydrogen to water. Each of the remaining carbon atoms combines with one molecule of oxygen in the inspired air. These conditions do not hold for either fats or proteins since some of the inspired oxygen is used in the production of water or other metabolic products. Hence, the average R.Q. for dietary fat has been found to average 0.707, and for dietary protein the average is 0.801.

In calorimetry, the R.Q. data from protein are often omitted since the error produced by this action is so small that it has no serious effect on the results. Since a mammal receives less than 15 percent of its total energy from protein, and the R.Q. of protein lies about halfway between carbohydrate and fat, the protein R.Q. is inconsequential in an estimation of body utilization of fats and carbohydrates. For instance, in a test animal, if the R.Q. is 1.00, it is probable that the animal is metabolizing carbohydrate to the exclusion of fat and protein; or, if the R.Q. is 0.70, the animal is metabolizing fats to the exclusion of carbohydrate and protein; and, if the R.Q. is 0.85, the animal is oxidizing approximately equal amounts of carbohydrate and fat.

More accurate determinations must consider the amount of protein utilized. This is done by analyzing the daily urine output for nitrogen, since it has long been known that the average protein composition is 16 percent nitrogen, and that 90 percent of the nitrogen produced by protein metabolism will be excreted in the urine in the form of creatine, urea, uric acid, and other nitrogenous wastes. For instance, if 10 grams of nitrogen were recovered from urine, 68.75 grams of protein were metabolized by the body, e.g. (calculating by ratio and proportion):

$$10 : x :: 16 : 100$$

(where 10 = gm N_2 recovered in urine, 16:100 represents the percent of nitrogen in the average protein, and x indicates the amount of body protein metabolized as represented by urine N_2)

$$16\ x = 1000$$

x = 62.5 gm protein represented by urinary excretion.

Since urinary excretion is only 90 percent of N_2 excretion, another 10 percent or 6.25 grams of protein must be added to the results to give total body utilization:

62.5 + 6.25 = 68.75 gm protein metabolized by the body.

Once the amount of protein is determined, then the quantities of oxygen utilized and carbon dioxide released by protein metabolism can be calculated and subtracted from the total oxygen. Carbon dioxide val-

Analysis of the Oxidation of Mixtures of Carbohydrate and Fat

R. Q.	Percentage of Total Heat Produced by		Calories Per Liter of O_2
	Carbohydrate	Fat	
0.70	0	100.0	4.686
0.71	1.10	98.9	4.690
0.72	4.76	95.2	4.702
0.73	8.40	91.6	4.714
0.74	12.0	88.0	4.727
0.75	15.6	84.4	4.739
0.76	19.2	80.8	4.751
0.77	22.8	77.2	4.764
0.78	26.3	73.7	4.776
0.79	29.9	70.1	4.788
0.80	33.4	66.6	4.801
0.81	36.9	63.1	4.813
0.82	40.3	59.7	4.825
0.83	43.8	56.2	4.838
0.84	47.2	52.8	4.850
0.85	50.7	49.3	4.862
0.86	54.1	45.9	4.875
0.87	57.5	42.5	4.887
0.88	60.8	39.2	4.899
0.89	64.2	35.8	4.911
0.90	67.5	32.5	4.924
0.91	70.8	29.2	4.936
0.92	74.1	25.9	4.948
0.93	77.4	22.6	4.961
0.94	80.7	19.3	4.973
0.95	84.0	16.0	4.985
0.96	87.2	12.8	4.998
0.97	90.4	9.58	5.010
0.98	93.6	6.37	5.022
0.99	96.8	3.18	5.035
1.00	100.0	0	5.047

From Dukes: *Physiology of Domestic Animals.*

ues and the remainder, the nonprotein respiratory quotient, can be used to establish the values for carbohydrate and fat metabolism.

It has been determined by study of numerous R.Q.s that metabolism proceeds at different rates, on different substrates, at different times, for different reasons. Shortly after eating, the principal metabolite is carbohydrate, and the R.Q. approaches 1.00. A few hours later, fat becomes the principal metabolite, and the R.Q. drops toward 0.70. In animals rapidly converting carbohydrate (which is rich in oxygen) to fat (which is oxygen poor), the R.Q. may go over 1.00. In animals with diabetes mellitus, where little carbohydrate is utilized and body energy is mainly derived from fat, the R.Q. is in the neighborhood of 0.70. Starvation will give similar results as diabetes in the early stages when carbohydrate is depleted, and, in the late stages when body protein is being utilized for energy, the R.Q. will be in the vicinity of 0.80.

Tables such as the following ones have been developed for the easy determination of the caloric values of oxygen and the heat production of carbohydrates and fats once the nonprotein respiratory quotient has been determined. To these data, the heat production from protein metabolism should be added to determine total body heat production. The tabular data can also be used to calculate the amounts of carbohydrate and fat that have been oxidized.

Examination of herbivores by the respiratory quotient method is complicated by the presence of carbon dioxide derived from ruminal or cecal fermentation. The figure is approximately 10 percent, and dairy cows in full production can produce 300 liters of carbon dioxide per day, some of which is eliminated by eructation (belching). Allowance must be made for this excess during the active phases of rumination or cecal digestion. A better method of determining metabolic rate in ruminants is to use a closed circuit breathing apparatus in which oxygen consumption can be accurately calculated. Since in this latter method both fermentative and metabolic carbon dioxide are absorbed, the presence of the additional gas is not a disturbing factor in the calculations.

chapter 9

The Vascular System

The vascular system includes all the duct systems of the body that carry blood or lymph.

The blood vascular system is composed of two parts, the arteries and the veins together with a pumping organ, the heart. Arteries are those vessels that carry blood from the heart (efferent vessels). Veins are those vessels that carry blood to the heart (afferent vessels). The blood vascular system is further divided into three circulations: the pulmonary, which involves the lungs; the systemic, which involves the general body area; and the portal, which involves the intestines and the liver. The portal circulation is entirely venous, beginning as capillaries in the intestines and ending in the sinusoids of the liver.

The lymphatic system is similar in structure and function in both adult and fetus. It begins in tissue spaces and ends in the cranial vena cava. Lymph itself is similar to blood plasma, but lacks proteins and contains certain specialized cells called lymphocytes.

Fetal circulation is a separate study since the fetus is a parasite and receives oxygenated blood from the placenta through the umbilical vein. The environmental difference results in a functional and structural variation from the free-living animal and requires special mention.

BLOOD

Blood is the fluid carried by the arteries and veins. It is somewhat sticky and viscous, having about five times the viscosity of water and a specific gravity about 1/20 greater than water (1.06) and an average pH of about 7.4. In color, it ranges from the bright red of oxygenated blood to the dark red of unoxygenated blood. It is slightly alkaline in reaction, has a distinctive odor and a salty taste, and is 6 to 10 percent of the total body weight.

Blood and plasma volume can be estimated by determining the dilution that occurs after the injection of a known quantity of dye (Evans Blue) into the bloodstream, or by the injection of radioactive 131 I-labeled serum albumin. Blood volume is ordinarily held within narrow limits by a number of factors, i.e., adjustment of water intake and excretion, concentration of sodium ions (Na^+), maintenance of cell numbers through the action of erythropoietin and maintenance of the balance between plasma and interstitial fluid. Variations in number of cells can also be affected by altitude of the habitat (i.e., the higher the altitude the greater the number of erythrocytes per mm^3) age, physical condition, and pregnancy.

Blood Percent of Body Weight	
Horse	10
Cow	8
Sheep	8
Goat	6
Dog	7
Cat	6.5
Human	8

Blood is composed of two parts, the cellular (formed) elements and the fluid elements (plasma). Each of these parts can be further subdivided. The cellular elements consist of red blood cells (erythrocytes), white blood cells (platelets and leukocytes, i.e., neutrophils, eosinophils, basophils, monocytes, and lymphocytes). The fluid elements (plasma) are a mixture of serum and fibrinogen. Plasma is the transport medium for blood cells, metabolic wastes, and dissolved nutrients such as proteins, amino acids, carbohydrates, fats, and salts. It also carries antibodies, enzymes, hormones, and vitamins. It is a yellowish to pearl colored fluid when seen in quantity. In thin layers it appears colorless. Among the domestic animals the darker colors appear in the plasma of horses, cattle and rabbits. Sheep, goats, dogs, cats and small rodents generally have light colored plasma. The color is caused by the presence of bilirubin and carotene, and is more pronounced in animals with high blood carotene levels or in animals with bilirubinemia. This latter condition can be measured by the icterus index where plasma is compared with standard dilutions of potassium dichromate. In most domestic animals, except cattle and horses where carotene and carotenoid enter into plasma color, the icterus index is a useful diagnostic tool to determine kidney and liver disease.

Plasma proteins serve as sources of amino acids for tissue protein anabolism, as substrates for antibody production, as a transport medium for drugs with protein affinity, and as one of the constituents that determine blood viscosity and pressure. The various protein fractions include fibrinogen, albumin and albuminoids, and globulins. Fibrinogen is essential in the clotting process. Albumin and albuminoids help maintain blood pressure and prevent edema through their osmotic effects. Globulins enter into immune reactions and antibody formation. The relative and absolute percentages of plasma protein change with advancing age; the percentage of plasma proteins rises and that of albumins and albuminoids declines.

Like whole blood, plasma coagulates, and as the plasma clot shrinks it squeezes out a watery fluid called serum. Serum contains virtually all of the dissolved substances in plasma except fibrinogen. It is, therefore, common practice to say that plasma is composed of serum and fibrinogen.

The relative volume of plasma is somewhat greater than the volume of cells. Generally about 60 percent of the total blood volume is plasma. These values can easily be determined by centrifuging blood in hematocrit tubes at 3,000 revolutions per minute for 30 minutes and directly measuring the plasma and the packed cell volume. Considerable variation in plasma percent will occur between species. As a result, the values in the accompanying table represent averages, rather than exact figures, and are midpoints in a fairly wide range. The specific gravity, however, is remarkably constant

	Hematocrit Percent of Packed Cells		Specific Gravity of Blood
	Average	Range	
Horse	35	24-44	1.052
Horse (thoroughbred)	42	32-52	1.053
Cow	35	24-46	1.052
Sheep	38	24-50	1.051
Goat	28	19-38	1.051
Pig	42	32-50	1.046
Dog	46	37-55	1.052
Cat	37	24-45	1.051
Rabbit	42	30-55	1.053

Blood transports nutrients from the digestive tracts to tissues or storage depots, transports oxygen and carbon dioxide between the lungs and body tissues, transports metabolic wastes from tissues to the kidneys, transports hormones, regulates pH and electrolyte balance in the body, regulates body temperature, and defends against disease-producing organisms or foreign matter.

The Erythrocytes

Mammalian erythrocytes average from four to seven and one-half microns in diameter. In their fully developed form they are non-nucleated biconcave discs with a thick rounded rim and a thin translucent center. In addition to the lack of a nucleus, the erythrocytes also lack mitochondria; hence, they can carry oxygen, but cannot utilize it in intracellular reactions. The red blood cell composition is about 65 percent water and about 33 percent hemoglobin, an iron containing pigment with a loose affinity for oxygen and carbon dioxide. The remaining two percent consists of proteins, lipids, vitamins, minerals, and other cell inclusions. Unstained erythrocytes appear yellowish- or greenish-red, but *en masse* impart the red color to the blood. These cells exist in large numbers in the circulating blood.

Diameter of Erythrocytes (Microns)	
Horse	5.6
Cow	5.6
Sheep	5.0
Goat	4.1
Pig	6.2
Dog	7.3
Cat	6.5
Human	7.5

A cubic millimeter of whole blood will contain from five to 13 million red blood cells. The size of the red blood cells is quite constant for each species. Number, however, may vary with age, sex, exercise, environment, altitude, diet, and climate. The total surface area of the red blood cells is enormous. It averages from 27 to 36 square meters per kilogram (2.2 pounds) of body weight. To illustrate the area involved, a 1,200-pound cow has about 50 square feet of skin surface, but the corpuscular surface of her red blood cells is in the neighborhood of 160,000 square feet.

The function of red blood cells is to carry oxygen to the tissues, to assist in carrying carbon dioxide from the tissues, to help maintain the normal pH of body fluids, and to help maintain the viscosity and specific gravity of blood.

In normal mammals there can always be found a low percentage (0.2–0.3 percent) of erythrocytes that have a network of bluish threads within the cell. These are reticulocytes and are immature cells that have been put into the bloodstream by the blood forming tissues. In some diseases or in the event of excessive erythrocyte loss or destruction, the number of reticulocytes increases.

Hemoglobin (Hb.) is the principal oxygen transport medium of the red blood cells. It is a polypeptide (globin) combined with an iron-containing pigment (hematin). Under normal conditions, hemoglobin values range from 10–17 grams/100 milliliters of blood (p. 247). Variations in normal hemoglobin levels occur between species. Within a species appreciable variations in hemoglobin levels occur as the result of physiological adaptations to high altitude, or as pathological changes arising from disease, metabolic disorders or hemorrhage.

Hemoglobin combines with oxygen or carbon dioxide to form oxyhemoglobin or carbohemoglobin. These combinations are relatively loose. Under abnormal conditions the bloodstream may be exposed to certain molecules, such as carbon monoxide, which combine tightly with the hemoglobin molecule and alter its character. Such molecules oxidize the ferrous ion in the hemoglobin molecule to ferric iron, and produce a compound known generically as methemoglobin. Since it is ferrous (i.e., two-valence) iron that functions in oxygen transport, its replacement by ferric (three-valence) iron results in a molecule that is unable to transport or release oxygen. Animals with severe methemoglobinemia will die of tissue anoxia, or more accurately cellular suffocation.

Hemolysis

Hemolysis is essentially a loss of erythrocyte contents into the plasma. This may result from cell rupture or from loss of cell contents through an intact cell membrane. Cell rupture (laking) usually occurs from the exposure of cells to such damaging agents as trauma (shaking or whipping), hypotonic solutions, parasites, soaps and wetting agents, fat solvents, venoms, toxins, and mixing bloods from different species of animals. Loss of erythrocyte contents without rupture of the cell wall occurs mainly from exposure of the cells to hypertonic solutions. In this latter event, the osmotic gradient toward the surrounding media literally pulls the cells' contents through the cell membrane, leaving behind shrunken, wrinkled and collapsed (crenated) cells.

The functional effect of hemolysis is the same, regardless of the cause. Hemolyzed blood is incapable of performing normal functions. The hemolyzed blood changes from opaque to translucent or transparent red. Hemolyzed blood is useful in the manufacture of microbiological media, but for little else.

Life Cycle of the Erythrocyte

Red blood cells are formed in specific places in the body. In the fetus and in the newborn, the bone marrow of long bones, the liver, and the spleen, as well as the marrow of cancellous bones, all have blood forming functions. In the adult, the red bone marrow of the ribs and sternum, and occasionally of the vertebral column and pelvis, will normally form all of the blood an animal will require. Under conditions of stress, hemorrhage, and certain types of anemia, the bone marrow of the long bones may revert to blood-forming organs. However, this will only occur in animals where the degradation of the marrow has not progressed to such a point that irreversible changes have occurred. Generally, mature adults have some blood-forming capabilities left in the bone marrow of the long bones but old animals do not. A condition known as extramedullary hemopoiesis, i.e., blood formation outside the marrow cavity, may sometimes occur in severely stressed animals. It is a normal process in the fetus and in the newborn. This usually occurs in the kidney, liver, or spleen. A hormone, erythropoietin (Erythropoietic Factor or EPF), secreted by the kidneys, controls the rate of red blood cell formation.

The red blood cells are apparently formed from undifferentiated cells known as hemohistioblasts, which pass through a number of changes culminating with the loss of the nucleus and the formation of the erythrocyte. In mammals the red blood cell is composed only of cytoplasm and the cell membrane. The last two development stages prior to erythrocyte formation are called nucleated erythrocytes and reticulocytes. Ordinarily there are no nucleated red cells and only a few reticulocytes in the circulation of a normal mammal. In some conditions of hemolytic or anemic disease where the capacity of the bone marrow is not sufficient to overcome destruction of the red cells, nucleated erythrocytes and reticulocytes may be seen in appreciable numbers. These cells, however, do not possess the hemoglobin concentration or carrying capacity of the mature red cell.

Percentage of Reticulocytes in Blood	
Dog	0.3
Cat	0.2
Human	0.6

The lifespan of red blood cells varies in domestic animals. Estimates vary from four to 120 days in different species. The average will probably be close to three months. During their life the red blood cells travel hundreds of miles through the blood vascular system. Ultimately, by a combination of wear against the walls of the blood vessels and squeezing and deformation in order to pass through capillaries, the cell membrane becomes weakened and the cell disintegrates. Enormous numbers of red cells disintegrate and are replaced daily. In a horse it has been estimated that three trillion red blood cells are formed every day or about 35 million every second. Disintegrated red blood cells are destroyed by cells of the reticuloendothelial system. These cells, known as macrophages, have the capability of engulfing and digesting foreign matter. According to one concept, the red blood corpuscles in the bloodstream break into progressively smaller fragments which still retain hemoglobin. When small enough they are ingested by the reticuloendothelial cells and the hemoglobin is broken down into an iron containing fraction and a protein fraction. The iron containing fraction is then split into hematin and pigment. The hematin is used to produce new red blood cells. The pigment portion is converted into the bile pigments bilirubin and biliverdin. These are excreted via the bile duct through the digestive system, and by the kidneys as the bile conjugate urobilinogen.

The exact fate of the protein fraction is unknown. It may be used in the metabolic processes of the macrophages, or may be returned to the bloodstream where it forms a part of the protein fraction of the plasma and enters into the formation of new cells or tissues.

Anemia and Polycythemia

The two commonest pathological processes that affect red blood cells are anemia and polycythemia. Anemia is an abnormality in which red blood cells are reduced in number and/or hemoglobin content. This results in poor oxygen and waste transport, weakness, prostration, and occasionally death. Anemia can be produced by excessive blood destruction (hemorrhage, hemolytic agents) or the presence of excessive numbers of white cells which phagocytose the reds (leukemia). Anemia may also result from defective production of red cells which may be due to one of three causes, i.e., bone marrow defects, nutritional deficiencies and disease. Defective production of red cells by the bone marrow may be the result of tumors or lead poisoning. Nutritional deficiencies are a common source of anemia and are associated with a lack of protein, iron, copper, or vitamins. In these cases the bone marrow is essentially normal, but lacks the raw materials necessary to produce red blood cells. Some hemorrhagic or hemolytic diseases may also produce anemia.

Pernicious anemia, which affects man but apparently does not affect other animals, may also occur. It is a condition in which the bone marrow lacks hemopoietic factor (also called antianemic principle), which is formed and stored in the liver. The antianemic principle is derived from two factors: one is known as "extrinsic factor," which is composed principally of vitamin B-12, and the other is known as "intrinsic factor," which is produced in the fundus of the stomach and is a specific component of gastric juice. These two factors pass through the intestines and into the liver where they are formed into the antianemic principle. A lack of either or both will result in a physiological anemia which can be suppressed by feeding the missing factor. Generally the extrinsic factor is the one which is lacking, but occasionally stomach fractions must be included in the diet. Rarely will both factors need to be supplied.

Polycythemia is an abnormality in which excessive numbers of red blood cells are found in the circulating blood (up to three or four times the normal number). It may be due to one of two factors: excessive production of red blood cells by the bone marrow or failure of the destruction mechanism. The latter is probably not important since physical destruction of red cells by normal wear and tear is constantly occurring. Excessive production may be due to abnormal stimulation of the bone marrow, often by tumors (myelocytomas).

The Leukocytes (White Blood Cells, Phagocytes)

White blood cells are divided into two groups: agranulocytes and granulocytes. Granulocytes are those cells of the leukocyte series that contain granular material within their cytoplasm. All white cells are complete organoids and contain both a nucleus and cytoplasm. They are all capable of some independent movement and have the capacity to engulf and digest foreign or unwanted materials.

They are much less numerous than erythrocytes, numbering only thousands per cubic millimeter of whole blood. The number varies with the species ranging from about 6,000 in man to 17,000 in swine (sheep, 7.5; cattle, 8; horse, 10; dog, 12; in thousands). The average number of white cells is reasonably constant for a species, but the range of normal variation is so great in some species of animals that the average white cell count is not too valuable in diagnosing disease. In general, the number of leukocytes per cubic millimeter of blood rises to meet specific emergencies such as disease and infection, but increase also occurs in response to certain physiological processes such as exercise, digestion of food, and pregnancy. Leukopenia is a decrease in white blood cell numbers. This is usually characteristic of the initial stages of viral infection. Leukocytosis is an increase in the number of leukocytes, principally neutrophils, and usually occurs as a response to bacterial infection. Leukemia is a pathological increase in number and the term is used today to denote one type of neoplastic disease. In general, except for lymphocytes, leukocytes are much shorter lived than erythrocytes. Estimates of their lifespan range from hours to approximately two weeks. It is probable that hours is a more accurate estimate than days. Old or worn out leukocytes are removed from the blood by the cells of the reticuloendothelial system.

Agranulocytes

These are cells that do not contain (or contain very little) granular material in their cytoplasm. There are two routinely encountered cells in the leukocyte series that fit this classification: lymphocytes and monocytes. The plasmacyte is sometimes included in this category.

Lymphocytes

Lymphocytes are small spherical cells with a relatively large nucleus and a variable amount of blue-staining homogenous cytoplasm.* There are two types of these cells: large lymphocytes and small lymphocytes. The essential morphological difference between the two varieties is the amount of cytoplasm. Lymphocytes are formed in the lymph nodes, spleen, bone marrow, thymus, and in other lymphoid tissue. They are believed to produce antibodies, neutralize or fix toxins, and aid in fat absorption from the intestines. They form most of the cellular material of the lymph. They are not greatly phagocytic and are capable of jerky independent movement. They appear to be able to move readily between the circulations of blood and lymph and the body tissues, and are often lost in large numbers by migration through mucous membranes to the outside of the body.

In recent years, with the development of the field of immunology, the lymphocyte has steadily increased in importance, and lymph cells have been described as "the fourth circulation."† It has been theorized that lymph cells are derived from two different embryonal sources at different times in embryonal development. These sources are considered to be the primitive gut and the thymus, which respectively appear early and late in development within the uterus (or egg). This theory postulates that the body is populated by two lymphocyte families of different origins. Its significance will be discussed later.

Lymphocytes are produced in the generative tissues from primitive reticular cells, large primitive cells with abundant cytoplasm. These cells pass through a number of rapid mitotic divisions (the usual number is six or eight) where the cytoplasm diminishes (and becomes increasingly rich in RNA). The end product of these divisions is the small lymphocyte, which apparently cannot divide for some time after its formation. Until recently the small lymphocyte was considered to be a mature cell with no known function, but many agreed that small lymphocytes can be considered to be a form of cellular "spore" reduced to the smallest size that is still capable of being conveniently restored to function. It is the large lymphocyte, not the small, that carries the bulk (if not all) of the

---

*The staining reactions referred to in this and subsequent paragraphs in this chapter are reactions to Wright Stain, a common dye mixture used in most laboratories and clinics.
†The other three fluid circulations are blood, lymph, and cerebrospinal fluid.

lymphocyte activity. It has been demonstrated by *in vitro* and *in vivo* studies that small lymphocytes can enlarge rapidly to form large lymphocytes, and may form primitive cells (lymphoblasts) that can form new lymphocytes. Eight mitoses of a primitive reticular cell could produce 256 descendants of a single ancestor. Each final descendant could conceivably become a primitive reticular cell (prolymphoblast) that is capable of repeating the entire process. Obviously, if this is the case, enormous numbers of lymph cells can be lost outside the body without lowering the lymphocyte numbers within the body.

The small lymphocyte, while biologically inactive, is quite active physically. It passes readily from the terminal capillaries of the blood circulation into the tissues, and from the tissue spaces into the lymph circulation, or it can pass from the tissues back into the bloodstream. Therefore, lymphocytes produced by lymphoid tissue can enter the bloodstream in two different ways: (1) through entry into the lymph vessels and from the lymph vessels via the thoracic duct or right lymphatic duct to the bloodstream and (2) through direct entry into the blood capillaries of the lymph node. Lymphocytes may enter lymph nodes either by the usual route from capillaries to tissue spaces, to the peripheral lymph vessels to the node, or by escape from the bloodstream directly into the node.

The development of the lymphocyte into an important part of the blood vascular system needed a specific biological catalyst, and this function was performed by an extract of red beans known as phytohemagglutinin (*phyto*—plant; *hemo*—blood; *agglutinin*—clumping or aggregating substance) commonly referred to as PHA. Originally this substance was employed in the culture of leukocytes to remove red cells from the culture media, but it was soon found that PHA not only agglutinated the erythrocytes, but also stimulated the lymphocytes to enlarge, develop considerable amounts of RNA-rich cytoplasm, and then to divide mitotically. PHA is now routinely used for this purpose, and the study of lymphocyte activity is one of the more eagerly pursued aspects of biological research.

One of the more interesting applications of this research has been the discovery that the lymphocyte is also the longest lived of the formed elements of the blood. Although their normal lifespan in the lymphocyte form ranges from 15 to 100 days, there is good evidence, based upon studies of survivors of the Hiroshima and Nagaskaki atom bomb explosions and from laboratory studies of irradiated animals and cancer patients, that lymphocytes can remain viable in the small or inactive state for up to 15 years and can revert to the active (large lymphocyte) form at any time during this period. This may also possibly explain the persistence of certain immunological reactions, such as graft rejections and so-called "permanent immunity." It may also explain the persistence and recurrence of lymphoid cancers that appear totally eradicated from the body by treatment with radiation or anticancer drugs.

All of these phenomena involve the functional aspect of lymphocytes in body defense mechanisms. Ever since the transplantation of organs and tissues has become an accepted surgical exercise to prolong life in patients whose own tissues and organs have become diseased or ineffective, the medical and paramedical professions have been concerned with a frustrating phenomenon known as the rejection mechanism. It appears that each body "recognizes" that which is its own and rejects foreign substances, or foreign tissues, even though they may be taken from animals of the same species. Peculiarly, some tissues of the body are not greatly influenced by this mechanism, notably the cornea of the eye and blood of similar types. It was, however, the study of blood and the mechanisms of blood transfusion that developed most of the early concepts of the rejection mechanism. Neither of these subjects receive a detailed study in this book. It is enough to mention that, with these exceptions, an animal will routinely reject unmodified tissues which are not its own, and the rejection is always accompanied by a severe and extensive lymphocyte and plasma cell reaction around the rejected tissue.

It was not hard to determine that lymphocytes had an important role in the rejection process, as well as a protective function against disease organisms, but it was not until the discovery of the effects of PHA that this could be proved. It was known that if a suspension of killed bacteria was injected into a rabbit's foot, some minor changes would occur in the regional lymph nodes through which the lymph from the foot passed. After a month, however, if a second injection was made, a severe reaction would take place within the regional lymph nodes, which would swell and become highly reactive. The lymph cells in the nodes would enlarge and mitose and a new type of cell, the plasma cell, which is structurally different from the lymphocyte, would appear. This is known as the Arthus phenomenon (p. 353). However, these plasma cells (plasmacytes), which produce antibodies, were not known for certain to develop from activation and mitosis of small lymphocytes until the development of PHA. Now it is commonly accepted that the stimulation of previously sensitized lymphocytes, which can remain in the body for years and their metamorphosis into plasmacytes is one of the bases for long term acquired immunity.

The lymphocytes also appear to have an important role in blood formation (remember that they apparently come from two different embryonal origins, although morphologically they cannot be told apart).

Blood is formed in red bone marrow. Several types of cells are formed in the marrow (probably all types except possibly plasma cells) but the two major groups are the granulocytes and the erythrocytes. In each case, the mature cells are produced (theoretically) from stem cells which are undifferentiated precursor cells. There has been a great deal of argument as to whether or not there is more than one type of stem cell, but this is not important in this discussion. The important thing is that at the stem cell level, it is impossible to distinguish between stem cells and small lymphocytes. Until PHA showed that lymphocytes could metamorphose and divide mitotically, the concept that stem cells and lymphocytes were the same could not be accepted. But even with acceptance of the hypothesis there must be proof, and proof was not long in coming.

It has been known for some time that an animal given a lethal dose of radiation dies, mainly because the cells of the bone marrow are destroyed and do not regenerate. A celebrated case some years ago recognized this in humans. It involved several Iron Curtain nuclear scientists who fled to Paris after their reactor had exploded and showered them with lethal doses of radiation. The scientists were given heavy intravenous doses of live bone marrow cells, and subsequently all recovered. With sublethal doses of radiation where the marrow is not completely destroyed, regeneration will occur, and before the new red blood cells and granulocytes appear, large numbers of lymphocytes are present in the regenerating marrow. After blood formation again gets under way, the number of lymphocytes are reduced. It has also been known for some time that with lethal doses of radiation, not only the bone marrow cells, but the cells of the thymus, spleen and lymph nodes are also destroyed.

As was mentioned earlier, there are two sources of lymphocytes. If a lethally irradiated animal is given a suspension of bone marrow cells, the animal recovers, and all other tissues involved in lymphocyte circulation (thymus, lymph nodes and spleen) also recover. But if lymphocytes derived only from thymus and lymph nodes are injected, the thymus, lymph nodes, and other lymphoid tissue will regenerate, but the marrow will not. However, if the lymphocytes are taken from the spleen or Peyer's patches in the gut, the marrow will recover just as though marrow cells had been injected. The conclusion is obvious that there is a difference between thymic (T) and bone marrow (B) lymphocytes and that both thymic and bone marrow stem cells are found in Peyer's patches and in the spleen. B-lymphocytes, incidentally, do not derive the "B" designation from bone marrow, but from bursal lymphocytes which were obtained from the bursa of Fabricius (p. 525) of chickens on which this work was originally done.

To clarify the difference between B and T lymphocytes a number of shielding experiments were devised to protect the spleen from radiation. The irradiated animals with shielded spleens recovered. Even when the spleen was removed from the irradiated animal an hour after radiation, the animal recovered. As short a time as ten minutes of spleen retention after irradiation was sufficient to produce significant recovery effects in some animals. Variants of this procedure, such as shielding the popliteal lymph nodes, gave rapid recovery of thymus, spleen, and other lymphoid tissues but did not result in restoration of the bone marrow, which indicates that peripheral lymph nodes are probably of thymic origin.

These experiments indicate that the two families of lymphocytes differ functionally from each other. This has promoted the theory of the dual origin of lymphocytes in the embryo, a theory which appears to be valid. Further experiments have shown that there is a basic difference in the mode of action of each lymphocyte family in the body. The bone marrow (B) lymphocytes are involved in antibody production, but apparently have little or no phagocytic powers and are not particularly reactive to foreign tissue. They metamorphose into plasma cells which produce antibodies against specific antigens. The thymic (T) lymphocytes, on the contrary, are moderately phagocytic, have considerable sensitivity to foreign tissue, and neither metamorphose into plasma cells nor produce antibodies.

Thymic lymphocytes are apparently the principal reason for the existence of the thymus, which is one of the more puzzling glands in the body. A discussion of the thymus is found in the section dealing with endocrine structures, although in light of recent discoveries it is possible that the gland may not belong among the endocrines. The thymus has long been known to be an active organ for lymphocyte production. However, the removal of the thymus apparently had no effect in reducing lymphocyte numbers or activity. With the recognition of the dual origin of lymphocytes and lymphocyte circulation the present consensus is that the thymus exerts its main effect before birth or shortly thereafter, and in a normal animal thymic lymphocytes migrate to and proliferate in all lymphoid tissues of the body except (perhaps) the bone marrow. Surgical removal of the thymus in newborn mice does cause failure of lymphoid tissues to develop properly and may result in early death from a general physical deterioration syndrome known as "wasting disease." It has been suggested that thymic lymphocytes migrate to regional lymph nodes and establish them as depots of thymic lymphocytes. This may well be true since some experiments have shown that early removal of the thymus suppresses (but does not completely eliminate) the rejection mechanism for homograft transplants. However, the problem is not completely solved, since thymectomy in adult mice results in a fall in number of thoracic duct lymphocytes, and after a time may also result in "wasting disease." The disease can be arrested by thymic extracts. It should be noted here that although the thymus persists functionally in adult mice, it shrinks and virtually disappears in many other large (and longer-lived) species of mammals. The mouse may therefore be a special case from which data applicable to other species cannot be drawn. Nevertheless, the problem of thymic generation of lymphocytes is receiving a great deal of attention and experimental work.

Plasmacytes

Plasmacytes are apparently a special form of lymphocyte and seem to be connected with rejection mechanism processes and/or antibody formation.

Monocytes

The third member of the agranulocyte series is the monocyte. Monocytes are large cells 11–13 microns in diameter. They possess an indented nucleus and abundant cytoplasm, which may or may not contain colored granules. They are formed in the liver, spleen, and bone marrow and comprise approximately two to ten percent of the total number of white cells in the circulating blood. These cells are motile, actively phagocytic, and are most active in acid environments (pH ranges below seven). In general, they are found around walled-off abscesses and infections which have been brought under control. They are also found in tissues, and in these locations are called macrophages. Their principal function appears to be janitorial, to remove foreign bodies and cellular debris produced by infection.

Recent studies indicate that "pure" suspensions of lymphocytes which have been incubated with antigens *in vitro* respond rather poorly in antibody competence compared to total leukocyte suspensions. This, and other experiments which have selectively removed granulocytes, suggest strongly that the immune body (antibody) production of lymphocytes is greatly increased by the presence of monocytes or macrophages. This opens another Pandora's box of discovery, which is still virtually untouched.

Granulocytes

The granulocyte series contains three members: neutrophils, eosinophils, and basophils. These cells are produced in the red bone marrow and are approximately 10–12 microns in diameter (about twice the diameter of an erythrocyte). The most numerous of the granulocytes are the neutrophils which are medium-sized cells with abundant, finely granular cytoplasm which stains a faint, grayish pink. The nucleus varies from roughly oval to rod-shape in young neutrophils, while in older cells the nucleus is lobulated, with four or five clumps of nuclear material connected together by thin threads.

Neutrophils are capable of independent ameboid movement and are actively phagocytic. Their optimum pH is above seven. They increase rapidly in certain infections and migrate through the walls of capillaries to the affected area. The act of migration of a white cell through the capillary endothelium is known as diapedesis. The neutrophil moves in an ameboid fashion through the interstices of the capillary endothelium. These cells are one of the principal agents in the formation of pus. Immature forms of neutrophils are called metamyelocytes and rod nuclear (stab) cells. These do not have the phagocytic power of the mature neutrophil, but are produced and put into the bloodstream in times of stress.

Eosinophils are large cells containing a bilobed nucleus and a cytoplasm filled with large red staining spheroidal granules. In certain parasitic infestations, such as trichinosis and stomach worms, their number will increase in the circulating blood, and they tend to increase in numbers in specific diseases such as eosinophilic myositis and mast cell tumors. They are mildly phagocytic and apparently inactivate histamine and histaminic substances. They are also capable of ameboid movement.

Basophils are similar to eosinophils except that their cytoplasm usually contains blue or purple staining granules. These occur infrequently in the bloodstream and compose less than one percent of the circulating white blood cells. Their function is thought to be phagocytic and similar to that of mast cells in the body tissues. However, there is much disagreement about this.

In general, granulocyte numbers (particularly neutrophils) increase in the presence of disease. Severity of the disease can be estimated by the kind and number of neutrophils present in the circulating blood.

Formed Element and Hemoglobin Content of Blood

Animal	Erythrocytes (Millions/mm^3)	Leukocytes (Thousands/mm^3)	Hemoglobin (gm/100 ml)
Horse	6.9	10.3	11
Cow	6.3	7.9	12
Sheep	8.1	7.4	11
Goat	13.9	8.9	11
Pig	7.4	17.1	12
Dog	6.2	12.6	13
Cat	7.2	12.5	10.5
Man	5.4	5.5	17
Woman	4.8	5.5	15.5

Leukocyte Distribution Percentage

Animal	Lymphocytes	Monocytes	Neutrophils	Eosinophils	Basophils
Horse	38	4	54	4	<1
Cow	64	10	21	5	<1
Sheep	48	6	42	4	<1
Goat	48	5	45	2	<1
Pig	47	8	42	3	<1
Dog	25	8	57	10	<1
Cat	27	10	58	5	<1
Human	23	7	67	3	<1

Blood Platelets

Blood platelets (thrombocytes) are the smallest of the formed elements in the circulating blood. These are oval, disc-like, or angular particles ranging from one to four microns in diameter and occurring in large numbers (400,000 per cubic millimeter of blood). They apparently originate from fragments of a large cell known as a megakaryocyte which is found in the bone marrow. They function to produce thrombokinase which is necessary in the clotting process.

Blood platelets are among the shorter-lived formed elements, having a nine- to 11-day lifespan in most domestic mammals. Certain hemorrhagic syndromes can be triggered by a deficiency of blood platelets (thrombocytopenia) or by an excess of platelets (thrombocytosis). Still other hemorrhagic disease can result when a normal number of platelets are present but their function is impaired (thrombasthenia). Two-thirds of all platelets in a healthy individual are in circulation; the remaining third are in the spleen for emergency use. Surgical removal of the spleen results in persistent thrombocytosis. The number of platelets increases in infections, iron deficiencies, traumatic injury, and certain cancers. Normal platelet counts are in the neighborhood of 100–200,000 per cubic centimeter of blood. Counts up to 500,000 are not considered abnormal, but are usually an indication of hemorrhagic disease. The possibility of a hormonal regulation mechanism for thrombocyte or platelet formation is suggested by the drop in platelet numbers following platelet transfusion, and the prompt increase in platelet numbers in conditions where clotting activity is necessary.

THE FLUID ELEMENT OF THE BLOOD

Blood plasma (p. 234) makes up about 60 percent of the blood volume. It maintains a remarkably constant composition considering that at one time or another it contains every product which cells use and also all substances produced by them. Water forms 90 percent of the plasma. Inorganic salts form 0.9 percent; over half the total of salt is sodium chloride. The salts form a buffered isotonic solution which maintains water and electrolyte balance between the body cells and the interstitial fluid. Glucose forms 0.08 to 0.14 percent of the plasma. It is important as a source of energy and an indispensable constituent of the internal environment.

Plasma proteins constitute from seven to nine percent of the plasma. They occur in colloidal suspension. The exact origin of these proteins is unknown, except for fibrinogen which is formed in the liver. They consist of fibrinogen, albumin and albuminoids, globulins and euglobulins and contribute directly to the viscosity of the blood and indirectly to maintaining normal blood pressure. The plasma proteins also exert a low, but constant osmotic pressure in blood, and are partially responsible for the osmotic exchange that takes place between the blood in the capillaries and the body cells. In addition, fibrinogen forms an essential part of blood clots, and other proteins (particularly the globulins) enter into antibody formation. The special plasma substances include a great variety of useful organic compounds (hormones, enzymes, antibodies), waste substances and dissolved respiratory gases.

Clotting

One of the outstanding characteristics of blood is its ability to clot and, within a short time after exposure to air, form an impervious cover for damaged tissue to prevent blood loss and keep out foreign matter. Clotting is a phenomenon of blood plasma, and is an elaborate mechanism that results in the change of blood from a fluid to a semi-solid gelatinous mass. Microscopically, needle-like fibrin threads appear in clotting blood. These increase in length and number to form an entangled network. Formed elements (principally erythrocytes) trapped in the fibrin meshes produce a clot's red color. Ultimately, the fibrin network shrinks, squeezing out a straw-colored, non-clotting liquid (serum). The time between stimulation and clot formation is less than five minutes for all domestic animals except the cow and horse. The average times are as follows:

Horse — 11½ minutes   Cow — 6½ minutes
Human — 5 minutes   Pig — 3½ minutes
Sheep — 2½ minutes   Dog — 2½ minutes

The substances and the mechanism of clotting involve a sequential series of enzymatic actions upon substrates. Thrombokinase (thromboplastin) is the enzyme that initiates the clotting process. It is liberated when blood contacts injured tissue cells or comes in contact with a foreign surface, and is released by platelet disintegration. Combined with ionized calcium, one of the inorganic elements always present in normal blood, thrombokinase attacks prothrombin, a carbohydrate containing protein that is probably formed in the liver and is always present in the blood as one of the plasma proteins. Prothrombin is broken down to form thrombin. This in turn converts fibrinogen (another plasma protein which is normally present as a sol) to a gel called fibrin. Fibrin then forms the network or foundation upon which the clot is based. This idea was advanced by William H. Howell several decades ago.

Summarizing: Platelets or damaged tissue →Thrombokinase
Thrombokinase + Ca^{++} + Prothrombin → Thrombin
Thrombin + Fibrinogen →Fibrin

Figure 9-1 represents a recent idea about the blood coagulation mechanism. One should note that even in this scheme there are named factors which have not yet been determined or otherwise identified although their presence in the scheme has been ascertained.

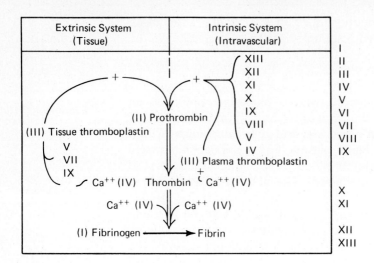

FIGURE 9-1   The clotting mechanism.

In essence, the new idea is an elaboration of Howell's original theory. Whether it gives a better understanding of the clotting process is a moot question. In any event, it is far more detailed and is a good illustration of the direction in which biological research is traveling in its effort to gain total comprehension of the processes involved in the functioning of complex life forms.

External factors may hasten or retard the clotting process. Clot formation may be speeded by chemicals such as adrenalin, malonic acid, thrombin, and thromboplastin; contact with rough-textured foreign matter such as gauze or lint; increasing the temperature of the blood; or gently agitating the blood (in a container). Of these, only the chemical and contact factors can be successfully applied in practice. Increasing temperature, while it does promote clotting, will also dilate the vascular bed and result in an increased flow of blood to the damaged tissue, which defeats the purpose of raising the temperature. Hemorrhaging animals should be kept quiet as movement tends to disrupt clots that have formed in the wound.

A number of factors can delay or prevent clotting. Decreasing temperature of blood will prevent clotting, but low temperature will also cause constriction of blood vessels and reduce blood flow. In animals, this is more important than increased clotting time in the treatment of wounds. Chemical factors that may interdict or delay clot formation include snake venoms, heparin, hirudin, or dicoumarol in living animals, or adding EDTA (ethylene diamine tetraacetic acid), sodium citrate, or potassium oxalate to collected blood. Deficiency of vitamin K which is required by the liver to produce prothrombin, or damage to metabolic processes in the liver, which produce fibrinogen can halt or retard the clotting process. Hereditary abnormalities or traits that result in deficiencies or abnormalities in the blood platelets will produce a condition known as hemophilia.

Clotting not only checks hemorrhage but also attempts to prevent potential leaks. Injury to the endothelium of blood vessels by parasites, trauma or bacteria will cause clots to form in an effort to strengthen the damaged part. Such clots which remain fixed are known as thrombi. A thrombus may develop to such an extent that it will fill the lumen of a blood vessel and obstruct the flow of blood. This will cause a local necrosis of the part distal to the vessel if the clot happens to form in an artery. If the clot becomes loosened from its attachment and floats freely in the bloodstream, it is known as an embolus. Emboli also can cause local necrosis, shock, and sometimes death by blocking small arterioles or capillaries.

INFLAMMATION

Although the phenomenon of inflammation more properly belongs in medicine rather than in functional anatomy, the process is so fundamental to living and homeostasis that a brief description of it should not be omitted. Inflammatory response can occur in a number of situations, ranging from inconsequential to severe, and usually consists of two phases, a primary and a secondary response. Secondary response does not ordinarily occur when stimuli are inconsequential. Primary response can be triggered by almost anything.

In man, and usually in animals, the four cardinal signs of inflammation, i.e., redness, swelling, heat and pain (first reported by Celsus about 34 A.D.), are associated with a number of physiological changes in the affected area. Blood flow initially accelerates and, as the vessels distend, the flow becomes sluggish. This accounts for the redness and heat. Following the appearance of hyperemia, the capillaries become more permeable, and plasma, erythrocytes, and leukocytes move from the vessels into the site of injury. Initially, water, salts, and glucose move through the capillary walls into the tissue spaces. These substances are followed by the higher molecular weight plasma proteins such as albumin and albuminoids, globulins, and fibrinogen (in roughly that order). White blood cells, commencing with neutrophils and ending with monocytes, then move into the area. Neutrophils that enter the affected area do not survive very long and on their death release a proteolytic enzyme which helps dispose of debris and lowers the pH of the inflamed tissue. The movement of leukocytes from the capillaries to the damaged tissues is called diapedesis and is essentially an ameboid movement of the leukocytes through interstices in the capillary walls. The fluid and soluble plasma constituents account for the swelling; the phagocytic and enzymatic activity, pH changes in the tissue, and pressure accounts for the pain.

The exact mechanism by which the white blood cells reach the site of an injury is not known, although experiments indicate that the attractant is probably chemical, at least in the case of neutrophils. Extracts of burned skin injected into rats produce a prompt mobilization of neutrophils at the site of injection. Apparently lowered pH is responsible for the attraction of monocytes and perhaps of lymphocytes to the affected area. Beyond these simple statements there is a vast amount of unknown territory.

What, for instance, causes the vascular permeability that allows plasma, white blood cells, and some red blood cells to escape from the vessels adjacent to injured tissue? Histamine is considered to be the substance principally involved, but its exact role is still not clear. It was not until 1963 that researchers discovered that histamine is released by a connective tissue phagocyte called a mast cell. However, it has since been indicated that histamine is not the only—and perhaps not the primary—substance involved in the inflammatory process. Antihistamines can delay but cannot prevent inflammation, and injury insufficient to cause mast cells to release histamine may still be followed by a delayed inflammatory response. Enzymes that produce histamine and others that break down epinephrine (whose effect is opposite to that of histamine) are found in second stage inflammatory processes where primary histamine release did not take place.

Another vascular permeability enzyme, bradykinin, has been found in appreciable amounts at inflammation sites, and may be involved in second stage inflammation. It is a chain of nine amino acids and appears to be produced by the action of enzymes on the globulin fraction of plasma that is a part of the inflammatory exudate. The entire story is still unknown.

OTHER DEFENSE MECHANISMS

In the most superficial study of bodily defenses against disease it becomes quickly apparent that there is something more than inflammation and direct cellular response involved. The granulocytes, agranulocytes, macrophages, monocytes, and lymphocytes all belong to the cellular police force that rids the body of foreign matter and invading organisms, but they do not do their work alone. They are aided by another defensive scheme called the immune mechanism, which is composed of antibodies, complement, and other defensive substances.

Antibodies are a product of something called an antigen. An antigen is practically impossible to define in terms of what it is, because it can be virtually anything—in any physical state of matter. It can only be described by what it does: It stimulates the formation of antibodies.

Antibodies are always present in the bloodstream as a part of the plasma protein fraction of the blood. Usually they are produced in response to the presence of an antigen. All antibodies appear to be proteins and to belong to a group of plasma proteins called globulins. The globulins with antibody properties are called immunoglobulins to distinguish them from other proteins which do not have immunologic capabilities.

In mammals there are usually several immunoglobulins present in the circulating blood. Their essential function is to combine with the antigen and either render it harmless or make it vulnerable to attack. The specific antibodies have long been thought to be produced by structures such as lymph nodes, spleen and liver, or by the specific cells or cell groups which are under attack (cellular antibodies). Antibodies can be specific or nonspecific. A specific antibody has a molecular configuration that combines with a specific antigen. A nonspecific antibody is one that is capable of being modified to connect with any number of antigens, i.e., its structure is not fixed to attack a specific organism or toxin. At all times, except possibly in case of severe disease, nonspecific antibodies are present in the bloodstream ready for instant modification to combat the presence of foreign substances in the body. Nonspecific antibodies can be produced in response to injection of foreign protein that is not in itself dangerous to life, and this tends to increase the body's capacity to combat disease.

Antibodies are passive. Antitoxins, for instance, detoxify poisons or combine them with a blood element called "complement." In themselves they do nothing, but they create an environment that is deadly to whatever substance they attack.

The deadly activitiy is carried out by the complement. Complement is manufactured in the liver and is also a normal component of the blood plasma. There are a number of different kinds of complement. In the human bloodstream there are nine, which are labeled C_1, C_2, C_3, etc. In animals, the number is similar but much work yet remains to be done before this subject is as well known in veterinary medicine as in human medicine.

Complement causes cells to lyse. Essentially what happens is that the combination of complement plus antibody destroys the ability of the attacked cell to keep out excess water. The water in the tissue fluid surrounding the attacked cell leaks in, the cell swells, ruptures and is destroyed.

It appears that when the antibody attaches to the antigen, it becomes changed in molecular conformation so that it will accept C_1. The C_1 complement then accepts C_2 and the process continues until all the complement fractions are linked together serially to the antibody. It is only then that the lytic changes can begin. The entire process is quick, usually taking place within three-tenths of a second.

Some of the cellular and all of the chemical portions of the defensive mechanisms of the body are under the control of the lymphatic system.

The two basic families of lymphocytes, the thymic or T-lymphocytes, and the bone marrow or B-lymphocytes have already been mentioned (p. 244). Their function in the immune mechanism, however, can stand a little more amplification.

Assume that a foreign organism to which the body is not already sensitized gains entrance. Shortly thereafter, it will come into contact with one or more lymphocytes. These lymphocytes apparently collect antigen from the invader's surface and carry this antigen to the nearest lymph node. In the lymph nodes (which contain both T- and B-stem cells) the appropriate stem cells are stimulated. B-stem cells produce plasmacytes, which are specific antibody producers. T-stem cells produce large active T-lymphocytes which are not antibody producers, but are moderately phagocytic and wall off infected areas. These active lymphocytes appear to act independently without support of granulocytes, macrophages, monocytes, antibodies, complement or any other defensive system. However, the T-lymphocytes aid the activity of the B-lymphocytes through the production of a soluble factor, which is as yet uncategorized. Apparently monocytes and tissue macrophages also enter into this reaction.

Other forms of immunity that do not deal with the blood vascular system and the formed elements of the blood include hypersensitivity which is still not well understood, and cellular immunity which is mediated through T-lymphocytes (p. 244) that are sensitized by exposure to appropriate antigens. Hypersensitivity is a condition that properly belongs to the field of immunology and will not be discussed here other than to state that it appears to be an overly active immune response to an antigen, which is often more detrimental than beneficial.

A somewhat better understood form of cellular immunity is embodied in the substance called interferon, which is produced in the animal body by cells that have been infected by a virus. The cells produce a defensive protein which apparently interferes with viral replication, hence its name. Moreover, the interferon can prevent or ameliorate virus infection of other cells, or infection with another virus and has been shown to have a beneficial effect on certain cancers. Recent research involving a genetically modified intestinal bacillus called *Escherichia coli* has resulted in an organism which has been hopefully called an "interferon factory." And this brings interferon to the borderline of what we call antibiotics—and so it goes. The sharp distinctions of yesterday become the fuzzy borderlines of today.

It is obvious that a great deal of work remains to be done before all the defensive schemes, their triggers, and their methods of operations are known. Yet it seems logical to assume that all parts of the mechanism must function properly to keep an individual at ease in its environment. For the environment is essentially inimical and an active, complete, and continuous defense is the individual's best hope for survival.

The Rejection Mechanism

The principles of the normal rejection mechanism are discussed under the heading of "lymphocytes" earlier in this chapter. However, recent developments seem to indicate that the rejection mechanism can be bypassed by exposing tissues to be heterografted to suitable preparation techniques. For instance, human skin from completely unrelated donors can be successfully grafted to a recipient without the usual rejection syndrome, provided that the donor skin has been held in tissue culture from four to six weeks. Presumably the tissue culture solutions remove some or all of the histocompatibility triggering apparatus. Skin from donors, or from cadavers up to 48 hours postmortem, can be used and can be held for six to eight months in tissue culture. Experimental skin transplants of human material have been 100 percent successful. Other organs are being tried, with work now being done in the tissue culture phase. Mouse skin grafts have not been so successful; they have had a 40 percent rejection rate.

A number of questions on the use of this technique remain to be answered, particularly how long skin can be stored, how long the grafts will remain on the recipient, and why apparently "dead" skin in tissue culture becomes "rejuvenated" when applied to a living recipient.

In clinical practice, grafts, until recently, were totally successful only if they were autografts (i.e., from other areas of the patient). With immunosuppressant drugs, homografts (i.e., from other members of the same species) could be used with some hope of success but until recently heterografts (i.e., from different species) could not be used except in the specific case of autoclaved bone which was used as a bridge for regrowth of lost bone or restoration of comminuted fractures. However, reports of an arterial graft made of tissue derived from cattle is perhaps the first successful application of heterografts in human medicine.

The graft is composed of selected bovine arteries, treated to render them nonantigenic so that they do not stimulate the production of antibodies when implanted in the human body. Chemists and microbiologists have succeeded in removing the immunologically active proteins from the arteries through chemical modification. The result is a nonantigenic artery that retains its natural structure and strength, but does not excite a rejection reaction by the recipient.

Clinical testing of the arterial heterograft has been carried out by implantation in the peripheral arterial system of nearly 300 patients with various types of obstructive arterial disease. Medical observations have extended to more than five years after the operation in some cases. At this time, however, the principal method of combating the rejection phenomenon is through so-called immunosuppressive drugs. The drugs commonly used include azothioprine, methotrexate, cyclophosphamide, and actinomycin C. A newer approach, involving immunology reactions rather than chemical suppression of lymphocyte activity, is offered by antilymphocyte globulin (ALG) derived from hyperimmunized horse serum. This material is still not cleared for human use, but is being extensively tested in laboratories.

The principal difficulty with all immunosuppressive drugs is that they not only keep the recipient from rejecting a tissue graft but that they also keep the recipient from rejecting invading bacteria, fungi, viruses, and other pathogens. Thus the individual on immunosuppressive drugs is much more susceptible to disease and has a measurable increase in susceptibility to cancer (particularly reticulosarcoma). Such great care must be taken of the patient that immunosuppressive therapy and transplant surgery are not going to be a normal veterinary procedure until simpler techniques are developed.

BLOOD TYPES

At one time it was thought that the only significant blood types occurred in humans, but with the expansion of research and increased delicacy of tests it is now known that blood types occur in the other animals and in some species have both commercial and medical importance.

Human erythrocytes at one time or another in their lives have surface antigens which are agglutinogenic and are responsible for the Landsteiner blood groups A, B, AB, and O as well as the Rh-positive and Rh-negative reactions. These are the routinely useful groupings of human blood typing, although a number of other blood derived antigens and antibodies are known to exist and are occasionally valuable to medicine. The agglutinins are specific plasma antibodies produced in response to the presence of a particular antigen. They are more active in man than in any of the domestic mammals.

The blood antigens are associated with one or more specific genes that may be inherited singly or as a group. This is helpful, particularly in cattle breeding, to identify parentage and determine whether twins of the same sex were produced from one or two ova.

The presence of incompatible antibodies or antigen in transfused blood can be determined by cross-matching, i.e., infusing the serum of the donor blood with cells of the recipient, and vice versa. The technique is not ordinarily necessary in domestic animals other than in purebred dogs and highly linebred or inbred animals, such as Thoroughbreds or Standardbreds. A single transfusion in domestic animals (but not in man) can be given with relative safety since natural antibodies are in low concentration. However, if several transfusions are contemplated, cross-matching should be done to preclude the possibility of the recipient developing antibodies against the donor's cells. If an animal has a history of isosensitization, great care should be taken to match the recipient's blood with the donor as death can result from transfusion of incompatible blood.

A word of explanation is indicated at this point. Blood identification has three basic categories of classification called "types," "systems," and "factors." A blood type is determined by a study of systems and factors. A system or "blood group" is designated by letter, as in the human A, B, AB, and O. Factors are generally distinguished by other means such as abbreviations, numbers, words, or a letter preceded or followed by the word "factor." In routine human blood identification the words "blood type" are used when "systems" are meant. This can lead to considerable confusion, but if one remembers that the systems are letter designated and biologically similar to human Landsteiner groups, and respond to similarly derived antigens, some of the problem will be clarified. Factors are subsidiary groups within the systems. Several of these may occur in a given antigen and can be detected by appropriate reagents.

The systems are determined by a single pair of allelic genes, one inherited from each parent. With few exceptions the alleles controlling a given system act as codominants. Since the systems and factors of an offspring must be found in one or both parents, the presence of a different system or factor than can be found in the parents indicates illegitimacy. In other words, we can tell from a study of blood types if an individual is not the parent of an offspring (but not if it is), and if there are sufficient numbers of subsidiary factors as there are in cattle, one can deduce with considerable accuracy whether a given bull is the sire of a given calf and whether sire's or dam's genetic characteristics are predominant in the calf.

In cattle particularly (and more rarely in sheep and humans) permanent modification of blood types can occur in dizygotic (fraternal) twins which are enclosed in a common chorioallantoic sac, and have a common chorioallantoic vascular system. In cattle, females born twin with males and share a chorioallantois are called freemartins (p. 440). Individuals with such modifications of blood genetics can properly be called "chimeric" or "mosaic" twins. This subject is far from completely investigated.

In blood typing, particularly for human transfusions, and in transfusions under certain conditions in horses and dogs, it is essential to distinguish between naturally occurring and immune (or induced) isoantibodies. Some of the reagents used in cattle, sheep, and horse tests are naturally occurring isoantibodies and resemble anti-A and anti-B sera in man, but the majority of the reagents in domestic animals are produced by deliberate isoimmunizing of donor animals. Some of these sera for cattle typing are now commercially available, but most are not.

Both agglutination and lytic techniques are used in typing. In humans, for instance, A, B, AB, and O types are agglutinative. In cattle and sheep virtually all techniques are lytic since cells of these species do not agglutinate even when sensitized by several doses of blood typing antibodies. Dog and chicken tests, on the other hand, are entirely agglutinative. Horses' blood reacts to both lytic and agglutination tests, depending on which factor is being assayed, so both are used; the choice must be determined by trial.

In dogs, at least six blood types are known, and are classified as A, A', B, C, D, and E. Of these only A seems to be of any importance in routine transfusions, although the other types have some important effects in certain types of transplantation experiments and in tissue rejection phenomena.

In cattle, there are more than 60 factors which are included in 10 types (systems), i.e., A, B, C, F-V, J, L, M, S-U, Z, and Z'. The exact reason for these peculiar letter nomenclatures is not readily apparent, and undoubtedly they will be systematized at some future date. Most of the cattle blood factors are found in the B system which involves over 200 alleles. The L, Z, and A systems are far less complicated: Each consists of but one blood factor and two alleles. The total recognizable blood variations in cattle, that are capable of being distinguished with present methods, number well over 100 billion. Blood typing is used in cattle as an analysis of parentage and has proved to be a valuable tool in determining an animal's ancestry, and for improving heritable traits or removing heritable defects.

In sheep, seven systems, A, B, C, D, M, R-O, and X-Z are known to exist. Systems B, C, and R-O are comparable to B, C, and J in cattle. Sheep blood analyses are being performed for the same purpose as in cattle.

In horses there are at least 19 blood factors, but classification of these into systems has only just commenced. The usefulness of these to determine parentage and cellular inheritance has not yet been established.

In swine there are probably 16 blood factors. Some of these have been systematized into three systems named A-O, E, and K, but the majority are still unclassified.

Five systems, A, B, C, D, and E, are found in chickens. The systems A and B are of interest not only because they are quite complex, but also because they are useful in identifying specific hybrid strains of poultry.

Blood studies are also recorded for cats, but while there are several known blood factors, there have been virtually no published studies of feline blood types.

Erythrolytic Syndromes

In certain individuals a peculiar syndrome involving a form of immunological destruction of erythrocytes may appear. Depending on the animal involved and on the triggering mechanism the disease is called the "Rh syndrome," "neonatal erythrolysis," and "isoimmune erythrolysis."

The "Rh factor" (*Rh* from the *rh*esus monkey, where the reaction was first noted) in human beings can result in a condition called isoimmune isoerythrolysis, which is a complex term used to describe a relatively simple antigen-antibody system. There are two Rh antigens, Rh^+ and Rh^-, either of which can be found in human blood (but not both at the same time). The Rh^+ is dominant; the Rh^- is recessive. Therefore, if an Rh^+ and an Rh^- are parents, the children will all be Rh^+. (Naturally, two Rh^- parents will produce Rh^- children, and two Rh^+ parents will produce Rh^+ children. However these need not be considered in the "Rh disease" syndrome.)

If the mother is Rh^+ and the father Rh^-, there is no problem insofar as "Rh disease" is concerned. The problem arises when the father is Rh^+ and the mother is Rh^-. In this situation, rarely with the first child, but often with the second and third child, sufficient interchange of fetal and maternal blood will have taken place to stimulate an Rh^+ immune body production in the mother. These Rh^+ antibodies attack the red blood cells of the fetus and destroy them, producing a fatal anemia. The anemia may kill the offspring prior to birth or shortly thereafter. Treatment once involved exsanguination and massive transfusion to remove the deadly antibodies from the infant. Today immuno-suppressive drugs have been developed to counteract the development of antibodies by the mother.

In horses and other animals a similar syndrome triggered by antibodies in the mother's milk has been recognized. The syndrome is called neonatal isoerythrolysis to distinguish it from the human disease.

In mares with a history of neonatal erythrolysis in their foals and in dogs with type A' blood sensitized by A donors, fatal reactions are possible. Compatible blood in horses and A' blood in dogs should be the only sort used if transfusions are needed. Previous sensitization in cattle can also lead to unfavorable reactions. This discovery extends the syndrome to another of a growing list of animals which includes humans, horses, pigs, rabbits, and dogs. It is probable that under experimental conditions the syndrome can be developed in a large number of animals which show no indication of the disease under natural conditions.

Cattle apparently can develop the syndrome under natural conditions, or from biological products containing erythrocytes, or from transfusion. Speculation exists that trauma from manual removal of placentas and caesarean section could also produce the condition in a subsequent pregnancy. In 1970 natural cases of neonatal isoerythrolysis were first reported in calves, and more reports have appeared since.

THE HEART

The blood vascular system begins and ends in the heart. This structure in mammals is a muscular, hollow organ divided into four compartments (chambers) by valves. The heart lies on the floor of the thoracic cavity in the middle mediastinal space. It is surrounded by a three-layered fibrous and serous sac, the pericardium, and is held in place by the great vessels at its base. The apex (pointed end) of the heart lies free within the pericardial sac and is oriented ventrally and caudally. The pericardial sac is attached to the sternal portion of the diaphragm by the pericardial ligament, and to the great vessels at the base of the heart. The sac is filled with tissue fluid which continuously bathes the heart.

Structure of the Heart

The heart is composed of a base and apex, two auricles and two ventricles, and a system of valves located and constructed to assist the flow of blood through the heart. These valves are the right and left atrioventricular valves and the pulmonary and aortic valves. The atrioventricular valves are located between the atria and ventricles and regulate the passage of blood from auricles to ventricles. The right atrioventricular valve has three cusps (valves), the left or mitral valve has two. The aortic valves and pulmonary valves have three cusps and are located at the origin of the pulmonary artery and aorta. They prevent backflow of blood from these vessels into the heart. The valves consist of sheets of connective tissue covered by epithelium. There are no valves in the large veins that enter the heart. The heart is regulated in its dilation by transverse muscular fibers known as moderator bands which pass across the ventricles. The heart is divided into four chambers by the interatrial septum which separates the two atria or auricles, the atrioventricular septa and the atrioventricular valves which separate the auricles from the ventricles, and the interventricular septum which separates the two ventricles. The right ventricle of the adult heart is larger and thinner walled than the left. The right side of the heart pumps venous blood. The left side pumps arterial blood. The heart is provided with its own circulation composed of the coronary arteries which branch from the aorta, and the coronary veins which empty into the right atrium.

The three tissue layers of the heart consist of the epicardium, myocardium, and endocardium. The epicardium is composed of squamous epithelial cells that lie upon a region of areolar tissue. It serves as the outer covering of the heart.

The myocardium (heart muscle) forms the bulk of the heart and is composed of two systems of branching muscle fibers. One system forms the musculature of the atria, the other the ventricles. The muscle fibers were once thought to form a true syncytium (a single mass) but electron microscope studies have shown them to be separate cytoplasmic units each with its own nucleus. The muscle, however, functions as a unit and obeys the "all or none law" of heart contraction, i.e., in response to a given stimulus the heart contracts totally or does not contract at all. Cardiac muscle fibers do not possess prominent sheaths and are separated by areolar tissue which forms a loose network between the branchings of the cardiac muscle. The muscles of the atria and ventricles are so arranged that when they contract they reduce the volume of the heart chambers which they enclose. These muscle systems are separated by connective tissue, but are functionally connected by Purkinje fibers which are muscle fibers adapted to conduct impulses. The connective tissue that separates the atria and the ventricles is a part of the central supporting structure of the heart to which most of the muscle fibers and the valves are attached. In most animals this cardiac skeleton is a dense, white, fibrous, connective tissue. In cattle, however, it contains bone (the os cordis), and in other animals it may contain cartilage.

The endocardium is the inner lining of the heart and is composed of squamous epithelium which is continuous with the endothelium of the vascular system. It is attached to the myocardium by a layer of elastic connective tissue.

The heart is essentially a self-contained organ which can function without the direct intervention of the voluntary or involuntary nervous system. This can be demonstrated by the fact that isolated hearts will continue to beat for a considerable period of time if they are kept in an appropriate environment. The intrinsic "nervous" system of the heart consists of a "pacemaker," the sino-atrial (S-A) node, and a number of modified cardiac muscle fibers (Purkinje fibers) which carry the contraction impulses in the ventricles.

The heartbeat originates at the S-A node, which is located in the right atrium between the entrance of the cranial vena cava and the coronary sinus. From the S-A node a wave of contraction passes outward through the atrial muscles. The speed of the contraction wave is very slow, about one meter per second. The atrioventricular (A-V) node in the interatrial septum then picks up the contraction impulse and conducts it through the A-V bundle of His to the Purkinje fibers which carry the impulse to the ventricular muscles. Since the Purkinje fibers transmit by contraction, the speed of the impulse (3–7 meters/second) is slower than it would be in a motor nerve, but is considerably faster than in the atria. Although not readily visible to the eye, the contractions of the heart are serial rather than simultaneous. The right atrium contracts first followed by the left, and subsequently by the virtually simultaneous contractions of the ventricles. The interval between the atrial and ventricular contractions can be observed in isolated hearts that have a slow heartbeat rate. Functionally this serial contraction of the heart muscle permits more efficient blood passage from the atria to the ventricles and from the ventricles to the vascular system.

The Cardiac Cycle

The cardiac cycle (a complete heartbeat) consists of two major phases, i.e., systole, the period of active contraction, and diastole, the period of relaxation and dilation (Figure 9-2). The cycle is considered to begin at

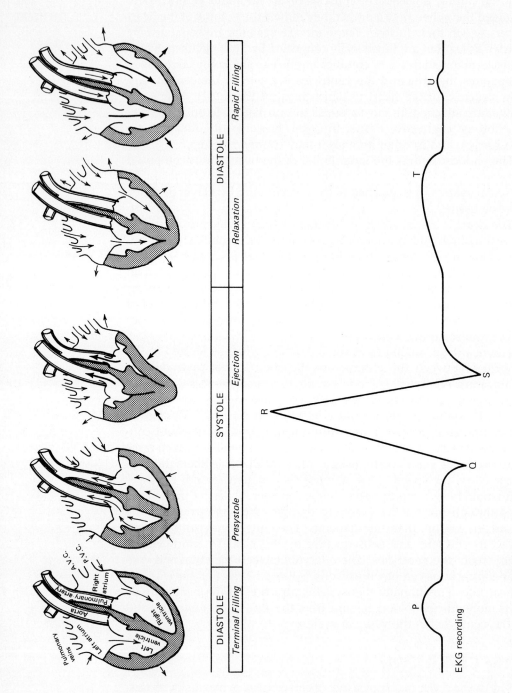

FIGURE 9-2 Phases of the cardiac cycle.

diastole. The pressure decreases in the ventricles as they begin to relax. The semilunar valves in the pulmonary arteries and aorta are forced shut by the higher pressure within the arteries. Ventricular pressure falls below the pressure in the atria and the A-V valves open allowing blood to enter the ventricles. The contraction impulse, acting upon the atria first, completely fills the ventricles. The ventricles then contract, building up pressure which closes the A-V valves and opens the semilunar valves in the pulmonary artery and aorta by exceeding the residual arterial pressure. By this time, diastole has already begun in the atria which fill with blood from the venae cavae and pulmonary veins. Diastole of the ventricles then begins, completing the heart cycle.

Heart Sounds

When an animal is examined with a stethoscope, two distinct sounds are heard in each cardiac cycle, a "lub," a low-pitched initial sound caused by the vibrations of contracting muscle fibers and the closing of the atrioventricular valves during systole, and a "dup," a second sound sharper and shorter and is caused by the closing of the aortic and pulmonary valves during late systole or early diastole. These two are called the audible heart sounds. There are two others which are called the inaudible sounds, or the third and fourth sounds. The third sound is produced early in diastole and comes from the opening of the atrioventricular valves and the rush of blood into the dilating ventricle. The vibrations are ordinarily below the level of audibility of the unaided ear, but can be detected with instruments. The fourth sound occurs after the P-wave and results from the terminal contractions of the atria to fill the ventricles with blood. The fourth sound can occasionally be detected by the unaided ear.

Regulation of Heart Rate

In spite of the fundamental automaticity of the heart, its rate is constantly regulated by the nervous system and by certain chemicals. This regulation exists to ensure proper coordination between the rate of heartbeat (and volume output) and the blood requirements in different parts of the body under different conditions. The nerves that modify the heartbeat rate are motor nerves from two different branches of the involuntary nervous system. They are defined as accelerator fibers of sympathetic origin which speed the heart, and inhibitory fibers of parasympathetic origin which slow the heart. The inhibitory fibers exert a continuous slight action upon the S-A node under normal conditions and, in consequence, severing the parasympathetic connections will result in acceleration of the heart.

The heart also possesses sensory nerves which are connected to the voluntary nervous system. These originate in two main areas: the wall of the right atrium and the wall of the aorta.

The sensory and motor fibers that supply the heart unite in the cardiac center which is located in the medulla oblongata of the brain. This center is influenced directly by changes in the chemical composition of the blood. Sensory stimulation and blood changes may result in stimulation of the cardiac center with subsequent effects on the heart.

There are at least three reflexes that exert an influence upon the heart rate. One of these is a general stimulation reflex that can be attributed to shock, stress, emotional states, or sudden insult to nerve plexuses such as the solar plexus. The other two are specific reflexes, triggered by blood pressure within the atrium or in the aortic arch. Sensory nerves in the right atrium stimulated by blood pressure result in a depressor reflex (Marey's Reflex) which slows the heart. Sensory nerves in the aortic arch stimulated by high blood pressure produce an accelerator reflex (Bainbridge's Reflex). These reflexes are essentially mechanisms for maintaining constant blood pressure.

The normal heart rate for a species is an average which is determined by numerous recordings of heartbeats per minute obtained from mature normal animals in a state of rest. This rate will vary considerably between individuals and, as a result, a normal animal will probably not have a "normal" heartbeat rate. The heart rate varies with age (young animals have faster rates than old), physical size (small animals have faster heart rates than large), size of the heart in proportion to the body (small hearts beat faster than large), amount of body fat (fat animals have faster rates), peripheral resistance, and sex. The pulse rate of the female is generally faster than that of the male of the same species.

Variations in the pulse rate may result from a number of causes. The most marked variations are those produced by temporary changes such as excitement or exercise. These result in accompanying variations in the minute volume of blood (the product of stroke volume times heartbeat rate). Minute volume can be effectively determined by Fick's method wherein simultaneous samples of arterial and venous blood are withdrawn from congruent vessels (e.g., jugular vein—carotid artery; cephalic vein—radial artery, etc.) The bloods are analyzed for oxygen content by the ferricyanide method, the results are expressed in volume percent (ml. O_2/100 ml. blood). The difference represents the milliliter percent of oxygen removed by the tissues. Since the brain has a high level of oxygen consumption it is advisable to draw one sample from the neck vessels and another from the periphery of the body (e.g., femoral vessels, brachial vessels) and average the data. The total oxygen consumption per minute now must be measured. This can be done by open circuit (Haldane's method) or closed circuit (Regnault and Reiset's method). With these data the minute volume can be calculated by simple ratio and proportion.

At work, skeletal muscles consume approximately 13 times as much oxygen as they do at rest. This necessitates an increased flow of blood to carry oxygen and remove waste products. With muscular exercise, the increased carbon dioxide produced raises the acidity of the blood which stimulates the cardiac center to increase the heart rate. The heart may also be stimulated by certain hormones or drugs. The commonest of these is adrenalin which exerts a direct stimulating action on the sino-atrial node. Thyroxin also has a stimulating effect upon the heart. This can be either direct or indirect. The indirect means is through the effect of thyroxin upon the adrenal gland. Sodium, potassium, and calcium ions also affect the heartbeat by varying the rate, force and amplitude of the heart muscle contractions.

Intrinsic Regulation of the Heart

There are also increases and decreases in stroke volume which are completely independent of outside regulatory effects. This has been shown through studies of "heart-lung" preparations where the two organs are completely removed from the body and kept alive (for a time) in an artificial environment. The heart is thus completely divorced from neural control and any known hormonal control. The blood values are held static, which removes the effect of mineral variations. Under these conditions, increased venous return results in increased amplitude and force of heart contractions which automatically keep inflow and outflow in a state of balance. This is important in the intact animal because if an imbalance between inflow and outflow exists for more than a brief period, circulatory stasis on the venous side of the systemic vascular tree and impaired pulmonary circulation will result.

The greater amplitude and force of "stretched" heart muscle contractions (the "stretch" is caused by the over dilation of the right heart by increased venous return) can be explained by the increase in size and surface area of the stretched heart muscle. Since the strength, and hence the energy release, of a muscle is proportional to the square of its linear dimension, the increased surface area would lead to a stronger contraction. This agrees with Starling's law of the heart which states, "the energy liberated by heart muscle fibers on contraction varies directly with the length of the fibers at diastole."

In an intact animal increased venous return is usually accompanied by increased stroke rate as well as force and amplitude, but the rate is influenced by neural and/or hormonal influences and not by "stretching" per se.

The heart compensates for increased load by three mechanisms: increased rate, increased amplitude and increase in size (hypertrophy). These three constitute the basis for the so-called "cardiac reserve." Cardiac reserve is the extent to which a heart can adjust to meet the demands placed upon it and still maintain equilibrium between blood flow into the right atrium and out of the left ventricle. If the reserve is inadequate, blood will accumulate in the pulmonary circulation, in the venous side of the systemic circulation and in the portal circulation. The condition is regularly seen in cases of congestive heart disease and "right heart failure." These are essentially decompensation phenomena caused by exhaustion of the cardiac reserve.

Electrocardiography

The electrocardiograph (EKG) is an instrument for recording the events of the cardiac cycle. The commonest of these instruments is the scalar electrocardiograph, which records the cardiac events as tracings upon a roll of paper. Recordings are called electrocardiograms.

To record the events of the cycle in man, electrodes are strapped to the surface of the skin. This technique cannot be successfully used in many animals due to the thickness of skin and hair coat and looseness of the underlying skin attachments; therefore, other means of attaching electrodes such as tissue clips or needles must be used. Since animals generally resent this sort of treatment, the EKG is not a usual technique in veterinary practice.

Since the location of the various electrodes (leads) influences the appearance and the sensitivity of the heart recording, a standard system has been adopted for human use and adapted with variable success for animals. Essentially the positionings of the electrodes form the points of a roughly equilateral triangle with the heart in the center. A number of variations of this procedure exist and cannot properly be described in the restricted boundaries of this text.

The tracings give information about the type and character of the events of the cardiac cycle (see Figure 9-2), heartbeat rate and rhythm, and show deviations which are normal occurrences in various species. For instance, resting sinus arrhythmia, the notching of the P-wave in horses, and the "dropped beats" (sinus arrest) in resting dogs and horses are recorded. For diagnosis of cardiac malfunction or disease, an EKG is a vital tool in human medicine. It is somewhat less useful in domestic animals.

Fibrillation

This pathologic condition does not properly belong in normal heart physiology, but it occurs so often in physiological experiments as well as in cardiac disease that it should be mentioned. Fibrillation occurs in two forms, atrial fibrillation, which appears as an unregulated rapid twitching of the atrial muscles accompanied by somewhat less rapid arrhythmic contractions of the ventricles, and ventricular fibrillation, which is characterized by incoordinate twitchings of the ventricles and failure of the ventricles to eject blood into the arteries.

Atrial fibrillation is apparently caused by defects in excitation of the atrial beat by the S-A node. The atrial contractions usually are very rapid and can be as many as several hundred per minute. However, the A-V node is not capable of responding that rapidly, and, in consequence, only one out of several atrial contraction impulses are passed to the ventricles at irregular intervals. The result is a twitching organ which is incapable of properly pumping blood. The EKG picture of atrial fibrillation typically has an absence of P-waves, which are replaced with an irregular wavy line, and an irregular QRS complex. Ventricular fibrillation most often occurs as a result of acute systemic shocks such as lightning stroke, toxemia, overdosage with chloroform, or thromboembolic disease.

Treatment for both types of fibrillation involves stopping the heart action with electric shock and then re-starting the heart, it is hoped in a normal rhythm. Ordinarily, cases of ventricular fibrillation will be dead long before such treatment can be applied.

Functions of the Pericardium

The pericardium functions to protect the heart against excessive dilation; to hold the heart in a relatively fixed position inside the thoracic cavity; to provide a sac to hold liquid (pericardial fluid), which acts as a homeostatic device and as a lubricant for the external surface of the heart; and to help the heart relax during diastole by exerting a small amount of negative pressure on the heart surface. The pericardium is not vital to heart function, since it can be removed surgically and the heart will continue to work.

In some disease conditions, excessive amounts of pericardial fluid can be secreted. This can interfere with heart function, primarily during diastole. Both ventricular output and peripheral blood pressure will decrease appreciably in such a situation.

Cardiac Circulation

Blood circulates through the heart in the following manner: venous blood enters the heart at the right auricle, passes through the right atrioventricular valve, and is pumped by the right ventricle through the pulmonary artery and into the lungs. The blood is oxygenated in the lungs and is returned to the heart by several pulmonary veins which deliver it to the left auricle. From the left auricle the blood passes through the left atrioventricular valve (or mitral valve) to the left ventricle which forces blood into the aorta.

THE VASCULAR TREE

At this point it should be mentioned that, while superficially similar, the vascular systems of mammals differ markedly between species and to a lesser extent between members of the same species. All vascular systems of mammals are built upon the same general plan. There is a heart, a lymphatic system, and a blood vascular system consisting of arteries and veins and composed of pulmonary, systemic, and portal circulations. Beyond that, there is only a strong similarity in the course of the vessels.

Analogous structures occur in different species, but details that apply to one species may not apply to another. Therefore, the following discussion of the vascular tree is deliberately general and related to the major vessels. For more detailed study of the vascular system, a number of excellent and detailed anatomy texts exist, some of which are listed in the bibliography.

THE ARTERIAL SYSTEM

The arterial system begins with the aorta, which is the largest artery in the body. It leaves the heart as the thoracic aorta and curves upward and backward to lie just below the thoracic vertebrae. The thoracic aorta passes through the diaphragm at the hiatus aorticus and is continued caudally as the abdominal aorta.

The Thoracic Aorta

The thoracic aorta gives off the following important branches: the coronary arteries, which supply blood to the heart, and the innominate artery (brachiocephalic trunk), which is the main trunk to the head, neck, and shoulder regions, the intercostal arteries, and the diaphragmatic (phrenic) artery. In some species the left subclavian (brachial) artery arises directly from the thoracic aorta immediately after the innominate artery is given off.

The brachiocephalic artery gives off the following major branches: the dorsal artery, which supplies the first three ribs and the deep shoulder region; the left subclavian, which supplies the left foreleg and shoulder region; and the common carotid, which gives off the right subclavian, which supplies the right foreleg. The common carotid then branches into the right and left carotid arteries, which supply the neck, face, and head; the vertebral artery, which supplies the neck and cervical vertebrae; and the deep cervical artery, which supplies the deep muscles of the neck.

The thoracic aorta, after giving off the brachiocephalic artery, curves dorsally and backward to lie under the thoracic vertebrae. It gives off a number of paired intercostal arteries (the number varies with the species and the number of ribs) and the diaphragmatic artery, which supplies the diaphragm. The thoracic aorta then passes through the diaphragm at the hiatus aorticus, and becomes the abdominal aorta.

The Abdominal Aorta

The abdominal aorta is the direct continuation of the thoracic aorta. It gives off the following important branches: the coeliac artery, which supplies the stomach, liver, spleen, and pancreas; the cranial mesenteric artery, which supplies the small intestine; the two (right and left) renal arteries,* which supply the kidneys; the caudal mesenteric artery, which supplies the cecum, colon, and part of the small intestine; two internal spermatic (or utero-ovarian) arteries, which in the male go to the testes and in the female to the uterus and ovaries; and a number of paired lumbar arteries, which go to the lumbar vertebrae.

The abdominal aorta ends at the pelvic region where it branches into five vessels: the internal iliac arteries (right and left), which supply the thigh muscles and genitalia; the external iliac arteries (right and left), which supply the hind legs and external parts of the abdomen and mammary glands in some animals; and the caudal artery, which supplies the tail. The internal iliac and caudal arteries have several branches of minor importance.

The external iliac artery gives off the following major branches: the femoral artery, which goes to the hind legs, and the prepubic artery, which supplies the abdomen and mammae (in those mammals with pelvic mammary glands). The femoral artery subdivides into several vessels which supply the lower portions of the hindlimb.

Structure of Arteries

Arteries are thick-walled with a small lumen. They contain considerable amounts of elastic tissue, maintain their shape without collapsing when empty, have no valves, and are deeply located.

The walls of arteries are composed of three coats or tissue layers: the tunica intima (inner coat), the tunica media (middle coat), and the tunica externa (adventitia or external coat).

The tunica intima consists of a layer of endothelial cells lying upon a layer of elastic tissue. It forms a relatively impervious lining that retains fluid and formed elements of the blood.

The tunica media consists of circularly arranged smooth muscle cells interspersed with elastic connective tissue. In the largest arteries the middle coat is composed almost entirely of elastic tissue. This decreases in amount as the arteries decrease in size until, in the smaller arterioles, elastic tissue is virtually absent.

The tunica externa is composed of loose collagenous and elastic connective tissue fibers that also contain some longitudinally arranged

---

*In dogs and cats, and probably in other animals, a paired phrenicoabdominal artery is found between the cranial mesenteric and the renal arteries. These vessels supply the diaphragm, the adrenals, muscles of the abdominal wall, and organs in the cranial portion of the abdominal cavity.

bundles of smooth muscle cells. This is essentially a protective layer which is strong and tough, and minimizes damage from cutting, tearing, and undue expansion.

Because of their structure, arteries are distensible and elastic, which allows them to accommodate their size to the increased pressure and volume of blood forced into them by each contraction of the heart. Their elastic qualities aid in maintaining blood pressure and act as an auxiliary blood pump by squeezing the blood onward as they return to their original diameter in time to receive a new volume of blood from the heart. This elastic structure, moreover, tends to smooth out and reduce the pulsation of arterial blood flow. By the time the arterial blood has reached the arterioles, most of the jetlike pulsation, so characteristic of the larger arteries, is lost.

Arteries hold their shape well and do not collapse when empty. Thus, a cut artery would remain open, except that the circular muscle near the cut contracts and the elastic fibers of the tunica media retract the artery a small distance inside its sheath, allowing a clot to plug the hole. In the larger arteries, however, the diameter of the lumen, the volume of blood, and force of the bloodstream are so great that the protective mechanisms ordinarily cannot function before the animal bleeds to death.

The largest arteries (pulmonary artery and aorta) can be quite impressive structures. In a large horse they may be about two inches in external diameter as they leave the heart. These large vessels decrease in size as they give off branches, until finally they blend indistinguishably into their terminal branches. The branches continue to divide and subdivide until they become the capillary bed. The smallest arteries are called arterioles (0.2 mm or smaller in diameter). The walls of the arterioles are composed almost entirely of smooth muscles, which by contraction, change the caliber of the lumen and thus regulate the flow of blood into the capillary network which the arterioles supply.

Capillaries

Capillaries are an anastomosing (interlocking) network of tiny vessels which are just large enough to permit the passage of red and white blood cells. They average about eight microns in diameter and ramify through most of the body tissues. The capillaries connect the smallest arteries (arterioles) with the smallest veins (venules).

The capillary walls consist of a single layer of endothelial cells that is continuous with the endothelium of the arteries, veins and heart. This continuous tissue is the structure that makes the vascular system a closed circulation. The capillary endothelium is quite permeable to gases and also permits the passage of a certain amount of fluid (and occasionally cellular) elements of the blood to and from the tissues. This movement of gas and fluid is responsible for nutrition of the cells; for transport of

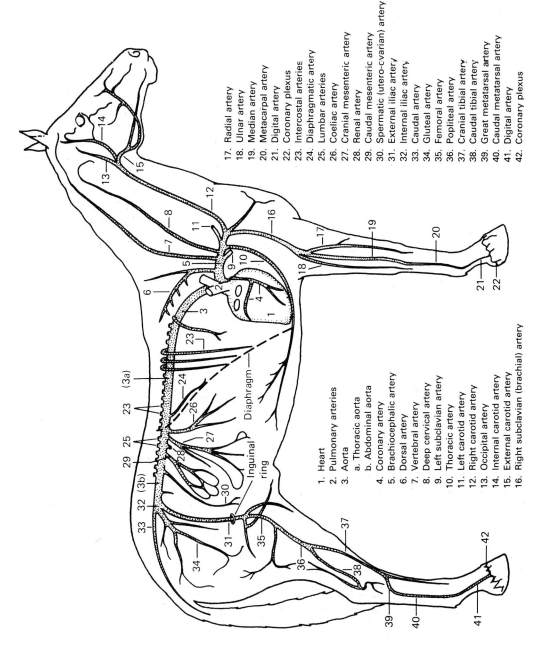

FIGURE 9-3 Schematic drawing of the arterial system of the horse.

cellular products and metabolic wastes; for transport of oxygen to maintain cellular reactions; and for the delivery of substances such as hormones and antibodies to places where they are needed.

The interlacing network of capillaries (the capillary bed) varies in shape and size in the various tissues of the body. Capillaries are not present in cartilage, hair, horn, hooves, or nails, the dense portions of the teeth and the cornea of the eye. All other tissues, however, possess capillary beds. In a one-square-millimeter cross section of skeletal muscle of the horse, there are some 1,000 muscle fibers and 3,000 capillaries. About 100 of these will be transporting blood at any given time; the remaining 2,900 will not be functioning. Blood transport shifts from one capillary to another in response to cellular need for oxygen or for removal of waste products. In dog or cat muscle there are about twice the number of capillaries per square millimeter as are in the horse. Tissues that require blood for purposes other than nutrition (secretion, absorption, excretion) have the most numerous capillaries and the most extensive capillary bed.

The remainder of the blood vascular system exists only to support the capillaries, for it is here that the functions of the blood are exercised, where the various reactions and interchanges between blood and cells that are essential for life take place. Although blood flow through individual sections of the capillary bed varies, it is probably never entirely cut off except in cases of disease or injury. Complete interruptions of capillary flow for any great length of time (the period varies with the tissues) can result in local necrosis and or loss of function.

The fluid pressure in the capillaries is variable and is regulated by neural and hormonal induced changes in caliber of the arterioles. Pressure is generally higher on the arterial side of the capillary bed than it is on the venous side. As the pressure decreases from arterial to venous side, the fluid elements of the blood are more slowly forced out of the interstices in the capillary walls and into the tissue spaces. Eventually the fluid pressure within the capillaries becomes less than the osmotic pressure in the tissue fluid, and a positive osmotic gradient is established in the direction of the capillary lumen. Fluid flow now proceeds into the capillaries. In general, outflow of fluid from the capillaries into the tissue spaces exceeds intake, and the excess interstitial fluid between the cells drains into the lymph system and is eventually returned to the bloodstream via the thoracic duct and the right lymphatic duct which empty into the cranial vena cava.

THE VENOUS SYSTEM

The vessels of the venous system are generally named the same as the arteries in the same area. There are exceptions to this rule. The principal exceptions in mammals are the saphenous vein (in some nomenclatures)

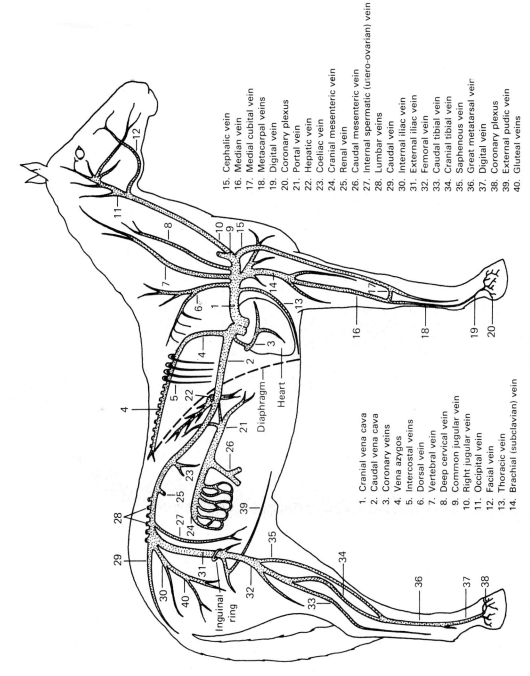

FIGURE 9-4 Schematic drawing of the venous system of the horse.

1. Cranial vena cava
2. Caudal vena cava
3. Coronary veins
4. Vena azygos
5. Intercostal veins
6. Dorsal vein
7. Vertebral vein
8. Deep cervical vein
9. Common jugular vein
10. Right jugular vein
11. Occipital vein
12. Facial vein
13. Thoracic vein
14. Brachial (subclavian) vein
15. Cephalic vein
16. Median vein
17. Medial cubital vein
18. Metacarpal veins
19. Digital vein
20. Coronary plexus
21. Portal vein
22. Hepatic vein
23. Coeliac vein
24. Cranial mesenteric vein
25. Renal vein
26. Caudal mesenteric vein
27. Internal spermatic (utero-ovarian) vein
28. Lumbar veins
29. Caudal vein
30. Internal iliac vein
31. External iliac vein
32. Femoral vein
33. Caudal tibial vein
34. Cranial tibial vein
35. Saphenous vein
36. Great metatarsal vein
37. Digital vein
38. Coronary plexus
39. External pudic vein
40. Gluteal veins

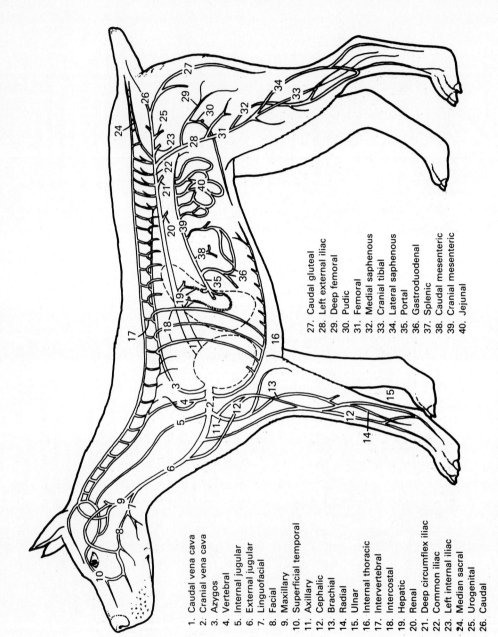

FIGURE 9-5 Schematic drawing of the venous system of the dog.

in the hind leg; the cephalic vein in the foreleg; the jugular vein in the neck; the azygos vein in the thorax; the portal vein in the abdomen, and the cranial and caudal venae cavae. Other nonhomonymous veins are found in various species. An example of such a vascular oddity is the subcutaneous abdominal vein ("milk vein") in cattle (Figure 15-5). Veins arise from capillaries and proceed toward the heart, becoming larger and fewer as they near this organ. Veins are thin-walled, collapse readily, possess valves, and in general are located more superficially than arteries. The large veins that enter the heart (i.e., the cranial and caudal venae cavae and the pulmonary veins) do not have valves.

Principal Venous Trunks

1. *Head, Neck, and Forelimb.* The jugular veins (right and left) drain the head and neck, and join together to form the common jugular vein. The common jugular then receives the cephalic vein which drains the forelimb. The subclavian veins (right and left) join the common jugular and form the cranial vena cava, which empties into the heart at the right atrium.

2. *Thorax.* The intercostal veins drain the ribs and the walls of the thorax, and empty into the caudal vena cava or into the azygos vein, both of which empty into the right atrium. The azygos vein is not present in some species.

3. *Abdomen and Hindlimb.* The caudal vena cava receives the hepatic vein from the liver, the renal veins from the kidneys; the spermatic (male) or utero-ovarian (female) veins from the gonads or uterus; the internal iliac veins from the thigh and pelvic region; the caudal vein from the tail; and the external iliac veins from the hindlimbs. The external iliac veins receive the femoral and the external pudic veins. The external pudics drain the mammary gland via the mammary veins. The caudal vena cava is the ultimate collecting vessel for all abdominal and caudal structures and empties into the right atrium of the heart. In cattle, the subcutaneous abdominal vein drains the cranial quarters of the udder and passes forward along the ventral part of the abdomen. It penetrates the abdominal musculature at an opening (the "milk well") that allows it to enter the caudal ventral portion of the thoracic cavity where it joins the thoracic vein and ultimately empties into the cranial vena cava. The hindlimb is drained by the femoral vein, which is continued directly as the external iliac vein once it passes through the inguinal canal. The femoral vein receives numerous branches among which are the external pudic vein, cranial and caudal tibial veins and the saphenous vein. This last vessel is a non-homonymous vein similar to the cephalic vein in the foreleg. It is a continuation of the great metatarsal vein and drains the anterior portion of the hock and thigh region before emptying into the femoral vein.

Structure of Veins

Veins return blood from the capillary beds to the heart. Their caliber and the thickness of their walls increase as they approach the heart, but their walls never attain the size or strength of their corresponding arteries. One or more veins of comparable capacity usually accompany each artery.

Although the aorta and the two venae cavae are approximately the same external diameter, the veins have a much greater capacity since the thinness of their walls results in a greater internal diameter. The walls of veins are also softer, and less elastic than those of arteries, but in most places consist of the same three coats that are present in arteries. Proportionately the tunica media is much less prominent, the elastic and muscle fibers are reduced, and the collagenous tissue fibers are relatively more numerous.

Veins, particularly in the periphery of the body, are supplied with valves that prevent backflow and keep the blood moving in the direction of the heart. These valves have only one cusp and do not completely close the lumen of the vessel. Their free edges point in the direction of blood flow. Valves are not present in the largest veins such as the pulmonary, portal, venae cavae, and jugulars, and are most numerous just beneath the skin and in the distal extremities of the limbs (but not in the digits). The number of valves is reduced as the veins approach the heart. In many of the larger vessels, valves occur where two veins join.

The Portal System

This portion of the circulatory system is venous only and arises in capillaries which are located in the stomach, spleen, pancreas, and intestinal tract. Veins pass from these organs toward the liver and coalesce into a single vessel (the hepatic portal vein). The portal vein enters the liver at the hilus and immediately branches into progressively smaller vessels that ultimately become sinusoids. The blood from the portal sinusoids is picked up by the sinusoids of the hepatic vein and is emptied into the caudal vena cava by that vessel. The hepatic portal system functions to transport nutrients and other substances from the digestive tract to the liver. Another portal system, the hypophyseal portal, occurs between the pituitary body and the hypothalamus (p. 346).

Blood Flow and Blood Pressure

Blood flows with considerable rapidity through the circulatory system of animals. Even for an animal as large as the horse, it requires only a little more than a minute to effect a complete circulation of blood. The rate of flow varies in the different parts of the circulation system. The velocity of the blood at any given point within the system is inversely proportional to the total cross-sectional diameter of all parts of the vascular system at that point. Thus the aorta of the horse, having a cross-sectional diameter of approximately two inches, will contain very rapidly flowing blood whereas the capillaries, whose total cross-sectional diameter is 800 times as great (or in the neighborhood of 1,600 inches) will have blood flowing through them very slowly (1/800 as fast). The velocity of the blood increases as it leaves the capillary bed and returns to the heart over the venous system, but in no case will the blood flow as fast within the veins as it will within comparable arteries.

Blood pressure has a direct relationship to the rate of flow. The pressure is highest in the aorta, and slowly diminishes through the arterial tree until the arterioles are reached. Here there occurs a sudden and marked pressure drop. The pressure of the capillaries is moderately low and decreases further in the veins, being highest in the venules and lowest in the venae cavae as they enter the heart. This gradually decreasing pressure is a physical phenomenon in which the energy of pressure is gradually dissipated as friction.

Summarizing: Blood, in order to perform its functions efficiently, must flow slowly and evenly through the capillary beds so that proper exchange of useful and waste products can take place. The initial pulsating flow through the arteries is gradually smoothed out by the elasticity of the arterial walls and the gradually increasing cross-sectional diameter of the total vascular tree as it leaves the heart. By the time the blood has reached the capillaries, these factors have reduced its pulsations and speed to a slow constant flow. In returning to the heart from the capillaries, blood moves slowly in the venules, more rapidly in the larger veins, but never as rapidly as in the corresponding arteries. The rate of flow in the venae cavae, which return blood to the right heart is only one-half that of the aorta. This is due principally to the fact that the cross section area of the veins entering the right heart is more than twice that of the arteries which leave it. The minute volume of blood entering and leaving the heart is identical; the only difference is the speed of flow.

Venous Flow

Since circulation in the veins can be a considerable distance from the direct action of the heart, a number of factors exist which tend to maintain venous blood flow. These are not necessary within the arteries since their circulation is accomplished by the direct pumping pressure of the heart. The factors that maintain venous circulation include some residual force from the heart, changes in pressure in the thorax and abdomen due to heart and respiratory movements, and contraction of skeletal muscles.

Arterial Flow

Arterial flow varies with the rate and amplitude of the heart contractions. The maximum average velocity in domestic animals is 350 millimeters per second; the minimum average is about 205 millimeters per second. This rapid flow is accompanied by a positive hydrostatic pressure within the vessels. The unqualified term "blood pressure" usually refers to intra-arterial pressure. It, of course, varies between contraction (systole) and relaxation (diastole) of the heart, thus giving a systolic and diastolic pressure. Systolic pressure is the highest pressure attained in the arteries; diastolic pressure is lowest, the difference between these two pressures is known as pulse pressure.

Function of Blood Pressure

Blood pressure is necessary to overcome gravity, to force blood through small vessels where resistance is great, to give a residual pressure to assist venous flow, and to provide sufficient pressure for adjustments of blood flow through various organs and tissues. At least five factors cooperate in the production and maintenance of blood pressure. These include force of the heartbeat, volume of blood, elasticity of the arterial walls, viscosity of the blood, and peripheral resistance. The only two that should require additional explanation are viscosity and peripheral resistance.

The relative viscosity of the blood is determined by the amount of plasma proteins as well as the blood cells themselves. Since viscosity is caused by friction between particles as they slide past one another in a moving stream, the colloidal particles of suspended protein in the plasma tend to make the blood more viscous. Any dilution of the blood by a less viscous substance, e.g., water, normal saline, and sugar solutions, does not help maintain blood pressure.

Peripheral resistance is affected partially by friction between the moving blood and the stationary vascular walls and partially by the diameter of the lumens of the vessels. Resistance is most noticeable in the region of the arterioles since these are able to constrict or dilate, thus increasing or decreasing peripheral resistance. Blood pressure varies directly with the degree of peripheral resistance. The arterioles to a great extent regulate blood pressure throughout the body by regulating peripheral resistance, and maintain pressure in the vascular system sufficiently high to insure adequate blood supply to all capillaries (particularly those of vital regions like the brain, heart, and kidney).

The arterioles also regulate blood distribution according to the demands of the body. Their function is controlled by a mechanism partly nervous, partly physical and partly chemical called the vasomotor mechanism. The nervous control of the vasometer mechanism functions through motor nerves from the vasomotor centers in the brain. These nerves are of two types: vasoconstrictor, which cause constriction of the muscles in the walls of the arterioles, and vasodilator, which cause a relaxation of the muscular wall. The vasomotor centers are located in the medulla oblongata of the brain and are influenced by stimulation from various sensory nerves and by chemical change of the blood which supplies the centers. The chemical control is similar to that which is exerted on the heart through the Marey's and Bainbridge's Reflexes. Chemical control can be affected by a number of circumstances; the chief of which is the amount of carbon dioxide in the circulating blood. Other chemicals are the hormone adrenalin, which causes a direct stimulation of the nerve endings of the arterioles, and the enzyme renin, which causes a stimulation of the heart. These three substances are typical of the variety that can enter into chemical control. Physical control is stimulated by intravascular pressure and acts through the nervous system. (See Marey's Reflex, Bainbridge's Reflex, p. 264).

THE LYMPHATIC SYSTEM

The lymphatic system begins in the tissue spaces between the capillaries of the blood vascular system. In the intestinal villi the capillaries are called lacteals. Interstitial fluid, the forerunner of lymph, is picked up and carried toward the heart by the lymph vessels, which are extremely thin-walled structures that collapse easily and possess a series of bicuspid valves along their length. The valves prevent backflow of lymph and keep the lymph moving toward the heart. This movement depends principally on external forces such as muscular contractions, gut movements, respiratory movements, and (minimally) suction pressures generated by the flow of blood past the openings of the terminal ducts.

Despite the great number and wide distribution of lymph vessels, lymph is a relatively small part of the total body fluid (two to three percent).

Lymph Nodes and Lymphoid Organs

Intercalated in the course of the lymph vessels are smooth, ovoid, pinkish bodies called lymph nodes, which are aggregations of lymphocyte-containing tissue, held together by a connective tissue interstitium or framework. The lymph nodes act as filters and phagocytic structures, removing foreign matter from the lymph stream before passing it on through other lymph vessels toward the heart. Several aggregations of lymph nodes may be passed through by the lymph before it enters the circulating blood at the anterior vena cava. Lymph nodes cluster at certain portions or regions of the body, and tend to check the spread of infection through the body. Infective matter is picked up and filtered by the regional nodes, and the lymph is clarified before it is sent to the

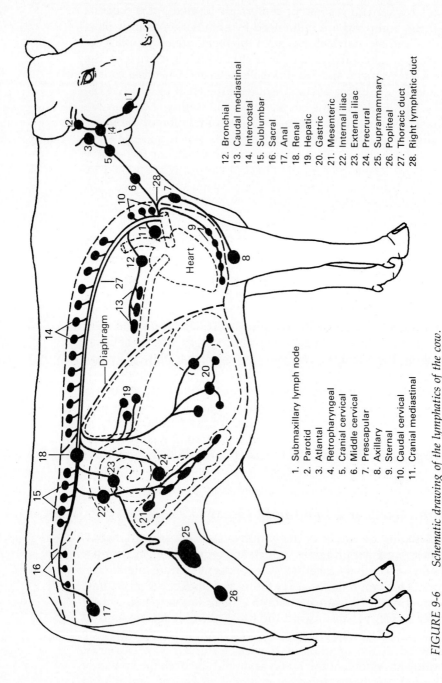

FIGURE 9-6   Schematic drawing of the lymphatics of the cow.

next series of nodes. If infection overcomes a lymph node, it tends to be a source of further infection. Lymph nodes are also thought to be the source of much of the antibody production of the body, and they also produce lymphocytes. Normal lymph is a clear to straw-colored fluid carrying lymphocytes in moderate numbers, but no other cellular constituents. The lymph systems of the caudal regions are passed through the diaphragm near the hiatus aorticus by a common structure, the cysterna chyli, which is continued toward the heart as the thoracic duct. The lymph systems of the head, neck, and forelimb usually unite to form the right lymphatic duct, which empties into the cranial vena cava along with the thoracic duct.

The term "lymphoid organ" has been used to label such diverse tissues as the spleen, thymus, bone marrow, tonsils, and Peyer's patches. These tissues do not conform structurally to the description of lymph nodes, and their primary functions tend to differ. The spleen is primarily an organ of storage, the thymus and bone marrow are lymphopoietic, and the tonsils and Peyer's patches function primarily as regional lymph nodes. Like lymph nodes, the lymphoid organs also have the capacity of producing lymphocytes.

Functions of the Lymphatic System

The lymphatic system has several major functions. It returns tissue fluid to the bloodstream from the extracellular spaces; filters out and phagocytoses bacteria or other foreign material which may have been picked up by the lymph; helps maintain homeostasis and tissue health; produces antibodies and lymphocytes; and absorbs and transports fat from the intestines to the bloodstream. The lymphatic circulation provides an avenue for drainage of some of the excess interstitial fluid and provides a mechanism for tissue fluid circulation, redistribution of body fluid and removal of foreign matter and waste products. Lymph mixes with the blood plasma upon its return to the bloodstream at the cranial vena cava, passes to the lungs and kidneys where waste products are removed, and subsequently returns to the tissues via the capillary bed. Blocking of lymph circulation can result in edema, or pooling of tissue fluids into a part of the body. Diseases that clog or poison the lymph nodes can result in severe edema. Simple standing on a hard floor without exercise can cause edema or "stocking" in the legs of a horse. Since edema interferes with tissue metabolism and can promote disease or degeneration lesions, it is always wise to reduce edema if possible. The ways are usually simple, such as elevating dependent parts and massage.

Interstitial Fluid

Since blood does not come into direct contact with tissue cells because it circulates in a closed system, tissue cells derive their nourishment from the interstitial (tissue) fluid which surrounds them. The fluid acts as a transfer system between the blood in the capillaries and the tissue cells. Interstitial fluid is the forerunner of lymph and is derived from the blood plasma elements which pass through the semipermeable capillary walls. Lymph and interstitial fluid are virtually identical, the only essential difference being that lymph is held within vessels while interstitial fluid lies between cells. Lymph and interstitial fluid are derived from, and are similar to, blood plasma, but have a lower protein content because of the filtering effect of the capillary endothelium which retains large protein molecules.

The movement of fluid and solutes between the blood vessels, interstitial regions and lymph vessels is vital to cellular nutrition and metabolism. This movement is accomplished in part by hydrostatic pressure and in part by an osmotic gradient. The hydrostatic pressure of the blood as it enters the capillary bed is positive (about 30 millimeters of mercury), while that of the interstitium is zero. The pressure differential forces fluids from the capillary bed into the tissue spaces. Large protein molecules, blood cells, and similar materials of high osmotic activity are left inside the capillaries. As the hydrostatic pressure of the blood drops, the osmotic effect of these retained substances increases until at the venule side of the capillary bed it has become about half as great as the hydrostatic pressure on the arteriole side. This osmotic pressure pulls tissue fluids back into the bloodstream together with dissolved waste products which are present. There is also a transference of excess interstitial fluid directly to the lymphatics. Between the blood vascular system and the lymphatic system the fluid equilibrium of the interstitium is maintained at a relatively constant level although circulation is continually going on and a constant transfer of nutrients and metabolic products is occurring.

The Spleen

The spleen is the largest of the lymphoid organs associated with the circulatory system. Because of its anatomic location, however, it is sometimes described along with the abdominal viscera although it is not actually a member of that group. The spleen varies in shape from triangular to oval depending on the species. It is somewhat purplish in color and lies entirely to the left of the median plane between the stomach and the left abdominal wall. It possesses a hilus, which contains blood and lymph vessels, and is held in place by the suspensory ligament and the gastrosplenic omentum. Its size varies greatly, depending chiefly upon the amount of blood that it contains. In the natural state the spleen is soft but not friable like the liver. It is not essential to life.

Structure and Function of the Spleen

The spleen is enclosed in a tough connective tissue capsule which extends into the body of the organ as trabeculae. The trabeculae form a fibrous network visible on the surface of the spleen and extend throughout the organ as the stroma, which supports the softer tissues or pulp. A considerable amount of smooth muscle fibers is present. The splenic pulp is composed of red and white pulp which surrounds the blood vessels that ramify through the spleen. The white pulp is essentially lymphoid tissue, and the red pulp forms the region of blood storage.

The spleen functions principally as a blood reservoir and contains the reserve supply of the body's blood which is needed in times of stress. It also functions as a lymphatic organ, as a blood-forming organ in young animals, as a place for destruction of senile red blood cells and storage of iron, and as a site for antibody production.

Fetal Circulation

In considering the mammalian fetal circulation, a new point of departure must be taken. The fetus is a parasitic growth which derives its protection and sustenance from its mother. It returns most of its excretory products to the mother's system. Therefore, in studying fetal circulation of mammals, one should start with the placenta, and proceed from there to the circulation within the fetal body.

At the placental attachment to the uterus, the fetal blood receives oxygen and nutrient matter from the mother and disposes of its accumulated metabolic wastes. The freshly oxygenated blood is returned to the fetus over the umbilical vein, which enters the fetal body via the navel (umbilicus). In the body the umbilical vein passes into the liver, where the nutrients in the blood are partly extracted and stored. At this point the "arterial" blood becomes mixed with "venous" blood from the portal system, and is then passed from the liver via the hepatic vein to the caudal vena cava, which empties into the right atrium of the heart. As the oxygenated blood enters the caudal vena cava, it becomes further mixed with the venous blood carried from the caudal parts of the body. In the fetal livers of some species, notably the bovine, a structure called the ductus venosus exists, which allows some of the blood from the umbilical vein to bypass the liver.

The right atrium of the heart receives venous blood from the cranial vena cava, and mixed venous and arterial blood from the caudal vena cava. This blood, instead of being passed directly to the right ventricle, may go in two directions, either to the right ventricle or to the left ventricle via the left atrium. This occurs because the interatrial septum is incomplete, and is perforated by a hole called the foramen ovale. Therefore, both the right and left heart receive the mixed venous-arterial blood sent to the right auricle.

The right ventricle pumps blood out through the pulmonary artery to the collapsed and nonfunctional fetal lungs, or over another bypass, the ductus arteriosus, which connects the pulmonary artery to the aorta at a point somewhat distal to the branching of the innominate artery.

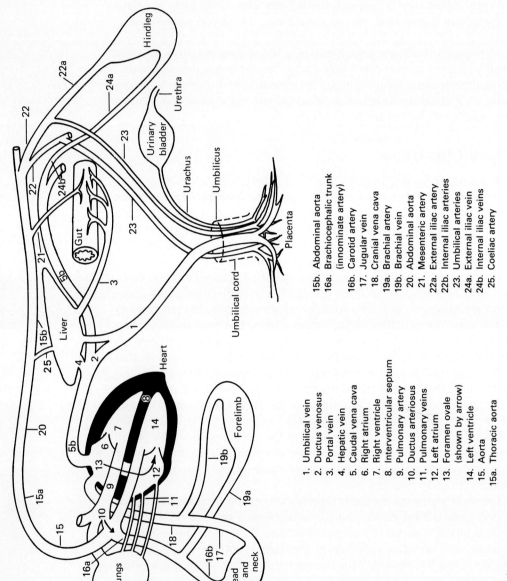

FIGURE 9-7 Fetal circulation.

The major portion of the fetal blood pumped by the right ventricle flows over the ductus arteriosus into the aorta and into the systemic circulation.

The left atrium receives a small amount of blood from the pulmonary veins, and a larger amount from the foramen ovale. This is sent to the left ventricle, and is pumped out into the aorta and into the systemic arterial circulation.

Significance of the Fetal Bypasses

This significance of these two bypasses, the foramen ovale and the ductus arteriosus, becomes apparent when one considers the fact that the fetal lungs are collapsed, semisolid, nonfunctional organs. The lungs, due to their structure at this time, are simply not sufficiently developed to handle the entire blood supply of the fetus; hence, the ductus arteriosus must be present to bypass this excess blood into the arterial system. Since the lungs cannot handle this large amount of blood, and as most of it is bypassed, the venous return to the heart over the pulmonary veins is a mere trickle, not sufficient to properly exercise the left heart, and allow it to develop along with the right. The foramen ovale must be present to allow an adequate amount of blood to enter the left heart and give it exercise and to provide for the physical demands of the heart and general body tissues.

It is noteworthy that during fetal life the right heart is much larger and better developed than the left, but at birth the left heart must take over the major portion of the work in supplying arterial blood. Without the presence of the foramen ovale the left heart would probably be too weak to do so.

At birth, or shortly thereafter, the ductus arteriosus and the foramen ovale close because of the major shift in circulation to the lungs and the left heart. No other significant variations between the vascular tree of the developed fetus and adult exist except for the two umbilical arteries. These leave the abdominal aorta at the region of the external iliac arteries, and form two trunks that pass ventrally and anteriorly along the midline inner surface of the abdomen, leaving the body at the umbilicus.

The two umbilical arteries, the single umbilical vein and an excretory duct from the urinary bladder called the urachus, plus a jelly-like binding substance called Wharton's jelly, form the umbilical cord. The umbilical arteries pass through the umbilical cord and carry venous blood to the placenta where it is purified, recharged with oxygen and nutrients, and returned to the fetus by the umbilical vein.

FETAL LYMPHATICS

No significant differences in the lymphatic system exist between the developed fetus and the adult.

chapter 10
The Urinary System

The principal function of the urinary system is the extraction and removal of waste products from the blood.

THE URINARY ORGANS

The urinary organs of mammals consist of the right and left kidneys and ureters, the urinary bladder, and the urethra. The urinary system begins in the kidneys, which are located in the dorsal cranial portion of the abdominal cavity on the right and left sides of the median plane and slightly caudal to the liver. The kidneys are retroperitoneal (i.e., outside the peritoneal cavity) and are connected to the blood vascular system by the right and left renal artery and vein which form short, direct connections between the kidneys and the abdominal aorta and the caudal vena cava.

The kidneys are held in position by the renal fascia, the perirenal fat, and the pressure of surrounding organs. The right kidney is usually slightly larger and more fixed in position than the left. Each kidney is covered with a tough transparent renal capsule that is derived from the renal fascia and is closely applied, but not adherent, to the outer layer (cortex) of the kidney. In diseased kidneys the capsule may be adherent.

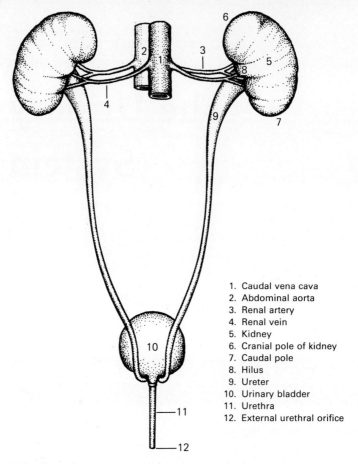

FIGURE 10-1 Schematic drawing of the urinary system.

1. Caudal vena cava
2. Abdominal aorta
3. Renal artery
4. Renal vein
5. Kidney
6. Cranial pole of kidney
7. Caudal pole
8. Hilus
9. Ureter
10. Urinary bladder
11. Urethra
12. External urethral orifice

Each kidney is drained by a single excretory duct, the ureter, which leaves the kidney at the hilus and extends retroperitoneally, caudally, and ventrally along the inner wall of the abdominal cavity and enters the neck of the urinary bladder cranial to the ducts of the prostate gland. The neck of the urinary bladder is usually located ventrally to the pelvic brim along the caudal floor of the peritoneal cavity. It connects to the outside of the body (the external urethral orifice) by a single duct, the urethra. The urethra is short, relatively straight, and solely urinary in function in the female. In the male it is relatively long, usually curved, and has both a urinary and a genital function.

THE KIDNEY

The kidneys of most animals are bean-shaped organs. The cranial curvature is called the cranial pole, the caudal curvature the caudal pole, and the indented medial portion is called the hilus. If the kidney is cut open through the hilus it will be seen that the renal hilus marks the external boundary of the renal pelvis. The renal pelvis is the funnel-shaped origin of the ureter, plus its various extensions which collect urine from the kidney lobules.

The lobule is composed of cortical (outer) and medullary (middle) portions. The cortex is usually granular and velvety in appearance whereas the medulla is characterized by radial striations. During fetal life the kidney is initially lobulated in practically all mammals, with each lobule containing its own cortex and medulla. In postfetal life, the kidneys of most mammals are not lobulated. Cattle, buffalo (and sometimes humans) are exceptions among the domestic animals. The lobulations are obliterated because the capsule is relatively small compared to the amount of gland substance. The basic lobular form of the gland can be recognized when the kidney is cut open. Small mammals such as rats and mice have only a single lobule in each kidney, while large animals like cattle may have as many as 25 or 30 lobules. The compression of the gland substance tends to distort the conical or pyramidal shape of the renal lobule but enough of the lobule persists for it to be recognized. The lobules are usually separated by bands of cortical substance called the renal columns.

The Renal Pelvis

In the kidneys of those animals where the lobular arrangement is not obliterated, the renal pelvis widens to form major outpouchings. These are known as the calyces major or major calyces. From these outpouchings, small collecting ducts go to the apices of each renal pyramid. These smaller collecting processes are known as the calyces minor, or minor calyces. The apical portion of the pyramids, which project into the minor calyces are known as papillae. On the face of each papilla are numerous small openings which form the area cribrosa. The area cribrosa is that region where the papillary ducts open into the collecting ends of the calyces minor. The equine kidney is unusual among the large animals in that the lobular form of the kidney is partially obliterated. The renal columns are greatly reduced, and the pyramids are fused into a band of medullary substance. Instead of numerous papillae, there is a renal crest composed of fused papillae. The cortical substance of the renal columns is indistinct. Neither major nor minor calyces are present in the pelvic extensions of the kidney of the horse. Instead, such outpouchings as occur are called diverticula, and the area cribrosa in the horse is located over the entire renal crest.

In dogs and other carnivores, and also in sheep, goats, deer, giraffe, and camels, the renal capsule so tightly compresses the kidney that the lobular arrangement is squeezed into what is, for all practical purposes, a single laterally compressed renal pyramid. In kidneys with this arrangement there are no apparent renal columns and no calyces major or minor. The area cribrosa is formed into a median longitudinal ridge (the common renal papilla, or renal crest) that concurs with a trough-like extension of the renal pelvis. In dogs, this trough bears a number of short flaps of tissue that fold inward to form grooves and incomplete tubular structures that encompass the transverse ridges of the renal papilla and the interlobular vessels (Figures 10-5, 10-6, 10-7). These buttress-like tubular flaps are not found in other carnivores or in the herbivores listed above. In small laboratory animals, such as rats, mice, and hamsters, the kidney forms a single renal pyramid, and the renal pelvis in these instances forms a cup-shaped extension that encloses the pointed apex of the pyramid.

The Medulla

The medullary part of any kidney is divided into two zones: an inner medullary zone, which possesses faint radial striations and varies from pinkish-yellow to red in color, and an outer medullary zone (Sisson's "intermediate zone"), which is dark red in color and is the darkest part of the kidney. Projecting radially from the outer medullary zone are striations of medullary substance which penetrate the cortical substance for a variable distance but never reach the outer surface of the kidney. These are known as medullary rays.

The Cortex

The cortex of the kidney is a band of reddish-brown granular tissue lying between the outer medullary zone and the capsule. Here are contained the majority of the secreting elements of the kidney.

The Capsule

The capsule is the outermost covering of the kidney and is composed of a thin, strong, transparent membrane.

The Perirenal Fat

A variable amount of fat surrounds the kidneys. It is most abundant in mature animals and tends to disappear with advanced age. However, it never disappears completely in normal animals. Under severe conditions of malnutrition, starvation, or emaciation resulting from pregnancy, stress, or disease, the perirenal fat will become minimal and will occasionally disappear. The fat is important as a shock absorbing device and as an insulating mechanism, and tends to give the kidneys a more stable external environment. It is interesting to note that the coronary fat in the heart, the periorbital fat in the eye, and the perirenal fat are the most persistent fat depots in the body and are the last to disappear in starving animals.

Arteries

The arterial blood supply to the kidney begins with the renal artery, which is a direct branch of the abdominal aorta.

Because of its direct and short connection to the abdominal aorta, the blood pressure in the kidney is high, about 70 percent of the aortic pressure. This high pressure helps filtration and improves kidney efficiency.

The renal artery enters the kidney at the hilus and branches in the renal pelvis to form a fan of interlobar arteries. The interlobar arteries penetrate the gland substance along the renal columns. At the juncture of the cortex and the medulla they branch repeatedly to form an anastomosing network of arcuate arteries. The arcuate arteries give off medullary branches called the arterioli rectae spuriae and more numerous cortical branches called intralobular arteries. In the cortex, the intralobular arteries branch to form the afferent glomerular arterioles, which enter Bowman's capsules and form anastomosing knots of capillaries called the glomeruli. The glomerular capillaries then coalesce to form the efferent glomerular arteriole. This leaves the glomerulus and branches repeatedly to form the medullary plexus, a network of capillaries which invest the uriniferous tubules and ramify through the cortical and medullary tissues.

Veins

The capillaries of the medullary plexus coalesce to form the efferent veins. The efferent veins empty into the intralobular veins, which unite to form the arcuate veins. These empty into the interlobar veins, which unite to form the renal vein. The renal vein empties into the caudal vena cava.

THE SECRETORY AND EXCRETORY MECHANISM OF THE KIDNEY

The gland substance of the kidney is composed of a parenchyma of uriniferous tubules, plus a certain amount of connective tissue stroma. The tubules have two functions which can conveniently be classed as secretory and excretory.

The Nephron

The nephron is the secretory portion of the uriniferous tubule. It begins with Bowman's capsule which empties into the proximal convoluted tubule. The proximal convoluted tubule is continued as the descending limb of Henle's loop, whose distal portion usually penetrates into the medulla of the kidney. The descending limb constricts near its termination and bends sharply backward to form the ascending limb. Near the terminus of the ascending portion, the tubule increases in diameter and becomes the distal convoluted tubule. The terminus of the distal convoluted tubule marks the end of the nephron. From this point onward, the function of the tubule is excretory.

The cellular aspects of the entire nephron have become most interesting since the advent of electron microscopy. Prior to this instrument it was known that there were such things as podocytes lining the urinary space in Bowman's capsule, and brush-bordered, striated cells lining the proximal convoluted tubule. Flattened cells were observed lining the descending limb of Henle's loop, and low columnar cells were

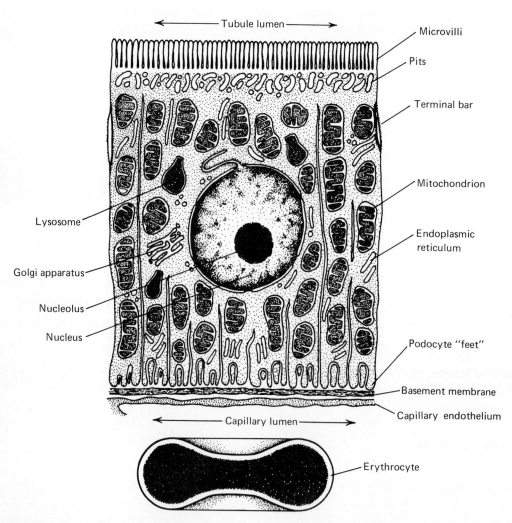

FIGURE 10-2 *Proximal convoluted tubule cell. (Adapted from Porter and Bonneville,* Fine Structures of Cells and Tissues.)

seen to form the lining of the ascending limb of Henle's loop and the distal convoluted tubule. The arched collecting tubule and the duct system, as far as the area cribrosa, were lined with cuboidal cells. However, except for some good guesses such as the brush border being microvilli and the striations being mitochondria, the fine structure (and consequently the details of function) were still unrecognized. Today a whole new field of microanatomy exerts profound influence on previously accepted theories of renal function.

One of the structures most affected is the proximal portion of the nephron. Analysis of the fine structure of the "brush-bordered" cells of the proximal convoluted tubule indicates that active secretion probably takes place in this region, as well as in other parts of the nephron.

Like the columnar cells lining the gut, the cells of the proximal convoluted tubule have numerous closely packed microvilli on their free surface which project into the lumen of the tubule. Materials in the lumen seem to be funneled between the microvilli into deep membrane-lined pits or wells, which terminate in tiny pouches, possibly designed for reception of large molecules (polypeptides-proteoses) in the protourine filtrate. These terminal pouchings apparently pinch off, and the packaged macromolecule migrates into the depths of the cell where it becomes incorporated into a lysosome and its contents are progressively hydrolized. What subsequently happens to the material is not yet known.

There are long tubular structures that may be intercellular membranes, which apparently connect the basement membrane of the cell with the bases of the microvilli. These tubular structures may represent the channels over which passive (osmotic) transfer of fluids from the tubule lumen to the interstitial fluid and the capillary plexus surrounding the tubule takes place. The intercellular borders appear to be sealed with desmosomes or terminal bars. However, spaces around these seals may conceivably exist.

The cells form protoplasmic "feet," similar to the podocytes in Bowman's capsule, and appear to interdigitate with the basement membrane. The mitochondria are numerous and are enclosed in deep folds of the cell membrane. The folds occasionally appear to pass almost completely through the cell, and can, in effect, make the cell function as a semipermeable membrane that will permit osmotic interchange between the lumen and the outer surface of the tubule. The large number of mitochondria indicate active participation of the cell in one or more secretory or excretory processes.

THE EXCRETORY PORTION OF THE KIDNEY

The excretory portion of the kidney begins with the arched collecting tubule, which is a continuation of the distal convoluted tubule. The arched collecting tubule empties into the straight collecting tubule, which passes through the cortex and medulla toward the area cribrosa. Several straight collecting tubules merge near the apex of the renal pyramid to become the papillary ducts (ducts of Bellini) which open into the area cribrosa. From this point, the urinary wastes are carried through the calyces minor, calyces major, renal pelvis, ureter, urinary bladder, urethra, and to the outside.

The straight collecting tubules and the distal portions of Henle's loop lie in close proximity to each other in the medulla, and when observed are called medullary rays. The glomeruli and Bowman's capsules (Malpighian corpuscles) are responsible for the granular, velvety appearance of the cut surface of the cortex.

The entire urinary tract, from the renal pelvis to the urethral orifice is capable of rapid changes in size and is lined with a remarkable arrangement of cells called transitional epithelium. Transitional epithelium is noteworthy for the rapidity with which it can accommodate to changes in area. Early light microscope studies of the epithelium developed a number of interesting theories about how the cells performed their function.

More recent electron microscope studies of transitional epithelium indicate that the cells do, in fact, slide over one another as was once postulated, but that this is not the only mechanism involved. In fact, the electron microscope has raised more questions than it has answered. However, it is now apparent that there is a basement membrane for transitional epithelium, which is a direct contradiction of earlier beliefs. The cellular anatomy has undergone some drastic revisions, which have in turn generated a number of hypotheses about function which are yet unproved.

The transitional epithelium lining the excretory portion of the urinary system ends with the termination of the urethra. In male domestic animals except for the ram, the urethral termination is not particularly noteworthy. The ram, however, has a long urethral process that extends beyond the rim of the penis and appears to be composed of excess duct tissue. In male pigs there is a round opening in the dorsal cranial end of the prepuce that leads to a small sausage-shaped sac called the preputial diverticulum, which extends caudally for a variable distance. The sac varies in size between individuals. The wall of the sac contains lymph glands and in uncastrated animals is usually filled with a foul-smelling semisolid mass of decomposed urine and shreds of epithelium, which give the unpleasant "sex odor" and taste to pork from mature uncastrated male swine.

In females the urethra usually ends in a longitudinal slit in the floor of the vagina. Noteworthy variations of this ending are found in cattle and dogs. In the cow, the termination of the urethra is a transverse slit in the anterior wall of a pouch called the suburethral diverticulum, which is found in the vaginal floor about four inches anterior to the vaginal opening. In the bitch, the urethra opens at the apex of a small

THE EXCRETORY
PORTION OF THE
KIDNEY

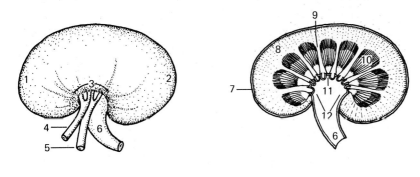

1. Cranial pole
2. Caudal pole
3. Hilus
4. Renal artery
5. Renal vein
6. Ureter
7. Capsule
8. Cortex
9. Renal columns
10. Medulla
11. Calyces
12. Renal pelvis
13. Outer medullary zone
14. Renal pyramid
15. Apex (area cribrosa)
16. Renal crest
17. Diverticula
18. Folds

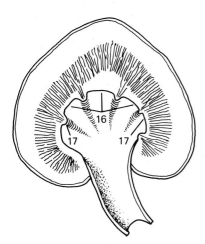

Kidney of Horse

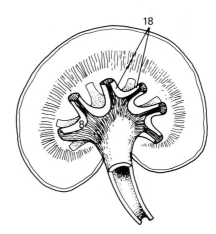

Kidney of Dog

FIGURE 10-3   Gross structures of the kidney.

THE URINARY SYSTEM

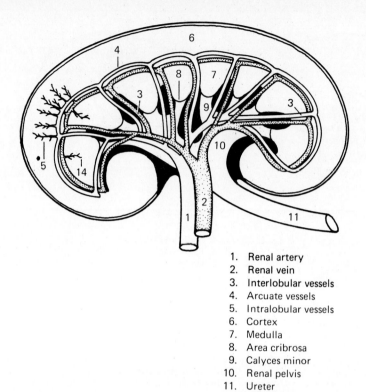

1. Renal artery
2. Renal vein
3. Interlobular vessels
4. Arcuate vessels
5. Intralobular vessels
6. Cortex
7. Medulla
8. Area cribrosa
9. Calyces minor
10. Renal pelvis
11. Ureter
12. Nephron
13. Renal columns
14. Arterioli rectae spuriae

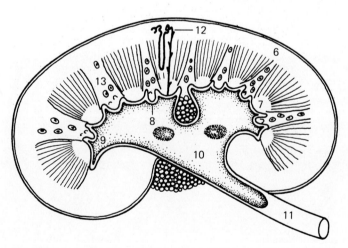

FIGURE 10-4   *Sagittal section of the kidney.*

THE EXCRETORY
PORTION OF THE
KIDNEY

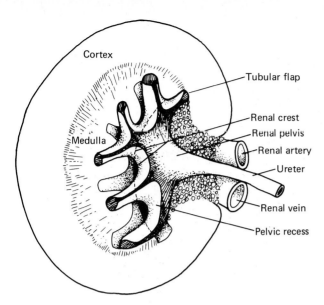

FIGURE 10-5   *Canine kidney, renal pelvis—dorsal view. The renal crest is curved to conform to the greater curvature (lateral border). Note the tubular flap extensions of the free borders. These enclose blood vessels and nerves.*

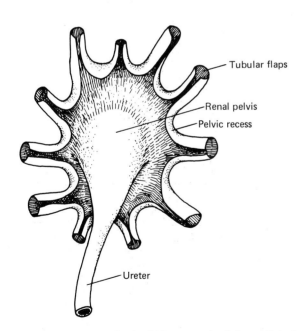

FIGURE 10-6   *Canine kidney, renal pelvis, medial view. The pelvis has been spread out to show the arrangement of the tubular flaps, which, in the intact kidney, enclose blood vessels and nerves.*

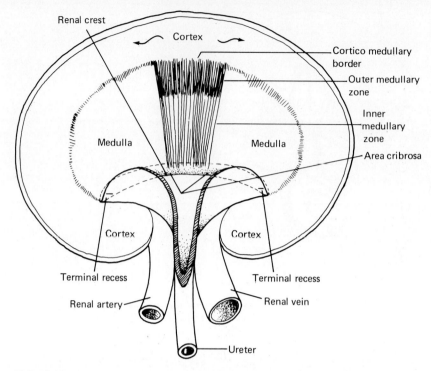

FIGURE 10-7   Ovine kidney.

mound of tissue, the urethral papilla, which is located on the floor of the vagina. These structures make catheterization of the urinary bladder of cows and bitches rather difficult.

RENAL PHYSIOLOGY

Essentially the kidney is a regulatory organ that tends to maintain stability of the internal body environment (homeostasis). It operates by regulating fluid balance and composition within the body by regulating the rate of fluid excretion and by regulating the character of fluid excretion. Renal function is ordinarily conservative, although the kidney can and does function rapidly at times to remove a great deal of excess water and waste products from the bloodstream, as anyone who has been on a beer drinking orgy can testify.

Kidneys can change size to accommodate for renal demand. Experiments with parabiotic (surgically created Siamese twin) rats have shown that kidneys can enlarge to handle increased excretory load and can return to "normal" size when the additional load is removed (Figure 10-8). Although this particular series of experiments would be difficult to perform on other species because of the problem of histocompatibility

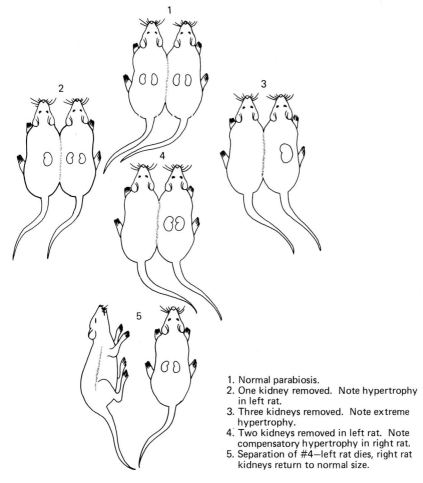

FIGURE 10-8   *Effects of kidney loss in parabiotic rats.*

1. Normal parabiosis.
2. One kidney removed. Note hypertrophy in left rat.
3. Three kidneys removed. Note extreme hypertrophy.
4. Two kidneys removed in left rat. Note compensatory hypertrophy in right rat.
5. Separation of #4—left rat dies, right rat kidneys return to normal size.

and rejection mechanisms (p. 243), it has been noted that animals with one kidney or a single functioning kidney will have appreciable adaptive or compensatory hypertrophy.

The formation of urine, which commences with the contact of blood plasma with the nephrons of the kidneys at the renal corpuscle, is the most important aspect of renal function. Urine is commonly considered to be solely an avenue for the disposal of metabolic wastes. Actually, this is but one aspect of the homeostatic process, which depends upon the constant adjustment of kidney function to body needs. A portion of this adjustment is physical, a portion is hormonal or neurohormonal, and a portion is enzymatic. The entire activity is constantly monitored by the nervous system which, either through direct action or indirectly through other structures, controls and adjusts kidney function.

PHYSICAL ASPECTS OF KIDNEY FUNCTION

The unit formed by Bowman's capsule and the glomerulus, i.e., the Malpighian (renal) corpuscle, acts as a simple filter for dissolved plasma elements other than protein. (Molecules with a weight of about 45,000 or more are normally retained in the bloodstream.) Considerable quantities (roughly 20 percent of the plasma content) of mineral salts, urea, water, sugar, and other products are filtered out of the blood and pass into the nephron.

As this fluid mixture passes through the tubules, selective reabsorption takes place through the walls of the nephron, and nutrients and liquids needed by the body are removed from the tubule and pass into the medullary plexus. The waste products pass into the excretory portion of the uriniferous tubule, from where they are ultimately excreted from the body in the form of urine. Most of the selective reabsorption takes place in the proximal convoluted tubule, although it is probable that Henle's loop exercises some selective reabsorption functions. From the ascending limb of Henle's loop to the arched collecting tubule, the cells forming the tubule walls actively transfer useful substances and water from the tubule lumen while retaining the waste material. Beginning with the arched collecting tubule and extending through the urethra, there is no further transfer of substance except water across the epithelium, and the fluid in the lumen is now definitive urine. The arched and straight collecting tubules and papillary ducts serve mainly to convey urine to the area cribrosa, from where it is carried to the outside by the ureters and the urethra. The urinary bladder is a distensible muscular sac lined with transitional epithelium, which stores the urine until such time that a quantity can be voided.

Since kidney function deals chiefly with fluid excretion, and since water is the principal solvent involved in renal function, it is advisable at this point to review briefly the function of water in the body (p. 179ff).

The fluid and solute interchange between cells, tissue fluids and the bloodstream results in the blood plasma taking on its share of the soluble wastes. Since only plasma and certain of its dissolved substances pass through the nephrons, the removal of water and soluble wastes from the body via the kidneys is basically a plasma function. The total function is a gradual process that does not cause abrupt changes in chemical composition within the body.

The Formation of Urine

Primarily the kidneys produce urine. The amount of renal excretion depends upon the flow of blood through the kidneys. The flow is surprisingly large. About one-fifth of the total cardiac output in any given minute passes through the kidneys; hence in five minutes, theoretically, the total blood volume of an animal will have passed through the renal

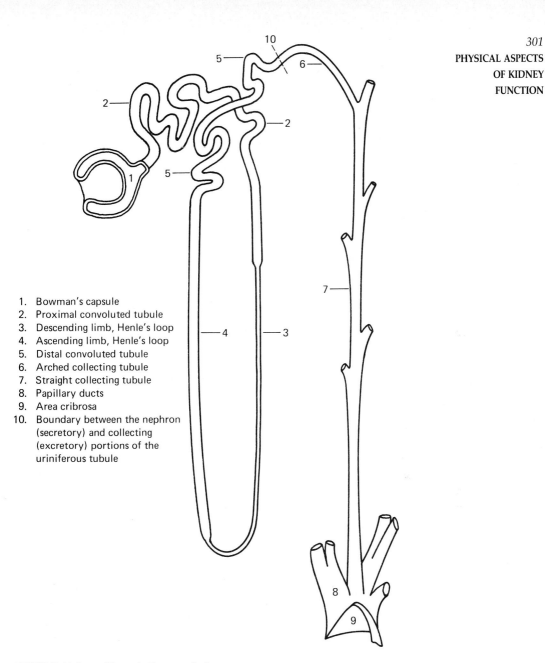

FIGURE 10-9   The uriniferous tubule.

1. Bowman's capsule
2. Proximal convoluted tubule
3. Descending limb, Henle's loop
4. Ascending limb, Henle's loop
5. Distal convoluted tubule
6. Arched collecting tubule
7. Straight collecting tubule
8. Papillary ducts
9. Area cribrosa
10. Boundary between the nephron (secretory) and collecting (excretory) portions of the uriniferous tubule

circulation, and within 15 minutes an entire new chemical equilibrium can be established in all of the extracellular fluid of the body. Since the kidneys are comparatively small (approximately one-half of 1 percent of the body weight), this disproportionately large volume of blood requires a direct and extensive system of arterial, capillary and venous channels.

Normally, more than 90 percent of renal blood flow is directed outward into the cortex and into the glomerular tufts. The remainder is diverted to nourish the supporting and nonexcretory tissues in the cortex, medulla, capsule, and pelvis. Nervous stimulation may change the distribution of renal blood flow. Blood may be virtually shut off from the cortex by constriction of the interlobar arteries and afferent arterioles, while the venous and medullary vessels remain open. This change in circulation has the effect of prohibiting blood from contacting the nephrons with the result that urine formation ceases or is markedly depressed. Prolongation of such a condition will result in a syndrome known as reflex anuria which may follow shock or sudden stress. Excessive continuation of this state will result in damage to the nephrons. Reflex anuria and danger of kidney failure may persist long after a normal blood distribution pattern has been restored in the stressed patient.

The formation of urine begins as the blood enters the glomerular tufts enclosed within Bowman's capsule. Each kidney contains a large number of these units (upwards of one million in the human), which are the vascular portions of the nephrons. The nephron continues as the tubular portion which performs the selective reabsorption and secretory functions necessary to form urine. Bowman's capsule is designed for the outward filtration from the plasma of water and substances dissolved in water. The energy for this filtration is derived from the blood pressure. The kidney, lying as it does beside the abdominal aorta, has a relatively short connection to the heart. As a result, about 70 percent of the aortic pressure reaches the glomeruli and a much higher pressure level is present in renal blood than in the blood supply of most other tissues and organs. The kidney capillaries, therefore, must sustain pressure which is two to two and one-half times that in other capillary beds. This resistance to pressure is accompanied by a greater retentiveness of the capillary membrane which, at least in the glomeruli, permits the ready passage of only water and solutes and retains more protein than any other capillary system in the body.

The high pressure in the glomerular capillaries tends to force water out of the plasma. This force is countered somewhat by pressures within the nephron that tend to prevent filtration. These include osmotic pressure of plasma proteins, interstitial pressure, and the pressure necessary to move the fluid through the tubule systems of the kidney. The initial force applied to the blood in the glomerulus is countered by opposing forces whose total pressures are about two-thirds as great. The difference represents effective filtration pressure.

The volume of glomerular filtrate depends on the levels of effective filtration pressure and the amount of renal blood flow. These functions are normally maintained at a relatively constant level even though the blood flow may be greatly changed in other areas of the body. This occurs because the renal arterioles and arteries have an ability to adjust their resistance rapidly and accurately during changes of arterial pressure.

Since filtration is the first step in forming urine, conditions that increase or impair filtration have corresponding effects on urine production. These may be one or a combination of the following three factors: hydrostatic change, which may produce increase or decrease in filtration by an increase or decrease in arterial pressure or in pressure opposing filtration; hemodynamic changes, which are ordinarily considered to be a decrease in blood flow due to vasoconstriction, shock or congestive heart disease; and glomerular changes in which disease of the glomerular tufts may partially or totally impair their function.

PHYSIOLOGY OF THE NEPHRON

The nephron begins with Bowman's capsule, which surrounds the glomerular tuft and continues as the proximal convoluted tubule, the loop of Henle and the distal convoluted tubule. Physiologically, the nephron has traditionally been divided into two different functioning regions. The original separation was made upon a basis of function. It was thought that the proximal convoluted tubule and the descending limb of Henle's loop functioned passively, through a difference in osmotic pressure between the fluids in the interstitial tissue surrounding the tubule, and the protourine held within the tubule. Electron microscope studies of the cells of the proximal convoluted tubule indicate that there is also an active secreting process, which requires energy other than osmotic pressure. The ascending limb of Henle's loop and the distal convoluted tubule are still considered to possess active secretory function, and to transport substances contained in the protourine back into the circulation. The fine structure of the epithelial cells of these regions tends to support this conclusion, since the cells contain large numbers of mitochondria, which are known to be involved in active transport and secretion.

At the end of the distal convoluted tubule the collecting duct system begins, and the fluid within the tubule (urine) is passed out of the kidneys into the excretory system and to the outside. From the end of the nephron onward there is little or no change in the chemical composition of urine of a normal animal, since only water interchange takes place through the walls of the excretory ducts. The secretory and selective functions apparently end in the nephron.

Each section of the nephron has relatively specific functions in the secretion or selective reabsorption of useful material by the body. For example, the resorption of glucose from the protourine is confined to the proximal convoluted tubule. This is a process that selectively takes glucose from the lumen of the tubule and discharges it into the interstitial fluid surrounding the tubule, where it is absorbed by the capillaries of the medullary plexus. If glucose levels in the nephron are extraordinarily high, the capacity of this transfer system can become overloaded and some sugar will remain within the tubule. The glucose that is not reabsorbed in the proximal convoluted tubule will pass, together with an increased water volume through the remainder of the nephron and into the excretory system, resulting in glycosuria. The resorptive capacity of the nephron is commonly known as the renal threshold, and has definite limits for specific substances. These limits are relatively constant within a species, but vary between species.

The selective return of nutrients and mineral salts to the body is an important aspect of kidney function. Since mineral salts are part of the buffering systems and electrolyte necessary for homeostasis, and since the diet of most animals contains an excess of acid or base which must be excreted if a stable bodily environment is to be maintained, the return must be precise. The fine adjustment aspect of mineral salt reabsorption is accomplished in the ascending limb of Henle's loop and in the distal convoluted tubule.

Reabsorption of water accounts for most of the difference between the large amount of filtrate that enters the nephron and the small amount (roughly 1 percent in volume) that leaves it. Unlike the reabsorption of glucose, which depends upon a transfer mechanism that is confined to the proximal portions of the nephron, the movement of water takes place throughout the entire structure and is largely dependent upon the difference in osmotic pressure within the tubule and in the interstitial fluid which surrounds it, and is associated principally with the concentration of sodium ions (Na^+) in the interstitial fluid. Sodium excretion from the nephron is an active process that requires energy, but osmotic activity is passive.

About two-thirds of the water that enters the nephron at the glomerulus is returned to the capillaries of the medullary plexus by the time the filtrate has reached the end of the proximal convoluted tubule. The terminal portion of the proximal convoluted tubule tends to be straight and lies close to the cortico-medullary border. This portion forms the thick part of the descending limb of Henle's loop. In the inner medullary zone, the tubule constricts to form the thin portion of Henle's loop. In mammals, the length of the loop seems to have some correlation with the concentration of urine. In desert rats, for instance, there are only long loops, and in humans, cattle, and swine—species noted for voiding large amounts of relatively dilute urine—there are relatively few long loops. Dogs, cats, camels, sheep, and goats, which all produce relatively concentrated urine, have a large proportion of long Henle's loops. However, birds have only short loops, yet produce a highly concentrated urate paste.

The thin segment of Henle's loop is permeable to water. It has about half the diameter and one-fourth the number of cells in cross section as the thick parts of the loop. It appears that this thin portion of the nephron has some function in passive fluid exchange. The medullary plexus (vasa recta) that surrounds the thin portion has a blood flow counter to the direction of the fluid flow within Henle's loop. This

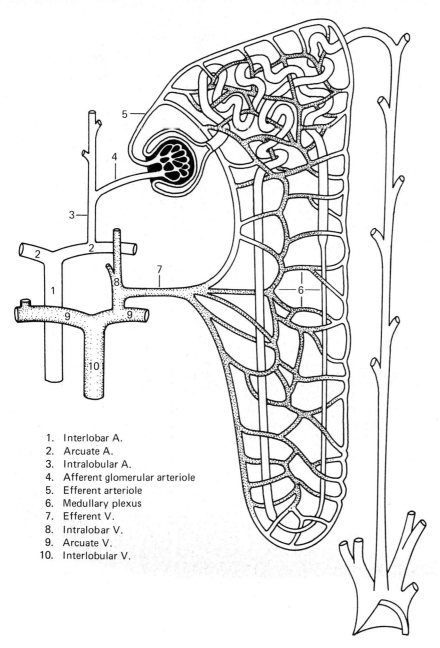

1. Interlobar A.
2. Arcuate A.
3. Intralobular A.
4. Afferent glomerular arteriole
5. Efferent arteriole
6. Medullary plexus
7. Efferent V.
8. Intralobar V.
9. Arcuate V.
10. Interlobular V.

FIGURE 10-10   *Circulation of the kidney tubule.*

countercurrent tends to promote the passage of fluid (principally water) from the lumen of the tubule to the capillaries. At the termination of the thin segment the osmotic pressure of the fluid in the nephron is higher than it is anywhere else in the kidney.

As Henle's loop becomes the ascending limb, the walls of the nephron thicken and the number of cells per cross section increase. These cells differ morphologically and functionally from those that form the thick portion of the descending limb. They are permeable to sodium ions (Na^+) but are relatively impermeable to water, and actively secrete Na^+ into the interstitial fluid outside the tubule. At the level of the distal convoluted tubule the osmotic pressures inside and outside the nephron are about equal.

The distal convoluted tubule is permeable to water and tends to absorb water from the interstitium. This causes the fluid inside the tubule to become more dilute and establishes an osmotic gradient toward the interstitium, which assists the cells in this portion of the nephron to return useful mineral salts to the bloodstream.

The fluid that leaves the distal convoluted tubule is of lower osmotic pressure than the interstitial fluid, and as the protourine passes through the collecting tubules, water diffuses outward into the interstitium. The waste products, excess minerals, and nutrients are retained.

Electrolyte and Nitrogenous Waste Excretion

The kidneys maintain the electrolyte composition and chemical balance of body fluid at a remarkably constant level in the normal mammal. There does not seem to be any differentiation made by the kidney between an excess electrolyte and a waste product. Once the protourine has passed the proximal convoluted tubule and Henle's loop, excess electrolyte in the nephron is treated no differently than other chemical wastes.

The ability of healthy animals to endure stress is dependent upon the proper functioning of the kidneys. Animals with impaired renal function can be maintained in comparative health only as long as they can be protected from stress that would seriously alter body chemistry or place excessive demands upon damaged kidneys. In practice, impaired renal function may not appear until some strain such as injury, excessive exercise, or surgery imposes a demand on the kidneys that cannot be corrected in the process of normal functioning.

Neural and Hormonal Regulation of the Kidney

The rate of urine production is dependent upon the number of nephrons involved in the formation of urine and the activity of the epithelial cells of the nephrons. Control of these factors is exerted by hormonal and neural action. Neural control functions principally via the juxtaglomerular apparatus, which regulates the amount of blood that passes through the glomerulus. Hormonal control is exerted principally by antidiuretic hormone (ADH) which is stored in the posterior pituitary. ADH (pitressin) is released in response to increased osmotic pressure of the circulating blood. It stimulates increased reabsorption of water, which in turn produces decreased fluid excretion (although the quantity of chemical wastes other than water is not changed). The increased amount of water returned to the bloodstream decreases the blood osmotic pressure, which removes the stimulus for antidiuretic production.

Possibly, there is a direct inhibitory action by the nerve cells of the hypothalamus on antidiuretic hormone production in the event of excessively hydrated blood with low osmotic pressure. However, this question has not been adequately answered, and from a conservative standpoint such a mechanism is probably not necessary. However, tumors involving both the pituitary and the hypothalamus have caused a condition known as diabetes insipidus, and have been used as evidence to indicate that inhibitory impulses exist. The problem is not yet resolved.

Enzymatic Regulation of Kidney Function

An aspect of renal function is concerned with changes in arterial pressure. Here a defect in kidney function or a deficiency in the flow of blood through the kidney results in an increase in arterial pressure. This is caused by a series of events called the renin-angiotensin-aldosterone mechanism, which is triggered by the release into the blood of an enzyme called renin, which is formed in the renal cortex.

The details of operation of the enzymatic mechanism have undergone changes over the years since they were first discovered. At the moment the general consensus is that it operates as follows:

1. Renin is secreted in the cortex of the kidney by the secretory cells of the juxtaglomerular apparatus and is picked up by the bloodstream.
2. The proenzyme angiotensinogen (hypertensinogen or renal substrate) is a glycoprotein produced in the liver as a normal part of liver function and released into the bloodstream. It is considered to be a part of the globin fraction of the plasma.
3. Renin reacts with angiotensinogen, activating it to form the polypeptide angiotensin I (hypertensinogen). Angiotensin I reacts in part with a blood-borne convertase to produce angiotensin II.
4. Angiotensin I and II are potent vasopressors which
   a. act directly on the smooth muscle of arteries to produce vasoconstriction. (The action may be effectuated by the myoepithelial cells in the juxtaglomerular apparatus);
   b. increase heart rate and amplitude;
   c. stimulate the adrenal cortex to produce aldosterone.

5. Aldosterone is a hormone that promotes sodium retention by stimulating cells of the ascending portion of Henle's loop and the distal convoluted tubule to return more sodium to the bloodstream. Sodium affects the walls of the arteries enhancing the direct vasopressor action of angiotensin I and II.

The pressor enzymes are ultimately destroyed by enzyme systems (angiotensinase) in the body and the mechanism subsides, providing the stimulus is transitory. The function of the mechanism essentially is to produce an increased flow of blood, and hence an increased formation of urine in response to increased stress or abnormal levels of metabolic waste within the bloodstream. In normal animals this mechanism functions to good effect. However, in animals with kidney disorders or disease, a constant stimulation is present which results in a cyclic condition called the renal hypertension syndrome. This tends to produce further kidney damage which may lead to ultimate destruction of kidney function and death.

The Juxtaglomerular Apparatus

The juxtaglomerular apparatus is apparently another mechanism that controls kidney function and the regulation of body water. It includes (1) a group of secretory cells associated with the glomerular tuft; (2) certain cell groups between arterioles; (3) a peculiar type of myoepithelial (muscle-epithelial) cell that lines the afferent and efferent glomerular arterioles; (4) a structure called the macula densa, which is composed of densely nucleated cells of the distal convoluted tubule (not necessarily from the same nephron) that make contact with the vascular pole of the glomerulus and are closely associated with the juxtaglomerular cells of the afferent arteriole; and (5) an associated nerve supply.

There is still a considerable amount of argument about the relationship of the apparatus to kidney function. The controversy covers tonus and renin production. A majority opinion holds that the principal function of the juxtaglomerular apparatus is to maintain tonus of the afferent and efferent glomerular vessels and to bypass specific glomeruli by shutting off blood supply. It has been suggested that the secretory and myoepithelial cells under the influence of nervous stimulation (e.g., sympathetic-vasoconstriction; parasympathetic-vasodilation) change the caliber of the arterioles and in so doing produce a rise in glomerular pressure and increased filtration pressure by constricting the efferent arteriole, a drop in glomerular pressure and decreased filtration pressure by constricting the afferent arteriole, and a complete shutoff of the glomerulus and diversion of blood to other glomeruli. This theory postulates a direct neural effect on the kidney. Another theory states that the juxtaglomerular apparatus is the initiating organ of the renin-angiotensin-aldosterone system described above. According to this theory, the secretory cells of the juxtaglomerular apparatus secrete renin and the apparatus acts as an intraneural volume regulator.

ENDOCRINE ASPECTS OF THE KIDNEY

The kidney also functions as an endocrine organ. A hormone or hormone-like substance called urogastrone, which inhibits gastric motility, has been isolated from urine. Whether this substance has genuine systemic activity or is a compound (with incidental hormone attributes) that does not enter into normal body function has not yet been determined. Another hormonal substance, erythropoietin, is a red blood cell stimulating factor (ESF). It originates in the kidney and stimulates the formation of erythrocytes by the bone marrow, and also functions to increase the amount of iron in the blood cells.

There are still many puzzling anatomic and physiologic problems about the kidneys that have not yet been adequately explained. Work in this area is continuing and from time to time new discoveries about kidney structure and function are published. It is a field that still furnishes rewarding results for detailed research.

chapter 11
The Nervous System

MAJOR DIVISIONS OF THE NERVOUS SYSTEM

The nervous system is the main control and organizing system of the body and is divided anatomically and functionally into three main parts: the central nervous system, the peripheral nervous system, and the autonomic nervous system.*

The central (voluntary) nervous system includes the brain and spinal cord. The peripheral nervous system is composed of the 12 pairs of cranial nerves and a variable number (depending upon the number of vertebrae) of paired spinal nerves which pass through the intervertebral foramina to supply the body and the limbs.

The autonomic (involuntary) nervous system is composed of two major and two minor parts. The major parts are the sympathetic (thoracolumbar) and parasympathetic (craniosacral) divisions. The minor parts are the more or less completely separated divisions of the heart (cardiac)

---

*For the purpose of clarification, "system" is used to categorize the major parts and "division" used to identify secondary structures. As of this date the hierarchial terminology is still confused, with words such as "system," "division," "part," and "complex" being used by various authors and educators to categorize major, minor, and subsidiary units indiscriminately without any particular regard for order or for the confusion that results from such usage. It is hoped that this may be clarified in the near future.

and of the intestines (myenteric). Of these latter two, the cardiac division is not precisely nervous, since the excitant part (the S-A node) and the conduction part (muscle fibers and Purkinje fibers) are not nerves, although they do transmit contraction impulses.

THE NERVE FIBER OR NEURON

Irrespective of whether the systems are central or autonomic, nerve impulses are carried by neurons. All neurons have a similar structure and function in a similar manner. A neuron is the structural and functional basis of the nervous system. Neurons are the most highly specialized cells in the body and have no centrioles and no powers of regeneration. An animal is born with its full complement of neurons and produces no more during its life.* A neuron is composed of three basic parts, a cyton or nerve cell, and two types of processes, i.e., axons or efferent fibers, and dendrites or afferent fibers, which respectively carry impulses away from or toward the nerve cell nucleus. The axon is usually single, while dendrites are usually multiple. The terminal ends of both axons and dendrites may have a number of branchings. The term axon is often used to describe any nerve fiber. It will not be used that way here. The general term will be "fiber," and axon and dendrite will specifically apply to efferent and afferent processes of a cyton.

The cyton is the life center of the neuron. It consists of the nucleus and a variable amount of cytoplasm called the perikaryon. This is filled with organoids and a number of granular aggregations of ribosomes and endoplasmic reticulum called Nissl bodies. The cyton is responsible for the maintenance of its axons and dendrites in a functional condition. When the cyton is destroyed, the processes die and become nonfunctional.

Each nerve cell contains a nucleus. The cell body, i.e., that portion of the cell which surrounds the nucleus, contains in addition to the nucleus at least two major structures which are visible under the light microscope. These are the Nissl bodies and neurofibrils. The Nissl bodies are composed of ribonucleoprotein and appear as fine granules in the cytoplasm of the cyton. They are associated with protein synthesis. Their numbers vary with the health of the cell. Fatigue, anoxia, and various toxic substances cause their breakdown and disappearance (chromatolysis). Nissl bodies are not found in axons, although they extend into the dendrites. Neurofibrils are often seen in prepared specimens. There is still argument as to whether they actually exist or are artifacts. Cells stained with the Golgi silver impregnation method show these fibrils plainly. Cells are stained with toluidine blue to demonstrate Nissl bodies.

---

*Recent work indicates that the receptor cells of the olfactory epithelium can proliferate and replace nerve fibers that have died or have been destroyed. This work, done at Florida State University, may herald a breakthrough in neurophysiology.

Ganglia and Glial Cells

Outside of the brain and spinal cord, cytons are usually found in clumps or aggregations called ganglia. Ganglia are distributed throughout the body and serve as cell and reflex centers for the autonomic nervous system, and cell centers for the sensory portion of the central nervous system. With the cytons are several types of supporting cells which, in aggregate, are called glial cells.

Glial cells fill the extracellular spaces around the neurons of the central nervous system and the cytons in the ganglia. They are about ten times more numerous than cytons and form about half the volume of the brain. Glial cells are further classified as macroglia (Gr. *macro*—large: *glia*—glue) and microglia (Gr. *micro*—small). The macroglia are composed of astrocytes (L. *astra*—star) and oligodendroglia (Gr. *oligo*—few; *dendro*—branches). The macroglia are by far the most numerous, with the oligodendroglia occurring in the greater number. The oligodendroglia cells probably enter into biochemical reactions involving neu-

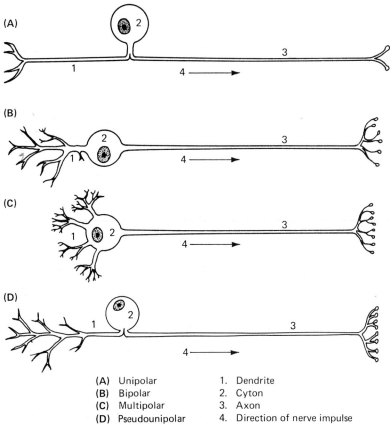

(A) Unipolar
(B) Bipolar
(C) Multipolar
(D) Pseudounipolar

1. Dendrite
2. Cyton
3. Axon
4. Direction of nerve impulse

FIGURE 11-1   *Types of neurons.*

rons and appear to have something to do with the formation of myelin sheaths (p. 315). Astrocytes are less than half as numerous as oligodendroglia, and are much larger with many branching processes. They are apparently a part of the "blood-brain barrier" that keeps many foreign substances out of the brain tissue and helps regulate neuron metabolism. The microglia compose about 10 percent of the glial cells and appear to have a phagocytic or protective function, since they tend to accumulate in areas of damage or infection.

Nerve Endings

At its beginning and end, a neuron has structures that allow it to receive stimuli and pass the stimuli on to other nerves or to organs and structures. Commonly, these endings are called receptors and effectors. Receptor endings are always found on the free ends of dendrites, effectors are always found on the free (terminal) ends of the axons.

Receptors are more complex in both structure and function than effectors. Receptors have a great variety of forms, ranging from simple free nerve endings and terminal arborizations to highly developed structures such as Meissner's corpuscles, Pacinian corpuscles, Merkel's corpuscles, Krause's corpuscles, Ruffini's end organs, Hederiform terminations, tactile hair terminations, muscle spindles, Golgi tendon apparatus, hair cells, rod and cone cells, olfactory cells, taste buds, and cells of the carotid and aortic bodies. The large number of different receptors correlates with a large number of different sensory stimuli that operate to give the brain an awareness of the environment and to initiate various reflexes.

Effector endings are relatively simple and consist mainly of free endings, terminal arborizations and motor end plates. This correlates with motor functions that principally involve muscle tissue and glands.

Preganglionic and Postganglionic Fibers

Two terms, used in place of axons and dendrites, are common in discussions of the autonomic and peripheral nervous systems. These are preganglionic and postganglionic fibers. Preganglionic fibers carry nerve impulses toward a ganglion. Postganglionic fibers carry impulses away from a ganglion.

Microscopic evidence reveals that there are many more postganglionic fibers on the effector side of the autonomic nervous system, which indicates that ganglia are probably instrumental in spreading stimuli to effectors. Functionally, sympathetic ganglia apparently prolong the effect of preganglionic stimuli during an excitement episode. Neither sympathetic nor parasympathetic ganglia serve as reflex centers. Such activity takes place either in the spinal cord or in the peripheral ramifications of the postganglionic fibers that involve such structures as the heart and the myenteric nervous system.

Nerve Centers

The aggregation of cytons that can be seen in the ganglia can also be found in the brain and spinal cord. In these structures, the cytons tend to be arranged in groups that have special configurations and functions. These groups of nerve cells form nerve centers or basal nuclei (Some authors call these structures "basal ganglia.") Their identity is an identity of function, since component parts of a nerve center may be located some distance from one another. Until quite recently, the central nervous system was thought to be composed of a group of discrete areas wherein definite functions were located. It was on this basis that Figure 11-2 was constructed. It has more recently been demonstrated that cells with similar functions are not so neatly definable, although the mechanisms through which they operate generally appear to be as shown in the sketch. However, there well may be a hierarchy of importance where some parts of the centers exert control functions over others. Many of the major nerve centers or basal nuclei particularly in cats, monkeys, and humans have been well mapped, so well that persons with little knowledge of either neuroanatomy or neurosurgery can perform probe implantations for neurologic studies.

Sheaths

All nerve fibers are covered by a close fitting delicate membrane, the endoneurium (neurilemma). Outside the neurilemma is a second membrane, the sheath of Schwann, which is formed by the membranous cytoplasm of a series of specialized cells (Schwann cells). Several fibers may be covered by a single sheath of Schwann. If this occurs, the fibers will not be myelinated. Usually the peripheral nerves, which carry sensations of heat, cold and pain, association neurons, and postganglionic fibers of the sympathetic and parasympathetic are unmyelinated. Masses of unmyelinated neurons are found in the gray matter of the brain, spinal cord, and in the gray ramus communicans of the sympathetic nervous system.

If the neuron possesses a fatty (myelin) sheath, the fat will lie between the sheath of Schwann and the endoneurium. The neuron will be called myelinated or medullated. Myelinated neurons form the white matter of the brain and spinal cord, and the preganglionic fibers of the sympathetic and parasympathetic nervous system. They differ somewhat from unmyelinated neurons. Schwann cells only invest single fibers of myelinated neurons, and at intervals the cytoplasm of a Schwann cell ends and constricts around the nerve fiber it covers. A tiny gap, the node of Ranvier, is left between successive sheaths. This gives the myelinated neuron the appearance of a string of beads.

The tiny gap of unmyelinated nerve fiber at the node of Ranvier leaves the neuron exposed to penetration by interstitial fluid which can diffuse through the nodal area and provide nutrition to the nerve fiber.

Moreover, and more important, the node of Ranvier is the site where depolarization of a myelinated nerve takes place when the neuron is stimulated (p. 335). Myelin apparently increases the speed of a neural volley. According to one popular theory, the depolarization at the node of Ranvier creates a potential great enough to "jump" along the myelin bead to the next node and bypass the sequential depolarizations that occur in unmyelinated nerves.

The Synapse

Obviously nerve fibers must connect with each other or else a nerve impulse would have no place to go, and all nerve reflexes would be merely feedback circuits between a point receptor and a point effector. The connections that link the nervous system into an integrated whole are called synapses. A synapse is that point where the axon of one neuron meets the dendrite of another and over which nerve impulses can pass. The axon and dendrite do not contact each other directly. There is a tiny gap between the two fibers which cannot normally be bridged unless there is a conducting medium present. In the synapses of the central nervous system and in the parasympathetic portion of the autonomic nervous system, this conduction medium is acetylcholine. In the sympathetic, the conduction medium is usually noradrenalin. These media have resulted in the names adrenergic and cholinergic, which are often used to categorize portions of the nervous system. The chemical compounds are formed as needed and are destroyed or inactivated by specific enzymes as soon as the need has passed. There are literally billions of synapses in the bodies of animals that function to connect and interconnect the entire nervous system into a functional unit.

The development of new techniques and chemicals in the past few years has greatly enlarged the scope of understanding about the function and chemical mediators of synaptic function, and the use of chemotherapy on human mental patients has added to the knowledge. Tranquilizers, dopamine, prostaglandins, histamine, and other compounds have increased our knowledge of nerve transmission and inhibition. The specific techniques, unfortunately, are too complex to be described here.

In 1969, scanning electron photomicrographs of nerve trunks of marine snails, revealed bundles of fibers and knobs located at spots where two major nerve trunks meet. At 40,000 diameters magnification the knobs and fibers appear to link up with each other and form a physical connection between axons and dendrites. If these pictures are indeed the synapse as the researchers (E.R. Lewis and Y.Y. Zeevi at University of California, Berkeley) believe, there may be something different as far as marine snails are concerned. The statement that the axon and dendrite do not contact each other directly may have to be modified. However, the researchers do not challenge the statement that transmission of the nerve impulse in mammals is over the synapse and the transmission is across a gap bridged by a conducting medium.

Reflex Arcs

THE NERVE FIBER OR NEURON

The passage of a nerve impulse from receptor to sensory nerve to synapse to motor nerve to effector is known as the reflex arc. Reflex arcs are the principal means by which the nervous system performs its functions. These arcs vary in complexity from the simple reflex arc described above to exceedingly complex ones involving memory, association, and

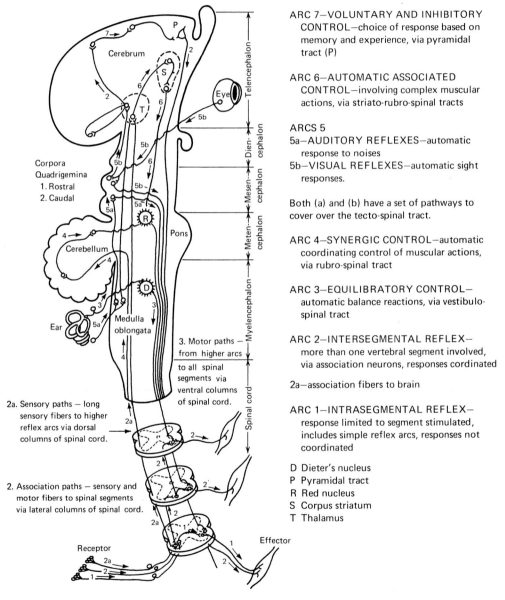

ARC 7—VOLUNTARY AND INHIBITORY CONTROL—choice of response based on memory and experience, via pyramidal tract (P)

ARC 6—AUTOMATIC ASSOCIATED CONTROL—involving complex muscular actions, via striato-rubro-spinal tracts

ARCS 5
5a—AUDITORY REFLEXES—automatic response to noises
5b—VISUAL REFLEXES—automatic sight responses.

Both (a) and (b) have a set of pathways to cover over the tecto-spinal tract.

ARC 4—SYNERGIC CONTROL—automatic coordinating control of muscular actions, via rubro-spinal tract

ARC 3—EQUILIBRATORY CONTROL—automatic balance reactions, via vestibulo-spinal tract

ARC 2—INTERSEGMENTAL REFLEX—more than one vertebral segment involved, via association neurons, responses cordinated

2a—association fibers to brain

ARC 1—INTRASEGMENTAL REFLEX—response limited to segment stimulated, includes simple reflex arcs, responses not coordinated

D Dieter's nucleus
P Pyramidal tract
R Red nucleus
S Corpus striatum
T Thalamus

FIGURE 11-2   *Reflex arcs C.N.S. (Adapted from Patten,* Embryology *of the Pig.)*

inhibitory control, where the voluntary portions of the brain control and modify response. The accompanying figure (Figure 11-2) gives a graphic picture of the seven basic reflex arcs involving the central and peripheral nervous systems. A similar system of reflex arcs exists for the autonomic nervous system, but the principal business of the autonomic is conducted at the level of reflex arcs one and two. The higher arcs seldom enter into autonomic activity.

THE BRAIN

The brain is the most highly specialized tissue in the body. It is almost entirely composed of nervous tissue and is the control center for all voluntary nervous reactions. It is a hollow organ, composed of a central system of ducts and cavities, surrounded by myelinated nerve fibers or white matter. Surrounding this white matter is a layer of nerve cells and unmyelinated processes called gray matter. The brain is enclosed in the meninges, a triple layer of fibrous tissue, and is bathed constantly in a lymph-like fluid (cerebrospinal fluid) which fills the central duct system and the space between the meninges and the brain.

The brain is developed from the cranial part of the neural tube of the embryo. The cranial portion of the tube is first divided into three main parts by two constrictions, thus forming the forebrain, midbrain, and hindbrain. The hindbrain gives rise to three secondary segments, and the forebrain to two, while the midbrain remains without further subdivision.

Regions of the Brain

Primary Segments	Secondary Segments	Derivatives
Rhombencephalon (hindbrain)	Myelencephalon	Medulla oblongata
	Metencephalon	Pons and cerebellum
	Isthmus rhombencephalon	{ Ant. cerebellar peduncle { Ant. medullary velum
Mesencephalon (midbrain)	Mesencephalon	{ Corpora quadrigemina { Cerebral peduncles
Prosencephalon (forebrain)	Diencephalon	{ Thalamus, Pineal body { Pituitary body, Optic nerve
	Telencephalon	{ Cerebral hemispheres { Olfactory tract and bulb

The Rhombencephalon

The medulla oblongata is sometimes called the brain stem. It is the caudal portion of the brain and extends from the foramen magnum to the pons. It lies on the basilar part of the occipital bone, is quadrilateral in outline, wider in front than behind, and compressed dorsoventrally. Its length, measured from the root of the first cervical nerve to the pons, is about 2 inches in the horse. It is the area of origin of the glossopharyngeal, vagus, spinal accessory, and hypoglossal nerves. Located in the medulla oblongata are the cardiac and respiratory centers, plus several nuclei that function to control equilibrium and touch sensations.

The pons is the part of the brain stem that lies between the medulla and the cerebral peduncles; it is marked off from these ventrally by rostral and caudal grooves. Viewed ventrally it is wider than it is long and is convex in both directions. Laterally a large part of its mass curves dorsally and backward into the base of the cerebellum. It is the area of origin of the trochlear, trigeminal, abducens and facial nerves, and is the area of entrance of the acoustic nerve axons. It contains the pontine nucleus of the pneumotaxic center and areas involved in postural reflexes.

The cerebellum is situated in the caudal cavity of the cranium rostral to the medulla oblongata and dorsal to the pons. It is separated from the cerebral hemispheres by a transverse fissure. It is separated from the pons and the greater part of the medulla by the fourth ventricle. Its shape is approximately globular but is very irregular and complexly wrinkled on its surface. It is considered to be composed of two halves or hemispheres, separated by a middle portion called the vermis, possibly because of its resemblance to a coiled and folded worm (Figure 11-3). In cross section the cerebellum has a tree-like appearance which is responsible for the poetic title "tree of life" or arbor vitae, which is sometimes applied to it. It is somewhat compressed dorsoventrally and its transverse diameter is the greatest. It functions primarily in maintaining tonus, and in posture and equilibrium reflexes.

The rostral cerebellar peduncles pass forward on either side of the dorsal surface of the pons to form the lateral boundaries of the forepart of the fourth ventricle.

The rostral medullary velum is a thin lamina which forms the rostral part of the roof of the fourth ventricle.

The Mesencephalon

The corpora quadrigemina are four (two in birds) grayish hemispherical eminences which are also called the rostral and caudal colliculi. They lie under the caudal part of the cerebral hemisphere. They consist of two pairs separated by a transverse groove. The rostral pair is the larger. This region contains the nuclei of the trochlear and oculomotor nerves and the neural paths connected with ocular reflexes. The red nucleus in the area is the origin of the rubrospinal tract that deals with automatic associated reflexes. The substantia nigra which is also found in this area is associated with Parkinson's disease in humans and yellow star thistle poisoning in horses.

The cerebral peduncles are two large rope-like stalks that appear at the base of the telencephalon. They emerge from the pons close together and divide as they go forward to enter the cerebrum. They disappear into the cerebrum at a point near the crossing of the optic nerves (the optic chiasma).

The Prosencephalon

The prosencephalon is composed of two major subdivisions, the diencephalon and the telencephalon. The diencephalon or inner brain includes the thalamus, hypothalamus and a number of other structures grouped around the third ventricle, which is the central cavity of this division of the forebrain.

The thalamus is the largest structure in the diencephalon. It is composed of two ovoid masses lying obliquely across the dorsal surfaces of the cerebral peduncles in such a manner that the long axes of the two thalami meet at about right angles. Around the area of fusion of the right and left thalamus is a sagittal space called the third ventricle. The thalamus contains the medial and lateral geniculate nuclei which are, respectively, major relay links for the optic and acoustic nerves. The hypothalamus is the lower (ventral) area of the thalamus at the point of fusion and lies immediately dorsal to the pituitary body. It is a geographic rather than anatomic division, but it differs functionally from the remainder of the thalamus in that it secretes hormones and control substances (p. 347). The pineal and pituitary bodies are discussed in the chapter on hormones (p. 346ff).

The optic nerves run from the caudal part of the eye, cross at the optic chiasma, and disappear into the brain in the direction of the rostral corpora quadrigemina.

The cerebral hemispheres form the greater part of the fully developed mammalian brain and are the outstanding structures of the telencephalon. In the aggregate they are often called the cerebral cortex, although this term should properly apply only to the outer layer of gray matter of the hemispheres. Viewed from above, the cerebral hemispheres form an ovoid mass of which the broader end, or base, is caudal and the greatest transverse diameter is just rostral to the base. The two hemispheres are separated by a deep median cleft. When the hemispheres are gently drawn apart, it can be seen that the cleft is interrupted in its middle by a white mass, the corpus callosum. This structure connects the two hemispheres for about one-half their length. The surfaces of the cerebral hemispheres are marked by numerous complex convolutions composed of folds (gyri) and grooves (sulci). In some mammals, notably rats, mice and other small rodents, the cerebrum is lissencephalic, i.e., "smooth brain." There are no gyri and sulci. This does not seem to impair the efficiency of their thinking apparatus. It has been postulated that the gyri and sulci exist to provide extra cortical area for

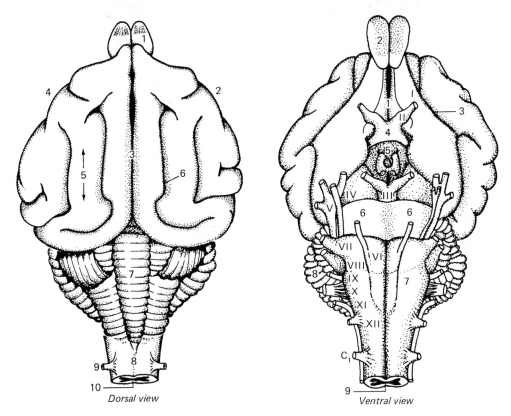

FIGURE 11-3   *Brain of the cat.*

1. Olfactory bulb
2. Right cerebral hemisphere
3. Longitudinal fissure
4. Left cerebral hemisphere
5. Gyrus
6. Sulcus
7. Cerebellum
8. Medulla oblongata
9. Spinal nerve
10. Spinal cord

I. Olfactory (S)
II. Optic (S)
III. Oculomotor (M)
IV. Trochlear (M)
V. Trigeminal (SM)
VI. Abducens (M)
VII. Facial (SM)
VIII. Vestibulocochlear (acoustic) (S)
IX. Glossopharyngeal (SM)
X. Vagus (SM)
XI. Spinal accessory (M)
XII. Hypoglossal (M)
C_1. First cervical (spinal) nerve

1. Longitudinal fissure
2. Olfactory bulb
3. Sulcus rhinalis
4. Optic chiasma
5. Tuber cinereum and pituitary body
6. Pons
7. Medulla oblongata
8. Cerebellum (right hemisphere)
9. Spinal cord

the brain and to improve reasoning capacity. While it is true that most of the more intelligent mammals have remarkably convoluted brains, the ratio of brain to spinal cord weight as a measurement of relative intelligence seems to be a better criterion.

The hippocampus lies in the caudolateral area of the cerebral hemispheres. Its exact function is not known. It serves as a key location for the diagnosis of rabies in animals.

The olfactory tracts and bulbs form the rostrolateral and rostral

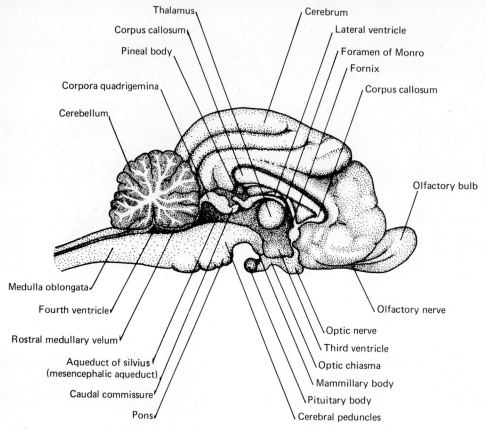

FIGURE 11-4   Brain of the cat—sagittal section.

portions of the telencephalon. The olfactory bulbs are hollow oval enlargements that are found rostroventrally to the cerebral hemispheres. Their convex rostral ends fit into the ethmoidal fossa and receive numerous olfactory nerve fibers through the cribriform plate; hence it is very difficult to remove the bulb intact. The olfactory tracts are bands of white matter which extend from the olfactory bulbs to the brain. They each contain a canal that connects the central cavity in the bulb to the lateral ventricles of the brain.

The Ventricles

The ventricles form the central cavity of the brain and are divided for convenience into four ventricles, two foramina, and one duct. The ventricles are I (left), II (right), III (third), and IV (fourth). The foramina*

---
*The medial foramen of Magendie is present in man and primates but apparently does not occur in other animals.

are the foramina of Monro and Luschka. The duct is the aqueduct of Silvius (mesencephalic aqueduct). The first and second ventricles receive the central canals from the olfactory tracts and expand to fill the central portion of the left and right cerebral hemispheres. They join along the midline to form the constricted foramen of Monro (interventricular foramen) which connects them to the third ventricle located in the dien-

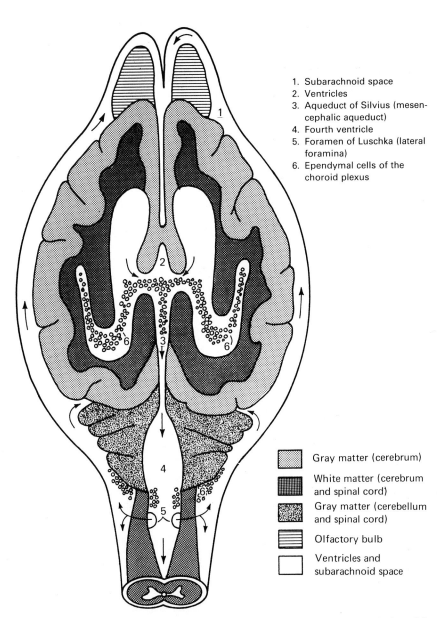

1. Subarachnoid space
2. Ventricles
3. Aqueduct of Silvius (mesencephalic aqueduct)
4. Fourth ventricle
5. Foramen of Luschka (lateral foramina)
6. Ependymal cells of the choroid plexus

- Gray matter (cerebrum)
- White matter (cerebrum and spinal cord)
- Gray matter (cerebellum and spinal cord)
- Olfactory bulb
- Ventricles and subarachnoid space

FIGURE 11-5 Brain—schematic—showing flow of cerebrospinal fluid. (Adapted from Innes and Saunders, Comparative Neuropathology.)

cephalon. The third ventricle constricts into a communicating duct, the aqueduct of Silvius (mesencephalic aqueduct), which empties into the fourth ventricle. The fourth ventricle underlies the cerebellum and forms a triangular area, dorsally located in the anterior of the medulla oblongata. This area is drained by the foramina of Luschka (lateral apertures of the fourth ventricle) which empties into the subarachnoid space surrounding the brain and cord.

A small amount of fluid passes into the central canal of the spinal cord. The ventricles are filled with a fluid similar to lymph called cerebrospinal fluid which is secreted by the ependymal cells of the choroid plexus. The choroid plexus is an aggregation of capillaries, which is found in all of the ventricles of the brain. Its principal function, other than supplying nutrition and oxygen to the deep tissues of the brain, is to produce the cerebrospinal fluid which is derived from the fluid elements of the blood. Measurements of cerebrospinal fluid pressure in various species indicate that mean pressure is about .110-.115 mm of normal saline solution. In animals with vitamin A deficiency, the pressure is higher.

Cerebrospinal Fluid

Cerebrospinal fluid, in contrast to blood plasma from which it is derived, is of a constant and uniform composition. This argues that the choroid plexus has a true secretory function. Studies with radioisotopes indicate that there is a constant inflow and outflow of electrolytes and water from the bloodstream to the ventricular cavities, and that the cells of the choroid plexus are involved in this movement.

The cerebrospinal fluid fills the ventricles of the brain and the central canal of the spinal cord, and leaves these cavities over two routes: the foramina of Luschka (and Magendie in primates), located in the posterior medullary velum, and from the ventricles through the nervous tissue to the surface of the brain. The fluid tends to remain in the subarachnoid spaces of the brain and spinal cord and to collect in pockets (or cisterns) at the base of the brain and in the cisterna magna between the cerebrum and cerebellum. Probably the majority of the cerebrospinal fluid is reabsorbed into the bloodstream via the arachnoid villi, the dural sinuses and the venous circulation. A certain portion, however, flows along the nerve sheaths of the spinal nerve into the tissue spaces and is picked up by the peripheral lymphatics.

The cerebrospinal fluid has both defensive and nutritive functions in addition to being a shock absorbing and homeostatic mechanism. It has antibacterial and antitoxic properties and carries antibodies. It is low in proteins, but contains considerable amounts of mineral salts, and carbohydrates. It is watery in consistency and normally does not contain cellular elements or constituents which are not present in the blood from which it is derived.

The Meninges

The brain is surrounded by three tissue layers called meninges. These consist of a tough outer layer called the dura mater, which is composed of two tissue layers and adheres closely to the inner walls of the cranium. The dura mater is separated by a small subdural space from the middle tissue layer or arachnoid. The arachnoid is a lace-like membrane bearing numerous villi, blood and lymph vessels, and is

THE BRAIN

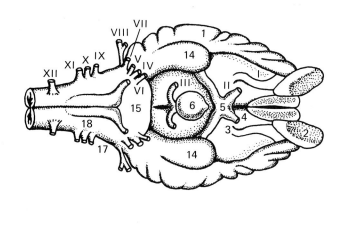

I. Olfactory
II. Optic
III. Oculomotor
IV. Trochlear
V. Trigeminal
VI. Abducens
VII. Facial
VIII. Auditory (vestibulocochlear)
IX. Glossopharyngeal
X. Vagus
XI. Spinal accessory
XII. Hypoglossal

1. Cerebrum
2. Olfactory bulb
3. Olfactory nerve
4. Optic nerve
5. Optic chiasma
6. Pituitary body
7. Corpus callosum
8. Fornix
9. Pineal body
10. Thalamus
11. Caudal commissure
12. Corpora quadrigemina
13. Mammillary body
14. Cerebral peduncles
15. Pons
16. Rostral medullary velum
17. Cerebellum
18. Medulla oblongata
19. Lateral ventricle
20. Foramen of Monro (interventricular foramen)
21. Third ventricle
22. Aqueduct of Silvius (mesencephalic aqueduct)
23. Fourth ventricle

FIGURE 11-6   Brain of the horse—ventral view and sagittal section. (Adapted from Sisson and Grossman, The Anatomy of the Domestic Animals.)

separated from the inner layer or pia mater by the subarachnoid space. The pia mater is a thin membrane which adheres closely to the brain and follows the fissures and sulci deep into the brain tissue.

The meninges of the brain are directly continued as the meninges of the spinal cord. They contain the cerebrospinal fluid and afford some additional mechanical protection of the brain and spinal cord against injury. The dura mater of the spinal cord, unlike that of the brain, is not two layered, nor is it adherent to the surrounding bone. Instead, there is an appreciable gap between the spinal dura and the wall of the vertebral canal. The gap is called the epidural space.

THE SPINAL CORD

The spinal cord is the direct continuation of the brain into the vertebral canal. There is no distinct line of demarcation between brain and cord, but for practical purposes it is assumed that the foramen magnum represents the point of division. Structurally, the cord is somewhat similar to the brain, being surrounded by meninges, composed of gray and white matter, and giving off a number of nerves. However, outside of this basic similarity, there is little comparable between brain and cord. Essentially the cord is a communicating link between brain and body; the main trunk line for nervous impulses passing to and from the brain and between the various segments of the spinal cord.

The arrangement of white and gray matter in the cord is the opposite of its arrangement in the brain. In the spinal cord the gray matter is centrally located and the white matter is peripheral. The spinal nerves (p. 329) which connect to the dorsolateral and ventrolateral portions of the spinal cord along well-defined lines pass from the vertebral canal through intervertebral foramina which ordinarily lie between two adjacent vertebrae.

A cross section of the spinal cord reveals that it is an oval structure, almost completely divided into right and left portions by two deep median folds, the dorsal septum and the ventral fissure. A small canal, which connects with the ventricles of the brain, partially fills the space between the ends of the folds. The central canal is surrounded by gray matter which extends into the right and left portions by forming an H-shaped mass. The gray matter of the cord, like that of the brain, is composed of cytons and unmyelinated fibers. The crossbar is known as the gray commissure and is separated from the dorsal end of the ventral fissure by a narrow band of white matter called the white commissure (#17 in Figure 11-7). The vertical bars of the H are called dorsal and ventral horns, depending upon whether the parts are located above or below the gray commissure. The gray commissure, horns, and the dorsal and ventral roots of the spinal nerves divide the surrounding white

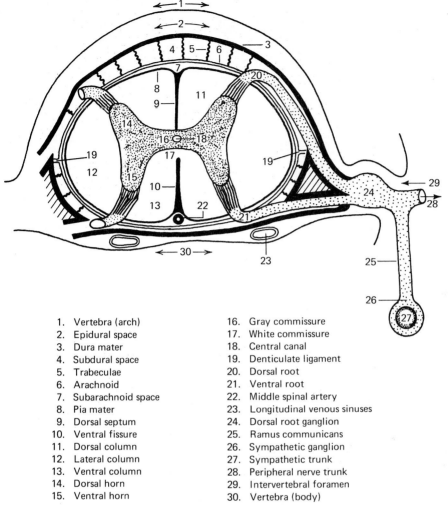

1.	Vertebra (arch)	16.	Gray commissure
2.	Epidural space	17.	White commissure
3.	Dura mater	18.	Central canal
4.	Subdural space	19.	Denticulate ligament
5.	Trabeculae	20.	Dorsal root
6.	Arachnoid	21.	Ventral root
7.	Subarachnoid space	22.	Middle spinal artery
8.	Pia mater	23.	Longitudinal venous sinuses
9.	Dorsal septum	24.	Dorsal root ganglion
10.	Ventral fissure	25.	Ramus communicans
11.	Dorsal column	26.	Sympathetic ganglion
12.	Lateral column	27.	Sympathetic trunk
13.	Ventral column	28.	Peripheral nerve trunk
14.	Dorsal horn	29.	Intervertebral foramen
15.	Ventral horn	30.	Vertebra (body)

FIGURE 11-7 *Structures of the spinal cord (schematic).*

matter on each side into three columns (dorsal, ventral, and lateral). These columns are composed mainly of myelinated nerve fibers and contain tracts that are somewhat specialized in function: the dorsal column mainly carries sensory impulses, the ventral column is largely motor, and the lateral columns carry mixed sensory, motor, and association fibers which connect to the brain or to other segments of the spinal cord.

The white matter is enclosed by the innermost of the meninges, the pia mater, which closely invests the surface of the cord. Surrounding the pia mater at a little distance is the arachnoid. The space between the pia mater and the arachnoid is called the subarachnoid space, and

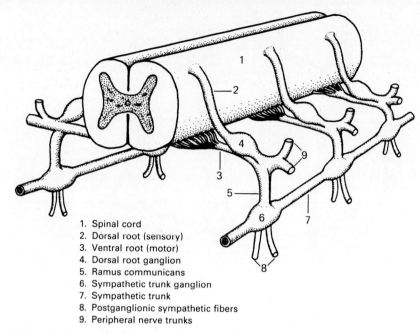

1. Spinal cord
2. Dorsal root (sensory)
3. Ventral root (motor)
4. Dorsal root ganglion
5. Ramus communicans
6. Sympathetic trunk ganglion
7. Sympathetic trunk
8. Postganglionic sympathetic fibers
9. Peripheral nerve trunks

FIGURE 11-8   *Spinal cord and sympathetic trunk.*

contains cerebrospinal fluid. In the subarachnoid space at the level of the intervertebral foramina a thin band of tissue, the denticulate ligament (ligamentum denticulum), runs from the pia to the dura mater and divides the subarachnoid space into dorsal and ventral parts. The dura mater is the tough outer membrane which encloses the cord and closely invests the spinal nerves at the inner openings of the intervertebral foramina and extends laterally a short distance into the intervertebral foramina. In the epidural space of the ventral part of the spinal cord are found two longitudinal venous sinuses which provide blood drainage for the cord and meninges. The larger nerve trunks outside the central nervous system are covered by a protective sheath of connective tissue called the epineureum which usually contains a network of capillaries that provide nutrients and remove wastes from the enclosed nerves.

Length of the Spinal Cord

The true spinal cord extends along the vertebral column from the foramen magnum to a point in the lumbar or sacral vertebrae that varies with the species of animal. There the terminal portion tapers to a point, called the conus medullaris, which splits to form a bundle of hairlike fibers, the cauda equina. Beyond this point the vertebral canal contains nerve fibers of the cauda equina which pass out of the canal through the remainder of the intervertebral foramina. The extreme caudal portion of the cord ends in a terminal filament which is composed of connective tissue derived from the pia mater.

Termination of Cord—Comparative Anatomy

Horse	1st Sacral vertebra	Pig	3rd Sacral vertebra
Cow	1st Sacral vertebra	Dog	6th Lumbar vertebra
Sheep	1st Sacral vertebra	Cat	3rd Sacral vertebra

THE PERIPHERAL NERVOUS SYSTEM

The outstanding feature of the peripheral nervous system is its intimate association with the central nervous system. It is composed of sensory and motor nerves that are connected to the brain or spinal cord. The nerve trunks that connect directly to the brain may be purely sensory or motor, or they may be mixed. The nerves that connect with the spinal cord are purely sensory or motor where they enter or leave the cord, but shortly thereafter they become united to form a mixed nerve.

The Cranial and Spinal Nerves

The peripheral nervous system begins in the brain with the terminal nerve (nervus terminalis). This nerve was first reported by Pinkus in 1894. Since then it has been found in a large number of animals including cats, dogs, and rodents. It has been classed as a cranial nerve by zoologists but not by veterinarians and physicians. It is closely associated with the olfactory nerves, leaves the brain at the olfactory bulb, passes through the cribriform plate of the ethmoid, and terminates in the nasal epithelium in fibrils which pass to blood vessels in the nasal mucosa, where it regulates blood flow. The 12 cranial nerves are specific structures which are similar in composition, origin, and function in all mammals. Although the nerves are similar in different animal species, they are not identical. The various drawings of the brains (Figures 11-3, 11-4, 11-5, 11-6) show some of the differences. The table of cranial nerves that follows this section will not be accurate for all mammals, but should serve as a guide and a memory aid for the general characteristics and nomenclature that should be remembered.

The paired spinal nerves that comprise the terminations and origins of the remainder of the peripheral nervous system form the dorsal and ventral roots of the spinal cord. The dorsal roots are sensory; the ventral roots are motor. The roots coalesce at the intervertebral foramen to form a single mixed nerve which contains peripheral sensory and motor fibers, and fibers that connect to the autonomic nervous system. Just prior to coalescence the cytons of the sensory roots are aggregated into the dorsal root ganglia which are located in the intervertebral foramina. The cytons of the ventral roots are mainly in the ventral horns of gray matter within the spinal cord. The sensory and motor fibers ramify through the body tissues, terminating respectively in receptor and effector endings.

The Cranial Nerves

No.	Name	S/M*	Origin	Termination
I	Olfactory	S	Nasal Epithelium	Olfactory bulb (diencephalon)
II	Optic	S	Retina of eye	Optic chiasma (diencephalon)
III	Oculomotor	M	Cerebral peduncle	Ventral oblique, dorsal, medial, and ventral rectus muscles of eye
IV	Trochlear	M	Rostral medullary velum	Dorsal oblique muscle of eye
V	Trigeminal	SM	Caudal portion of pons	Muscles and receptors of face, oral, nasal, and buccal region
VI	Abducens (Abducent)	M	Rostral medulla oblongata	Lateral rectus muscle of eye and retractor oculi muscle
VII	Facial	SM	Medulla oblongata lateral surface	Muscles of head and face; nasal, palatine, and lacrimal glands; salivary glands
VIII	Acoustic (auditory, vestibulocochlear)	S	Internal ear	Medulla oblongata, lateral surface caudal to VII
IX	Glossopharyngeal	SM	Medulla oblongata lateral surface caudal to VIII	Tongue, tonsils, palate, pharynx
X	Vagus	SM	Medulla oblongata caudal to IX	Pharynx, larynx, trachea, esophagus, heart, lungs, stomach, intestines, thyroid, liver, spleen, and kidney
XI	Spinal accessory (accessory)	M	Medulla oblongata lateral surface & several roots in the cervical region	Musculature of neck and shoulders
XII	Hypoglossal	M	Medulla oblongata lateral surface caudal to X	Hyoid muscles and tongue

*S = sensory, M = motor, SM = mixed

THE AUTONOMIC NERVOUS SYSTEM

The term "autonomic" was coined to indicate that this portion of the nervous system has autonomy, or the ability to function independently of the central nervous system. It is that part of the nervous system which transmits stimuli to the heart, smooth muscle, and glands, and conducts impulses from the viscera to the central nervous system.

It is impossible to completely separate the autonomic from the central and peripheral nervous systems anatomically, since autonomic fibers are frequently carried in nerve trunks along with peripheral fibers. The only method of division is physiological, by observing what the autonomic system does and how it functions. Through extensive studies it has been shown that the autonomic system supplies the visceral muscles and glands and conducts the major part of its operations through the means of reflex arcs which do not necessarily involve the higher nerve centers.

The autonomic nervous system is composed of two major divisions, the sympathetic (adrenergic) and the parasympathetic (cholinergic). These two divisions are sometimes referred to respectively as thoracolumbar and craniosacral, a terminology based upon their anatomical location. The terms "adrenergic" and "cholinergic" refer to the chemical compounds noradrenalin and acetylcholine which are necessary for the transmission of nerve impulses across the synaptical gaps separating the nerve fibers of the respective systems. Functionally the autonomic nervous system differs from the peripheral and central nervous systems in that the nerves appear to be less irritable and more sensitive to stimuli of long duration. The speed of conduction along the autonomic nerves is considerably slower than it is in the central or peripheral nervous system. A minor division, the myenteric, is also considered to exist.

Sympathetic Division

This portion of the autonomic nervous system is frequently called the thoracolumbar system because its fibers connect to the spinal cord only in the thoracic and lumbar regions.

Sympathetic fibers pass to the body through the intervertebral foramina, along with fibers from the sensory and motor roots of the peripheral nervous system. After emerging from the foramina most of the sympathetic fibers leave the main trunk as the white ramus communicans, which joins the sympathetic trunk ganglia.

The sympathetic trunk is composed of two parallel bundles of autonomic nerve fibers running along the ventrolateral portions of the bodies of the thoracic and lumbar vertebrae. At each vertebral junction, the sympathetic trunk forms a sympathetic trunk ganglion which receives fibers from the cord via the white ramus communicans. Here the sympathetic fibers from the cord form synapses and proceed in three directions: (1) along the sympathetic trunk to other sympathetic trunk ganglia; (2) directly into visceral plexuses or ganglia, or to visceral receptors and effectors; (3) to somatic glands (sweat, sebaceous, ceruminous, mammary) and the arrectores pili muscles. These last fibers return to the spinal nerve as postganglionic axons over the gray ramus communicans and proceed along with the peripheral nerves to their proper destinations.

Sympathetic fibers to the head and neck arise from the caudal cervical (stellate, cervicothoracic) ganglion which receives fibers from the sympathetic trunk and the first thoracic segment. These fibers pass cranially along the sides of the cervical vertebrae and enter the cranial cervical ganglia which lie near the base of the skull. From the anterior cervical ganglia the fibers pass to such structures as the salivary and lacrimal glands and the intrinsic muscles of the eye (dilator iridis and ciliary muscles).

In a similar manner sympathetic innervation is carried to structures caudal to the lumbar region. The fibers pass from the lumbar ganglia and the caudal end of the sympathetic trunk to the caudal mesenteric ganglion, where postganglionic fibers pass to the mammary glands (where these are located pelvically or abdominally), the urogenital apparatus, and the glands of the terminal portions of the colon, rectum, and anus.

Sympathetic innervation to the heart is effected by postganglionic fibers from the stellate ganglion and from preganglionic fibers from other thoracic segments. The fibers terminate in the cardiac ganglion at the base of the heart and send postganglionic fibers to the atria.

Sympathetic innervation to the lungs is doubtful. Since lung action and function is controlled by the respiratory center in the brain, these neural connections are apparently unnecessary.

Sympathetic innervation to the intestines is mainly over long postganglionic fibers from the celiac, cranial mesenteric, and caudal mesenteric ganglia. The celiac ganglion is the main junctional region for sympathetic innervation to cranial abdominal viscera. Postganglionic fibers from the celiac and cranial mesenteric ganglia and from the sympathetic trunk ganglia in the caudal thoracic region pass to the stomach, liver, spleen, kidneys, adrenals, and the cranial part of the small intestine. They terminate mainly in the smooth muscle walls of the smaller arterioles and the muscularis mucosa and probably in the intestinal glands. Their adrenergic effects result in vaso-constriction and muscular relaxation, which stops or inhibits digestive activity and gut motility.

Sympathetic innervation to the caudal portions of the small intestine and the cecum is principally derived from the cranial mesenteric ganglion and postganglionic fibers from the last thoracic and first two or three lumbar vertebrae, depending on the species of animal. Connecting postganglionic linkages to and from the celiac ganglion permit dissemination of major impulses over a wide area. Connecting links to the caudal mesenteric ganglion, and impulses arising from the lumbar spinal segments and sympathetic trunk ganglia give sympathetic innervation to the large and small colon, the rectum, the sex organs, and the urinary bladder. Depending upon their location, the mammary glands receive sympathetic fibers from the celiac, cranial mesenteric, and caudal mesenteric ganglia, and from spinal segments and sympathetic trunk ganglia extending from the mid-thoracic through the lumbar region. Animals with long functional milk lines, such as rats, may have all spinal segments in the thoracic and lumbar region involved in mammary innervation and secondary involvement of all the abdominal ganglia.

Parasympathetic Division

This portion of the autonomic nervous system is sometimes referred to as the craniosacral system because its nerve fibers arise from cranial nerves III, VII, IX, and X, and from the sacral portion of the spinal cord. The fibers generally go to the same structures as are supplied by the sympathetic. Normally two nerve fibers are involved; the cyton of the first fiber lies in the brain or the sacral region, while the terminal cyton is located in a ganglion near or on the organ supplied. The postganglionic fibers of the terminal portions of the parasympathetic are usually short.

Parasympathetic fibers from cranial nerves III, VII, and IX are distributed only to the regions of the head, to effectors antagonistic to those supplied by the cranial cervical ganglion of the sympathetic. The vagus nerve, however, carries parasympathetic innervation to the esophagus, heart, lungs, stomach, small intestine, and the cranial portions of the large intestine.

The parasympathetic fibers from the sacral portion of the system innervate the urogenital apparatus and the structures caudal to the first third of the large intestine. However, there is no parasympathetic innervation of the mammary glands or the skin, and it is doubtful if there is any appreciable parasympathetic control of the gonads.

In general, the sympathetic is an emergency system and is concerned with rapid release of energy to meet critical situations, while the parasympathetic takes care of the normal demands of living. The parasympathetic also functions to restore energy and produce the relaxation necessary to prepare for sudden energy release.

To some extent the parasympathetic and sympathetic are antagonistic to each other, as shown in the following table:

Sympathetic	*Parasympathetic*
Accelerates heartbeat rate and amplitude	Slows heartbeat and amplitude
Dilates pupil of eye	Contracts pupil of eye
Constricts blood vessels	Dilates blood vessels
Dilates bronchi and bronchioles	Constricts bronchi and bronchioles
Inhibits gastrointestinal motility	Excites gastrointestinal motility

Myenteric Division

The myenteric division of the autonomic nervous system is confined to the intestinal tract. It arises in structures called Auerbach's plexuses which are tiny ganglia and unmyelinated nerve fibers that lie between the muscular layers of the gut and can be found scattered along its length. The nerve fibers connect the ganglia to Meissner's plexuses in the submucosa of the intestine, and since the Meissner's plexuses are responsible for stimulating intestinal movements, the gut can thus be motivated without either sympathetic or parasympathetic innervation. This argues that the Auerbach-Meissner arrangement should be classed as an independent or quasi-independent division of the nervous system rather than a subsidiary part.

Without carrying this presently unresolved argument any farther, the information is presented here without comment except to state that there are protagonists and antagonists—and, at the moment, the antagonists are the majority.

Cardiac Division

The principle that a structure which has autonomous reflex capability and functional independence of both the voluntary and other divisions of the involuntary nervous system should be classed as a separate division is also illustrated in the heart. The heart will continue to beat after all extrinsic nerves are removed, but the usual heartbeat stimulus is not of nervous origin or transmission. Since this mechanism has already been described (p. 261) it will not be included here.

GENERAL PHYSIOLOGY OF THE NERVOUS SYSTEM

The principal function of a nerve fiber is to conduct electrochemical impulses along its length. A synapse serves to connect one nerve fiber (axon) with another (dendrite). Stimuli are furnished by specialized dendrite endings known as receptors. Actions are accomplished by specialized axon endings known as effectors.

Although in the laboratory it can be shown that a nerve impulse can be conducted in either direction along an isolated neuron, in the body the impulse in a single neuron is unidirectional, i.e., from dendrite to axon. This phenomenon results from the fact the synapses conduct nerve impulses in only one direction, i.e., from axon to dendrite.

The principal properties of nerve fibers are excitability and conductivity. These are electrochemical in nature. When a stimulus is applied to a nerve (in the laboratory these are usually weak electrical currents which do not damage the nerve tissue) chemical and electrical changes appear throughout the entire nerve. These changes have been shown to be completely independent of the primary stimulus and do not reflect its nature, which may be anything (e.g., electrical, chemical, thermal, mechanical) capable of producing a response. Furthermore, a nerve obeys the "all or none law": Either it fires or it does not.

The firing (the excitation of the nerve impulse) results from an electrochemical reaction that produces an effect called the action potential, which passes along the nerve fiber. Before it can be excited, however, a neuron must have a resting potential difference of some 50 to 100 millivolts between the inside and the outside of the nerve fiber. The resting nerve cell accomplishes this by removing sodium ions (Na^+) from the inside of the neuron. Once removed, the Na^+ are kept out by the impermeable cell membrane. This builds a negative charge within the cell and a positive charge outside it.

When the nerve is fired, the stimulated local area of the cell becomes momentarily permeable to sodium ions which rush through the cell membrane to satisfy the demand for ions that was created during the resting phase. This reverses the polarity of the cell in this area and the inside of the cell becomes 25 to 50 millivolts positive. This in turn increases the permeability of the cell membrane in the adjoining negatively polarized region and another change in polarity occurs. These successive changes in electrical potential (action potential) pass along the nerve fiber resulting in the phenomenon of "progressive depolarization" or a chain of propagated action potentials. The summation of these effects causes a wave of electrical energy to flow down the nerve. This is a nerve impulse.

The cell then passes into a refractory period during which the resting potential is reestablished.

The resting potential is maintained at an effective level by the nerve cell mitochondria which keep a high concentration of potassium and a low concentration of sodium inside the cell processes (compared to that which exists in the extracellular medium).

The fact that the interior of a nerve cell process is electrically negative as compared with its exterior was established using cannulated nerves of squid which possess so-called giant axons. Into and around these large fibers, investigators inserted electrodes, and then connected the electrode from the periphery to the one inside the axon through an amplifier and an oscillograph. This registered the resting potential. When the nerve was stimulated, an action potential peak was generated that was far greater than the resting potential. Other investigations showed that this rise in action potential was due to depolarization of the resting potential. Still other experiments demonstrated that action potentials along a nerve have a characteristic and constant value under identical conditions. The action potentials are propagated at the same speed as the nerve impulse and are present in every part of the nerve during the period it is conducting an impulse.

These facts gave rise to the theory of progressive depolarization, which is now generally accepted and which states essentially that a wave of depolarization flows along the surface of the nerve fiber and produces the action potential. This activity, while rapid, does not result in particularly rapid transmission of a nerve impulse over a naked axon. Nerve fibers have a number of modifying characteristics that affect the speed of transmission, i.e., the cross-sectional diameter of the fiber, the presence of a myelin sheath, and the location in the neural network. Fibers of larger diameter conduct impulses faster than smaller ones. Medullated fibers conduct impulses faster than unmyelinated ones, and generally, the large myelinated (group A) fibers, and the small to medium (group B) fibers of the peripheral nervous system will conduct impulses faster than comparable neurons of the autonomic nervous system. Unmyelinated fibers (group C) have the slowest transmission speeds. Moreover, sensory transmission appears to be slower than motor in both systems. The significance of these varying speeds is still far from clear.

Nerve transmission speeds vary considerably between species as well as within the individual. In mammals, motor nerve speeds range from 90 to 120 meters/second. In frogs, similar nerves transmit impulses at about 25 meters/second. Recorded speeds of small to medium sized myelinated sensory nerves in mammals run from 25 to 50 meters/second, and similar autonomic fibers conduct at 3 to 14 meters/second. Unmyelinated peripheral and association neurons transmit at from 3 to 10 meters/second while similar autonomic fibers range from 0.3 to 2.3 meters/second, which is hardly more than a crawl.

This all goes to show that the speed of thought is slow at best since the gray matter of the brain is unmyelinated and the speed of conductivity probably does not exceed three meters per second—about the speed of a jog trot—seven miles per hour. It is the difference between the pedestrian activity of the brain and the transmission speed of an electronic computer (186,000 miles per second) that gives the computer such an enormous advantage in solving a properly programmed problem. With such a difference in relative speeds the computer can afford to go through an insanely repetitive sequential examination of data bits and still come in considerably ahead of a human even though the human has the capability of intuitive jumps that can bypass repetition.

The speed of transmission along myelinated fibers is accomplished because the depolarization impulse apparently jumps from one node of Ranvier to the next, and the intensity of the action potential at one node is sufficient to jump along the myelin sheath to the next node. Experiments have demonstrated a higher action potential at the nodes than can be found elsewhere along the myelinated axon, a potential appreciably higher than that which occurs in unmyelinated axons of comparable size.

While it has been known since 1937 that acetylcholine is an essential element in the transmission of most nerve impulses, its role is still not completely understood. Theoretically, depolarization of the nerve fiber may be due to the presence of acetylcholine which is deactivated by acetylcholinesterase once the impulse has passed. At any rate, acetylcholine and acetylcholinesterase are very important in a majority of the synapses where they form the "on-off" mechanism of the uniting medium between nerves.

Although the reflex arcs have been previously described (p. 317) their physiological properties were not enumerated. Essentially these are included under two major headings, categories and aspects.

1. Categories
   (a) Exteroceptive reflexes: those that arise outside the body and involve
      (1) Cutaneous receptors: those which involve stimulation of receptors in the skin.
      (2) Chemoreceptors: those involved in taste and smell.
      (3) Wave receptors: those involved in sight and hearing.
   (b) Enteroceptive reflexes: those which arise inside the body and affect
      (1) Visceroceptors: sense organs in the various parts of the internal organs such as the intestinal tract, heart, lungs, and liver.
      (2) Proprioceptors: sense organs in the equilibratory organs and in muscles and muscle-tendon junctions stimulated by body movements and changes in spatial orientations.
2. Aspects
   These include all of the inherent properties of the nerve impulse through the reflex arcs.
   (a) Forward conduction: This is possible in one direction only because of the mediating activity of the synapse which permits travel only from dendrite to axon in a nerve fiber and from axon to dendrite across a synapse.
   (b) Facilitation: If a stimulus is repeated frequently, but not so often as to become tiring or tetanic, the reflex becomes increasingly easier to elicit. Some authors refer to facilitation as the requirement that more than one neural volley is necessary to excite a postsynaptic neuron. In such cases facilitation would be similar to summation.
   (c) Fatigue: This occurs when stimuli are too frequently repeated at too short an interval for recovery to take place in the effector structure or organ.
   (d) Summation: The repetition of subliminal stimuli at sufficiently short intervals will finally produce excitation of the neuron.
   (e) Synaptic delay or resistance: Transmission is less effective across a synapse; weak transmissions may be interrupted, others can be delayed. The delay is about 0.002 second per synapse.
   (f) Inhibition: One reflex may inhibit another, or there may be reciprocal stimulation. Inhibition occurs where one pathway may have two opposite functions. Simultaneous stimuli may call for contrary action, which will ordinarily result in the one

which is more important overriding and suppressing the other. The "more important" action is the one that has been determined by the most facilitation, rather than the action which may have the greatest survival value to the individual.

(g) After discharge: The activity initiated by a reflex does not stop immediately upon cessation of the stimulus.

(h) Rebound: If during a reflex contraction of a muscle the stimulation is suddenly interrupted, there may be a short pause followed by an after discharge; i.e., the muscle resumes contracting after a brief pause in the contraction cycle.

(i) Long cycle rhythms: occur over a prolonged period (about 24 hours) and involve habit patterns and cyclic motor activity correlated with light and temperature adjustments.

Many aspects of the functioning of the nervous system have been obtained from deliberate surgical mutilation of experimental animals. These techniques, while necessary to the development of understanding of the functioning of the nervous system, will always have a group of detractors who state that once is enough and that successive generations of students need not go through the inherently unpleasant process of pithing frogs, decerebrating cats, and experimentation upon other animals. Nevertheless, the best way of learning is to perform an experiment. More is learned by the student who stimulates a pithed frog than from any book that can be read on the subject. It is regrettable but necessary that such exercises must be conducted if students are to acquire useful knowledge and skills in the practical application of physiology. The following statements, therefore, are in the nature of a guide and serve to indicate what may occur. For a more practical and more memorable process, experiments are advisable and should be conducted, under appropriate supervision, if there is any opportunity to do so.

CENTRAL NERVOUS SYSTEM

The spinal cord may be transected at a number of levels. It may also be divided longitudinally or a portion of one or the other side of the cord may be removed. Sections of one side or the other may be removed at different vertebral segments. Dorsal or ventral roots may be transected. Sympathectomies may be performed, or the brain itself may be completely or partly destroyed by pithing, decerebration or hemidecerebration. Results, in general, will conform to the following patterns:

Spinal Cord

1. *Transection of Spinal Cord*
   Sensory, motor, and association impulses arising proximally (i.e., toward the brain) will be interrupted at the point of transection.

Segments distal (posterior) to the transected portion will show only responses associated with reflex arcs 1 and 2 and the response from arc 2 will extend no further cranially than the transection. Voluntary control movements distal to the transection will be eliminated. Sensation distal to the transection will be eliminated.

2. *Partial Transection of the Spinal Cord*
   a. Transection of dorsal columns of the cord:

   The sensory impulses from segments distal to the transection will be (mostly) absent. However, some will be present if the lateral columns are left intact.

   b. Transection of both lateral columns of the cord:

   Coordinate motions involving segments distal to the point of transection will be impaired and occasionally abolished.

   c. Transection of one lateral column of the cord:

   Coordinate motions involving segments distal to the point of transection and on the same side as the transection will be impaired.

   d. Transection of the ventral columns of the cord:

   Motor impulses originating in the higher centers and those arising cranial to the transection will be interrupted. Segmental and intersegmental motor reflexes distal to the transection will not be impaired.

   e. Transection of the ventral roots of one or more segments:

   Areas supplied by the transected roots will have motor paralysis on the side where the transection has taken place.

   f. Transection of the dorsal roots of one or more segments:

   Areas supplied by the transected roots will have a loss of sensory function.

   g. Longitudinal division of the cord:

   Areas supplied by the segments where the cord is divided will have no coordinate function in animals where simultaneous coordination is not a vital aspect, e.g., in birds such a division will have no effect on the coordination of the wing strokes, but will affect the coordination involved in walking. In mammals, there will be a loss of motor coordination in hopping species.

   h. Hemisection of the cord:

   If the hemisection involves one or more spinal segments both sensory and motor elements supplied by and to that section will be absent. Areas distal to the hemisection will have a relatively complete loss of sensory and motor function on the side where the hemisection is performed. Sensory and motor responses will at first be totally absent. Later a more or less effective motor function will be restored although the sensory function will continue to be absent.

i. Alternate hemisection of the cord:
Areas distal to the complete alternate hemisection will be devoid of proximal or higher central control. A few association reflexes may be present which will bypass both transections. These will be motor. Sensory function to the higher nervous centers will be absent.

3. *Sympathectomy*
Sympathectomy results in a complete absence of excitement and loss of the "fight or flight" syndrome. In birds it causes erection of feathers and dilation of cutaneous blood vessels.

4. *Parasympathectomy*
Parasympathectomy of the caudal portion results in no grossly observable somatic changes. Technically, the performance of a cranial parasympathectomy is a surgical exercise of such magnitude that it is not performed. Although certain parasympathetic nerves can be isolated and transected, the technique is not for students. Since most parasympathetic fibers run together with visceral motor and sensory fibers or with cranial nerves, the damage done to the other nerve fibers clouds the results, and makes acquisition of useful information problematical.

Brain

1. *Pithing*
A pithed animal has no higher center reflexes, although all intersegmental and intrasegmental reflexes will be present. Pithing results in what is commonly called a "spinal reflex preparation."

2. *Extirpation or Partial Extirpation*
This involves removal or partial removal of one or more of the divisions of the brain: the cerebellum, the optic and olfactory lobes, the thalamus, and hypothalamus. Structures that are essentially pathways such as the medulla oblongata, diencephalon, and mesencephalon cannot be extirpated without producing other damage. Removal of the medulla oblongata would produce effects equivalent to pithing except that the functions of cranial nerves I through VI would still be preserved. General somatic responses distal to the medulla oblongata would be absent.

Transecting or removing the diencephalon or mesencephalon would probably destroy the functioning of cranial nerves I through IV, and VI, plus a part of the function of V. The somatic behavior would be that of a decerebrate, or pithed animal.

a. Complete extirpation of the cerebellum would affect movement and balance and, in birds, would destroy the ability to fly. Partial extirpation would have less effect and in general the signs would tend to disappear, although locomotion would remain somewhat impaired as would complex muscular acts. Forms of dietary extirpation or impairment of the cerebellum and cerebrum exist, and the condition of microcerebellum or cerebellar hypoplasia in domestic animals as a secondary effect of viral infections of the fetus or of the mother have also been recorded. Hereditary loss or failure of cerebellar development also exists. The dietary form results from the ingestion of methylazoxymethanol (MAM), an aglycone of cycasin, a glycoside found in the nuts of the cycad tree. MAM interferes with the division, multiplication and differentiation of the developing nervous system. It is also carcinogenic and mutagenic.

b. Destruction of one optic lobe in birds results in blindness in the contralateral eye as it does in some lower mammals. In higher mammals this effect does not occur, particularly in those with binocular vision where complete decussation of the optic nerve does not occur. More delicate operative techniques have shown that there are layers of perception in the optic lobes. For instance, cortical damage to the lobe results in no loss of reaction to light, but a loss of discriminating function. Damage to the optic lobes also results in motor disturbances such as wryneck and circling.

c. Thalamic destruction can only be evaluated by comparison with animals whose forebrain has been so carefully removed that the thalamus is left intact.

d. Destruction of the hypothalamus causes a drop in body temperature, a decline in pituitary function and damage to a number of other functions including smell. Removal of the forebrain and the diencephalon causes mammals to become poikilothermous. Removal of only the forebrain does not affect maintenance of body temperature. Concomitantly with a drop in body temperature, hypothalamic destruction also results in a drop in blood pressure and marked diuresis.

Since the hypothalamus has a vascular linkage with the rostral pituitary and the rostral lobe has no neural control, it follows that the hypothalamus exerts control over rostral pituitary activity through "releasing factor." Additionally, the pressor hormones, i.e., oxytocin and pitressin in mammals and oxytocin and vasotocin in birds, are produced in the neurosecretory cells of the supraoptic and paraventricular nuclei of the hypothalamus (p. 349). Electron microscopy of these nuclei demonstrates the presence of secretory granules in the cells. Surgical removal of the nuclei produces concomitant loss of pressor hormones.

3. *Decerebration*

Elimination of the cerebral cortex apparently removes discriminatory and psychic function. Reflexes appear to be unimpaired. This effect, at least in birds, is most notable in the feeding mechanism. Birds will peck or seize, but will not ingest. The birds have the necessary visual sensory perception of food. They have also a functioning motor apparatus for swallowing, but the associative function that exists between "seeing and doing" is absent in decerebrate individuals.

4. *Electrical Stimulation*

   More recently, electrical stimulation of various regions of the brain has been conducted, and highly complex "maps" have been made of cranial functions. It is not the purpose of this book to attempt to record these findings. However, foci and centers have been recorded that control emotional stability, temperature, phonation, complex body movements, and sexual activity, to name a few.

5. *Electroencephalography*

   A great deal has been and is being done in this field with both human and animal subjects, since an electroencephalogram (EEG) can be made without the dangerous and complex procedure of placing electrodes in brain tissue. Actually in mammals the correct term for the electronic recording of brain activity would be an electrocorticogram (ECG) since only the cerebral cortex is involved, whereas in birds, the entire encephalon seems to be involved in the electronic recording and hence the record would be an EEG. Again, the EEG and ECG phenomena are so numerous and detailed that they have no place in this work. It is enough to say that although a great deal has been and is being done, there is a great deal of work yet to do in this area.

   One could write a book on these two aspects alone, but the subjects, while extremely interesting, are ones which, of necessity, have to be kept short. At the present state of the art, the practical applications of EEG to the understanding and management of domestic and wild animals are almost nil. Nevertheless it is felt that these techniques should be mentioned even if they are not explored in detail.

INTELLIGENCE, INTELLECT AND MEMORY

The large brain is a peculiarly mammalian structure that reaches its highest development in humans and certain of the cetaceae such as dolphins and toothed whales. A corollary of brain development appears to be intelligence, which is the ability to acquire and apply knowledge, and intellect, which involves the ability to reason and to transmit the products of such reasoning to successive generations. Intellect, therefore, requires both communication and comprehension. The problem of intelligence and intellect in aquatic mammals is presently under intensive investigation and is being conducted along the lines developed by studies in man and terrestrial animals. At the moment, the results are inconclusive. Intelligence is undoubtedly present, but intellect commensurate with intelligence has not been proved.

Paleontological evidence indicates that the human brain began to enlarge several million years ago, after the physical and cultural basics of the hominid group had been fairly well developed (i.e., upright stance, opposable thumbs, terrestrial rather than arboreal habitat, binocular vision, tribal or group organization, ability to pronate, bipedal locomotion, tool using, and language). The cultural imperative of communication was a survival trait and operated to select individuals who could remember and discriminate better than their fellows. In consequence, those with better functioning brains had a better chance to reproduce (all other attributes being relatively equal). Through application of Romer's Rule, i.e., a trait must be present in order to be developed, it is fairly certain that the protohumanoids had achieved language and the basic elements of a culture prior to the expansion and enlargement of their brains.

Some recent experiments with chimpanzees which, like other apes and monkeys, have no speech centers or vocal cords capable of duplicating human speech sounds, indicate that communication and reasoning is not the sole property of the human race. Chimpanzees taught deaf and dumb sign language can communicate on a fairly respectable level and use about 150 combinations that indicate sounds. Some of the experimental animals also appear to show a capacity to reason and think in the sign language and to coin new "word" combinations to explain new experiences. Reasoning backward from the highly developed brains of the dolphins and toothed whales it would appear that these species must have a well-developed language and possibly a culture. This still awaits proof, but it is highly probable that both intelligence and intellect are not the sole property of the human race, although civilization may be.

The function of the brain is twofold, i.e., it stores data and applies stored data to problems. Little is known about either mechanism. Indeed, the problem of data storage, let alone the complex combinations and recombinations of data involved in thinking (as distinguished from memory) is still a fertile field for investigation.

It has long been known that the operation of the brain is associated with electrochemical activity, and that unconscious, active, and pathological brain functioning give markedly different recordings on an electroencephalogram. Research into DNA and RNA function has produced some breakthroughs in understanding the data storage mechanism of the brain. It now appears that data is stored in brain cells by alterations of the chemical structure of RNA in a manner similar to that used by DNA to store genetic information. The concept that data storage is chemical rather than physical should open entirely new fields to research. At the moment, the evidence is mostly circumstantial, but it does suggest that specific changes in neural RNA are the basis of memory, and since RNA is a constituent of all cells, it might also clarify the concept of "cellular memory," which is periodically discredited and revived.

Work with a chemical called 8-azoguanine and learning ability in rats has shown that the chemical has an inhibiting effect on the learning process. The chemical is so similar to one of the RNA bases that if it is injected into rats, the enzyme systems responsible for RNA manufacture will use it, and produce an RNA quite different from the natural product. Tests showed that injected rats remembered previously learned skills but were only half as efficient as uninjected rats in solving new problems.

Other tests on planaria, a primitive flatworm, which can regenerate when cut in half, indicate that conditioned worms cut transversely in half and allowed to regenerate in a normal environment retain their conditioning in both head and tail pieces. But, if the two cut portions are regenerated in the presence of ribonuclease, an enzyme that destroys RNA, the head ends remember but the tail ends do not. The newly formed head of the tail half was apparently wiped clean of memory by the RNA-ase, which affected the regenerating tissue but not the already formed structures in the old head.

A classical experiment where conditioned planaria were ground into fragments and fed to unconditioned planaria has given some additional corroboration to the chemical theory of memory. The unconditioned cannibals fed conditioned planaria were twice as smart at solving problems as were unconditioned planaria fed a normal diet, providing that the problems were the same as those to which the triturated worms were conditioned, prior to being converted into feed. Unconditioned cannibals fed unconditioned worms were as inept at problem solving as planaria fed a normal diet.

Cellular extraction techniques involving single nerve cells have demonstrated that nerve cells contain 10 times more RNA than surrounding glial and stromal cells, and have a higher rate of RNA synthesis and breakdown. There also appears to be a different relationship between the four compounds that enter into the formation of the RNA molecule (cytosine, adenine, guanine and uracil). Other studies have indicated that problem solving in rats is accompanied by a rise in adenine and a fall in cytosine content of neural RNA.

All of these examples are merely indications rather than proof. Before RNA can be pinpointed as the chemical involved in memory (if indeed memory is a chemical process), the entire neuron, not just nuclear or cytoplasmic RNA, will have to be defined and its intrinsic and extrinsic relationships understood. This may be a long way off, but in the meantime the search will go on.

chapter 12
The Endocrine System

GENERAL CONSIDERATIONS

Endocrine (Gr. *endo*—internal; *crine*—secretion) glands are ductless glands which empty into the bloodstream. They produce substances called hormones that exert either a stimulating or inhibitory effect upon the development or functioning of body organs and structures. Most hormones possess names ending in "in" or "one," e.g., pitocin, parathormone, insulin, enterogastrone. Recently the "one" ending is the usual one used. A hormone is usually defined as the product of an endocrine gland which is carried by the bloodstream to some other part of the body where it exercises specific (and observable) effects. This definition may need revision since some recently discovered hormones and hormone-like substances do not quite fit.

Endocrine glands themselves may be divided into two types: those that are only endocrine in function, i.e., pituitary, parathyroid, thyroid, and adrenal, and those that possess other functions besides producing hormones, e.g., pancreas, gonads, uterus, gastric epithelium, intestinal epithelium, and kidneys.

A few years ago, the entire endocrine system of the mammalian body was considered to be composed of six definitive glands, plus an additional six organs which were not purely endocrine structures. This is no longer true, and the discovery of new hormones is a fairly common occurrence. Among the more recent discoveries are secondary hormones of the thyroid and the pancreas, a hormone produced by the kidneys and a hormone-like substance produced by a number of structures not heretofore known to have endocrine functions. The end of hormone identification is not yet in sight.

THE PITUITARY

The pituitary body (hypophysis) occupies a unique position among the endocrine organs in that many of its hormones function chiefly to regulate the activity of other endocrine glands. It is this regulation of the endocrine system that has caused the pituitary to be called the "master gland" of the body. Despite its broad controlling functions, the gland is not essential for life, as hypophysectomized animals can survive for long periods of time.

The pituitary body (hypophysis) is a small round or oval structure attached by a stalk of nervous tissue to the floor of the brain in the region of the tuber cinereum. It is composed of three parts: the rostral pituitary (adenohypophysis), the caudal pituitary (neurohypophysis), and the pituitary stalk (infundibulum). A cleft of connective tissue usually separates the rostral and caudal pituitary. The infundibulum attaches the pituitary to the brain.

The gland is enclosed in a bony pocket (the sella turcica) in the basisphenoid bone, and is surrounded by a dense meshwork of small arterioles and capillaries. These, in turn, are enclosed in a fibrous connective tissue capsule that adheres to the walls of the sella turcica and forms an incomplete membrane across the mouth of the bony pocket.

The rostral pituitary includes the pars tuberalis and the pars distalis (rostral lobe), and the pars intermedia. The caudal pituitary includes the infundibulum and the pars nervosa. The infundibulum can be divided into three parts: the median eminence (labrum infundibularis), the stalk (pediculus infundibularis), and the infundibular process (bulbus infundibularis). The infundibular process when united with the pars intermedia is usually called the pars nervosa. The pars intermedia, although originating together with the other portions of the rostral pituitary from the embryonal Rathke's pouch, develops caudal to the cleft and unites with the bulb (infundibular process) of the infundibulum. The following sketch illustrates these rather complex relationships.

The pituitary receives blood from two sources, the caudal hypophyseal artery, which mainly supplies the caudal pituitary, and the cranial hypophyseal artery, which supplies the cranial pituitary. Both of these vessels are branches of the internal carotid artery. Other branches of the internal carotid ramify through the tuber cinereum and the hypothalamic nuclei and form extensive capillary beds in these structures.

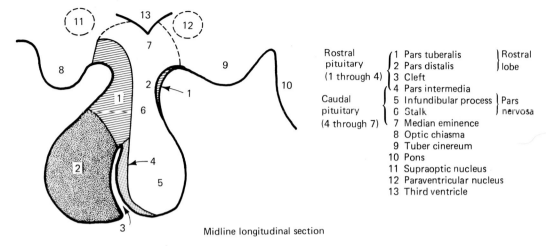

FIGURE 12-1   Pituitary body (hypophysis cerebri).

From the capillary beds of the hypothalamic nuclei, blood drains into the long hypophyseal portal veins that extend into the pars nervosa via the infundibulum. From the capillary bed in the infundibular process short portal veins enter the rostral pituitary.

These vessels form the hypophyseal portal system, and are the vascular connecting links between the hypothalamus and the rostral pituitary. It was the presence of these vessels that first gave rise to the idea that the hypothalamic nuclei might have an endocrine function. "Releasing factors" secreted in the hypothalamus are carried to the rostral pituitary via the portal veins and regulate the hormone output of this portion of the gland.

The Rostral Pituitary

The rostral pituitary produces the majority of the pituitary hormones. Six separate hormones have been identified:

1. Growth Hormone (GH) also called Somatotrophic Hormone (STH).
2. Follicle Stimulating Hormone (FSH) in females; Spermatogenesis Stimulating Hormone (SSH) in males.
3. Luteinizing Hormone (LH) in females; Interstitial Cell Stimulating Hormone (ICSH) in males.
4. Prolactin.
5. Thyrotrophic Hormone (TTH).
6. Adrenocorticotrophic Hormone (ACTH).

As its name implies, the growth hormone regulates the size of the

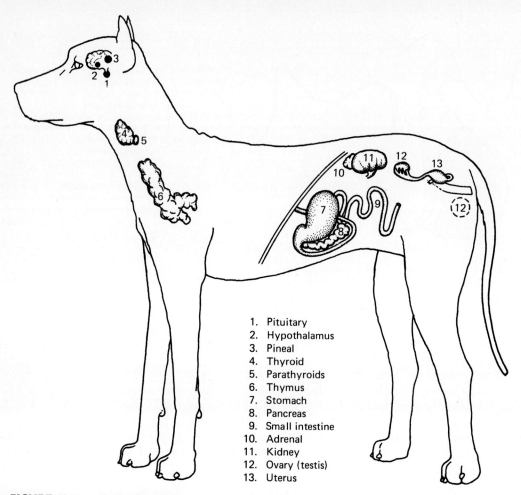

FIGURE 12-2   Endocrine system.

individual. Excess growth hormone can result in unusually large and occasionally malformed individuals. Absence or impaired production of the hormone results in dwarfism.

The follicle stimulating hormone (FSH) stimulates the growth of Graafian follicles in the ovary and the production of spermatozoa.

Luteinizing hormone (LH) stimulates ovulation and the development of the corpus luteum in the ovary, and causes enlargement of the interstitial cells in the testes.

Prolactin maintains the secretion of the corpus luteum and the interstitial cells and also functions to maintain lactation.

Thyrotrophic hormone (TTH) is necessary for the efficient functioning of the thyroid gland, and aids in regulating metabolism.

Adrenocorticotrophic hormone (ACTH) is essential for the functioning of the adrenal cortex. Absence of this hormone results in gradual degeneration of the adrenals.

Hypothalamic Pituitary Relationship

The function of the hypothalamus (and particularly the supraoptic and paraventricular nuclei of this organ) in rostral pituitary activity has become increasingly interesting. It has been demonstrated that there are neurosecretory cells in both of the above nuclei, and experiments indicate that the hypothalamus exerts regulatory control over the rostral pituitary via the vascular linkage of the hypophyseal portal system. The control is exerted by "releasing factors" which stimulate secretion in the rostral pituitary.

In 1969 a team of researchers at Baylor University in Waco, Texas, succeeded in isolating and synthesizing a hypothalamic product that apparently controlled the functions of the pituitary and thyroid glands. The substance was named "thyroid releasing factor" (TRF) and caused the pituitary to release thyrotropin which in turn stimulated the thyroid to release thyroxin. This was the key that opened insight into the function of the hypothalamus. TRF is a relatively simple compound composed of pyroglutamic acid, histidine, and proline and can be of medical value in treating thyroid disorders. Releasing factors have now been identified for all the rostral pituitary secretions.

Caudal Pituitary (Neurohypophysis)

Two separate hormones have been isolated from the caudal pituitary of mammals. These are usually called pitressin and oxytocin (pitocin). The hormones appear to originate in the supraoptic and paraventricular nuclei of the hypothalamus, and are carried to the caudal pituitary along nerve channels. The caudal pituitary is thus an organ of storage rather than an active hormone producing structure.

Pitressin causes constriction of smooth muscles. It exercises a pressor effect upon the blood vascular system, and can produce a rise in arterial blood pressure. It also interrupts smooth muscle activity in the gut and produces an antidiuretic effect in the kidneys that results in the term antidiuretic hormone (ADH) being applied to it.

Oxytocin increases the power of smooth muscle contractions by increasing irritability. Oxytocin stimulates the muscular coats of the uterus, gut and urinary bladder, and in addition stimulates the letdown of milk in the lactating animal.

Pars Intermedia

A hormone called intermedin is produced by the pars intermedia. In warm blooded animals it apparently exerts no significant effect, although it is thought to affect water metabolism. In cold blooded animals, particularly frogs and chameleons, it affects response of the color cells of the skin to the color of the environment and is called melanocyte stimulating hormone (MSH). The secretion of this hormone is thought to be influenced by melatonin, a substance produced by the pineal body. Some authors consider the pars intermedia to be a portion of the rostral pituitary because of its embryonal origin, and MSH is therefore listed by them as a rostral pituitary hormone. However, the strong anatomic relationship that exists between the pars intermedia and the caudal pituitary (resulting in the term pars nervosa) indicates that the hormone intermedin (MSH) should probably be considered separately from the hormones of the pars distalis.

The Infundibulum

The infundibulum is a connecting stalk of nervous tissue that attaches the pituitary body to the brain. It is often called the pituitary stalk. It produces no known hormones. It acts as a transfer or transport mechanism for the pressor hormones produced by the hypothalamic nuclei and is considered to be composed of three parts: the median eminence (labrum infundibularis), the stalk (pediculus infundibularis), and the infundibular process (bulbus infundibularis). The infundibular process and the pars intermedia are often considered together as the pars nervosa.

This terribly confused and confusing nomenclature may be clarified some day, but this day does not appear to be imminent.

THE PINEAL BODY

The pineal body has long been considered to be an endocrine gland, principally because of its structure. It lies buried along the midline of the brain above the corpora quadrigemina, immediately rostral to the cerebellum. It is bounded laterally by the caudal ends of the thalamus and dorsally by the caudal rim of the corpus callosum. It is connected by a hollow stalk to the roof of the third ventricle and is surrounded by a network of blood vessels. René Descartes once stated that he thought the pineal body was the "eye of the soul." More and more, this view seems to be correct insofar as the first word of the phrase is concerned. Studies in reptiles and amphibians, and several species of night-active lizards, have demonstrated beyond reasonable doubt that in these animals the pineal body is part of a light receptor apparatus and functions to aid activity at night. In mammals, the "third eye" effect is lacking, and the presence of a parietal nerve connecting the organ to the brain has not been demonstrated. Tumors of the pineal have occasionally been associated with precocious sexual maturation, which together with its structure, has promoted ideas that it may have an endocrine function. The pineal body is not necessary for life nor does its removal (in mammals) seem to have any bad effects. Pineal extracts produce no significant results when injected into experimental animals although melatonin (melanone), a pineal extract, causes skin blanching in frogs. Some evidence has been developed that melatonin has activity in promoting seasonal estrous cycles in some mammals by blocking secretion of pituitary gonadotropins which inhibit gonad activity. Since melatonin production is inhibited by light, long days would reduce its suppressive effects and permit sexual activity. The evidence to date is somewhat "iffy," and it is possible that the pineal body is merely an evolutionary remnant of a "third eye." However it is also possible that it is an endocrine gland.

At the moment, it is best to leave pineal function in mammals open to question.

THE THYROID

In mammals the thyroid gland consists of two lobes lying along the lateral walls of the trachea close to its junction with the larynx. In many species an isthmus, or band of glandular tissue, extends across the ventral surface of the trachea and connects the two lobes. This isthmus is not routinely present in dogs and cats, and is often absent in man.

The thyroid gland is oval, reddish brown in color, and covered with a fibrous tissue capsule that is moderately adherent. The gland is essential for life. It is composed of a large number of hollow spherical follicles filled with a substance called colloid. The colloid contains the hormone thyroxin mixed with a protein. Thyroxin is one of the hormones whose chemical structure is known. It is unique in that it contains iodine. It functions to control metabolism, and is secreted from the gland into the blood and lymph stream. It also serves as a concentrator and storage depot for iodine. A second hormone, calcitonin, has been isolated from thyroid tissue. It functions to control calcium metabolism and bone development. In laboratory animals calcitonin appears to be produced by the ultimobranchial bodies, which become incorporated into the thyroid late in fetal life and lose their structural identity. This hormone may have considerable importance in correcting the "brittle bone" syndrome in aged mammals.

The thyroid has a close interrelationship through hormones with other endocrine glands, notably the pituitary, adrenals, and gonads. Oddly enough, there is little thyroid-parathyroid relationship although anatomically the two organs are next door neighbors, and functionally have a certain amount of overlap.

THE PARATHYROID GLANDS

The parathyroid glands are small brownish bodies found either on or near the thyroid gland. Commonly, they occur as two pairs, although their location and number is inconstant. The cranial pair are generally found buried in the thyroid gland while the caudal pair may be some distance away. In some species, notably swine, the parathyroids are entirely separate from the thyroid gland. Occasionally accessory parathyroids may be found in the thorax embedded in the thymus or along the course of the carotid artery. In some individuals all of the parathyroid bodies may be found embedded in the thyroid gland. The gland is essential to life.

One hormone, parathormone, is produced. The hormone, an 84-amino acid polymer, was first synthesized in 1972. It functions to control calcium levels of the blood. An increase of parathormone results in an increase of blood calcium. The exact mechanism of action is not entirely clear. The hormone apparently produces this effect by controlling the rate of phosphorus excretion by the kidneys. Since the calcium-phosphorus ratio of the blood is held within rather narrow limits, a decrease or increase of circulating blood phosphorus will result in a corresponding decrease or increase in calcium under normal conditions. However, variations in calcium levels can also be affected by diet and calcitonin (see thyroid gland). High protein dietary levels can increase calcium excretion, and low protein diets favor calcium retention.

Apparently there is a relationship between parathormone and vitamin D although this is not completely understood. The vitamin can partially protect the body against the effects of parathormone deficiency by improving calcium metabolism.

THE THYMUS

The thymus, like the pineal gland, has been considered to be endocrine. The status of this structure, however, has waffled between endocrine and nonendocrine in recent years. For a while the lymphoid nature of the gland and the exciting discoveries of lymphoid function involving the thymus tended to make any belief in thymic hormone questionable. However, new evidence indicates a probable hormonal function although no definite thymic hormone has yet been isolated. The gland is apparently not essential to life, providing it is not removed at birth. In mice, neonatal thymectomized individuals die of "wasting disease" (p. 245) after a few weeks or months of life. More recent investigations reveal that thymectomized young adult mice can also develop a form of "wasting disease" which can be arrested with thymic extracts which are considered by some investigators to be hormones. More research indicates that the thymus' major importance may be in thymic lymphocyte production.

The thymus is a triangular or Y-shaped organ composed of lymphoid tissue and is pinkish to yellowish-white in color. Its divided base rests upon the pericardium and its upper portion extends cranially along the trachea as far as the larynx. It is composed of lobes loosely joined together by connective tissue, and contains characteristic thymic corpuscles (Hassall's corpuscles).

The thymus is remarkable for its changes in size. In most mammals from birth to the time of sexual maturity the gland enlarges. After this period it shrinks, retreating into the thoracic cavity where it remains as small masses of glandular tissue surrounding the base of the heart.

The hormonal function of the thymus was for a time overshadowed by its lymphatic aspects (p. 244ff). New data indicate the gland is necessary for the production of tissue immunity and for the proper development of immunologic competence at the cellular level, and that the gland accomplishes these acts through what is probably a hormonal factor.

The early studies on thymectomized neonatal mice revealed that these animals usually did not reject heterografts and that sensitizing and shocking doses of antigen failed to produce either sensitivity or the Arthus phenomenon (see p. 243). These results prompted further studies that revealed that thymectomized mice had normal humoral immunity but lacked cellular immune response. Other studies of thymus transplants from one strain of inbred mice to neonatally thymectomized mice of another strain caused the recipient to accept tissue transplants from the donor strain and produce lymphocytes that reacted to antigens and produced antibodies. These results were obtained even when the transplanted thymic tissue was enclosed in a compartment, which prevented the migration of thymic cells into the body of the recipient.

Recent work has isolated a steroid from thymus tissue that inhibits lymphocyte formation and also inhibits tumor growth.

Present thinking among the proponents of a hormonal function for the thymus is that the hormonal factor is transported to stem cells in the lymphoid tissues and programs them to produce lymphocytes which are immunologically competent and are compatible with the donor strain's tissues. These acts meet the criteria for hormones, and once again the thymus is at least a peripheral member of the family of endocrine glands. It is to be hoped that it will stay that way.

THE ADRENAL GLANDS

The adrenal bodies are two flattened, brownish-pink structures located dorsally and cranially to the kidneys and embedded in the perirenal fat. Each is composed of two parts: a dark brownish outer layer or cortex and a granular middle portion, or medulla. In young animals, so-called accessory adrenals are occasionally found. These may consist of either cortical or medullary tissues, or both. They generally are located between the kidneys and ventral to the abdominal aorta in the region of the renal arteries.

The cortex and medulla of the adrenal are two separate and distinct structures, each producing hormones. The cortex is essential to life. The medulla is not, although its function is broader than was thought not too long ago.

Cortical Hormones

There is still considerable confusion about the exact hormone status of the adrenal cortex. It produces three families of hormones, i.e., glucocorticoids, mineralocorticoids, and androgenic hormones. All are modifications of a steroid molecule closely related to cholesterol and have the generic classification of corticosteroids. A crude extract of the adrenal cortex is called cortin, which is useful in the treatment of Addison's Disease in humans.

The glucocorticoids consist principally of cortisol, cortisone, and corticosterone. The names tend to be confusing, since the compounds can be synthesized and a number of analogs with hormone effect can be constructed in the laboratory. There are also natural analogs. Among these latter substances is deoxycorticosterone which is virtually identical in effect to corticosterone. Glucocorticoids promote gluconeogenesis (p. 394), decrease inflammatory reaction, affect muscle tone and excitability of nerves, inhibit cartilage growth and development, and have roles in fat and water metabolism. Glucocorticoids are used principally in the treatment of ketosis in animals and rheumatoid arthritis in man and to suppress inflammatory response in certain infections.

Aldosterone is the principal mineralocorticoid. It is an electrolyte regulating hormone which has a potent effect on water metabolism. It controls the reabsorption of sodium from the kidney tubules. Its precise mechanism of operation is unknown. There are other mineralocorticoids, and deoxycorticosterone, which have a similar but considerably less potent action. Aldosterone is principally used in the treatment of water metabolism dysfunction associated with excessive electrolyte loss and in treatment of the renal hypertension syndrome (p. 308).

The androgenic hormones are virtually identical to those produced by the testicles (p. 357) and have the same physiological effect. However, the amount of these hormones is extremely small and their effect is minimal. In some texts these androgens have been called "accessory sex hormones." They can occur in females of certain species (mainly humans) in sufficient quantity to cause an appreciable degree of masculinization. Tumors of the adrenals can result in sexual precocity in the male and masculine characteristics in the female. Other relationships are with the pituitary and thyroid.

Medullary Hormones

Two hormones, adrenalin (epinephrine) and noradrenalin (norepinephrine), are produced by the medulla. Since they both are emergency stimulators of the so-called "fight or flight" reaction, and accomplish much the same effects, there is no particular reason to separate them in this basic discussion. Both hormones stimulate the sympathetic nervous system and result in accelerated pulse, higher blood pressure, contraction of the stomach and sphincters, suppression of intestinal movements and activity, glandular stimulation, and increased conversion and utilization of liver glycogen. They find their principal use in medicine as stimulants, decongestants, hemostatic agents and as adjuvants given together with local anesthetics to delay their absorption and prolong their effect.

THE PANCREAS

The pancreas is a pinkish, lobulated gland located in the first loop of the duodenum after it leaves the stomach. The endocrine portions of this organ are the Isles of Langerhans, which are small masses of cells occurring between the acini. The islet tissue is essential to life and secretes the hormone insulin, and a second hormone called glucagon. The tissue is composed of a stroma and two types of secreting cells called alpha and beta. The beta cells are more numerous and secrete insulin. The alpha cells produce glucagon.

Insulin functions to regulate blood sugar levels by acting as a catalyst that lowers the threshold resistance of body cells to glucose. The presence of insulin allows normal intermediate metabolic processes to function in the presence of low levels of glucose. In other words, insulin makes glucose in the blood and tissues more readily available for use. Without insulin, blood glucose levels rise abnormally high to overcome the threshold resistance of the cells. The effect of abnormally high sugar levels is a disease called diabetes mellitus, which is characterized by excess sugar in the blood (hyperglycemia), sugar in the urine (glycosuria), ketosis, fatty degeneration of the liver, and excessive thirst and urination.

Glucagon functions in the metabolism of liver glycogen. It stimulates activation of the enzyme phosphorylase which catalyzes the breakdown of glycogen to glucose (p. 373ff).

Diabetes

It might be of interest to examine insulin in a bit more historical depth, since this hormone has preserved literally millions of human lives and threatens to become a racial disaster.

On July 6, 1921, Frederick G. Banting, a young surgeon, and Charles H. Best, a graduate student, discovered that a pancreatic extract could lower blood sugar levels in a diabetic dog. They had discovered an effective treatment for diabetes mellitus in humans, a feat which won Banting the Nobel Prize for Medicine in 1923.

Diabetes is a disease that, if untreated, kills a sufferer fairly early in life. Its symptomatology was recorded over 2,000 years ago. In the 16th century, Paracelsus mistakenly identified the foreign substance in the urine of diabetics as salt. In the 17th century, Thomas Willis thought the condition was stomach disease: He tasted diabetic urine and found it to be sweet (hence *mellitus*—honey). In the 19th century, Claude Bernard isolated glycogen from liver, identified the process of blood glucose excretion by the kidney when blood levels rose above the renal threshold, and described diabetes as a liver dysfunction, i.e., overproduction of sugar.

In 1867, Paul Langerhans identified islet tissue in the pancreas, and Gustav Laguesse suggested that it might be an endocrine gland. Oscar Minkowski accidentally made a dog diabetic by removing its pancreas, and, in the last decade of the 19th century, diabetes was known to be a pancreatic dysfunction.

In the early 20th century the difference between the acinar (enzyme-forming) structures and the islet (endocrine) tissue of the pancreas was recognized. Robert Bensley, a histologist, established the independence of the islet tissue, and one of his students described the details of the alpha and beta cells in the islets. The Fehling test for urine sugar, later modified by Benedict, gave researchers a convenient tool for identification of diabetics.

All this information was available to Banting and Best when they started their work at the University of Toronto Medical School. Banting had an idea that the islet tissue was the key to carbohydrate metabolism and possibly to diabetes, and after a series of failures caused by pancreatic enzymes which digested the islet secretions, the two men prepared an extract from frozen tissue that caused temporary recovery of a diabetic dog. Further discoveries led to the regimen of continuously repeated low-level injections, and by January 1922 they were ready to try their extract on a human diabetic. It worked.

Since that time, insulin was isolated, and analogs of insulin, sulfonurea compounds, dietary controls, and early detection tests have made diabetes a controllable (albeit incurable) disease. Insulin, in fact, is a treatment that less than one-fourth of all diabetic patients require.

To avoid diabetes, one must have nondiabetic ancestors, and that, of course, is the racial disaster referred to earlier. For diabetes is heritable, and with the enormous number of surviving diabetics and their increased reproduction, the percentage as well as the absolute number of people afflicted with this incurable disease rises with each passing year.

It is now known that some dogs develop diabetes mellitus and the Egyptian sand rat *Psammomys obesus*, the New Zealand white rat, *Rattus norvegicus* var. (NZW), and the spiny mouse *Acomys cahirinus* also may develop the disease. There is no lack of animal models for research, and with all this going for the investigators, perhaps a cure for diabetes mellitus can be found.

Spinoff of insulin research indicates the hormone may function in hibernation. Studies of hibernating animals such as groundhogs and hedgehogs have shown that insulin induces hypoglycemia that results in a state similar to hibernation. However, reversal of the hypoglycemia results in fairly rapid awakening, which does not occur in arousal from true hibernation where the awakening period is prolonged and characterized by periods of fever and cardiac, respiratory and glandular hyperactivity.

THE TESTICLES

The male gonads or testes not only produce male germ cells but also produce male sex hormone or androgens. Androgens are steroid compounds produced by the interstitial cells (Leydig cells) of the testicle, and are also produced in small quantities by the adrenal cortex. The principal male sex hormone is testosterone in most mammals. Rats, however, do not have this hormone but instead produce an analog called androstenedione. Other analogs of testosterone are androsterone, epiandrosterone, and dehydroandrosterone, which are the result of certain modifications of the testosterone molecule as it passes through the kidneys. These are less potent and less biologically active than testosterone. Testosterone and its analogs are not essential to life. They circulate in the bloodstream bound to plasma protein and are either rapidly utilized or are degraded by the liver and/or kidneys, and are excreted through the bile duct into the intestine or are excreted as part of the urine.

Testosterone and related androgens are responsible for the male secondary sex characteristics, body conformation, muscular development and libido or sex drive. They inhibit pituitary production of ICSH and when injected into the female delay or suppress estrus. They also stimulate growth, development and activity of the male accessory sex glands, activate spermatogenesis and the duct system of the testicles, maintain viability of the spermatozoa in the testicle duct system and stimulate growth of the penis.

A cyclic pattern of pituitary-gonad relationship similar to that in the female also exists in the male. The pituitary hormones SSH (FSH) and ICSH (LH) respectively stimulate sperm production and Leydig cell activity.

THE OVARIES

The ovaries produce several female sex hormones (estrogens); i.e., estradiole, estrone, progesterone, and a number of related compounds in addition to producing female germ cells (ova). A third hormone called relaxin appears near the end of pregnancy. Estradiole (which has been called estriol, estrin, and theelin) and estrone are produced by the Graafian follicles, (Figure 14-5), and progesterone and its analogs (progestins) are produced by the corpus luteum.

Ultramicroscopic evidence indicates that the cells of the theca interna produce most, if not all, the estrogenic hormone of the ovarian follicle. Their structures, although superficially resembling fibrocytes, differ ultramicroscopically in that they contain lipoid material, and their mitochondria strongly resemble those found in steroid secreting cells of the testicular interstitium and the adrenal cortex. Evidence that the follicular cells produce estrogen is less apparent, since the mitochondria are somewhat different and lipid is not as frequently encountered. Without proof to the contrary it is probable that both follicular and thecal cells are involved in estrogen production. Both theca and follicular cells appear to enter into progesterone production during the luteal phase of the ovarian cycle, with the follicular cells playing the major role.

When the Graafian follicle ruptures it undergoes some rather remarkable changes, and becomes transformed into the corpus luteum or "yellow body." The corpus luteum is a second endocrine structure. It produces the hormone progesterone and at least two analogs which are presently called progestins. All appear to have similar effects. Extracts of ovaries obtained from pregnant animals contain a water soluble polypeptide hormone called relaxin. Relaxin is presumably produced in quantity just before parturition and functions to relax the pelvic ligaments, cervix, and vaginal musculature. The exact site of production of this hormone in the ovary is not known, although the boundary cells between the cortex and medulla are thought to be involved.

Estrogens are responsible for the development of the female secondary sexual characteristics and body conformation and, in addition, enter into the complex hormonal interrelationships of the estrous cycle, which will be described later.

Progesterone and its analogs prepare the uterus for reception of a fertilized ovum, suppress the development of new Graafian follicles, prepare the mammaries for lactation, and suppress heat or estrous. Neither estrogen nor progesterone is essential to life.

THE UTERUS

Normally the uterus of the nonpregnant female is not an endocrine structure, but in some species during pregnancy a number of changes take place in the lining of the uterus (endometrium) that cause it to become an endocrine organ. With the development of the placenta (p. 416) the pregnant (gravid) uterus may develop endocrine activity. This has been demonstrated in horses, rats and primates, and probably occurs in all species where the corpus luteum does not persist functionally until term (birth). The placental hormone, chorionic gonadotropin, helps maintain the uterus in a condition that will support growth and development of a fetus. In the early stages of pregnancy before significant placental development takes place progesterone handles this function, but if the corpus luteum of pregnancy regresses, the placental hormone takes over.

In certain animals, notably the mare and the human female, the production of chorionic gonadotropin is so high that a detectable amount is found in the urine and/or blood serum. Tests for the presence of this substance have been developed to a high degree of accuracy and are used in the early diagnosis of equine and human pregnancy. Cows, ewes, bitches, and sows either do not produce this hormone or produce so little that it cannot be detected by the analytic methods which have been used in mares and women.

THE STOMACH

The wall of the stomach produces the hormone gastrin. The hormone is produced by the epithelial cells in the pyloric and possibly the fundic region. It is somewhat peculiar insofar as hormones are concerned because it is produced by the stomach wall and affects the stomach wall. The hormone functions to stimulate gastric secretion and activity, and causes more rapid gastric digestion of food.

THE SMALL INTESTINES

Outside of the pituitary body, no single structure produces as many hormones as the small intestines. All of the intestinal hormones affect the digestive processes in one way or another. Five known hormones are produced and others are postulated. The known hormones are secretin, enterocrinin, enterogastrone, pancreozymin, and cholecystokinin.

Secretin stimulates pancreatic, bile, and duodenal secretion. It functions to raise the fluid levels in the gut but has little or no effect upon enzyme production. It also appears to have an inhibitory effect on stomach activity.

Enterocrinin stimulates secretion of digestive juice and enzymes by the small intestine.

Enterogastrone inhibits gastric secretion and activity. It is produced in response to the presence of fats and fatty acids in the intestine.

Pancreozymin stimulates the pancreas to produce enzymes. It does not apparently affect fluid production of the gland. It is produced in response to the presence of protein derivatives, fats and fatty acids in the small intestine.

Cholecystokinin stimulates the gall bladder to contract and empty bile into the small intestine. Its function in animals that do not possess gall bladders (horse, deer, rat) is not clear.

Intestinal gastrin is suspected to exist, but its existence is not yet proven. It is similar in action to gastrin.

THE KIDNEY

The kidney produces two hormones, urogastrone and erythropoietin. Urogastrone exerts an inhibitory effect upon gastric secretion. There is some question whether or not this hormone is actually produced by the kidneys, as it is found in the urine and might have originated elsewhere, or it may be an accidental result of chemical combinations involved in the urinary process. A second hormone, erythropoietin (Erythrocyte Stimulating Factor or ESF), is apparently produced or activated in the kidney. It is responsible for the regulation of erythrocyte production in normal animals. Here is an example of the newer techniques in hormone assay since the classical methods obviously cannot work insofar as the kidneys are concerned, since these organs are essential for life.

THE LIVER

The liver has long been suspected, but has never been proved, to be a site for hormone production, since the liver is essential for life and mammals will not live more than 48 hours if deprived of this organ and not supported by ancillary treatment. New techniques, however, give some promise that the hormone status of this organ may yet be clarified.

PROSTAGLANDINS

According to the definition of a hormone (p. 345) prostaglandins does not precisely qualify, yet in some respects it has a hormone-like action. As knowledge of bodily function increases, the old definitions lose meaning when faced with the new realities of physiology and biology. The "gray areas" keep increasing with advances in knowledge about the mechanisms of life.

The problem surfaced many years ago with choline (p. 179ff) which was not precisely a vitamin according to definition, but certainly had some vitamin-like functions. It is being carried on today by substances which have hormone-like functions. Where the process will end makes interesting speculation, but also causes considerable trouble to classifiers and textbook authors since there is no convenient pigeonhole into which these discoveries can be placed.

As long ago as 1934 it was found that human semen and extracts of fluid from sheep seminal vesicles contained a fat soluble fraction that lowered blood pressure and stimulated intestinal muscular activity. But it was not until 1960 that the active principle was isolated, determined to have hormonal activity, and named prostaglandins because of its supposed origin in the prostate. Since that time prostaglandins has been isolated from a large number of tissues including gut, kidney, brain, lungs, eye and sex organs, and four basic forms of the substance (E, F, A, B) have been described together with a number of subsidiary forms that result from minor alterations of the molecule.

Prostaglandins has had its most useful application in cases of reproductive disorders, particularly the synchronization of estrus and the treatment of anestrus caused by persistent corpora lutea. Prostaglandins also has effects upon circulation that vary with the basic and subsidiary forms of the material. Types E_1 and E_2 are peripheral vasodilators. In some species Type F_2 is a vasoconstrictor. Type F_{2a} is particularly involved in estrus activity and can be used to terminate early pregnancy. A number of other possible uses include inhibition of stomach secretion, dilation (type E) and constriction (type F) of bronchial muscles, and increase of inflammatory action.

The hormone is readily deactivated by the liver and, under natural conditions, is presumed to exert its effects in the tissues where it is formed. It is also inhibited by antiinflammatory drugs such as aspirin.

HORMONE INTERRELATIONSHIPS

Hormones serve to regulate the action of other parts of the body. The endocrine glands are sensitive to the needs of the body and adjust their production according to these needs. Being closely related to blood supply, their activity depends on the concentration of chemicals in the tissue fluid surrounding them. Thus a change in the composition (e.g., ions, hormones, enzymes) of the serum adjacent to the secreting cells may stimulate them to produce or inhibit production of hormones and make the adjustments necessary to correct imbalances. Other forms of stimulation such as neural or mechanical also apply to hormone production.

Hormones frequently have complex interrelationships, many of which are not completely understood. One of the best known of these is the pituitary-gonad relationship. Its results are expressed as anatomic as well as physiologic changes and serve to show in a limited way the effect of hormones upon body structure and function (Figure 12-3) and upon each other.

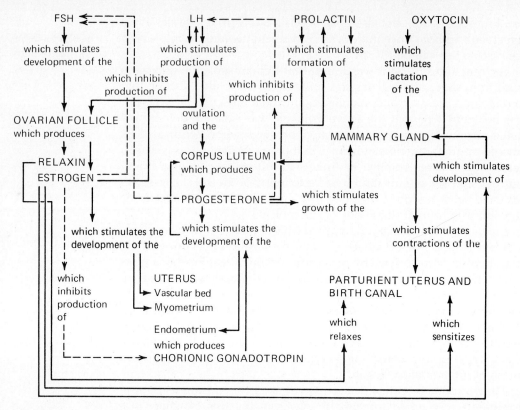

FIGURE 12-3   Hormone relationships of pituitary and female reproductive organs.

REGULATION OF REPRODUCTIVE AND ACCESSORY SEX STRUCTURES

In the female a complex hormonal interaction exists which is responsible for the maintenance of the normal ovarian cycle, the adaptation of the reproductive structures to contain and hold the embryo and fetus during the gestation period, and the development and functioning of the mammary glands. Four structures enter into this relationship: the pituitary, the ovary, the uterus and the mammary glands. The first three of these four structures produce hormones.

The Nonpregnant Female Hormone Interrelations

In the nonpregnant female a cycle of alternating estrus (heat) and anestrus is present. This may occur once a year (monestrus), twice a year (diestrus), at regular intervals during certain seasons of the year (seasonally polyestrus), and at a regular cycle which extends throughout the year (polyestrus). Regardless of the length and period of the heat cycle (which appears to be a species function and differs markedly between species), a similar complex of hormones is involved in the control and maintenance of the estrous cycle. Two hormones produced by the pituitary, i.e., FSH (follicle stimulating hormone) and LH (luteinizing hormone), and two hormones produced by the ovary, i.e., follicular hormone (estrogen) and lutein hormone (progesterone) are principally concerned in the maintenance of the estrous cycle.

Follicle stimulating hormone produced by the pituitary stimulates the ovarian follicles to develop, ripen and mature. As the ovarian follicle matures, it produces female sex hormones or estrogens which are apparently secreted by the theca interna and granulosa layers within the follicle. Estrogens produce the physical manifestations of heat and, in addition, exert two effects upon the pituitary, one of which inhibits further production of follicle stimulating hormone; the other stimulates production of luteinizing hormone. Luteinizing hormone from the pituitary stimulates ovulation and formation of the corpus luteum within the ovary. This structure produces a hormone (progesterone) which inhibits further production of luteinizing hormone and also inhibits follicle stimulating hormone. When the Graafian follicle ruptures and its cells change to form the corpus luteum, the inhibition of FSH exerted by estrogen production is removed from the pituitary body. Progesterone, which also inhibits FSH production, gradually exerts a weaker effect as the corpus luteum matures and becomes nonfunctional. After a varying period (two weeks in large polyestrous animals, two days in mice and hamsters), the progesterone effects have declined to such an extent that follicle stimulating hormone can initiate a new ovarian cycle.

The Pregnant Female Hormone Interrelationships

In the pregnant animal the initial steps of follicle production and production of the corpus luteum are the same as in nonpregnant animals. However, in this case, another factor enters; the ovum that has been discharged has become fertilized. The fertile ovum introduces an entirely new factor into the cycle. During the development of the Graafian follicle the estrogens produced in the follicle stimulate the development of the muscular wall and the vascular bed of the uterus. This prepares the uterus for the reception of a fertile egg. The progesterone produced by the corpus luteum stimulates the development of the epithelial lining (endometrium) of the uterus rendering it capable of receiving and supporting a fertile egg.

At implantation the uterine wall undergoes some marked functional changes. The endothelium is stimulated to produce chorionic gonadotropin which has two effects upon the ovary. It inhibits follicle development and stimulates the corpus luteum to continue in an active condition beyond its normal involution time. A cyclical mechanism is set up between the wall of the uterus and the corpus luteum in which the chorionic gonadotropin stimulates the production of progesterone, which in turn stimulates the endometrium of the uterus. This keeps the uterine lining in such a condition that the initial development of the fetus is maintained.

In addition, the further production of progesterone acting through FSH of the pituitary suppresses further follicle development. The other effect of progesterone (i.e., inhibition of LH) is counteracted by the presence of chorionic gonadotropin in ever increasing amounts.

A third function of progesterone now enters the picture. Prolonged progesterone stimulation causes mammary growth and stimulates the pituitary to produce prolactin, which in turn stimulates the formation of the secreting cells of the mammary glands. This accounts (at least in part) for the enlargement, inflammatory processes, and increased physical size of the mammary glands prior to birth. Prolactin also functions to stimulate maintenance of the corpus luteum, and thus furnishes a second hormone to replace the functioning of LH which is inhibited by the production of progesterone.

As gestation progresses, the corpus luteum becomes physiologically exhausted, but the fetus and endometrium by this time require no further stimulation and maintain themselves without appreciable outside assistance. During this period the fetus increases markedly in size and ultimately, probably through physical outgrowth of placental capacity, becomes either a foreign body or a strain upon the mother which cannot be adequately handled by the reproductive structures and maternal metabolism. At this time relaxin is produced by the ovary. It has been discovered in ovarian extracts, and is probably the result of an increasing foreign body reaction between the fetus and the mother. It causes relaxation of the pelvic ligaments and the birth canal, and sensitizes the uterus and cervix. Since the corpus luteum of pregnancy is virtually nonfunctional immediately prior to parturition, follicle stimulating hormone again appears to act upon the ovary resulting in the abortive production of follicles and the release of low levels of estrogen. This estrogen tends to inhibit chorionic gonadotropin production, sensitizes cervical and vaginal musculature and also stimulates the development of mammary tissues. At this time the hormone oxytocin, which is constantly present in the body of the animal, stimulates contractions of the sensitized uterus and promotes dilation of the sensitized cervix. These contractions help initiate parturition. Later, the presence of the fetus in the cervix and vagina causes a release of adrenalin which helps excite more forcible contractions of the uterine and vaginal musculature and intensifies the expulsive movements of birth. There is no proof that the act of birth results in increased production of oxytocin.

A probable neurohormonal relationship exists between the lactating mammary and follicle stimulating hormone of the pituitary, but this has not been proved. Yet it is known that among mammals in early lactation the production of Graafian follicles by the ovary is inhibited and estrum normally does not occur for several weeks to several months after birth. An exception to this is found in the mare and the jenny (and possibly in other equines) which about 4-14 days after birth will have a single heat period followed by suspension of estrum for several months. After the mother's body has become adapted to the strain of lactation, ovulation and estrus will again appear and the gestation cycle is capable of being repeated.

Hormonal Functions in the Male

In the male the hormonal relationship between the gonads and other endocrine structures is more simple. Testicular maturation and spermatozoa production are influenced by pituitary gonadotropin (LH). The hormone testosterone produced by the interstitial cells of the testicles stimulates gonadotropin production in the pituitary.

Since sperm production in the male is virtually a constant phenomenon from maturity to senescence, the cyclical organization of the male gonad is not pronounced. Increase in ambient temperature, particularly during the summer months, has the effect of lowering fertility in some species. This, however, does not appear to be related to hormone production but is merely a physical effect of temperature on spermatozoa production. Certain species have seasonal periods of fertility but these may be affected by artificial changes in the environment. In general, male libido in mammals seems to be directly affected by female receptivity. In both male and female, certain "accessory sexual hormones" are produced, which stimulate or enhance the effects of the sex hormones produced by the gonads. These substances are apparently derived from the still imperfectly understood adrenal cortex.

REGULATION OF THE INTESTINAL TRACT

The stomach and small intestine hormones modify the actions of the alimentary tract. Secretin maintains the composition of the intestinal contents by adjusting the pancreatic, duodenal, and bile secretions to conform to the moisture content and type of food ingested. Secretin, however, does not stimulate enzyme production. The digestion of foods is accomplished by enzymes derived from the stomach, pancreas and duodenum. These catalysts are produced only in the amount necessary to digest the food present.

The presence of incompletely digested food in the intestine triggers a hormone secreting mechanism in the intestinal wall which produces enterocrinin and pancreozymin. These hormones regulate the production of needed enzymes and inhibit further discharge of chyme from the stomach. When the enzyme concentration in the duodenum is inadequate, these hormones stimulate the proper structures to produce the required enzymes. The reaction is fairly precise. For instance, a deficiency in pancreatic enzymes will result in the production of pancreozymin, which stimulates the pancreas to produce more trypsin and chymotrypsin.

Stomach contractions may also be stimulated or inhibited by hormones produced from the stomach and small intestine and possibly a hormone produced by the kidney. These hormones are gastrin, intestinal gastrin, enterogastrone, and urogastrone. Gastrin and intestinal gastrin stimulate stomach secretion and motility, while the two gastrones inhibit gastric function.

REGULATION OF WATER AND ELECTROLYTE BALANCE

To maintain an ionic environment suitable for the functioning of the various cells of the body, the concentration of the ions of sodium, potassium, phosphorus and calcium in blood and tissue fluids must be maintained within relatively narrow limits. This is accomplished by the action of the pituitary, the adrenal cortex and parathyroid glands upon the kidney. The level of calcium ions in the blood is controlled by the parathyroid glands through parathormone, which controls the excretion of phosphorus by the kidney. An increase in parathormone results in decreased excretion of phosphorus, a raising of blood phosphorus levels, and a concomitant increase of blood calcium and of calcium and phosphorus in the bones. The rostral pituitary controls sodium and potassium levels and, through ACTH, influences the production of adrenal cortical hormone that in turn regulates the excretion of sodium by the kidney. The water balance of the body is regulated to some degree by the adrenal cortex. Cortical hormones control, to some extent, the excretion of water through the kidney. This particular homeostatic mechanism is not designed to control large intake or output of fluids; it is more of a "fine tuning" effect that maintains the terminal degree of the body's fluid balance.

REGULATION OF METABOLISM, BODY TEMPERATURE, AND HIBERNATION

The rostral pituitary, acting through thyrotropic hormone, regulates the production of thyroxin by the thyroid, which controls the general level of metabolism. The pituitary also affects the production of the adrenal hormones which affect metabolism by the action of ACTH. Insulin production by the Isles of Langerhans serves to lower the threshold of glucose needed for utilization by the cells. This also increases metabolism.

Metabolic regulation affects the ability of mammals and birds to maintain their body temperature at a fairly constant state regardless of the external temperature. Indeed, some mammals do this so adeptly that they can maintain their body temperature within one-half degree Celsius throughout the day and night in every season of the year. In Arctic and Antarctic habitats this is quite a feat and requires an enormous expenditure of energy, which in turn requires a high food intake and a reasonably high metabolic rate. Although such structures as the rete mirabile can allow the appendages of the body to become quite cold in arctic dwellers, the keeping of the central body warm is still an enormous task. Temperature regulation, or thermostasis, is best developed in mammals, although birds have this ability to a lesser extent. Those animals that can achieve thermostasis are called homothermous animals.

The opposite numbers on the temperature scale are the poikilothermous animals whose bodily temperatures vary with the ambient temperature. Reptiles, amphibia, fishes, and other animate life lower on the evolutionary scale are poikilothermous, and can endure only brief exposure to heat, and are paralyzed or immobilized by cold.

There are, within five of the 18 orders of living mammals, species that have rejected both the poikilothermous and homothermous adaptation, or it might be better said that they have selected and developed some of the qualities of both. Such animals are classed as hibernators. Marsupiala, Insectivora, Chiroptera, Rodentia, and Primates all contain species that hibernate. Bears, however, do not have this talent despite popular tales to the contrary. They sleep, but their body temperature does not fall below 90°F. Moreover, they awaken relatively quickly. Some species of birds also possess the ability to hibernate.

Hibernation is not sleep. It is an interruption of the metabolic processes. The hibernating animal is similar to the estivating one except that cold rather than heat triggers the interruption of active metabolic processes. These conditions are not like the stasis that results in amphibians or reptiles when they are exposed to cold, since metabolic and body processes continue although at a much reduced rate. Hibernation appears to be ordinarily a function of cold or stress (except in ground squirrels where it is a cyclic phenomenon) and can be triggered experimentally in those animals capable of hibernation by reduction of the ambient temperature.

One characteristic apparently common to most hibernators is a suppression of thyroid and adrenal cortex activity usually on exposure to cold, which is the exact reversal of the process in nonhibernators.

A structure called the "hibernating gland" is a mass of brown fat found dorsally between the scapulae of groundhogs, hibernating rodents, and, for that matter, in nonhibernating rodents such as rats and mice. It is supposedly a metabolism-reducing structure and has been found to contain steroids similar to androgens. It does not seem to be of any significance except in the groundhog, nor does it seem reasonable, at least at this time, to class the structure among the endocrine organs.

The process of hibernation may be slow or rapid depending on the species, but eventually the animal loses its homothermous capability and the internal body temperature falls to a few degrees higher than the ambient temperature. Oddly enough, the various body centers of metabolic activity retain their sensitivity, arterial blood is well oxygenated and acid-base balance and blood electrolytes are normal. Blood sugar may be low but is never dangerously so. Body temperature falls or rises with ambient temperature but always remains above freezing.

Hibernators tend to awaken periodically during the winter months. Bats probably have the longest hibernating period and can remain in a suspended metabolic state for over a month.

Arousal requires a great physiologic effort. Heart rate accelerates a hundredfold over the hibernation rate. Metabolic rate jumps to abnormally high peaks. Respiration is rapid and irregular. Shivering and erected hairs mark the waking process as the animal warms itself by expending muscle energy.

Regulation of the Neuromuscular System

The sensitivity of muscle is affected by the level of pitressin released by the caudal pituitary and in some instances by the presence of oxytocin. Pitressin causes constriction of the smooth muscle. Oxytocin produces contractions in sensitized muscle of the female reproductive tract and in the lactating mammary gland. Nervous regulation of the sympathetic nervous system and concomitant muscular response is in part accomplished by the release of epinephrine (adrenaline) by the adrenal medulla.

Although progress has been made in understanding the hormone relationships in animals, a great deal remains to be done before understanding of the endocrine system is complete.

chapter 13
Intermediate Metabolism

INTRODUCTION

Intermediate metabolism, insofar as this text is concerned, will deal mainly with carbohydrates, fats, and proteins. The subject is of considerably broader scope, but it would serve no useful purpose for a book of this nature to attempt to give more than brief coverage of what is an extraordinarily complex field. Intermediate metabolism is principally biochemistry and is an area where the classical physiologist tends to become lost in a morass of enzymatic reactions and reversible equations.

The entire field of intermediate metabolism is still not completely understood. There are blank areas where knowledge is absent, and gray ones filled with educated guesses that are not proved. There is much left to learn in the field and there are considerable opportunities for meaningful research.

THE DYNAMIC STATE OF BODY CONSTITUENTS

The detonator for this particular aspect of the knowledge explosion was produced in 1942 when R. Schönheimer published a monograph entitled "The Dynamic State of Body Constituents." Schönheimer proposed that the blood vascular systems, lymphatics, and tissue fluids functioned as a metabolic pool into which cells dumped their wastes and from which they extracted the necessary elements and molecules to carry on their life processes. The essential difference between this idea and older ones was the concept of flux, a dynamic state wherein nothing was static and where there was constant turnover and replacement of cellular components. Today the dynamic state is considered to be basic to (and one of the most important concepts of) the entire field of intermediate metabolism.

Proof of the concept has long since been completed, mainly through the use of natural and radioactive isotopes of such elements as carbon, hydrogen, oxygen, nitrogen, sulfur, and phosphorus. Since isotopes cannot be differentiated from ordinary elements by the body but can be recognized in the laboratory, researchers had a potent tool to acquire knowledge. The body metabolizes isotopes precisely as it does "normal" substances; hence, a compound labeled with isotopes can be traced through metabolic pathways in an animal and the processes which lead to the compound's fate can be determined.

PHASES OF METABOLISM

Metabolism has two broad phases, catabolism and anabolism. Catabolism involves the breakdown of substances to (or toward) their end products, usually with a production of energy that either can be stored, or utilized immediately in other biochemical reactions, or dissipated as heat. Anabolism involves the building of substances from different or less complex precursors. This usually involves an expenditure of energy. Quite often, since a majority of biochemical reactions are reversible, a reaction can go in either direction. The direction is dependent upon the relative concentration of reacting substances, their energy relationships and pH. If the energy level and the reaction product are markedly different from its precursors, the reaction is irreversible, but if there are strong similarities between precursor and product, and energy levels are similar, the reactions are usually reversible. Thus different directions of the same reaction pathways can result in catabolism and anabolism, although not concurrently.

In the aggregate, the metabolic reactions of cells are the manifestation of life itself and produce the other vital phenomena of reproduction, growth, and irritability. By far the largest part of intermediate metabolism is the making of food energy and food products available to body cells, and that is the portion of the total metabolic scheme that will be discussed here.

All of the energy producing foods (carbohydrates, fats, proteins) can be altered chemically. If they are oxidized they can be made to release energy. However, this is not done rapidly, since desirable bodily energy output is not heat energy (which occurs in rapid oxidation such as burning), but the sort of energy that promotes muscular contraction, glandular secretion, low level chemical reactions, and formation of resting potentials in nerves. This more gradual energy release is obtained from intracellular sequential enzymatic reactions that proceed at a relatively low temperature and which involve energy transfer systems.

Since normal nutrient substances are quite stable, they must be destabilized before they can be metabolized. To do this, a sufficient amount of energy must be released into the nutrient (or into the reaction) to start the mechanism. If catabolism is the direction, the substance will yield more energy than the amount needed to start and perpetuate the reaction. If anabolism is the direction, the energy input will be greater than the output.

Energy Sources for Metabolic Reactions

The energy is obtained from a substance called adenosine triphosphate (ATP) which contains high energy compounds between the last two phosphate molecules. There are other high energy compounds, such as phosphocreatine and phosphoenolpyruvate which appear in metabolic reactions, but by far the bulk of the energy storage of the body is ATP. Adenosine triphosphate has the general formula:

$$R - C - O - P - O \sim P - O \sim P - O^-$$

(with O, O, O above and O⁻, O⁻, O⁻ below the phosphates)

where R represents adenine and ribose. The last two of the terminal phosphates are connected to the rest of the molecule and to each other by high energy bonds ($\sim$). The amount of energy in each of these bonds is about 7,000 small (gram) calories. Loss of one PO_4 radicle from ATP yields adenosine diphosphate (ADP) and 7,000 calories. A second PO_4 loss can convert ADP to adenosine monophosphate (AMP) and another 7,000 calories. However, this second reaction seldom occurs, and ADP and ATP are the phosphates ordinarily found in the body.

Adenosine triphosphate is present in both nucleoplasm and cytoplasm of every living cell in the body and furnishes the vast majority of the energy required by physiological mechanisms. In the process ATP loses a terminal high energy phosphate and becomes ADP. ADP is restored to ATP by the oxidation of hydrogen in the cells. Since the phosphate that forms the high energy bond is always available as free radicles from metabolic sources, the only problem in the restoration of ATP is its acquisition of seven kilocalories. This is usually done during a resting state when the cells are not particularly active.

CARBOHYDRATE METABOLISM

There are, in mammals, two fundamentally different methods of carbohydrate metabolism, the nonruminant and the ruminant method. The two should be considered as distinct entities. The ruminant (or acetate) method is discussed on p. 199ff. In this section the more conventional carbohydrate metabolic routes will be covered.

In omnivores, carnivores, and nonruminant hervibores, some of the higher molecular weight carbohydrates such as cellulose, hemicellulose, and pentosans may be converted in the gut to fatty acids by microbial fermentation. This is principally a cecal activity and occurs so late in the process of digestion that it does not have the profound effect upon general metabolism that occurs in ruminants.

In nonruminants, starch, glycogen, and the common disaccharides are hydrolyzed through enzymatic processes to hexose (6-carbon) monosaccharides in the small intestine. These simple sugars are passed across the epithelial barrier of the gut and enter the lacteals and blood vessels of the villi. The rate is controlled directly by the presence or absence of insulin. Disaccharides that inadvertently enter the system are excreted in the urine. Somewhat over 60 percent of most of the absorbed monosaccharides enter the portal circulation: the remainder are picked up by the lymph.

The absorbed hexose monosaccharides are converted into either glucose or fructose molecules in the liver and then may pass in two directions. They can be transported as glucose or fructose in the bloodstream to body cells where the molecules may be used at once, or stored as tissue (principally muscle) glycogen. They can be stored in the liver as liver glycogen, or they can be converted into fat and stored in fat depots. The anabolic reactions that convert glucose to glycogen are called glycogenesis. The conversion reactions that change hexose monosaccharides into fat are called lipogenesis.

Carbohydrates are found in all tissues of the body and can be rapidly augmented by diet and from reserve stores of liver glycogen or through glucogenesis. Muscle glycogen is not used to supply blood glucose. Carbohydrates are the primary source of energy for various metabolic processes, and their role is vital to normal functioning of the general metabolic scheme. At least two-thirds of all energy utilized by nonruminants comes from oxidation of carbohydrate.

Carbohydrate levels vary with the diet and the time after eating. Immediately after a meal, blood levels of hexose monosaccharides are high and will contain detectable quantities of fructose and galactose in addition to glucose. A short while later the only simple hexose sugar present in the blood is glucose. Either the liver has converted the other sugars to glucose or the fructose has been utilized in various glycolysis schemes. Although galactose can be absorbed over the gut epithelium, it must be converted to either fructose or glucose in the liver. Glucose is the main—and usually the only—hexose monosaccharide found in the blood. Normal blood concentration of glucose three or more hours after a meal will be 90 percent, and except in cases of diabetes the blood glucose level will seldom exceed 140 milligrams percent.

Carbohydrates are depleted in the blood by oxidation within the cells, by storage as muscle or liver glycogen, by conversion into fat by the liver and adipose tissue, and by excretion through the kidneys when the threshold level is exceeded. In fasting animals blood glucose levels are maintained by the liver and kidneys. The liver is the principal site for conversion of glycogen to blood glucose, and is the principal organ involved in the maintenance of blood glucose levels, although glucose may also be supplied to the bloodstream by cells of the intestinal mucosa and the kidneys. The liver, moreover, is the principal storage organ for blood glucose.

In the liver appropriate enzymes called convertases are available to convert hexose monosaccharides from one form to another. Although glucose can interconvert in virtually all body cells, galactose is able to interconvert only in the liver, since the appropriate convertase enzymes for this reaction are present only in liver cells.

Muscle Glycogen

Somewhat over 50 percent of total body glucose in nonruminants is present in muscle tissue, but little of this gets into the bloodstream. Muscle glycogen is structurally complex and the molecules are so large that they are unavailable for blood transport. Moreover, the glycolysis reactions in muscle produce a glucose compound that satisfies energy demands of muscle tissue but which cannot leave the muscle cells: hence no muscle glucose (and only a small amount of glycolysis by-products) can enter the bloodstream. A small amount of the breakdown products of muscle glycogen (notably lactic and pyruvic acids) may enter the bloodstream and be transported to the liver where they may be resynthesized into blood glucose or liver glycogen. The sequence of events that form the cycle from blood glucose to muscle glycogen to lactate and pyruvate to blood glucose is known as the Cori cycle, and is one of the cyclic schemes that eventually led to Schönheimer's concept of the dynamic state of body constituents.

The Glycolysis Scheme and the Tricarboxylic Cycle

The catabolic reactions that convert glycogen to hexose monosaccharides (simple sugars) and then convert the monosaccharides to energy, carbon dioxide, and water are known as the glycolysis scheme and the tricarboxylic (Krebs) cycle. Both glucose-6-phosphate and fructose-6-phosphate can be utilized as basal energy sources in the glycolysis scheme and the tricarboxylic cycle. Galactose, however, cannot be so used, and must first be converted to glucose. Moreover, fructose can be utilized only in the glycolysis scheme, since glucose is the basis of the maltose units that are polymerized into glycogen. The result is that glucose forms the main carbohydrate utilized for energy in the body, and is the sole precursor of the complex carbohydrate called glycogen (animal starch), which is the storage form of hexose carbohydrates.

In glycolysis, successive molecules of glucose in the polymer chain are split off the main body by a process called phosphorylation. The enzyme phosphorylase is the catalyst of this process. The enzyme is activated only when needed, and its absence permits glycogen to be stored. The hormones glucagon and adrenalin (p. 355, p. 354) both activate phosphorylase, and although they operate through somewhat different mechanisms, the end result is the same.

Once the phosphorylated glucose molecule (glucose-6-phosphate) is split off the glycogen molecule it can go two ways, either down the glycolysis scheme which commences with glucose-6-phosphate (Figure 13-2), or the phosphate (PO_4) molecule can be removed by the enzyme phosphatase, which is found in cells of the kidney, intestine and liver. The dephosphorylated glucose molecule is not trapped within these cells, but can diffuse out and be transported by the blood to other body cells. In other body cells, particularly muscle, glycogen is degraded to glucose-6-PO_4 by phosphorylase but phosphatase is not available; hence glucose-6-PO_4 cannot be dephosphorylated to glucose in these locations and is not able to leave the cells and enter the vascular system. Instead there is local glycolysis and virtually complete catabolism of carbohydrate. The following figures (Figures 13-1 and 13-2) show the fundamental reactions through which the metabolism of carbohydrate and other substances is accomplished.

While glycolysis is relatively complete in the muscles, only a small portion of liver glycogen or glucose is completely catabolized. Most liver glycogen is converted to glucose which is sent to the blood. The reaction is powered by the oxidation of hydrogen atoms released by glycolysis, which are incorporated into the restoration of ATP.

Note that the reactions follow circular or cyclical patterns. This is commonplace in biochemical reactions, which differ from usual chemical reactions not only because they tend to be circular and/or reversible, but because they operate in a sequential fashion at low temperatures and relatively low levels of energy. The reactions in the glycolysis scheme are each catalyzed by a specific enzyme or enzymes, and the energy release in any one step is usually too small to form ATP molecules from ADP. This energy is dissipated as heat and serves no primary metabolic function. However, between diphosphogylceric acid and phosphoglyceric acid, and between the enolphosphopyruvic acid and pyruvic acid stages the energy release is great enough (7,000 calories) to form ATP. This results in the formation of four molecules of ATP for each molecule of glucose-6-phosphate that enters the scheme. However, the net gain

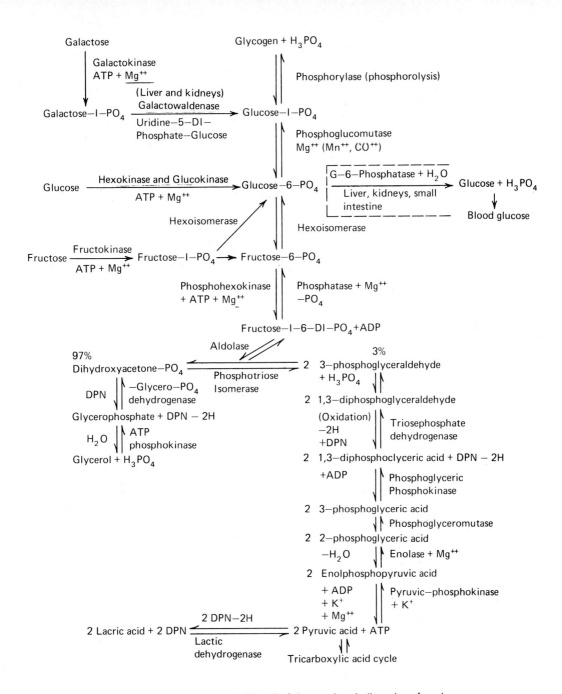

A composite diagram of the glycolytic cycle. Not all of the reactions indicated are found in all tissues and organs; for example, only liver, kidney, and possibly intestinal mucosa possess a glucose–6–phosphatase that produces glycogenolysis. Specific enzymes are shown.

FIGURE 13-1   *The glycolytic cycle. (Adapted from Dukes,* Physiology of Domestic Animals.*)*

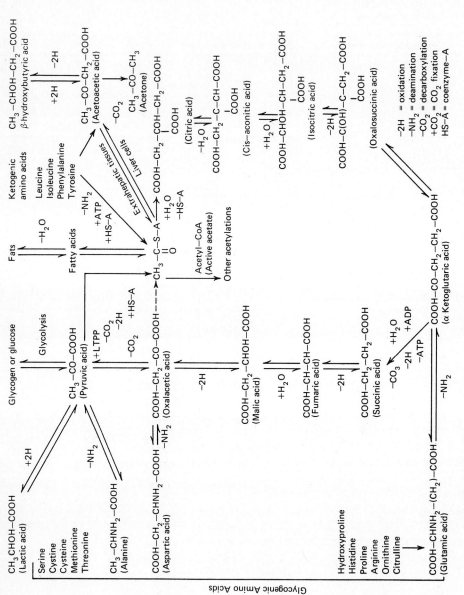

FIGURE 13-2 The tricarboxylic (Krebs) cycle. (Adapted from Dukes, Physiology of Domestic Animals.)

is only two molecules of ATP, or about 14,000 calories, since two ATP molecules are required to initiate the reaction. However, some 56,000 calories of energy are lost as heat during the reaction which makes this particular reaction only 25 percent effective. Additionally, there are four hydrogen atoms lost in the transformation of diphosphoglyceraldehyde to phosphoglyceric acid. These, plus eight more hydrogen atoms released in the tricarboxylic acid cycle, are eventually oxidized by respiratory oxygen to provide most of the energy needed to synthesize ATP.

The Tricarboxylic (Krebs) Cycle

The tricarboxylic cycle (Figure 13-2) commences with the conversion of the molecules of pyruvic acid, which are formed from the breakdown of fructose 1-6 diphosphate in the glycolysis scheme. The pyruvic acid molecules convert to two molecules of acetyl Co-A (active acetate) with a release of one molecule of CO_2 and two atoms of hydrogen for each molecule of acetyl Co-A formed. This conversion does not form ATP although it gives off enough energy to do so, but six molecules of ATP are later formed when the subsequent oxidation of the four hydrogen atoms provides energy for ATP formation. In the tricarboxylic cycle the acetyl portion of acetyl Co-A is converted into CO_2 and hydrogen. The hydrogen atoms are later oxidized and provide still more energy to form ATP.

It is theorized that the enzymes involved in the tricarboxylic cycle and the formation of ATP are contained in the mitochondria of the cells. The enzymes are possibly attached to the cristae of the mitochondria in sequential order so that successive products can be shunted from one enzyme to another with a minimum of delay. The theory is based upon the known fact that mitochondria provide ATP to the cell.

The net product of the tricarboxylic cycle is two CO_2 molecules, eight hydrogen atoms, one Co-A molecule, and one ATP molecule for each molecule of acetyl Co-A that enters the scheme. A net of three molecules of water, one molecule of ADP, and one acetyl Co-A must enter the scheme for the reaction to proceed. The cycle begins and ends with oxaloacetic acid and theoretically could go on forever, accepting acetyl Co-A, ADP and water, and giving off Co-A, hydrogen, CO_2, and ATP.

The release of hydrogen is catalyzed by the enzyme dehydrogenase, and in the entire glycolysis-tricarboxylic cycle there are six points where hydrogen is released; hence, six different dehydrogenases are necessary to catalyze the reaction at those points. At four of the six points (the ones between ketoglutaric acid and fumaric acid are not included) diphosphopyridine nucleotide (DPN) attaches to the dehydrogenase, which is catalyzing the substrate. Then the complex breaks apart, leaving the substrate with two less hydrogen atoms, the dehydrogenase, the DPN plus one hydrogen atom, and a free hydrogen atom.

The reaction will not occur without initial dehydrogenase activity and available DPN to act as a hydrogen carrier. Ultimately both hydrogen atoms are oxidized and enter into ATP formation. The hydrogen atoms given off in the formation of succinic and fumaric acids combine with a specific dehydrogenase that does not require DPN, and pass directly into the oxidation process.

The tricarboxylic cycle is of considerably greater functional importance in carbohydrate metabolism than glycolysis. It is the liberation of hydrogen rather than the formation of ATP that marks the importance of the cycle. If the entire glycolytic reaction is considered from the viewpoint of the glucose molecule, a molecule of glucose produces six molecules of ATP and uses two for a net gain of four in the entire glycolysis-tricarboxylic scheme. However, the scheme also produces a total of 24 hydrogen atoms. Twenty of these are oxidized through the oxidative phosphorylation scheme, which produces three ATP molecules for every two atoms of hydrogen that are metabolized, or a total of 30 ATP molecules. The remaining four hydrogen atoms, which are released in the succinic and fumaric conversions in the tricarboxylic cycle, enter the oxidative phosphorylation scheme after the initial DPN→DPNH reaction, and produce two ATP molecules for each two hydrogen atoms metabolized for a total of four. The combined total ATP molecule production from the complete oxidation of one molecule of glucose-6-PO_4 is 38.

The molecules of ATP store 266,000 small calories of the 686,000 that are released during the total oxidation of each gram molecule of glucose. The storage transfer efficiency is 39 percent. The remaining 61 percent (420,000 gram calories) is lost as heat and cannot be specifically utilized by cells. However, this heat loss is utilized by the body to prevent hypothermia in cold environments. In humans this is the reason why knowledgeable inhabitants or travellers in arctic environments eat high energy foods such as fats and chocolate bars, which are quick suppliers of glucose.

Rarely, the oxygen supply to cells becomes so poor that cellular oxidation cannot take place, but this does not entirely interrupt the production of energy from glucose. The glycolysis scheme will continue to operate, since the conversion of glucose to pyruvic acid does not require oxygen. This, however, is a wasteful process since it converts less than 2 percent of the glucose to energy, compared with 39 percent if the glucose passes through the tricarboxylic cycle. Moreover, it can be dangerous since considerable lactic acid is formed from pyruvate and this material can be toxic in excess. However, the reaction may be life-saving in certain circumstances, and the lactic acid can readily be converted back to pyruvic acid once adequate oxygen becomes available. The pyruvic acid can then either go through the tricarboxylic scheme or be converted back to glucose.

It is interesting to note that while excess lactic acid in skeletal muscle can result in skeletal muscle damage in horses, the heart muscle frequently utilizes lactic acid for energy when adequate supplies of glucose are temporarily unavailable.

Oxidative Phosphorylation

The biochemical reactions that involve the oxidation of the hydrogen atoms released in the various glycolytic reactions and the tricarboxylic cycle are not completely known, and are one of the more important holes in our skein of knowledge of the metabolic processes. However, there are some educated guesses that can be made about the process based upon what we already know. The oxidation reactions are theorized to be a sequential series of enzyme-catalyzed reactions that convert the hydrogen atoms into hydrogen ions (H^+) and change the dissolved oxygen in the intracellular fluid of the reacting cells into hydroxyl ions (OH^-), and then unite the hydrogen and hydroxyl into water (H_2O). During the course of the reactions large amounts of energy are released to form ATP. A theoretical structure of the oxidative phosphorylation reaction scheme is shown in Figure 13-3.

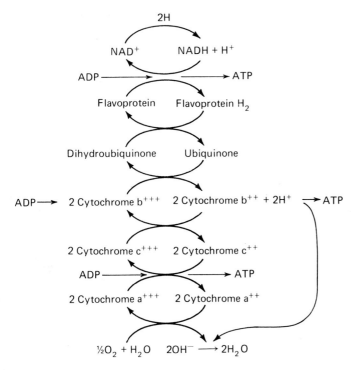

FIGURE 13-3   *Theoretical scheme of the oxidative phosphorylation reaction. (Adapted from Guyton,* Textbook of Medical Physiology.*)*

The simplified reaction would read

$$2H^+ + O^+ + 3ADP \rightarrow H_2O + 3ATP$$

Since this reaction uses none of the enzymes of the tricarboxylic cycle, it is an independent source of energy. In the first four steps of the reaction one molecule of carbon dioxide and four atoms of hydrogen are released. The remaining steps are devoted to the formation of various monosaccharide sugars that can be converted into glucose. The reaction leaves five molecules of glucose for every six that enter the scheme. The single molecule of glucose that is lost is combined with six molecules of metabolic water and is converted into six carbon dioxide molecules and 24 atoms of hydrogen. As the reaction continues and other molecules of glucose and water are injected into it, a continuous flow of carbon dioxide and hydrogen will be given off.

The oxidative phosphorylation scheme involves ionization reactions and electron transport. The two hydrogen atoms removed from the substrate during the breakdown of glucose are finally released into the interstitial (intercellular) fluid as hydrogen ions. The removed electrons are passed to cytochrome-b, which in turn passes them to cytochrome-a. Here four electrons are transferred from four cytochrome-a molecules to a molecule of dissolved oxygen and two molecules of water to form four hydroxyl ions. The hydrogen ions previously formed in the reaction by removal of electrons, and the hydroxyl ions created by the addition of electrons to water and oxygen, now unite to form water, and the energy released is used to convert ADP to ATP.

Each reaction should produce three molecules of ATP for every two atoms of hydrogen which pass through the scheme.

The Phosphogluconate Reaction

In addition to the glycolysis scheme, the tricarboxylic acid cycle, and the oxidative phosphorylation reaction, energy can be obtained from glucose by yet another method, the phosphogluconate reaction (also called the hexose monophosphate shunt). Although this reaction converts only about 5 percent of total body glucose into energy, it is more important than it appears. The reaction takes place primarily in the liver and over a period of time will convert about one-third of the liver glucose to energy. Since liver functions in synthesis and detoxification are of vital importance to life and health, the energy released by the phosphogluconate reaction has a disproportionately large survival value for the individual. The reaction is shown in Figure 13-4.

Unlike the hydrogen produced in the glycolytic scheme, the phosphogluconate hydrogen reacts with TPN (triphosphoridine nucleotide), rather than DPN (diphosphoridine nucleotide) since the enzymes in the phosphogluconate reaction do not react with DPN. However, the result

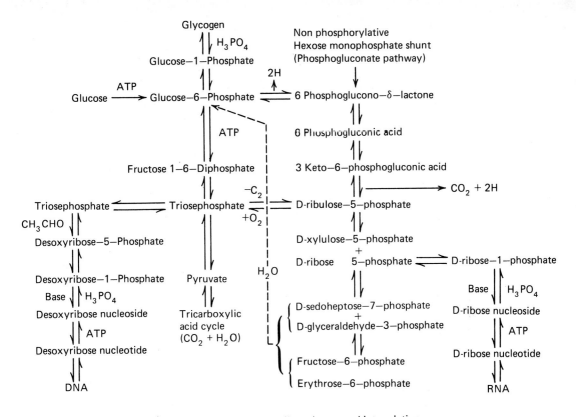

An illustration of the complex metabolic pathways and interrelationships which exist between the glycolysis scheme, the hexose monophosphate shunt, and the metabolism of pentose sugars in the synthesis and catabolism of nucleosides, nucleotides and nucleic acids. The enzymes involved in these reactions are not shown.

FIGURE 13-4   *Metabolic interrelationships—hexose glycolysis and pentose syntheses. (Adapted from Dukes,* Physiology of Domestic Animals.*)*

is the same. Hydrogen ions are produced. A few of these react with DPN to enter into the conversion of ADP to ATP, but most remain bound to TPN as TPNH, and in this form are used to synthesize fats from carbohydrates.

The phosphogluconate pathway remains operative whether or not the glycolysis scheme is operating, and in the event excess carbohydrate is present, operates through TPNH to convert the excess into fat which is stored in fat depots. It also functions in the metabolism of pentose sugars and the formation of RNA and DNA.

The continual release of energy from glucose when energy is not required by the body would be wasteful. Therefore, a regulating mechanism is required. This mechanism involves ATP. If all the cells' ADP has been changed to ATP, the entire glycolysis scheme and oxidative phosphorylation mechanism will stop since they both require conversion of ADP to ATP as part of the step sequence of their reaction schemes. As a result, glucose will not be metabolized until ATP is utilized by different energy demanding functions and sufficient ADP is formed to make the energy reactions go.

There is a principle here that should be noted: *Unless all elements of an intermediate metabolism reaction are present in the proper quantity, the reaction will not occur.*

High levels of blood glucose (hyperglycemia) are usually controlled by insulin through the mechanism described on p. 355. This promotes sugar transport into cells, resulting in lowered blood glucose levels and raised cellular levels of glucose and glycogen. As blood levels drop, the output of insulin decreases.

If blood glucose levels fall much below normal (hypoglycemia) the sympathetic nervous system becomes stimulated and releases adrenalin and noradrenalin from the adrenal medulla. These hormones go to the liver and activate phosphorylase, which catalyzes glucose from liver glycogen. The glucose thus produced can rapidly elevate blood levels and correct hypoglycemia. As the blood concentration rises to normal levels, adrenalin and noradrenalin production is reduced.

FAT METABOLISM

Fat (lipids) occur in three major forms and a few others of minor significance. The important forms are triglycerides (neutral fats), phospholipids, and cholesterol.

The basic compound in both triglycerides and phospholipids is the fatty acid. This is essentially a long chain saturated or partially saturated organic acid with the general formula, CH_3-R-COOH, where R is a variable number of CH_2 or CH groups. The more CH groups that are present, the less saturated (and more fluid) the fatty acid will be. Highly unsaturated fats are oils, highly saturated fats are waxes.

In the general metabolic scheme, the utilization of neutral fat is by far the most common, if not the most important function. Three fatty acids, stearic, oleic, and palmitic, are the most common and are fully saturated except for oleic acid, which has one double bond in the middle of the chain. As has been mentioned previously in the chapter on digestion, fats are split in the stomach and the anterior part of the small intestine into glycerol, monoglycerides, and fatty acids, which pass across the intestinal epithelium. Once across the epithelial barrier the fats are resynthesized into triglycerides and pass into the lacteals.

The tiny fat droplets, sometimes called chylomicrons, adsorb an equally minute amount of protein, which coats them and increases their suspension stability. The lymph transports the fat from the intestine to the bloodstream via the intestinal lymphatics, the cisterna chyli and the thoracic duct which empties into the cranial vena cava. Immediately after eating, lipid concentration in the blood plasma rises and the lacteals become engorged with chylomicrons. This phenomenon disappears in about three hours as the fat is absorbed by the body.

The largest portion of the absorbed fat is hydrolized into fatty acids and glycerol in the blood via an enzyme called lipoprotein lipase. Lipoprotein lipase is scant in the blood, abundant in fatty tissues and heart muscle, and moderately abundant in skeletal muscle.

The amount of lipoprotein lipase can be increased by injection of heparin, a liver anti-clotting factor. Either heparin is a part of the enzyme or it is a mobilizing agent. A rise in blood-borne lipoprotein lipase increases the rate of hydrolyzation of the chylomicrons in the plasma and rapidly clears the plasma of the turbidity which they impart. The enzyme was once called "clearing factor" because of this ability.

A lesser portion of the chyle is transported to the liver by the portal and systemic circulation, where the liver cells absorb it directly and utilize it as energy or in synthesis reactions. The liver synthesizes a major part of the triglycerides and virtually all of the phospholipid and cholesterol found in the body.

Unesterified Fatty Acids

A small and inconstant quantity of fatty acids is always present in the circulating blood combined with albumin. Fatty acids ionize readily in water and the ionized acid combines with the albumin fraction of the plasma. Ordinarily about three molecules of fatty acid combine with one molecule of albumin, but this can rise to a 30-to-one ratio when the need for fat transport is increased. This is the "unesterified fatty acid" fraction of the blood plasma, so named because all the other lipids in the bloodstream are transported as esters. Although the blood concentration of fatty acids is small (about 10 milligrams percent) it represents the major transport system for lipids since its turnover is extremely rapid. Over 50 percent of the unesterified fatty acids in the blood are replaced every two or three minutes. At this turnover rate about half the energy demand of the body is satisfied under normal conditions. Under abnormal conditions such as starvation, or where carbohydrates cannot be utilized for energy, the quantity of unesterized fat in the blood increases several times, which indicates that these fats are probably being utilized to satisfy energy demands.

Lipoproteins

Once absorption of the fats from food is completed and the enzymatic hydrolysis of the chylomicrons is finished, over 95 percent of all the lipids in the plasma are in the form of lipoproteins, i.e., mixtures of triglycerides, phospholipids, cholesterol and proteins. These are formed primarily or totally in the liver. Protein represents about 25–35 percent of this mixture. The relationships, expressed as milligrams percent of the plasma, are roughly

triglycerides	150
phospholipid	150
cholesterol	175
lipid-bound protein	200

The percentages vary with the species of animal and the diet. These can be further classified as alpha (high density) or beta (low density) lipoproteins. High density lipoproteins are small quantities of lipids combined with alpha (α) globulins. Low density lipoproteins represent relatively large quantities of lipid combined with beta (β) globulins. Each of these two major classifications can be further subdivided.

The precise functions of lipoproteins are unclear. They are thought to be the means whereby lipids can be transported from one part of the body to another for energy utilization, fat storage, or special purposes such as myelin formation.

Unutilized fats are stored for the most part as energy reserves in fat depots or adipose tissue. Small amounts of fat around the kidneys, the coronary portion of the heart, and in the orbit of the eye apparently have homeostatic functions, and peripheral fat, especially in swine, seals, whales, and dolphins, serves as an insulating device to conserve body heat.

Fat cells are specialized connective tissue cells which are capable of storing triglycerides. Depending on the species and degree of unsaturation of dietary fat, the fat within the cells may be semisolid or liquid. In sheep and cattle the fat is notably firm. In horses, dogs, and cats it is more liquid.

Large amounts of lipoprotein lipase are present in fat depots and apparently function to maintain the unesterified fatty acid blood levels. During periods of utilization of fats for energy, the rate of utilization of body fat increases and the fat depots decrease in size.

Triglyceride Metabolism

In metabolizing fats for energy, the liver plays a major role. It functions to split triglycerides into fatty acids and glycerol, to break down or desaturate long chain fatty acids, to synthesize or resynthesize triglycerides and to form specialized fats or fatty acids needed by the body. In times of stress, impaired or damaged carbohydrate utilization, or during starvation, the liver functions as a reception and conversion center where triglycerides from fat depots are either converted into glucose or converted into acetyl Co-A by a process called beta oxidation.

The first state of the beta oxidation process is the hydrolyzing of triglyceride into fatty acids and glycerol. This can occur in other body locations than the liver, but to a far less significant extent. The glycerol is promptly converted into glyceraldehyde by intracellular enzymes, passes through the phosphogluconate pathway (Figure 13-4) and is used for energy.

The fatty acids require more complex treatment. In a step reaction the long chain fatty acids are split into acetyl Co-A, which then enters the tricarboxylic cycle (Figure 13-2) and goes through the same process as the final breakdown of glucose.

A large number of ATP molecules can be produced from ADP by the breakdown of a single molecule of fatty acid. Conversion to acetyl Co-A by beta oxidation releases four hydrogen atoms from the fatty acid molecule each time a molecule of acetyl Co-A is formed. By using stearic acid as an example, nine molecules of acetyl Co-A can be formed. Eight of these will be involved in hydrogen atom release for a total of 32 hydrogen atoms. The tricarboxylic cycle yields eight hydrogen atoms for every molecule of acetyl Co-A that passes through it. The nine acetyl Co-A molecules will thus produce 72 more hydrogen atoms for a total of 104.

Of this total, 34 atoms are removed by flavoproteins, and 70 by DPN. Since these two groups enter the oxidative phosphorylation reaction at different points, their ATP production differs. Flavoprotein-bound hydrogen produces one molecule of ATP for each hydrogen atom, but DPN-bound hydrogen produces one and one-half molecules of ATP for each hydrogen atom. In addition, each of the nine molecules of acetyl Co-A that pass through the tricarboxylic cycle forms one molecule of ATP. A total of 148 ATP molecules can thus be derived from the degradation of one molecule of stearic acid. This entire reaction occurs in the liver, but the liver is incapable of handling all the fat conversion that may be required in times of severe need. In consequence, other tissues of the body enter into the scheme. To accomplish this shift of the tricarboxylic cycle from liver cells to elsewhere, the liver combines two molecules of excess acetyl Co-A into one molecule of acetoacetic acid, which is transported by the bloodstream to other metabolic locations. In the new location the reaction is reversed and two molecules of acetyl Co-A are formed from the acetoacetic acid and enter into the tricarboxylic acid cycle.

Ordinarily the acetoacetic acid is transported rapidly and causes no noteworthy variation in the blood chemistry. However, in certain situations, notably starvation, early lactation, diabetes or pregnancy toxemia, when there is increased demand for glucose beyond the ability of carbohydrate and normal fat metabolism to satisfy, an abnormal amount of acetoacetic acid may enter the blood and tissue fluids.

Ketone Bodies

Acetoacetic acid can be readily converted into two other compounds, β-hydroxybutric acid and acetone. For example,

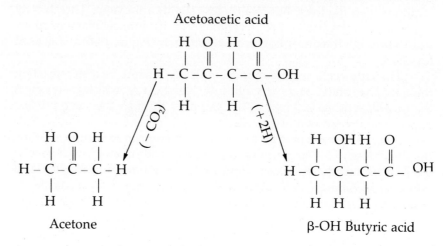

These three substances are the so-called ketone bodies that are present in ketosis. Ordinarily they will be found in urinalysis. In severe cases the odor of acetone will be present in expired air. In excess, the substances are metabolic poisons, which can produce severe acidosis by decreasing blood pH or by causing excessive sodium loss through the demands of urinary excretion. The sodium loss occurs because the ketone bodies are too acidic to be completely excreted in the urine in their acid form. Some must be combined to form basic salts, and extracellular sodium is the most readily available bodily substance for this conversion.

The sodium loss results in a major deficiency of alkali in the body's buffering systems, thus permitting further acidosis. If the condition is continued over a period of time, it can result in death.

Sources of Fat Other Than Triglyceride Ingestion

Fats can be formed from carbohydrates. The sites of synthesis are in the liver and adipose tissue. The basic process is a reversal of the beta oxidation scheme, converting acetyl Co-A to 16-, 17-, or 18-carbon fatty acids, which are conjugated with glycerol to form triglyceride. The enzymes that catalyze this recombination appear to be specific for fatty acids in the 14- to 20-carbon range and will not work on shorter chain fatty acids. The energy loss in the entire process is small, about 15 percent. This form of storage is important principally because glycogen storage is not only slight, but it is, with the exception of liver glycogen, a specialized storage for specific (e.g., muscle energy) purposes, and consequently the glucose derived from it is not available for general use.

Additionally, a gram of fat contains 2.25 times the energy of a gram of glycogen; hence, fat as an energy storage form is more concentrated.

Proteins, also via acetyl Co-A, can be synthesized into fatty acids and triglycerides, and this occurs when there is excess protein in the diet.

The rate of fat metabolism is regulated by a number of mechanisms, i.e., amount of available TPNH, rate of CHO metabolism, amount of acetyl Co-A, and various hormones. If there is excess TPNH, which usually happens when excess glucose is metabolized via the phosphogluconate pathway, the conversion of acetyl Co-A to fat is favored. If there is low glucose metabolism, fat will be mobilized from depots to compensate for the lack. Concentration of acetyl Co-A from any source affects fat metabolism. If the concentration is high, fat metabolism is reduced; if it is low, more fatty acids will be degraded and less fat will be stored.

Hormones involved in fat metabolism include insulin, which can directly increase or decrease fat metabolism by controlling cellular utilization of glucose. Thryoxin can cause increased mobilization of fats by increasing the general metabolic rate. Adrenal cortical hormones act directly upon fat cells to produce increased fat mobilization; in their absence fat utilization is low. Adrenalin also can stimulate fat metabolism. Pituitary hormones, i.e., ACTH and STH, both have fat mobilizing effects. ACTH probably functions more through stimulation of the adrenal cortex than directly on the fat cells.

Phospholipid Metabolism

The second group of fatty substances involved in fat metabolism are the phospholipids. These fall into three major groups, the lecithins, cephalins, and sphingomyelins. They ordinarily contain a fatty acid, a phosphoric acid radical, and a nitrogenous base. Their chemical structures vary but their physical properties are similar. They are all lipid soluble and are transported as lipoproteins in the blood plasma and appear to be utilized similarly throughout the body. Most body cells are capable of forming phospholipid although about 90 percent of blood phospholipid comes from liver synthesis. Virtually no phospholipids are excreted from the body, which indicates that they are broken down into other compounds and metabolized. Certain metabolites, notably choline (for lecithin) and inositol (for some cephalins), are necessary for specific phospholipid formation. Other metabolites such as biotin, cystine, and thiamin can depress phospholipid formation. Phospholipids are used to help transport fatty acids across the intestinal mucosa, to form thromboplastin, to form myelin, to donate phosphate radicals, to act as "carriers" on the cell membrane transport systems, and to form structural elements within cells.

Cholesterol Metabolism

Despite the furor over the possible relationship between cholesterol and atherosclerosis, it should be recognized that cholesterol is a normal dietary component and a necessary compound in the structure and function of the body. It is a part of cholic acid, and of bile salt formation which is necessary for fat digestion. It forms the nucleus for a number of adrenal and sex hormones. It functions to make the skin relatively

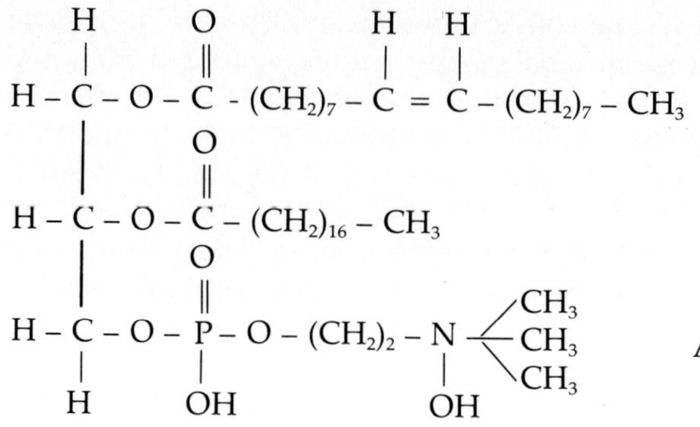

$$\begin{array}{l} H \\ | \\ H-C-O-\overset{O}{\underset{\|}{C}}-(CH_2)_7-\overset{H}{\underset{|}{C}}=\overset{H}{\underset{|}{C}}-(CH_2)_7-CH_3 \\ | \\ H-C-O-\overset{O}{\underset{\|}{C}}-(CH_2)_{16}-CH_3 \\ | \\ H-C-O-\overset{O}{\underset{|}{P}}-O-(CH_2)_2-N{\overset{\displaystyle CH_3}{\underset{\displaystyle CH_3}{-CH_3}}} \\ | \quad\quad | \\ H \quad\quad OH \quad\quad\quad\quad\quad\quad OH \end{array}$$

A lecithin

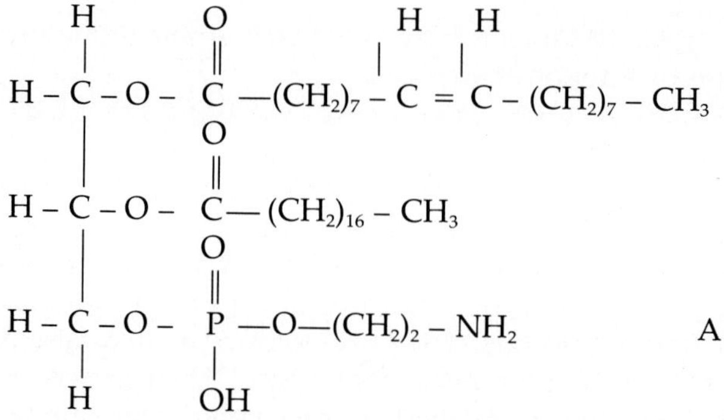

$$\begin{array}{l} H \\ | \\ H-C-O-\overset{O}{\underset{\|}{C}}-(CH_2)_7-\overset{H}{\underset{|}{C}}=\overset{H}{\underset{|}{C}}-(CH_2)_7-CH_3 \\ | \\ H-C-O-\overset{O}{\underset{\|}{C}}-(CH_2)_{16}-CH_3 \\ | \\ H-C-O-\overset{O}{\underset{|}{P}}-O-(CH_2)_2-NH_2 \\ | \quad\quad | \\ H \quad\quad OH \end{array}$$

A cephalin

$$\begin{array}{l} H \quad H \quad H \\ | \quad | \quad | \\ H-C-C=C-(CH_2)_{12}-CH_3 \\ | \quad\quad H \quad O \\ | \quad\quad | \quad\quad \| \\ H-C-N-C-(CH_2)_{16}-CH_3 \\ | \\ H-C-O-\overset{O}{\underset{|}{P}}-O-(CH_2)_2-N{\overset{\displaystyle CH_3}{\underset{\displaystyle CH_3}{-CH_3}}} \\ | \quad\quad | \\ H \quad\quad OH \quad\quad\quad\quad\quad\quad OH \end{array}$$

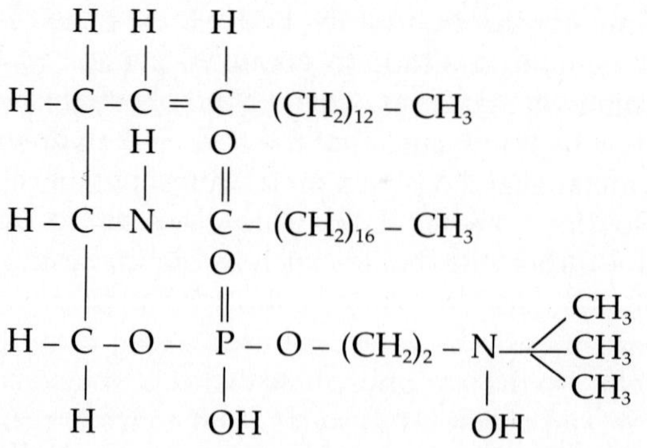

A sphingomyelin

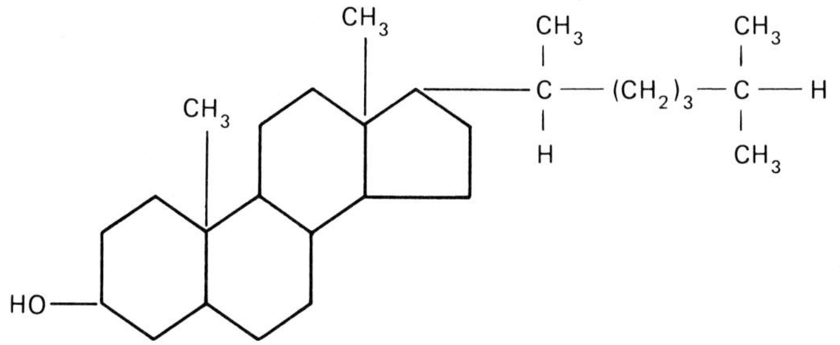

FIGURE 13-5   Cholesterol.

impervious to water, water soluble substances, and chemical agents such as acids and tends to prevent evaporation of water through the skin. This last feature is highly important for without cholesterol in the skin, evaporation would be about 60 times greater than it is, which could be fatal.

Both phospholipid and cholesterol appear to be a normal and necessary part of all membranes and probably have some effect upon their permeability. In this connection it should be remembered that there are more membranes in a cell than the outer limiting membrane and these other membranes probably also require cholesterol for normal functioning. At least three additional membranes are present in cells: the nuclear membrane, a nucleolar membrane, and a mitochondrial membrane. A requirement of these structures is that they are not soluble in water. Of the organic body components only lipids and some high molecular weight proteins qualify in this regard. Lipids, therefore, undoubtedly contribute to cell integrity. In addition, both phospholipids and cholesterol have a slow turnover rate and consequently would contribute to long-term integrity of structures.

Cholesterol is readily absorbed into the body over the intestinal epithelium. It requires no preliminary digestion, is readily fat soluble and is only slightly soluble in water. It can and does form esters with fatty acids. Ordinarily cholesterol comes from exogenous and endogenous sources. Endogenous cholesterol is formed mainly in the liver but can be made elsewhere, and probably all cells are to some extent concerned in its formation. Its basic structure is a sterol nucleus, which can be synthesized entirely from acetyl Co-A. Various side chains modify the basic nucleus and form a number of adrenal and sex hormones.

Plasma cholesterol levels can be influenced by dietary cholesterol. The influence is slight in a mixed diet, but high saturated fat diets can appreciably raise blood cholesterol levels. Unsaturated fat diets slightly depress blood levels of cholesterol. Lack of thyroxin increases blood cholesterol levels, excess thyroxin depresses them. Lack of insulin causes a rise in blood cholesterol.

PROTEIN METABOLISM

Proteins often have been called the building blocks of the body. They are, without doubt, the most complex molecules that occur in nature, and, as has been indicated earlier, are polymers of amino acids that can have a molecular weight of several million. The amino acids are joined together by a so-called peptide linkage that involves the amine (NH_2) group of one acid and the carboxyl (COOH) group of another. Amino acids are the form in which protein is transported across the intestinal epithelium. The majority of the amines are picked up by the blood capillaries of the intestinal villi and are transported to the liver via the portal circulation. Here they may undergo a number of processes or be directly transferred to the bloodstream.

It should be emphasized that there are always free amino acids in the blood, interstitial fluid, and lymph. Some of these come from dietary protein, others from tissue protein. Protein metabolism in mammals is essentially the metabolism of amino acids.

It has been the use of "tagged molecules," i.e., molecules carrying radioactive atoms (isotopes) such as carbon (C^{14}), deuterium (H^2), and nitrogen (N^{15}), that has made the study of intermediate metabolism of foods meaningful and accurate, since by intelligent use of these isotopes the metabolic paths and end products of metabolism can be discovered.

Insofar as protein metabolism is concerned, it has been shown that there are at least nine possible outcomes:

1. deamination with the formation of α-ketoacids and ammonia;
2. transamination;
3. ureogenesis, i.e., conversion of ammonia to urea and uric acid;
4. gluconeogenesis, i.e., conversion of α-ketoacid to glucose;
5. liponeogenesis, i.e., conversion of α-ketoacid to fat;
6. oxidation of α-ketoacids for energy;
7. synthesis of amino acids into body proteins;
8. other synthesis reactions that produce:
   a. hormones,
   b. enzymes,
   c. creatine and creatinine;
9. excretion of amines, proteins, and protein breakdown products in the urine.

Deamination

All amino acids have this characteristic terminal configuration:

$$R - \underset{\underset{NH_2}{|}}{\overset{\overset{H}{|}}{C}} - \overset{\overset{O}{\|}}{C} - OH$$

The terminal carbons with the attached carboxy (COOH) and amine (NH₂) groups are commonly known as the "amine radical." The R portion may be a number of chain (aliphatic) or ring-and-chain (aromatic) compounds. Other terminations, such as beta (β) amine or gamma (γ) amine radicals, are chemically possible but are rarely, if ever, encountered in biological systems, for example,

$$\overset{(\beta)\ (\alpha)}{H - \underset{\underset{NH_2}{|}}{\overset{\overset{H}{|}}{C}} - \underset{\underset{H}{|}}{\overset{\overset{H}{|}}{C}} - \overset{\overset{O}{\|}}{C} - OH} \qquad \beta \text{ amino propionic acid}$$

$$\overset{(\gamma)\ (\beta)\ (\alpha)}{H - \underset{\underset{NH_2}{|}}{\overset{\overset{H}{|}}{C}} - \underset{\underset{H}{|}}{\overset{\overset{H}{|}}{C}} - \underset{\underset{H}{|}}{\overset{\overset{H}{|}}{C}} - \overset{\overset{O}{\|}}{C} - OH} \qquad \gamma \text{ amino butyric acid}$$

and while diamine amino acids and other R-configurations that include amine groups, such as:

$$\overset{(\delta)\ (\gamma)\ (\beta)\ (\alpha)}{H - \underset{\underset{NH_2}{|}}{\overset{\overset{H}{|}}{C}} - \underset{\underset{H}{|}}{\overset{\overset{H}{|}}{C}} - \underset{\underset{H}{|}}{\overset{\overset{H}{|}}{C}} - \underset{\underset{NH_2}{|}}{\overset{\overset{H}{|}}{C}} - \overset{\overset{O}{\|}}{C} - OH} \qquad \begin{array}{l}\text{Ornithine}\\ \alpha, \delta \text{ diamino valeric acid}\end{array}$$

and

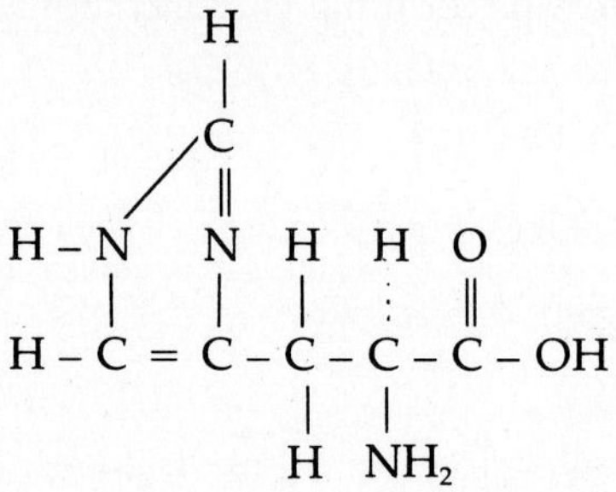

$$\underset{\text{Histidine}\atop\text{α amino imidazole propionic acid}}{H-C=C-C-C-C-OH}$$

(structure of histidine with imidazole ring, showing H–N, N, H, H, O and H, NH$_2$ substituents)

are not rare, the terminal amine group is an α amine. And since these compounds are all included under the classification of α amino acids the remainder of this section will deal with α amino acid reactions.

In deamination, the removal of two hydrogen atoms from an α amino acid can produce an imino acid, for example,

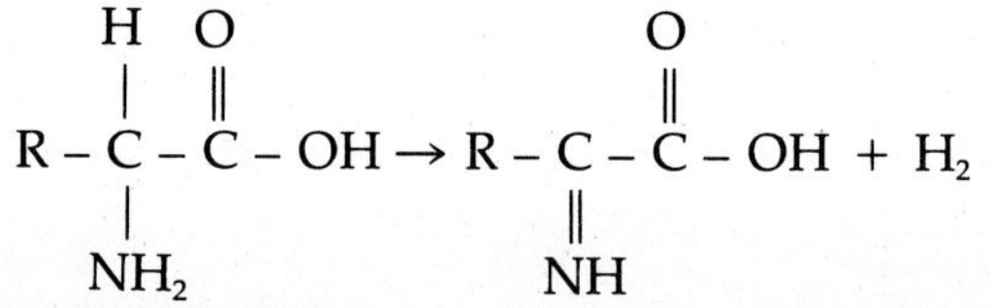

$$\underset{\text{(amino acid)}}{R-\underset{NH_2}{\underset{|}{C}}-\overset{O}{\overset{\|}{C}}-OH} \rightarrow \underset{\text{(imino acid)}}{R-\underset{NH}{\underset{\|}{C}}-\overset{O}{\overset{\|}{C}}-OH} + \underset{\text{(hydrogen)}}{H_2}$$

By the addition of a molecule of water to the imino acid an α ketoacid plus ammonia (NH$_3$) can be produced, for example,

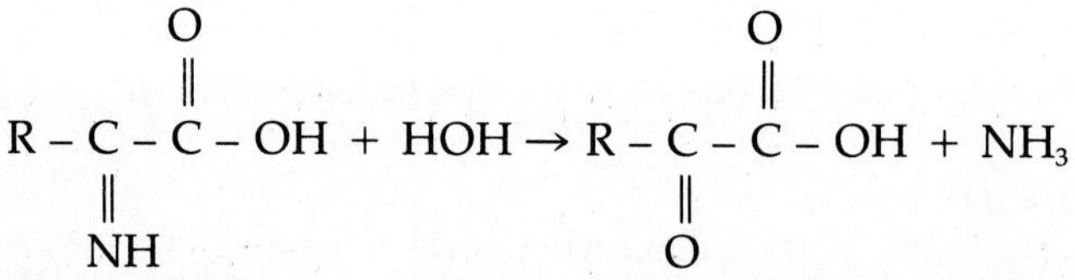

$$\underset{\text{(imino acid)}}{R-\underset{NH}{\underset{\|}{C}}-\overset{O}{\overset{\|}{C}}-OH} + \underset{\text{(water)}}{HOH} \rightarrow \underset{\text{(α ketoacid)}}{R-\underset{O}{\underset{\|}{C}}-\overset{O}{\overset{\|}{C}}-OH} + \underset{\text{(ammonia)}}{NH_3}$$

The amine group has now been removed from the original amino acid. The process requires several enzymes and is not as simple as shown. Like most low level biological reactions it is reversible, but the deamination process is a common reaction that occurs in the liver, kidneys, muscles, and other body tissues.

Transamination

The transamination reaction that converts pyruvic acid to alanine is shown here:

$$\text{H}_2\text{N}-\overset{\overset{\text{O}}{\|}}{\text{C}}-\text{CH}_2-\text{CH}_2-\text{CHNH}_2-\text{COOH} + \text{CH}_3-\overset{\overset{\text{O}}{\|}}{\text{C}}-\text{COOH} \xrightarrow{\text{(glutamic transaminase)}}$$

(glutamine)   (pyruvic acid)

$$\text{H}_2\text{N}-\overset{\overset{\text{O}}{\|}}{\text{C}}-\text{CH}_2-\overset{\overset{\text{O}}{\|}}{\text{C}}-\text{COOH} + \text{CH}_3-\text{CHNH}_2-\text{COOH}$$

(α keto glutamic acid)   (alanine)

This is a type reaction. There are many others like it that are carried out within cells. All involve the interchange of amine (CHNH_2) and ketone ($-\overset{\overset{\text{O}}{\|}}{\text{C}}-$) groups. Each reaction is specific and requires a particular enzyme for each different amino acid that enters into the reaction. The enzymes belong to a family called transaminases and all are derivatives of pyridoxine, one of the B-complex group of vitamins. The above reaction is catalyzed by glutamic transaminase. No other enzyme will work.

In the absence of pyridoxine, nonessential amino acids cannot be synthesized, and protein production of the body is seriously impaired. Transamination is an important mechanism that is the principal supplier of amine groups to ketoglutarates to form glutamic acid molecules which have wide use in protein syntheses.

Conversion of Ammonia to Urea and Uric Acid

Ammonia is toxic and must be either promptly excreted or converted to less toxic substances. In mammals, some ammonia is excreted in the urine, but most is converted into urea or uric acid. Herbivores excrete extremely small amounts of ammonia in the urine. Urea excretion, however, is large. The basics behind the conversion of ammonia to urea are simple. Two units of ammonia are conjugated with one unit of carbonic acid to form one unit of urea and two units of water.

$$2\text{NH}_3 + \text{H}_2\text{CO}_3 \rightarrow \text{CO(NH}_2)_2 + 2\text{H}_2\text{O}$$

The actual reactions are more complex. In biochemical shorthand, the reaction looks like this:

$$\text{Ornithine} \xrightarrow[{-\text{H}_2\text{O}}]{+\text{NH}_3 \atop +\text{CO}_2} \text{Citrulline} \xrightarrow[{-\text{H}_2\text{O}}]{+\text{NH}_3} \text{Arginine} \xrightarrow[{+\text{H}_2\text{O}}]{\text{(arginase)}} \text{Ornithine} + \text{Urea}$$

394
INTERMEDIATE
METABOLISM

The actual reaction probably is as follows in Figure 13-6.

The complexity of the above is why only a few complete reaction schemes have been included in this text and why, insofar as possible, explanations are kept simple. Urea is a normal body constituent and is formed mainly from tissue and dietary protein. The liver appears to be the site of urea formation, since hepatectomized animals lose all urea from their bloodstreams before they die.

Gluconeogenesis

Protein can be used as fuel. The process of converting protein to carbohydrate is called gluconeogenesis, and according to the metabolic

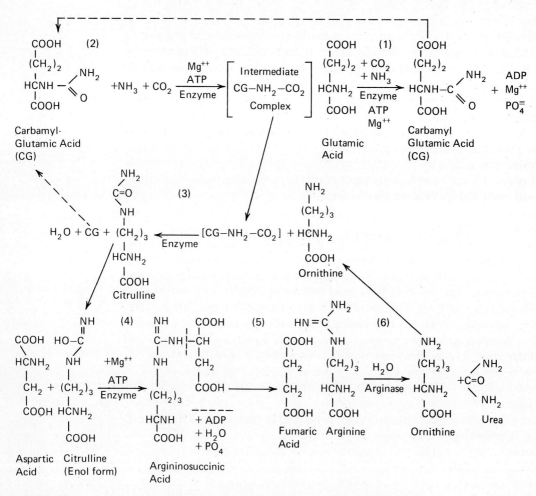

FIGURE 13-6   *Conversion of Ammonia to Urea. (Adapted from Dukes,* Physiology of Domestic Animals.)

pool concept, a certain amount of this activity goes on at all times in the living animal. Under normal conditions this sort of transformation is a relatively low level activity, but in cases of starvation it can become the major source of body energy, with a consequent "wasting away" of muscle to provide energy to keep the organism alive.

Proteins, to be converted into glucose, must first be broken down into amino acids and then deaminated. The appropriate fragments, the glycogenic amino acids, then enter the glycolysis scheme at the pyruvate stage of the tricarboxylic cycle (Figure 13-2). This, in turn, enters the reverse reaction path of the glycolysis scheme and subsequently forms glucose or glycogen.

Of the 23 amino acids useful in nutrition, 15 are known to reduce to so-called glycogenic (glucose forming) acids which yield three-carbon fragments; another four become ketogenic acids which yield two- or four-carbon fragments, while the fate of the remainder is unknown. Their breakdown occurs principally in the liver, and to a lesser extent in the kidneys.

The usual method of determining whether an amino acid is glycogenic is to feed it to a diabetic animal, or an animal made diabetic by removal of the pancreas, or by administration of the diabetogenic drugs alloxan or phlorizin. Alloxan destroys the insulin producing tissues of the pancreas, while phlorizin lowers the renal threshold for glucose. However the diabetes is caused, after several days of fasting, the carbohydrate reserve of the animal is virtually exhausted. If carbohydrate is still excreted in the urine, it is being derived from body protein. Now if measured amounts of the amino acid in question are fed to the animal, a study of glucose and nitrogen elimination through the urine will reveal whether or not the amino acid is glycogenic. If it is, both nitrogen and glucose elimination will be increased.

Liponeogenesis

The formation of fat from protein is difficult to demonstrate although there is proof that it occurs. Under normal nutritional conditions it is of doubtful importance since protein is more useful elsewhere. If there were energy needs that required synthesis of protein into energy substance, its conversion to glucose (gluconeogenesis) would be a simpler and more direct way of satisfying the energy demand.

There is, however, one condition where protein could be transformed into fat. If the glycogen reserves of an animal were completely filled and protein was subsequently fed at such high levels that the body could not utilize it either as amino acids or sugar, there would be storage of the protein as fat. Evidence of this effect can be obtained by studying tagged molecules and the respiratory quotient.

Oxidation of α Ketoacids for Energy

The deaminization of amino acids results in the production of α ketoacids as has been shown on p. 390. These α ketoacids can be converted to glucose by entrance into the tricarboxylic acid (Krebs) cycle and continuing along the reverse pathway to glucose or glycogen as shown in Figures 13-1 and 13-2.

Synthesis of Amino Acids into Protein

There should be no difficulty in visualizing the synthesis procedure. The appropriate amino acids are polymerized into proteins through the peptide linkages, for example,

$$\left.\begin{array}{c} \text{H} \quad \text{OH} \\ | \quad\quad | \\ \text{R} - \text{C} - \text{C} = \text{O} \\ \quad\quad\quad\backslash \\ \text{H} \quad \text{NH} \\ | \quad\quad | \\ \text{R} - \text{C} - \text{C} = \text{O} \\ | \\ \text{NH}_2 \end{array}\right\} \text{"peptide linkage"}$$

This continues for several hundred to several thousand linkages, which may involve all 23 amino acids. For this mechanism to operate, there must be adequate amounts of B-complex vitamins, adequate caloric intake, and adequate amounts of the essential amino acids. Since protein synthesis is an "all or nothing" mechanism, i.e., all the amino acids and all the catalysts necessary for the synthesis of a given protein must be present for synthesis to occur, there must be an adequate quality as well as quantity of protein present. It is here that the "essential" amino acids are so important, since there is virtually no storage of any amino acid in the body. The so-called "storage protein" is only a temporary thing. Therefore, there must be a daily dietary replacement of lost essential amino acids or serious protein deficiencies can occur.

PHYSICAL STRUCTURE OF PROTEINS

Proteins fall into two general physical categories once they are formed: they are either globular or fibrillar. Globular proteins have low molecular weight and a spherical or ovoid shape. They are kept in this shape by coiling of the peptide chains. Usually they are soluble in water or salt solutions. The more complex proteins are fibrillar and are usually shaped as long strands or fibrils bound together in bundles by cross linkages. These are the proteins involved in fibrin and collagen formation and in other structural proteins of the body. They are extremely strong and elastic, and have a tendency to stretch under constant pressure, but if the pressure is removed they will shrink to a much smaller length. In general they are insoluble in water and salt solutions, although they can be softened by this exposure.

There is no simple functional classification of proteins any more than there is a simple functional classification of words in a dictionary, and for the same reason. There are literally thousands of different functions performed by proteins in the living animal, and proteins of widely different structures may perform analogous functions. However, for rough classification, proteins can be lumped together into groups, mainly on the basis of their chemical properties as:

1. Albumins and albuminoids: low molecular weight, water soluble.
2. Globulins: low molecular weight, poorly soluble in water, soluble in salt solutions.
3. Scleroproteins: fibrous, high molecular weight, insoluble in water or salt solutions, resistant to chemical breakdown.
   a. Collagen: fibrous connective tissue.
   b. Elastin: elastic tissue, tendons.
   c. Keratin: hair, horns, hooves.
4. Histones: highly basic proteins, considered with nucleic acids as nucleoproteins.
5. Nucleoproteins: simple proteins plus nucleic acid, DNA, RNA, etc.
6. Mucoproteins: proteins plus complex polysaccharides, chemically resistant.
7. Lipoproteins: proteins plus lipids.
8. Glycoproteins: proteins plus sugars.
9. Chromoproteins: proteins plus coloring agents, e.g., cytochromes, hemoglobin.
10. Phosphoproteins: proteins plus phosphorus.
11. Metalloproteins: proteins plus copper, iron, zinc, magnesium, cobalt, etc.

Other Synthesis Reactions

A similar mechanism to that involved in the synthesis of body proteins is involved in the production of hormones, enzymes, and other substances which are dealt with in the chapters on endocrines and the digestive system. They will not be repeated here. However, the production of creatine and its anhydride creatinine is sufficiently dissimilar to be considered separately.

Creatine Synthesis

Creatine is an important constituent of muscle tissue and is apparently essential to muscle function. It appears to be formed in the kidney and liver from the amino acids glycine, arginine, and methionine. These reactions, unlike most bodily reactions, are irreversible. Creatine, therefore, must be eliminated either as creatine or its anhydride creatinine. About 98 percent of all body creatine is found in muscle tissue principally as energy-rich phosphocreatine. Most of the remainder is found in nervous tissue. Skeletal and cardiac muscle have higher creatine content than smooth muscle, and rapidly acting striated muscles contain more than slower acting muscles. Creatine is a normal excretory product in birds, but in mammals creatinuria occurs in unusual circumstances such as rapid growth, pregnancy, starvation, fever, exophthalmic goiter, diabetes mellitus, high protein diet and muscular dystrophy. In all of these conditions, creatinuria is associated with increased catabolism of protein. It also may occur as a sequel to amputation or surgical removal of large muscle masses, probably because less need for creatine exists.

The normal mammalian excretory product of creatine is creatinine. This substance is always present in the blood and urine of normal animals in small amounts, and increases in the abnormal conditions listed above. In normal mammals the daily output is quite constant—about 2 percent of the total creatine in the body—and is not affected by exercise, urine output, or dietary protein level. The daily excretion of milligrams of creatinine nitrogen per kilogram of body weight is called the creatinine coefficient. This amount has a normal range of from 7 to 11, and increased body weight will result in an increased level of creatine excretion. Creatine excretion is essentially a measurement of the metabolic activity of the tissues.

Excretion of Amines, Proteins, and Protein Breakdown

Within a few hours after ingestion of protein by an animal with a stable body weight (the time varies with the species of animal), a quantity of nitrogen equal to that which was absorbed will be excreted to maintain nitrogen balance in the body. The body contains labile proteins (storage proteins), which are readily subject to metabolism; these exert a sparing action on the more stable proteins incorporated into the body tissues, but there cannot be a net gain of protein or any other substance in a stable animal. The intake must be balanced by excretion. Through the use of heavy hydrogen (deuterium-H^2) to label the α ketoacid portion of the amino acid molecule, and N^{15} to identify the amine group, plus careful monitoring of all excreted nitrogen it has been shown that about 40 percent of absorbed dietary amino acids are deaminated immediately in the liver. Of the remaining 60 percent, about 50 percent are incorporated into storage protein, plasma protein and body tissues, and about 10 percent are present in the circulating blood and tissue fluid.

The ammonia and amines produced by deamination and other forms of amino acid metabolism are mainly excreted in urine and feces. Fecal nitrogenous compounds come principally from undigested or unabsorbed dietary substances, and secondarily from metabolic sources such as molecules of albumin and urea that pass from the body into the intestinal lumen from cells abraded from the intestinal mucosa and from passed intestinal enzymes. Although the material can be recycled by the body there is always some loss, and the loss is usually greater when there are appreciable amounts of fiber in the diet, which could be an argument against dietary fads.

Sulfur containing compounds enter the body principally in the amino acids cystine, cysteine and methionine, and less frequently in sulfoproteins, glycoproteins, and inorganic sulfur. The latter molecule cannot be utilized by the metabolic scheme, and methionine is the only one of the sulfur containing amines that is essential. Cystine and cysteine have a sparing action on methionine.

Usable sulfur is found in:

1. Normal body protein
2. Hair, hooves, horns, mucin and chondroitin sulfuric acid
3. Glycoprotein in various connective tissues
4. Taurocholic acid in bile
5. Ergothionine in erythrocytes, urochrome, and melanin
6. Sulfolipids in the nervous system
7. Inorganic sulfur in body fluids
8. Insulin
9. Thiamine, diphosphothiamine, cocarboxylase
10. Decarboxylation coenzymes
11. Coenzyme-A
12. Several dehydrogenase enzymes
13. Several sulfhydryl groups

Sulfur is lost in the urine, in bile, in saliva, in gastrointestinal secretions, in shedding of hair, through wearing away of nails, horns, and hooves.

VITAMINS

Vitamins of the B-complex have important roles in intermediate metabolism. For examples of these roles see p. 183ff.

chapter 14

The Reproductive System

INTRODUCTION

The class Mammalia is bisexual, which should come as no surprise to anyone. The sexes are male and female, which also should not be any earthshaking news. However, there are many gradations between the genotype-phenotype male and the genotype-phenotype female, since sex, which is determined at the moment of fertilization, passes through a number of stages of development in the embryo and juvenile before reaching its definite anatomical and physiological form in the adult. In the course of this development modifying factors can produce a number of variations from "purely" male or female. Since primordia of both sexes are present in every individual, a multiplicity of variants can, and do, exist. Fortunately their number is small, and with the possible exception of humanity, which perpetuates genetic errors through surgery and medicine, their effects make them self-limiting.

The function of the reproductive system, again with the exception of humans, who have shortened function to fun, is the production of offspring. This is a basic drive and one may say with some justice that animals are ruled by their sex hormones.

Anatomically, there is no place in the mammalian body that has undergone more modifications of the basic pattern than the reproductive system. If reproduction were music, the spectrum of structural and functional differences would certainly be called variations on a theme. The overriding purpose of all these various habits, techniques, structures, and appurtenances is to bring a male spermatozoon through the defenses of the ovum and into its cytoplasm and to provide a relatively secure and stable environment in which the fertilized egg can grow and develop until birth. After birth, the process provides maternal care and nutrition until the new individual can fend for itself.

Reproduction is a most interesting subject, and so it is perhaps appropriate to begin with the simpler and less interesting of its two primary manifestations, the male. It is mildly amusing from an etymological viewpoint to note that early students of the male reproductive apparatus considered the testicles to be of little importance, even though castration and eunuchs were known as early as the Persian Empire period, and probably before that. Still, the Latin "testis" has the same root as "testament" or "testify." It means "a witness" and a testiculus is a small witness. Obviously, those workers felt that the testicles were present merely to witness the more spectacular activities of the penis!

THE ANATOMY OF THE MALE GENITAL SYSTEM

Basically the male genital apparatus consists of the testicles plus a system of ducts, blood, and nerve supply, accessory glands, and penis. The functioning mammalian testicle is ordinarily external. While there are some mammals with internal testicles (e.g., beavers, elephants, hyrax, seals, porpoises, and whales), and a number of species (mainly rodents) that retain the testicles within the body during most of the year, during the breeding season the testicles of most of these latter species descend to the outside to become functional. As a rule, temperatures within the body are too high to permit development and maturation of mammalian spermatozoa (male germ cells) in most species, although hormone production is not impaired.

The testicle has two functions: the production, maturation, transportation, and storage of spermatozoa and the production of male sex hormones (androgens).

The Descent of the Testicles

The testicles are originally formed from two masses of germinal tissue in the embryo that are set apart early in gestation. Each mass lies outside the peritoneal cavity of the embryo and is connected by a cord of connective tissue, the gubernaculum testis, to a spot on the floor of the cavity where the scrotum will eventually form. As the embryo grows, the gubernaculum remains relatively constant in length. As a result, the testicular tissue is gradually moved from its original location backward and downward along the posterior wall of the abdominal cavity. This is an example of the differential growth rate that is a major characteristic of embryonic growth and development. Actually, the male embryo grows away from its testicles. As the embryo grows upward and forward the germinal tissues move backward and downward held in place by the gubernaculum. During this movement they slowly develop into definitive glands and pick up a covering fold of peritoneum. This fold, the tunica vaginalis propria, also encloses the blood vessels and nerves that supply the gland substance. As the testicle moves into the developing scrotum it picks up a second fold of peritoneum that covers the inner opening of the inguinal ring. This membrane becomes the tunica vaginalis communis. Once the descent of the testicle into the scrotum is completed, the gubernaculum becomes a layer of tissue called the scrotal ligament, which lies under the skin at the bottom of the scrotal sac and attaches the testes to the scrotum.

Descent of the testicles can be incomplete. Those individuals who have this defect involving both testicles are called cryptorchids. If only one testicle fails to descend completely, the animal is called a monorchid.

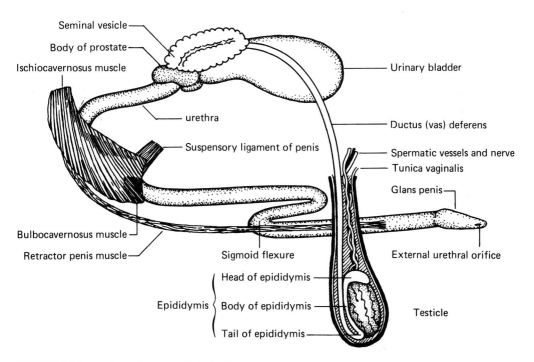

FIGURE 14-1   Genital organs of the bull.

The Scrotum

The testicles are enclosed in a sac called the scrotum, which is composed of skin, muscle, and fibrous tissue and is lined with the two layers of modified peritoneum, the tunica vaginalis communis and the tunica vaginalis propria.

The scrotum provides testicular temperatures from four to seven degrees Celsius cooler than the general body under normal summer conditions. The temperature of the testicles is important since sperm production in most mammals will not occur at normal body temperature. The scrotum, therefore, is constructed to radiate heat. The skin is thin, relatively hairless, lacks subcutaneous fat, and is supplied with sweat glands. In many animals the scrotal sac hangs below the general body surface. In cold weather, heat must be conserved. A muscle (cremaster) retracts the testicle against the body to protect it from excessive cold. Small scrotal muscles can pucker the skin of the scrotum, which helps keep the testicles close to the body and gives less radiating surface for heat loss through the skin. A thermal regulation reflex apparently arises from the scrotum and can elicit shivering or rapid breathing that can, respectively, raise or lower body temperature.

The Testicle

The testicle is egg-shaped and covered with a layer of whitish connective tissue called the tunica albugenia. The gland has a rich blood supply. It possesses a hilus from which connective tissue invades the gland substance and divides it into pyramid-shaped lobules, which have their bases at the periphery and apices at the hilus.

The Duct System

A system of semiferous tubules ramify through each lobule converging at the apices of the lobules to form the canaliculi recti, which penetrate the mediastinum testis and empty into a plexus of anastomosing tubes in the hilus, the rete testis. From the rete testis a lesser number of ductuli efferentes (efferent ducts) leave the hilus and lead to the epididymis. The epididymis lies upon the surface of the testicle and is composed of three parts: the head (caput epididymus), the body (corpus epididymis) and the tail (cauda epididymis).

A series of ducts, the coni vasculosi, lead from the ductuli efferentes and form the enlarged, highly convoluted caput epididymis. From the caput epididymis a single convoluted tube, the corpus epididymis, extends along one side of the testicle. This structure is continued as the cauda epididymis, which bends sharply backwards at the ventral pole of the testicle and becomes the relatively straight vas deferens.

The epididymis stores, concentrates, matures and transports spermatozoa. The epididymal tubules have both absorptive and secretory functions, and apparently have some specific effect upon spermatozoa maturation, since both preepididymal and epididymal sperm are infertile, but spermatozoa in the vas deferens and the ampulla are capable of fertilizing ova. The caudal portions of the epididymis appear to be the locale for this maturation process. Spermatozoa storage is mainly in the tail portion of the epididymis and during copulation, peristaltic contractions of the vas deferens propel the sperm into the urethra.

The Vas Deferens and the Spermatic Cord

The vas deferens, together with a small artery and a vein, is enclosed in one side of a fold of the tunica vaginalis propria. Parallel to it but enclosed on the opposite side of the fold are the spermatic artery, vein and nerve. These two sets of structures, physically separated by the folded tunica vaginalis propria, form the spermatic cord. The cord passes upward into the abdominal cavity through the inguinal canal and

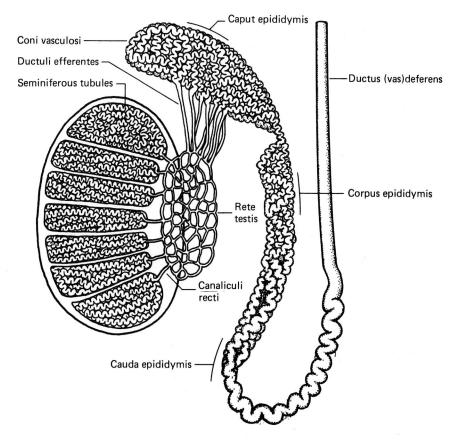

FIGURE 14-2   *Schematic drawing of the duct system of the testicle.*

lies between the wall of the abdomen and the peritoneum. At this location, the vas deferens leaves the spermatic cord and connects to the urethra near the neck of the urinary bladder. At the point of connection, or shortly before it, the vas deferens usually forms a spindle-shaped enlargement called the ampulla.

The Ampulla

The term, "ampulla," is something of a misnomer, since in many genera and species there is no enlargement of the lumen of the vas deferens. The thickened walls of the ampulla contain branched tubular (ampullary) glands that produce a serous fluid. The ampulla appears to be a place of temporary storage for spermatozoa. It may be either single or paired, and in some species (notably swine) it may be absent. Its functional significance is not well understood. The ampullae open into the urethra in the same area as the ducts of the seminal vesicles. The ampullary opening marks the termination of the vas deferens.

Blood, Nerve, and Lymph Supply

The blood to the testicle is mainly supplied by the internal spermatic (testicular) artery, which arises from the abdominal aorta caudal to the caudal mesenteric artery. The vessel, covered by a fold of the tunica vaginalis propria, descends into the testicle and ramifies in the gland substance. The testicular vein arises from capillaries and venules in the gland substance, and is carried back into the body in the same fold of the tunica vaginalis that encloses the testicular artery and accompanies that artery along its course until the vein enters the caudal vena cava. The testicular nerve is mixed and receives motor, sensory, sympathetic and parasympathetic fibers from the lumbar region and caudal mesenteric ganglion. The testicular nerve accompanies the testicular artery until it enters the testicle.

The lymph drainage arises from the interstitial tissues in the lobules, the tunica albugenia, and the epididymis. The route of the larger lymph vessels is generally the same as that of the testicular veins. The lymph system is neither large nor extensive, but on occasion when the regional lymphatics are damaged by disease, spectacular testicular and scrotal enlargements can occur as the result of blocked lymph drainage. Ordinarily the lymphatics function to drain interstitial fluids and transport androgens.

In the testicles of mammals the arterial blood is delivered in such a manner that it is appreciably cooler than the blood within the systemic circulation, and the blood flow through the testicle is smoothed to the point where it lacks any appreciable pulsation. The cooling effect is accomplished by a structure called the pampiniform plexus, a cone-shaped mass of anastomosing venules derived from the testicular vein. The plexus lies near the head of the epididymis.

The testicular artery, once it has left the body and enters the spermatic cord, becomes highly convoluted and intimately associated with the venules of the pampiniform plexus. This plexus forms a so-called vascular cone (which should not be confused with the coni vasculosi of the head of the epididymis). From this structure arterial branches are given off to the head of the epididymis and to the tunica albugenia.* The spermatic artery then passes to the testicle and forms complex surface patterns before entering the testicle at the hilus and branching in the mediastinum to supply the seminiferous tubules. The veins from the gland substance coalesce into a central vein, which passes from the hilus into the pampiniform plexus. This plexus consists of numerous anastomosing venules that surround the spermatic artery and act as a heat exchange device. The plexus in turn finally coalesces into the spermatic veins that ultimately enter the caudal vena cava.

ACCESSORY GLANDS OF THE MALE GENITAL SYSTEM

As the ductus deferens enters the urethra it dilates to form the ampulla or bulb. From this point onward the duct system of the testicle is incorporated into the urethra and loses its identity. The accessory glands of the male empty into the urethra and add their products to the spermatozoa released from the ampulla. The combination is known as semen. Commonly there are three different accessory glands: the prostate, the seminal vesicles, and the bulbourethral (Cowper's) glands. Some species may have fewer than three; the dog, for instance, has only the prostate, and the cat has only a prostate and bulbourethral glands. The size and functional significance of these glands also may vary greatly between species. In general, animals without (or with greatly reduced) seminal vesicles will have relatively prolonged copulating times, and animals with extraordinarily large seminal vesicles will also have prolonged copulating times. In the remainder, the duration of the mating act is relatively short. The significance of these variations is not understood. Cowper's glands serve two purposes; to provide nutrition for spermatozoa and to coagulate semen (in rodents they are often called coagulating glands). They open into the urethra via ducts that are located closer to the urethral orifice than the ducts of the seminal vesicles or the prostate.

---

*The tail of the epididymis is supplied by the artery of the vas (the deferential artery), which is a branch of the hypogastric artery and has an entirely different connection to the aorta than the testicular artery. A small deferential vein may also be present.

Comparative Anatomy of the Accessory Glands of the Male

Animal	Seminal Vesicle	Prostate	Bulbourethral Gland	Ampulla
Stallion	Large and bladder-like	Present	Present	Present
Bull	Lobulated, glandular and compact	Present	Very small	Present
Ram	Similar to bull	Present	Very small	Present
Boar	Very large, less compact than horse	Present	Very large	May be absent
Dog	Absent	Enormous	Absent	Present
Cat	Absent	Present	Present	May be absent

The Penis

The penis is the external copulating organ of the male. It contains the urethra, which performs double duty as a urinary and genital duct. The penis consists of the glans penis, urethra, erectile tissue, and prepuce. In some species muscle tissue may be present, and in others (dog, raccoon, mink, beaver, etc.,) a bony structure, the os penis (baculum) is found. It is an aid to support soft tissues of the penis where erectile tissue is reduced or absent, and may in some species (e.g., mink) be an essential structure in copulation.

The terminal glans contains sensory nerve endings and may or may not contain erectile tissue. In the horse, the glans contains a considerable amount of erectile tissue and may become so engorged and enlarged that an "overeager" stallion cannot enter the mare on breeding. In the bull and the ram the glans is small, contains a minimum amount of erectile tissue, and is only slightly distensible. The boar lacks a true glans penis. The cat possesses a glans, which is covered with short spines and is moderately distensible. The dog is a special case and will be discussed later.

The urethra lies along the ventral part of the penis. It is the common outlet for the products of the urinary and genital systems. In the ram the urethra is continued beyond the glans penis as a long terminal filament called the urethral process (processus urethrae), which apparently serves no useful function.

The erectile tissue consists of two parts, the corpus cavernosum penis, which lies dorsal to the urethra, and the corpus cavernosum urethrae (corpus spongiosum), which surrounds the urethra. These spaces become engorged with blood during sexual excitement, causing the entire organ to stiffen and become erect. The erectile tissues receive blood mainly from three sources; the deep artery of the penis (a branch of the obturator artery), the dorsal artery of the penis (a branch of the external pudic artery), and the artery of the bulb (a branch of the internal pudic artery). In the bull, ram, and boar, the penis is composed mainly of

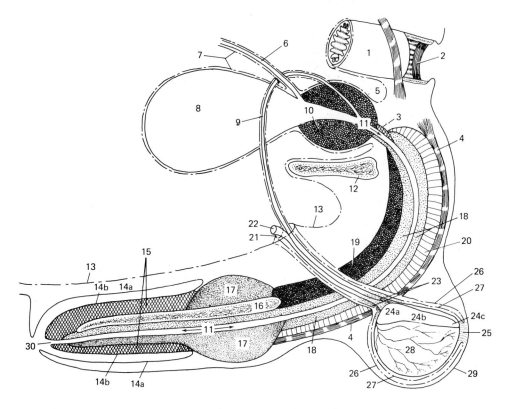

1. Colon
2. Internal Anal Sphincter
3. Urethral Muscle
4. Retractor Penis Muscle
5. Pararectal Fossa
6. Ureter
7. Visceral Peritoneum
8. Urinary Bladder
9. Ductus Deferens (Vas Deferens)
10. Prostate Gland
11. Urethra
12. Symphysis Pelvis (Ischium)
13. Parietal Peritoneum
14. a. Parietal b. Visceral Layer of Prepuce
15. Pars Longa Glandis (Glans Penis)
16. Os Penis (Baculum)
17. Bulbus Glandis (Penile Bulb)
18. Corpus Spongiosum Penis
19. Corpus Cavernosum Penis
20. Bulbospongiosus Muscle
21. Vaginal Ring
22. Testicular (Spermatic) A.V.&N.
23. Pampiniform Plexus
24. a. Head b. Body c. Tail of Epididymis
25. Ligament of the Tail of Epididymis (gubernaculum testis)
26. Parietal (Common) Vaginal Tunic
27. Visceral (Proper) Vaginal Tunic
28. Testicle
29. Scrotum
30. External Urethral Orifice

FIGURE 14-3   *Schematic drawing of the male genital system of the dog.*

fibrous connective tissue. The corpora cavernosa are small in extent, and are confined to a region just forward of the root of the penis. The penis is quite firm in the nonerect state, and is not greatly enlarged during erection. It merely becomes more firm. A sigmoid flexure or S-curve exists in the penis of these animals, which when straightened by pressure of the engorged erectile tissue provides for the necessary extension or protrusion.

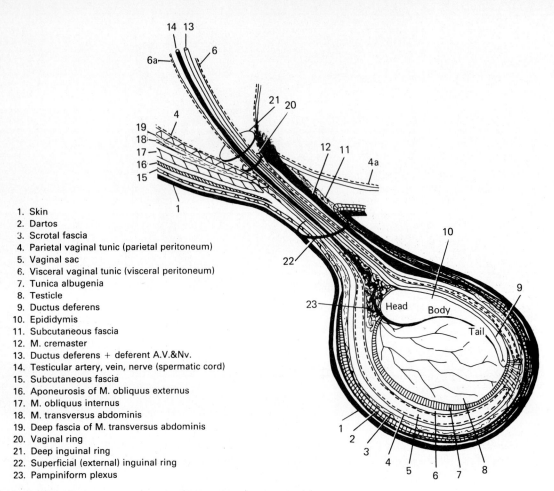

1. Skin
2. Dartos
3. Scrotal fascia
4. Parietal vaginal tunic (parietal peritoneum)
5. Vaginal sac
6. Visceral vaginal tunic (visceral peritoneum)
7. Tunica albugenia
8. Testicle
9. Ductus deferens
10. Epididymis
11. Subcutaneous fascia
12. M. cremaster
13. Ductus deferens + deferent A.V.&Nv.
14. Testicular artery, vein, nerve (spermatic cord)
15. Subcutaneous fascia
16. Aponeurosis of M. obliquus externus
17. M. obliquus internus
18. M. transversus abdominis
19. Deep fascia of M. transversus abdominis
20. Vaginal ring
21. Deep inguinal ring
22. Superficial (external) inguinal ring
23. Pampiniform plexus

FIGURE 14-4   *Canine testicle and inguinal canal (schematic).*

The prepuce is the sheath of skin covering the head of the penis. The boar possesses a preputial pouch, which is usually filled with decomposing urine and cellular debris and is in part responsible for the foul (sex) odor of the beast. The muscles of the penis are the extensor penis and the retractor penis, and are best developed in those animals that possess a sigmoid flexure.

The Penis of the Dog

The penis of the dog and other members of the canine genus differs considerably from other mammals and needs separate discussion. The prostate encircles the urethra and is the only accessory gland. The corpus spongiosum penis, which forms the caudal penile bulb, is quite long and extends from the root (radix) of the penis cranially to the bulbus glandis and becomes incorporated into the bulb except for a thin band lying dorsal to the urethra that extends almost to the external urethral orifice. The corpus is confined to the caudodorsal part of the penis and caudal to the bulbus glandis. The glans penis is very long and is supported by the os penis or baculum, a bony structure considered to be an extension of the corpus cavernosum. The glans penis (pars longa glandis) is divided into a cranial part, which is elongate and pointed and covers the cranial end of the penis, and a caudal part, which is bulbous (and may be called the bulbus glandis). The glans contains considerable erectile tissue supplied only by venous blood. Erection is slow and affects the long anterior portion first. The process of ejaculation in the dog occupies considerable time, mainly because of the lack of seminal vesicles and Cowper's glands. At copulation the bulbar part of the glans (bulbus glandis) becomes extremely dilated and should "lock" or "tie" the copulating pair together. The "tie" appears to be an essential process in the act of breeding, because the bitch will ordinarily not become pregnant or will produce only one or two pups if the "tie" does not occur. It is probable that the copulatory lock functions as a part of the ejaculatory mechanism in the dog, although all the details of this mechanism are not known.

ANATOMY OF THE FEMALE REPRODUCTIVE SYSTEM

Introduction

The more complex reproductive apparatus in mammals occurs in the female. Mammals are not unique in being live-bearing; some snakes and fish also have this capacity, but they are unique in the possession of a placenta, a definitive uterus, and, of course, the mammary glands.

Comparative Anatomy of the Female Reproductive Organs

	Mare	Cow	Ewe	Sow	Bitch
Ovary size	1 1/2" × 3 1/2"	1/2" × 1 1/2"	1/2" × 1"	1" × 1"	1/2" × 1"
Length of oviduct	8—12"	12—14"	no demarcation from uterus	6—12"	2—3"
Uterus—length of horn	10"	12"	4—5"	48—60"	5—6"
Shape of horn	straight	spiral	spiral	coiled	straight
Length of body	7—8"	1/2—1"	1"	0	1"
Comparative length of horn to body	5 to 4	10 to 1	6—7 to 1	horns only	6 to 1
Cervix length	2—3"	4"	1/2—1"	4"	1"
Shape of lumen	straight	very crooked	crooked	spiral	straight
Vagina length	6—8"	8—12"	3—4"	4—5"	relatively long

It is in the female mammal that the great variations in anatomy and physiology of the reproductive system are found. It would be a practical impossibility to discuss all of the known variants in this book, and even confining discussion to the domestic animals produces a large enough body of material to confuse rather than enlighten. As a result, this section has been kept general rather than specific, with an emphasis on principles rather than long digressions into peculiarities.

The female sex organs consist essentially of the ovaries, oviducts (Fallopian tubes), uterus, cervix, vagina, and vulva. Considerable variation in size shape, and position of the internal parts of the female reproductive system exists among the domestic animals. These differences are summarized in the table on p. 411.

The Ovaries

The ovaries are located in the caudal dorsal part of the abdominal cavity. The exact location varies with the species of animal. They are irregular, oval, retroperitoneal structures suspended from the dorsal abdominal wall by a ligament, the mesovarium, and are held in place by the pressure of surrounding organs. They produce the female germ cells (ova) and female sex hormones (estrogens), and are embryonically homologous with the testes of the male.

The ovary is composed of a loosely organized central portion or medulla and a denser outer portion or cortex, which is covered by a connective tissue envelope (tunica albugenia) and a thin germinal epithelium. The medulla is covered by the cortex, except at the hilus where the blood vessels and nerves enter the gland. The cortex and medulla are not separated by a distinct line of demarcation but blend together at the indistinct cortico-medullary border.

The cortex is the region where growth of ova and development of follicles and corpora lutea takes place. Each ovum (egg) as it ripens is surrounded by a gradually enlarging fluid-filled sac, or follicle. A mature follicle (Graafian follicle) ruptures at estrus and expels a mature ova into the dilated (pavilion) end of the oviduct. The ruptured follicle then passes through three stages, the corpus hemorrhagicum, the corpus luteum (either of estrus or of pregnancy), and finally the corpus albicans, which is nonfunctional scar tissue. Most mature, functioning, nonpregnant ovaries will show all the structures listed above, except for bovine ovaries, which do not normally have a corpus hemorrhagicum. The blood supply of the ovary is derived from the utero-ovarian artery. The nerve supply is derived from the utero-ovarian nerve, which contains sensory, motor, and sympathetic fibers arising from the posterior mesenteric ganglion and the lumbar spinal nerves.

The Oviducts

The Fallopian tubes (uterine ducts, oviducts) connect each ovary with the uterus and convey mature ova from the ovary to the uterus.

The oviducts are complexly convoluted tubes lying in a fold of the mesovarium. The ovarian end of the tube is dilated into a funnel-shaped infundibulum, whose free border possesses finger-like projections or fimbriae. The infundibulum is attached at one point on its free margin to the hilus of the ovary. The fimbriae completely surround the ovary. However, there is no direct tissue contact between the gland substance of the ovary and the remainder of the reproductive tract.

The caudal ends of the oviducts open into the tips of the horns of the uterus. (Humans and other primates are exceptions to this statement since their uteri usually do not possess horns. In these species the oviducts open into the body of the uterus.)

The lumen of the oviduct is lined with ciliated columnar epithelium, which assists the ovum in passing down the oviduct to the uterus. In most mammals fertilization of the ovum must occur in the oviduct for pregnancy to ensue (p. 432).

The Uterus

A complete uterus is a musculomembranous sac consisting of two horns (cornu), a body (corpus), and a neck (cervix). There are a number of variations on this arrangement with one portion or another having a greater development. Most domestic animals possess the bipartite uterus* (which is often called bicornuate). Primates possess a simplex uterus, marsupials, rabbits and monotremes possess duplex uteri and sows are considered to have bicornuate uteri.

The multiplicity of uterine types are not as clearcut as might be inferred from the accompanying illustrations (Figure 14-4). Actually, there is not only a great deal of variation between species but also within a species. Primates, for instance, may possess bipartite or even bicornuate uteri. The exact nomenclature of the various uterine forms has not been entirely clarified. Although the bovine uterus is anatomically bipartite, a considerable number of authors call such a uterus bicornuate, and one refers to it as didelphic, although a didelphic female reproductive system, which has two separate vaginas, cervices, and uteri, is not found in higher mammals. To cope with this latter arrangement, the male penis is forked. This situation apparently gave rise to the quaint, but widespread belief that copulation in the opossum (which has a didelphic uterus in the female and a forked penis in the male) is conducted via the nostrils.

Although none of the common laboratory animals normally has a

---

*The nomenclature of the bicornuate and bipartite uterus is confused. Properly, a bipartite uterus consists of two parts, i.e., body and horns, and a bicornuate uterus consists of two horns only. This is the classification used here. However, there is a considerable body of usage which employs bicornuate to describe a uterus with two horns and a body, and bipartite to indicate a uterus with two horns only. Pitfalls in nomenclature such as this are usually cleared up by a nomenclature committee. So far, this has not been done with these terms.

double vagina, double cervices are not rare, and some of these species (e.g., marsupials, rabbits, and some species of mice and other rodents) have duplex uteri. For that matter, double cervices are not uncommon in cattle, sheep, and humans. The additional cervix may, but need not necessarily, interfere with parturition and result in dystocia. In those domestic mammals where the double cervix is an abnormality, it is good policy to eliminate the animals that possess this structure.

The interior of the nonpregnant uterus differs greatly among domestic animals. In the nonpregnant uterus of the mare and sow the inner wall is folded. In the cow and ewe there are present about 100 "buttons" or caruncles. The uterus of the bitch and queen is smooth and undifferentiated. In all uteri except the simplex, the horns are the locations where the embryos develop.

Placental Types

During pregnancy the variations in internal structure of the uterus become more noticeable with the development of the fetal and maternal placentae.

Uterine Types and Structures

	Species of Animal	Structures & Comments
Didelphic uterus (uterus didelphys)	Opossums, kangaroos, most (or all) other marsupials	Essentially a double uterus, each containing one body and one horn, and connected to one ovary by an oviduct 2 cervices 2 vaginas and vulvar openings (n.b. to cope with this arrangement the male penis is forked or split)
Duplex uterus (uterus duplex)	Some mice and other small rodents, also lagomorphidae (rabbits).	Essentially a double uterus and cervix with a single vagina. (This explains why spacing of embryos throughout the uterus cannot occur in rabbits, and some rodent species)
Bicornuate uterus (uterus bicornus)	Swine	Two horns, no body, 1 cervix, 1 vagina and vulva
Bipartite uterus (uterus bipartis)	All domestic mammals except swine and rabbits. Most laboratory mammals except primates; lagomorphs, marsupials and monotremes. Most wild mammals except as above.	Two horns, 1 body, 1 cervix, 1 vagina and vulva
Simplex uterus (uterus simplex)	Most primates	No horns, 1 body, 1 cervix, 1 vagina and vulva

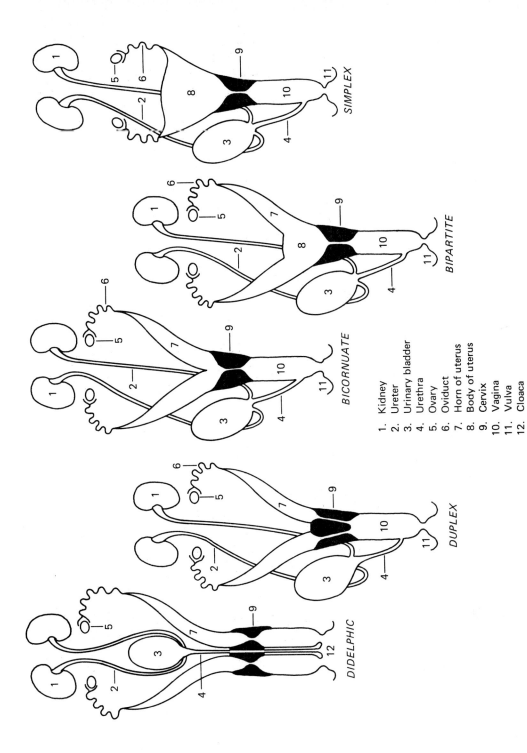

FIGURE 14-5 Types of uteri—schematic.

415

Two main types of placentation exist: nondeciduous (the uterine wall is not shed at birth) and deciduous (the placental portion of the uterine wall is shed at birth with resultant hemorrhage). Under nondeciduous placentaed animals are included two types: diffuse placentation, where the placenta is loosely attached over the entire wall (mares and sows have this type), and cotyledonary placentation, where the placenta is well attached to the uterus by means of caruncles (cows and ewes have this type, also deer, goats, antelope, and other ruminants). Under deciduous placentaed animals are also included two subtypes; zonary placentation, where the placenta is firmly and intimately attached to the uterus along a narrow cylindrical zone passing around the inner surface of the uterus (dogs, cats, and other carnivores have this type), and discoidal placentation, where the placenta is firmly and intimately attached to the uterus at a disc-shaped area (primates and rodents have this type).

The Cervix

The uterus is separated from the vagina by a neck, or cervix, which is normally closed. In most animals the cervix may be dilated by manipulation, but in the cow (and to a lesser extent in the sheep) it is tightly and firmly closed except at estrum and parturition. The cervix is capable of rapid dilation at parturition.

The Vagina, Vestibule, and Vulva

The vagina is a muscular tube lined with stratified squamous epithelium and contains a number of mucous glands (vestibular glands). It forms a channel for birth of the fetus and for the entrance of the male penis. The terminal portion of the vagina is known as the vestibule. A transverse fold of membrane called the hymen separates the vagina and vestibule. The vulva or lips of the vagina are divided into two portions: the inner lips, or labia minora, and the outer lips, or labia majora. The outer lips, together with the anus, form the escutcheon. In all domestic mammals the vagina is located ventral to the anus.

On the ventral wall or floor of the vagina, at its junction with the vestibule, is the urethral orifice which may lie in a crypt (as in the cow), or open directly into the floor (as in the mare), or open on a raised fleshy papilla (as in the dog). On the ventral wall of the vestibule, just anterior to the vulva, is a small mass of erectile tissue called the clitoris, which is homologous to the male penis. Here are located considerable numbers of sensory nerve endings. The clitoris functions to increase the sexual excitement of the female during copulation. On the lateral walls of the vagina, just inside the labia minora, are the vestibular glands, which are homologous to the bulbourethral glands of the male, and which secrete mucus.

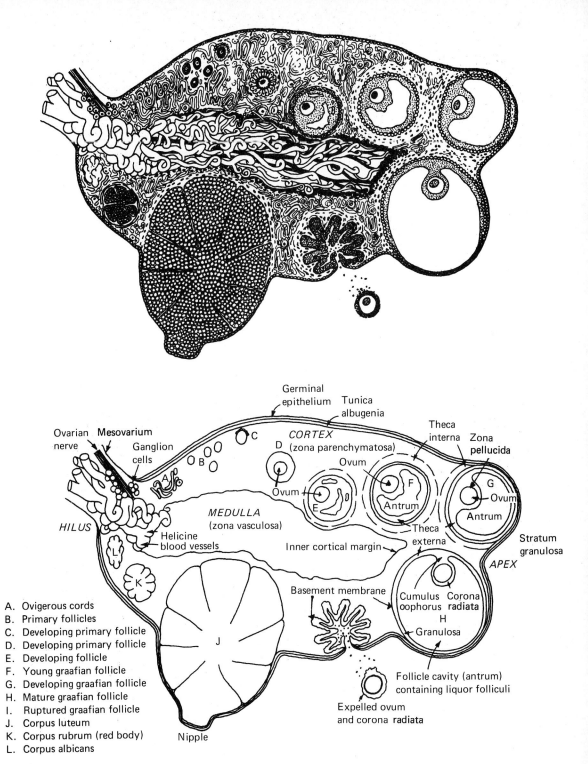

FIGURE 14-6 *Cross section of the bovine ovary.*

PHYSIOLOGY OF REPRODUCTION

Mitosis and Meiosis

In dealing with reproduction, it is advisable to review the two methods by which a cell divides, since both are involved in the reproductive schemes of complex animals. The two types of cell division are usually called mitotic and meiotic (maturation) division. Mitotic division is the normal method of cell reproduction and at one time or another during an animal's life involves every cell type in its body, although nerve cells (except for olfactory nerves—cf. p. 506) lose their capacity for reproduction prior to birth, along with the loss of their centrioles. Meiotic division, however, involves only germ cells.

In mitotic division there are four distinct stages and a resting state where the cell is not reproductively active.

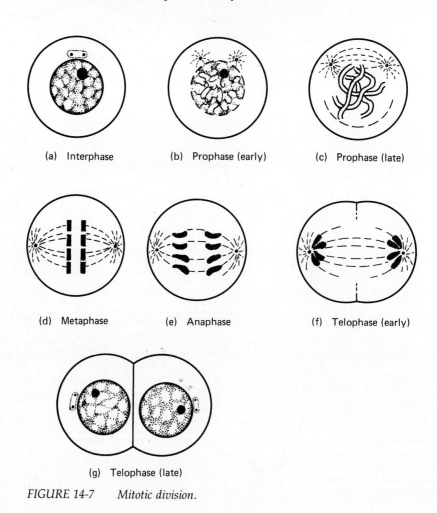

FIGURE 14-7   Mitotic division.

418

1. *Prophase.* The dispersed chromatin in the nucleus becomes a long slender thread that thickens and then forms a number of segments called chromosomes, which divide longitudinally. The nucleolus and the nuclear membrane disappear, the centrosome separates and astral rays appear. These radiate from each centrosome which migrates to opposite poles of the cell and forms the aster spindle, which extends from the centrosomes to the chromosomes in the center of the cell.
2. *Metaphase.* The longitudinally split chromosomes (still stuck together) line up along the equatorial plane of the cell and become attached to the aster spindles of the centrosomes. Each chromosome is then pulled apart along the longitudinal division by the spindle fibers.
3. *Anaphase.* The divided chromosomes are moved toward opposite poles of the cell apparently by contraction of the spindle fibers, so

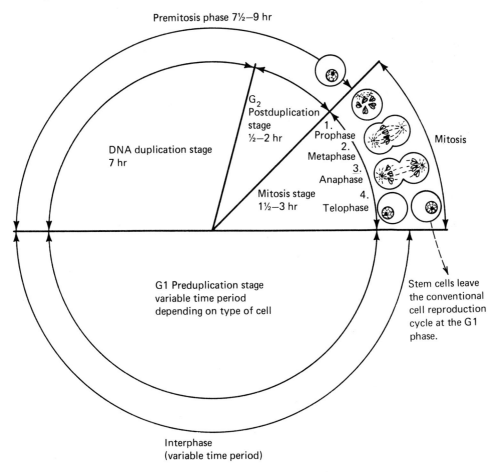

FIGURE 14-8 *Stages of the cell cycle. (Adapted from Ham,* Histology.*)*

that each pole contains a chromosome population qualitatively and quantitatively identical to the chromosomal pattern of the cell prior to prophase.

4. *Telophase.* A constriction develops along the equator, which quickly becomes a complete cell membrane that passes through the equatorial plane and separates the cell into two daughter cells, which are qualitatively identical in cytoplasm and nuclear content with their parent. The chromosomes become first threads and then nuclear chromatin. The nucleolus and nuclear membranes reappear.

5. *Resting stage or interphase.* During interphase the daughter cells separate from each other and a period of growth takes place until the daughter cells are quantitatively as well as qualitatively identical with their parent. The cells then assume their normal functional activity, which continues until another reproductive episode occurs.

In meiotic division there is a distinct difference in the mechanism. Meiotic division involves two stages. The first, a primary stage, is roughly similar to mitotic division in that there are four subsidiary stages: prophase, metaphase, anaphase, and telophase. However, what happens to the chromosomes is quite different. In meiotic division, the chromosomes form into pairs prior to the formation of the aster spindle, and then in late prophase of the first meiotic (I meiosis) division, each pair replicates to form a tetrad. The number of tetrads is one-half the chromosome number of the species, but since each tetrad is composed of four chromosomes, the total number of chromosomes in the cell has been doubled (Figures 14-10 and 14-11).

In metaphase the tetrads line up transversely along the equatorial plane and are pulled apart by the aster spindle fibers. The separated tetrads, called dyads, each contain two chromosomes, so the total chromosome number of the species is preserved, but the daughter cells do not have the qualitative identity of their parent; only one-half the chromosome variation of the parent goes into each daughter cell.

In anaphase and telophase of I meiosis the mechanics of cell restoration are similar to mitotic restoration, except that dyads do not change back into linin threads and then vanish into chromatin granules, but fade into chromatin without the intermediate change. The cells resulting from the primary division are called secondary spermatocytes or secondary oocytes.

In meiosis there is no appreciable resting state. The daughter cells (secondary spermatocytes or oocytes) promptly enter a second reproduction phase called the second maturation (reduction) division (II meiosis) where the dyads reform in prophase, pair up along the equatorial plane in metaphase, and are pulled apart transversely without longitudinal

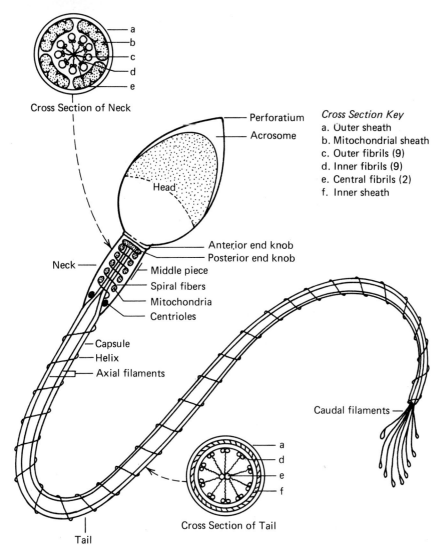

FIGURE 14-9   Spermatozoon anatomy. (Adapted from Svendsen, An Introduction to Animal Physiology.*)*

division occurring. This produces two daughter cells (spermatids or ootids) which are haploids, i.e., which contain one-half the species number of chromosomes. Hence in maturation (II meiotic) division, the cells divide twice, but the chromosomes divide only once. The end result is the formation of four haploid cells from a diploid primordium.

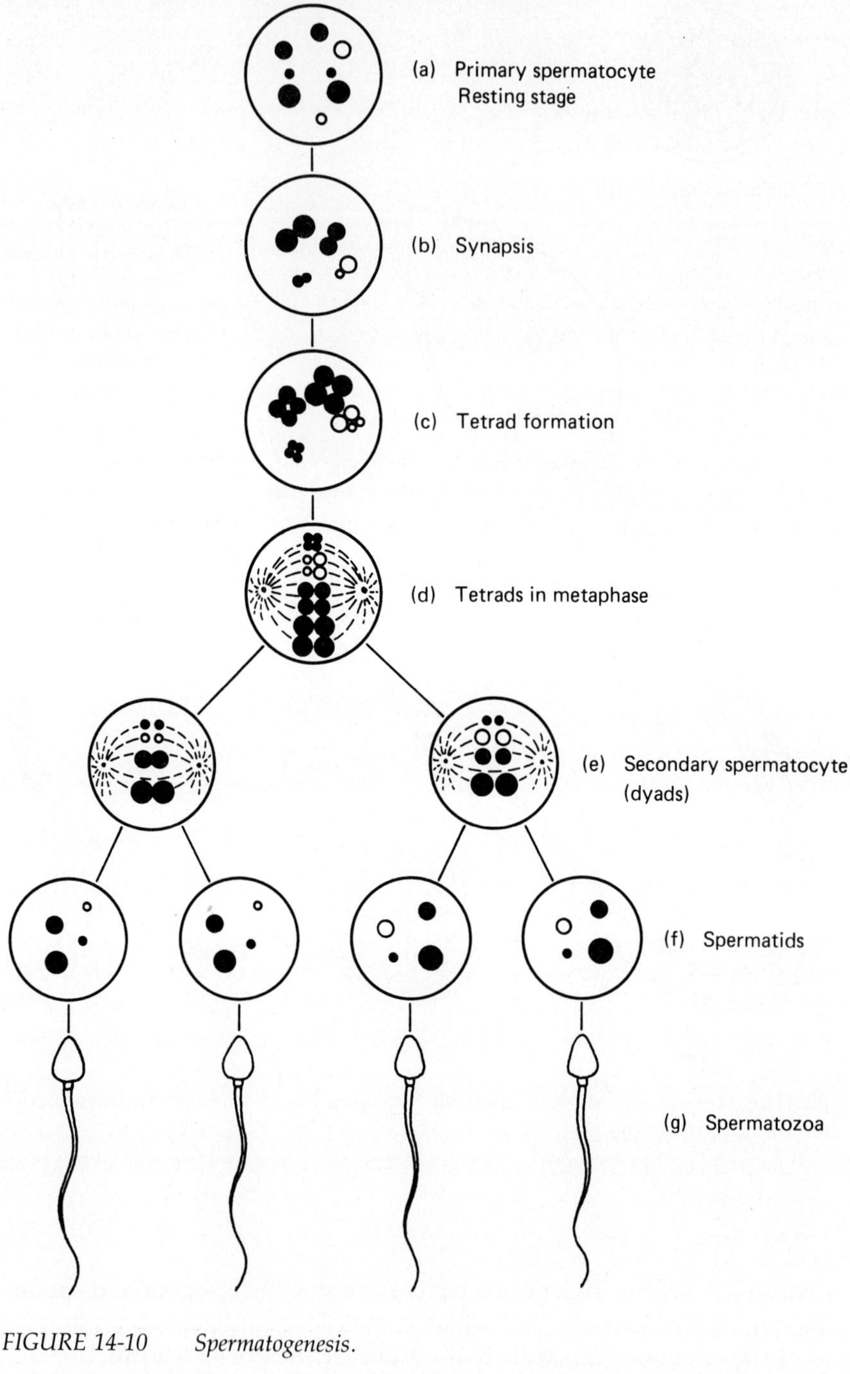

422
THE REPRODUCTIVE SYSTEM

(a) Primary spermatocyte
Resting stage

(b) Synapsis

(c) Tetrad formation

(d) Tetrads in metaphase

(e) Secondary spermatocyte (dyads)

(f) Spermatids

(g) Spermatozoa

FIGURE 14-10   *Spermatogenesis.*

SPERMATOGENESIS

Spermatogenesis has turned out to be a complex process, but the complexity is in detail rather than in method. The classical physiologists were correct in their conclusion that spermatozoa arise from spermatogonia, but more critical modern studies have complicated the process.

Spermatogonia develop from gametocytes (primordial spermatogonia) which are formed in the embryo and lie dormant in the seminiferous tubules of the testicle until puberty. At puberty the gametocytes multiply by mitosis and form spermatogonia, which congregate beneath the limiting membranes of the seminiferous tubules, and remain there in a dormant state.

Spermatogonia subsequently divide to form two daughter cells. One of these becomes another spermatogonium and returns to the dormant state. The other becomes a Type A or active spermatogonium. The TYpe A cell is polarized toward spermatozoa production. It divides four times, forming two generations of intermediate cells, a generation of Type B cells, and finally 16 primary spermatocytes (Figure 14-12). These undergo meiotic division to become 64 spermatids that ultimately metamorphose into spermatozoa as shown in Figure 14-10.

The terminal portion of the process (maturation division) does not vary appreciably from the classical concept except for the time involved. The end results are numerically different in that 64 spermatozoa can be produced from one Type A spermatogonium. However, the mechanism that causes the polarization of the Type A spermatogonium to form primary spermatocytes is still unknown.

With the formation of the spermatids the II meiosis (reduction division) phase of spermatogenesis is completed. The spermatids must now metamorphose into spermatozoa. This process involves both morphological and cell content alteration of the spermatid. There is a loss of most of the cytoplasm, the Golgi apparatus forms the acrosome, the mitochondria concentrate along the proximal part of the axial filament of the tail, the centriole becomes a part of the middle piece and forms the axial filament of the tail, and the nucleoplasm becomes condensed and packed into the head.

The entire process of formation of a spermatozoon from a dormant spermatogonium takes from one to two months, depending on the species of animal. This is considerably longer than what was once thought. One-fourth to one-half of the time is taken up by meiotic prophase, which requires about 15 days.

In stained sections of the seminiferous tubules, the relatively dark-staining spermatogonia and their descendants are interspersed with pale-staining cells. These are Sertoli or sustentacular cells, and in actively spermatogenic portions of the seminiferous tubules form a fairly numerous group that extends to the lumen. They furnish nutrition to the

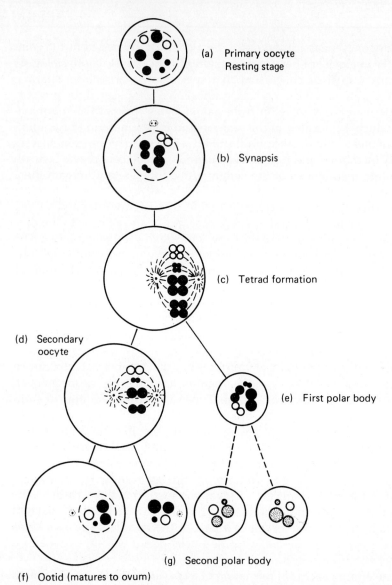

FIGURE 14-11 *Oogenesis.*

developing spermatids and spermatozoa, and produce the fluid that moves the spermatozoa toward the urethra.

Spermatozoa

Despite variations in shape (chiefly in the head), spermatozoa follow a generally similar plan. They are composed of a head capped with an acrosome, a neck, a middle piece, and a tail. The head is composed

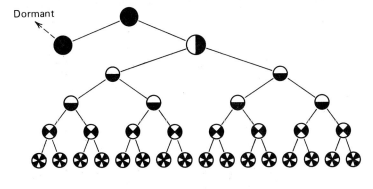

FIGURE 14-12 *Family tree of a primary spermatocyte.*

of condensed nuclear material and contains one-half the species number of chromosomes. Between the head and middle piece is the neck. A long filamentous thread of protoplasm, the axial filament, is attached to the neck and runs caudally through the middle piece into the tail. The middle piece consists of the axial filament and two sheaths, an inner nine-unit fibril sheath and an outer nine-unit fusiform sheath and centriole elements. Outside these structures is a spiral sheath of mitochondria. The tail, a long filamentous structure, is the posterior continuation of the spermatozoan. It is capable of rapid whip-like or undulant propulsive movements. Spermatozoa contain, in modified form, all the parts of a cell. However, there is no food storage mechanism within the sperm, and consequently they are dependent upon outside sources for nutrition. In the absence of these they quickly die.

Although sperm survive longest in an isotonic neutral medium, they are most active and most efficient in producing conception in a media sufficiently alkaline to neutralize the mildly acid uterine secretions that otherwise would kill or immobilize the sperm. In general, sperm are less affected by hypertonic than hypotonic media.

Except perhaps in swine and horses, semen contains many times the number of sperm per cubic milliliter required for efficient fertilization. Using bull semen as an example, an ejaculate contains 2.5 billion sperm per cubic milliliter, but fertility is not impaired when one-milliliter doses of diluted sperm are used in artificial insemination, until the sperm count is less than 10 million per cubic milliliter. Hence, bovine sperm can be diluted 250 times before the dilution factor (or the lack of sperm numbers) has an effect on fertility. It is this fact, above all else, that makes artificial insemination of cattle a profitable business.

Temperature has profound effects on spermatozoa. A 10°C rise in ambient temperature will double the metabolic rate of sperm and concomitantly increase their motility and shorten their lifespan. A 15°C rise above normal body temperature will produce permanent damage to sperm within five minutes. Sperm cooled rapidly to 0°C will die of cold shock. However, if they are cooled slowly from 15°C down, motility and metabolism slowly disappear, and if held in appropriate media (egg yolk citrate, glucose, glycerol, penicillin, streptomycin) can be kept in a suspended animation state for some time. Bull sperm in such a media can be deep frozen (to minus 196°C) and held indefinitely. Freezing does not work with boar or dog semen, but it will effectively preserve stallion, ram and goat semen. Artificial insemination in thoroughbreds and standardbreds is prohibited for a number of nontechnical reasons, and although its use in other horse breeds has not been prohibited, it is seldom employed.

There is a fairly good correlation between number of sperm produced and testicle size. Bulls and rams possess large testicles and can produce between four and eight billion sperm per day.

Changes in the microscopic structure of the testicles occur rapidly at puberty and slowly thereafter. Before puberty the seminiferous tubules have no lumens, and except for spermatogonia and interstitial cells, there are no recognizable reproductive cells. At puberty the gland becomes functional and remains so until old age when the seminiferous tubules progressively degenerate, followed by interstitial cell degeneration.

Defects of Spermatozoa

Among the domestic animals, bulls have been subjected to the most intensive sperm examination. Among the results of these studies is the information that bovine semen contains variable numbers of dead and malformed sperm, and that excessive numbers of these abnormal cells will result in poor reproductive capacity. A compilation of the various sperm defects resulted in a number of kinds of damage being classified as heritable that were once thought to be results of disease or overusage. Among the defects were lost or detached acrosomes, middle piece and tail deformities, curled tails, split or coiled tails, narrow head, and separated heads. Records indicated that these abnormal sperm were not only associated with low fertility but also that a tendency to produce these defective cells was passed on from sire to offspring.

It is unwise to extend data from one species to another without careful examination and testing, but the findings in cattle may probably be extended to other domestic animals if the semen examinations of low fertility males of other species are codified.

Hormones

See endocrine section on gonads, Chapter 12.

Spermatozoa Transport and Storage

It takes from several weeks to two months for sperm to travel from their place of formation to the region from which they can be ejaculated. They are moved along by a positive pressure of fluids in the tubules of the testes. The slow current carries the sperm at least as far as the tail of the epididymis and in some species to the termination of the vas deferens, where organs of temporary storage, the ampullae, hold the sperm for a time. During this period, the sperm are not motile. Motion is acquired during ejaculation when the sperm encounter the mildly alkaline fluids secreted by the prostate and the bulbourethral glands. The seminal vesicles produce the bulk of the fluid (semen) in which the sperm are held during ejaculation.

Ejaculation

Ejaculation is the process by which spermatozoa are propelled out of the male's body. This process involves a neuromuscular reaction that originates in the receptors of the glans penis and ends in the muscular walls of the urethra. The act of ejaculation involves a rhythmic peristalsis of the urethra which, together with the contraction of the seminal vesicles and the increased secretion of the accessory glands, results in the formation and expulsion of the mixed product known as semen. Semen is composed of spermatozoa and seminal fluid. The seminal fluid contains relatively high levels of choline, citric acid, fructose, bicarbonate and protein. The fluid is a natural buffer that resists abrupt changes in pH. Seminal fluid apparently counteracts the relative acidity of the female vagina, which would kill unprotected spermatozoa.

OOGENESIS

Ova, or female germ cells, are produced in the cortex of the ovary. The primordial germ cells are derived from the germinal epithelium in the embryo and migrate into the cortical connective tissue where they become grouped in clusters known as egg nests or ovigerous cords.

After this cellular migration takes place, the tunica albugenia develops beneath the germinal epithelium, effectively preventing further migration of the epithelial cells into the cortex. In many mammals the interdiction of the germinal epithelium takes place prior to birth; in some it is a postnatal phenomenon, and for awhile after birth in some species (notably cattle) there can be germinal epithelium cell migration into the cortex. There may or may not be further proliferation of the cells of the ovigerous cords. In any event, cell multiplication soon ceases and from this point onward the number of potential female germ cells is steadily reduced until eventually there will be none left.

Once the tunica albugenia is established, little happens until puberty, when follicular structures begin to appear in the cortex. These structures are apparently influenced to develop by good nutrition, favorable climatic conditions, and the presence of follicle stimulating hormone. The follicular estrogens start the cyclic phenomenon of ovulation and the development of secondary sexual characteristics. The ovulation cycles are called monestrous, diestrous, seasonally polyestrous, and polyestrous depending upon their occurrence during the year. The estrous cycle, once established, continues throughout the animal's fertile life, although it can be influenced somewhat by diet and confinement.

Through some presently unknown mechanism, one of the cells in an egg nest begins to change and develops into a functional ovum, and some of the remainder become nutritive or sustentacular cells which support and supply the ovum during its period of growth. During the development of the Graafian follicle the ovum usually passes through the first maturation division and forms two secondary oocytes. One of these receives virtually all of the cytoplasm while the other becomes the first polar body, which either degenerates or disappears. The surviving secondary oocyte then undergoes the second maturation division wherein the chromosome count is halved and two ootids are formed. One of the ootids receives virtually all of the cytoplasm, while the other becomes the second polar body, which usually disappears or degenerates. The surviving ootid matures to form the ovum. In some animals the final maturation of the ootid occurs in the Graafian follicle prior to rupture; in others (the bitch for example) no maturation division occurs until the ovum enters the oviduct. In still other animals (cows and ewes) the second maturation division takes place in the oviduct.

Ovum Anatomy

The mature ovum is a large spherical cell with a well-defined nucleus which contains a dark-staining nucleolus. The cytoplasm is granular, and the periphery is enclosed toward the end of the maturation process in a dense carbohydrate capsule, the zona pellucida, which appears to be secreted by the cells of the cumulus oophorus. The cell itself is undifferentiated and is not specialized to perform any specific function.

Preovulation and Postovulation Changes in the Ovary

As the ovum matures, a number of changes take place in the surrounding tissue. The granulosa or sustentacular cells multiply in number and change from a layer of flattened cells to a surrounding zone of distinctly columnar cells. During this period the ovum and its follicular cells are known as a primary follicle. The follicular cells continue to increase in number and become smaller and roughly spherical. Cell multiplication is faster on one side of the ovum than the other. This forces the egg into an eccentric position on the border of the growing follicle. During this time, by differential growth, the follicle migrates into the deeper layers of the cortex.

Clefts now appear between the follicular cells and become filled with a tissue fluid called liquor folliculi. At the same time, certain connective tissue cells form a definite band of tissue, the theca externa, which surrounds the follicle. At the junction of the follicle cells and the theca externa, a layer of mixed cellular elements develops which is known as the theca interna. The follicle cells are formed into a peripheral layer a few cells thick called the granulosa. At this point where the ovum rests against the follicle wall, a local proliferation of granulosa cells forms a small mound called the cumulus oophorus. A thin layer of granulosa cells called the corona radiata covers the free surfaces of the ovum. Structurally, the follicle, which is now called the Graafian follicle, is complete. All that is required now is growth, which is accomplished by proliferation of the follicular and theca cells, and an increase in the amount of follicular fluid. As the follicular fluid increases, the Graafian follicle becomes a cyst-like structure that bulges to the surface of the ovary.

Shortly prior to ovulation a small area on the outer surface of the Graafian follicle thins down to a few connective tissue fibers and one or two layers of granulosa cells. This "cell poor" area is where the follicle will rupture at ovulation. At ovulation the ovum is expelled from the follicle and the ruptured follicle passes through the series of cellular changes which form the corpus hemorrhagicum, corpus luteum, corpus rubrum and corpus albicans.

Hormones

See endocrine section, Chapter 12.

The Ovulation Cycle

The changes from initial development of the ovum to the formation of the corpus albicans form what is known as the ovulation cycle. The entire cycle takes place in a relatively limited period of time, usually over two estrous cycles in polyestrous animals. In monestrous or diestrous (one or two heats a year) animals, the entire cycle is completed in a relatively small fraction of the time between heats.

The Movement of the Ovum Outside the Ovary

Rupture of the Graafian follicle tears the ovum loose from the cumulus oophorus and expels it to the outside where it is caught in the dilated proximal portion of the oviduct (the infundibulum) and conveyed down the oviduct to the uterus. A combination of ciliary beating of the cells lining the oviduct, peristaltic movement of the oviduct and ameboid movement help the ovum move from the ovary to the uterus. In domestic animals movement of the ovum through the oviduct takes from three to five days. If fertilization does not occur the ovum will degenerate within a week. Occasionally an ovum may escape the oviduct and fall into the abdominal cavity. If such an ovum has been fertilized, ectopic pregnancy may result.

THE ESTROUS CYCLE

This is the rhythmic period that occurs in normal mature females, and is characterized by the physical change known as "heat" (estrum, estrus) which accompanies ovulation. Mammals fall into three groups as far as estrous cycles are concerned: monestrus, where ovulation occurs once a year; diestrus, where ovulation occurs twice a year; and polyestrus, where ovulation occurs many times a year. Of the commonly known animals, captive furbearers such as foxes and mink are monestrus; dogs are diestrus; cattle and swine polyestrus; and horses, sheep, and cats seasonally polyestrus. In polyestrous animals the rhythmic cycle varies from four days to four weeks in duration, but may be interrupted by pathological changes or pregnancy.

Phases of the Estrous Cycle

The estrous cycle can be divided into four more or less distinct phases, which are called proestrus, estrus, metestrus, and diestrus, plus a fifth phase, anestrus, in which no ovarian activity apparently occurs. Anestrus may be physiological or pathological. Physiological anestrus occurs in animals that are monestrus, diestrus, or seasonally polyestrus and is the nonfunctional period between periods of sexual activity. Pathologic anestrus results from disease or malfunctions involving the hormonal apparatus of the female reproductive organs.

Proestrus is the period of Graafian follicle growth. The growth involves the production of follicular fluid which is rich in estrogen, particularly estradiole. The hormones enter the bloodstream and cause a thickening of the vaginal wall and a cornification of the vaginal epithelium. The submucosa of the uterus becomes thickened and vascularized with concomitant proliferation of the epithelium. The epithelium of the oviduct forms an increased number of cells with more numerous cilia. The period is one of preparation for ovulation and coitus. In some dogs the uterine phase of this period is accompanied by bleeding from the cervix and vagina. Tests for proestrus involve vaginal smears which show large numbers of cornified cells when examined microscopically.

Estrus is the period of heat or "season." It involves marked behavioral alterations, willingness to breed, swelling or turgidity of the external genitalia (in female monkeys and baboons this involves the spectacular production of "sex skin") and postural and voice attitudes, which are easily recognized. In most mammals this period terminates with the rupture of the Graafian follicle and the expulsion of the ovum into the oviduct. At this time the signs of estrus are intensified. In some species, notably cats, rabbits, mink, ferrets, and ground squirrels, ovulation does not occur until the animal is bred. As a result, the signs of estrus will persist for a considerable period of time and cause particular problems to cat owners, since queens (particularly Siamese) are not the pleasantest companions when they are in season.

Metestrus is the period of functional development of the corpus luteum: The granulosa and theca interna fold into the blood-filled cavity of the corpus hemorrhagicum, and commence to proliferate and form the theca lutein and granulosa lutein cells. Blood vessels proliferate into the blood-filled cavity and the developing tissue overgrows the organized hemorrhage and becomes a large mass of yellowish tissue (the corpus luteum). The corpus luteum secretes progesterone, a hormone that suppresses further follicular growth, prepares the endometrium of the uterus to receive a fertilized ovum, and causes a loss of epithelial cells of the vagina together with a large number of leukocytes. Examination of vaginal smears easily detects this period.

Diestrus is the longest period of the estrus cycle in polyestrus or seasonally polyestrus animals. It is during this period that the corpus luteum exerts maximum effect. Two possibilities now can occur.

1. If fertilization has taken place, the corpus luteum will persist as the corpus luteum of pregnancy, and the estrus cycle will be more or less completely interrupted for the greater part of the pregnancy period.
2. If fertilization has not taken place, the corpus luteum becomes the corpus luteum of estrus and a number of characteristic changes will take place in the reproductive tract. The initial changes, i.e., thickening of the endometrium, proliferation of the uterine glands, and development of the uterine musculature, persist for awhile until the progesterone secreting activity of the corpus luteum declines.

    In polyestrus and seasonally polyestrus animals the corpus luteum regresses to a corpus rubrum or "red body" and a new cycle of follicle growth begins. The uterine epithelium sloughs with a variable production of debris and hemorrhage (menstruation). As a rule, the menstrual phenomenon is confined to primates.

    In monestrus and diestrus species, in the rabbit and occasionally in other species, if pregnancy does not take place, the corpus luteum may persist for awhile and produce the phenomenon known as pseudopregnancy. In pseudopregnancy the animal shows all the signs of pregnancy but has no offspring. The false gestation period is somewhat shorter than normal pregnancy.

COPULATION

With the exception of some primates and rabbits, female mammals are unwilling to breed except during the ovulation phase of the estrus cycle. The act of natural breeding in mammals involves conjugation or physical contact between male and female. Copulation usually results in ejaculation by the male and the deposition of semen in the vagina or uterus of the female.

Fertilization

The spermatozoa are normally deposited in the cervix or the lower part of the uterus. They ascend the female reproductive tract and pass into the oviduct. Normally fertilization takes place in the upper third of the oviduct although this is variable. Exceptions include Madagascar anteaters, moles, and shrews, where fertilization occurs within the Graafian follicle, and monotremes, where fertilization occurs in the uterus.

In fertilization, only one spermatozoon is involved in the restoration of the diploid number of chromosomes in the ovum, but considerably more than one are required to overcome the physical resistance of the ovum to penetration so that the head and middle piece of the fertilizing spermatozoon may enter the ovum's cytoplasm. Once this occurs a number of changes take place. The cell membrane of the ovum becomes refractory and refuses any further attempts at penetration by other sperm. The nuclear materials of the sperm and the ovum unite to form the zygote and restore the species number of chromosomes and centrioles, and the zygote shortly thereafter divides mitotically (cleavage) to form two daughter cells. The cells of the zygote continue to divide mitotically and increase in number as the growing mass of protoplasm makes its way down the oviduct and into the uterus.

Stages in Embryo Development

The cell mass stage. In the beginning of embryonal development, the dividing cells of the zygote form a mass of unspecialized cells, somewhat ball-shaped in appearance and only slightly larger than the original ovum from which they are derived. This is known as the cell mass. After it passes the 16-cell stage it is called the morula.

Implantation. The morula, still growing in number of cells, enters the uterus and comes to rest against the uterine wall after a period ranging from several hours to several days, depending on the species of animal. An enzyme secreted by the morula dissolves away a pocket of tissue in the endometrium into which the morula implants itself.

The blastula. As the morula develops in the wall of the endometrium it becomes transformed into a hollow, roughly spherical mass called the blastula.

The gastrula. The blastula, by continued growth, forms the gastrula, a cup-shaped mass of cells, thicker on the side opposite the rim. This mass of cells, by differential growth, works its way toward the lumen of the uterus.

The embryonal disc. As the base of the gastrula continues to thicken by further cell growth, it forms a roughly oval mass known as the embryonal disc. The edges of this disc begin to proliferate outward along the inner surface of the uterus.

The extra-embryonal membranes. The edges of the embryonal disc grow with great rapidity to form a thin transparent membrane which lines the uterus. A short while later a yolk sac placenta is developed from the embryonal disc and impinges against the endometrium. Primitive vascular channels arise in the yolk sac placenta and convey nutrient from the mother to the growing embryo. In some species the yolk sac placenta does not develop.

Hensen's node and the primitive streak. On the surface of the embryonal disc that protrudes into the uterine lumen, a raised area of extremely active cells appears. This is the growth center called Hensen's node. Extending cranially from this node is a dark line of cells called the primitive streak, which marks the future location of the spinal axis of the developing embryo.

The neural groove. Along either side of the primitive streak, columns of cells form, making a depression between them which contains the streak. This depression is called the neural groove.

The somites. The columns of cells that form the neural groove become differentiated into blocks of tissue called somites. These structures are the precursors of the muscular and skeletal tissues of the body.

Craniad precocity. Since virtually all of the body cells originate in Hensen's node and are pushed forward by a process of differential growth and formation of new cells by the node, the older (and more highly evolved) cells are found in the cranial portions of the embryo. Therefore the structures in the head and thoracic region tend to develop early, resulting in the phenomenon of craniad precocity.

Development of the chorion. Two membranous sacs, the allantois and the amnion, are developed from the ventral surface of the embryo. The allantois unites with the yolk sac placenta to form the allantochorion (or chorion), while the amnion closely surrounds the developing embryo. These structures vary considerably in size, extent, and development among animals. However, some form of them exists in all mammals. Even the human embryo, which does not have a functional amnionic vesicle, has a primordium of this structure. The amnion and allantochorion are fluid-filled sacs within which the fetus floats. Amniotic fluid is slightly alkaline and contains (in addition to water) protein, sugar, salts, fat, and traces of urea. The urea indicates a possible urinary origin of some of the amniotic fluid constituents. In general, however, amniotic fluid is thought to be derived from tissue fluid. The allantoic fluid, however, is composed mainly of excretory products since it is connected to the fetal kidneys by the urachus. The chorion produces the placental structures, which unite the fetus with the wall of the uterus via the umbilical stalk.

The placenta. This is a reasonably late development in the embryonal period and is the area of attachment and close conjunction between the maternal and fetal bloodstreams over which oxygen, nutrients, and metabolic waste pass in maternal-fetal interchange. Several types of placenta exist among animals, but all begin as epitheliochorial. As gestation proceeds, the membranes and tissues are lost until the species type of vascular relationship is attained.

The fetus. With the development of all definitive structures and the assumption of general species form, the embryo becomes the fetus. The fetus confines its major activity to growth and development rather than to the formation of new organs and structures.

Placentas and Vascular Relationships

Placental Type	Species	Number of Membranes	Vascular Relationship
Diffuse	Horse—pig	6	Epitheliochorial
Cotyledonary	Cow—sheep	5	Syndesmochorial*
Zonary	Dog—cat	4	Endotheliochorial
Discoidal	Man—apes	3	Hemochorial
Discoidal	Higher rodents, i.e., lagomorphs, guinea pig, rat.	1	Hemoendothelial

*There is evidence that this classification is incorrect and that six membranes separate the fetal and maternal bloodstreams, which would make the relationship epitheliochorial.

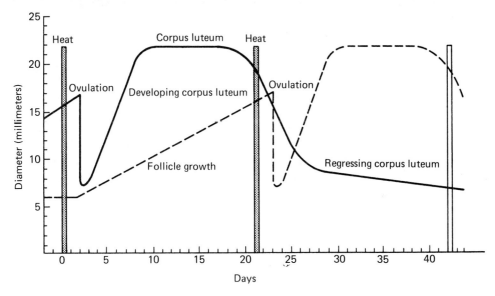

FIGURE 14-13   *The ovarian cycle of the cow.*

The Gestation Period

The gestation period or pregnancy period is the time from fertilization to parturition. It varies considerably between species and to a minor extent between individuals of the same species. For convenience, it is usually divided into three parts or trimesters. The first trimester is considered to be the embryonal period, the second the developing fetal period, and the third the growth period. In this sort of scale the trimesters are of unequal length. In cattle the first trimester would be about 80 days, the second about 100 days, and the third about 100 days. Fetuses in the third trimester may be born alive but require intensive care to keep them alive if they are born in the first two-thirds of this period. In human gestation the trimester is considered to be one-third of the ges-

Estrous and Gestation Periods

Animal	Estrous Cycle	Gestation True	Gestation Approximate
Horse	21 days	346 ± 28 days	11 months
Cow	21 days	282 ± 21 days	9 months
Sheep	18 days	150 ± 18 days	5 months
Pig	18 days	115 ± 5 days	4 months
Dog	6 months	63 ± 3 days	2 months
Cat	Variable	60 ± 9 days	2 months
Rabbit	None	30 ± 3 days	1 month
Man	28 days	274 ± 21 days	9 months

tation period and is only roughly comparable to embryonal and fetal development according to the scheme used in other mammals.

Parturition

At the end of the gestation period, parturition or birth takes place. The factors that initiate parturition are imperfectly known, but among them are probably size and weight of the uterus, size and weight of fetus and attachment to the placenta, hormone balance of fetus, and hormone balance of the mother. Possibly additional factors play cooperative parts in inducing parturition, or perhaps only one of these is the true initiating factor. The answer has not yet been found.

In addition to fetal hormone balance which may trigger the act of birth, at least three maternal hormones are known to function in the process; relaxin, oxytocin, and adrenalin. These have been discussed elsewhere (p. 364).

At birth the uterine muscles contract with gradually increasing force and push the fetus out of the uterus and into the birth canal. The contractions of the uterus are assisted by the pressure of the abdominal viscera, which are forced downward and backward upon the uterus by the contraction of the abdominal muscles (straining). The musculature of the cervix and vagina first relaxes, allowing both the cervical and vaginal canals to dilate, and after the fetus enters the birth canal, the musculature contracts rhythmically in slow peristaltic waves to assist the movement of the fetus through the canal to the outside.

Pressure stimulation by the physical presence of the fetus in the birth canal sets up a mechanical stimulus to the nerves of the pelvic plexus resulting in direct nervous stimulation of the vaginal and cervical musculature, and the majority of the labor pains. In addition, the pressures developed in the birth canal have a somewhat deadening effect upon the pain receptors in the walls of the birth canal.

Postparturient Changes in the Uterus and Birth Canal

After birth, the uterus and birth canal remain for a short period in a state of shock and are refractory to stimulus. During this relaxed period the placenta and fetal membranes (afterbirth) ordinarily are shed.

After the refractory period, which may vary from several seconds to several hours, the uterus and birth canal commence to involute or retun to their approximate original size. In the beginning this process is very rapid, but the whole process of involution may require from one to two months for the reproductive organs and several months for the abdominal musculature, skin, and subcutaneous fascia. Actually, involution is never complete, and the result is a uterus and abdomen slightly larger and with slightly less tone than it had prior to pregnancy.

The culmination of this process over a number of pregnancies results in a large flabby uterus with poor muscle tone.

Occasionally before birth, but usually afterward, milk secretion commences in the mammary glands. During gestation the glands increase in size and the glandular structure develops. At parturition milk flow is stimulated by the hormone oxytocin. In most animals the mammary glands involute following lactation. In some species lactation inhibits or stops female reproductive activites.

The cervix, vagina, and the pelvic ring form a composite structure called the birth canal. There are many variations in this structure, although functionally all birth canals are designed to form a passageway leading from the uterus to the outside through which a fetus can pass. The principal differences in structure involve the bony ring of the pelvis. This structure may be complete as in the horse and cow, incomplete as in seals and cetaceans, or partially complete as in mice and other small animals which bear comparatively large offspring.

In the domestic animals, the pelvic ring does not dilate appreciably, or at all, during parturition. Stories to the contrary are numerous and are probably inspired by the fact that small animals with partially complete pelvises such as mice and guinea pigs do have an appreciable dilation of the pelvic opening. In the larger animals such as man, horses, cattle, sheep, and swine, the pelvic girdle is complete and bound together by immovable joints at the sacrum and symphysis pelvis. Any appreciable dilation of the pelvic ring would separate these joints and cripple the animal.

The Menopause or Female Climacteric

Rarely, if ever, do wild animals live past their reproductive lives. Occasionally this occurs in domestic animals (mainly cats and dogs), and it commonly occurs in women. In domestic animals the cessation of female function is apparently a gradual process and individuals can become pregnant very late in life. In women, however, the menopause is a relatively sudden process, which usually occurs sometime after 40 years of age and may be accompanied by hormonal dysfunction, emotional upset and personality alterations.

Ectopic Pregnancy

Occasionally, and fortunately rarely, a pregnancy will take place outside the uterus. This is called ectopic pregnancy. The most common of these is where the embryo implants and begins to develop in the oviduct. This is tubal pregnancy, and by the time it is recognized the only recourse is surgery and removal of the involved structures.

False ectopic pregnancy occurs when a fertilized ovum implants, a placenta develops, and then because of a uterine defect or rupture the developing embryo passes through the uterine wall and into the abdominal cavity, remaining attached to the placenta which lies within the uterus.

True ectopic pregnancy results from the dropping of a fertilized ovum into the abdominal cavity and implantation and placenta formation in the abdominal cavity. This is extremely rare but has been reported in the literature. The outcome is almost invariably fatal due to extensive disruption of abdominal organs, adhesions, and development of abnormal blood channels to supply the displaced fetus.

SEX RATIO

In mammals under natural conditions there seems to be a slightly greater number of males born than females. This is the basis for what is called the sex ratio. There are, however, three sex ratios, i.e., primary, secondary, and tertiary, that are involved.

The primary ratio is the number of each sex conceived, the secondary ratio is the number of each sex born, and the tertiary ratio is the number of each sex that reach reproductive age. In animals that have been studied, males are usually more numerous at birth but steadily decline in number until, at reproductive age, the ratio is approximately equal between males and females.

The sex ratio can be varied by selection and breeding programs. In albino rats with a basic 108:100 male to female secondary ratio, a program of selection and inbreeding resulted in two strains, one with a 124:100 ratio and the other with an 82:100 ratio. The selection was for phenotype rather than genotype. Phenotype autosomes possibly exert an influence on which sex survives in the uterus, at least under the conditions of this experiment.

Other factors that apparently affect the sex ratio (at least as far as experimental data on a restricted number of species show) include

1. *Frequency of ejaculation of the male.* In horses, seldom-used stallions had a secondary sex ratio of 97 males to 100 females in their offspring. When used frequently they gave a 101:100 ratio.
2. *Seasonal factors.* In horses, the secondary sex ratio was 98:100 from March through June; 133:100 from July through October.
3. *Number of pregnancies.* In rats, the secondary sex ratio was 122:100 for first pregnancies; 103:100 for fourth. In mice it was 124:100 for first litters and 150:100 for fourth.

It seems possible that environment as well as genetic factors exert an influence on sex ratio, and it appears that certain selective functions have been applied to mammals. If the hypothesis that autosomal characteristics influence the sex ratio is correct, a number of problems may be explained, e.g., why the normal dog secondary sex ratio is 110–124 males to 100 females, and the fact that considerably greater numbers of male fetuses are aborted in man, cattle, sheep, and swine. In these species, there has been considerable conscious or unconscious selection for male offspring and "prepotent" males with a high masculine autosomal gene content. The primary sex ratios, at least in these species, are conceivably unequal, or if conception produces equal numbers of males and females, embryonal mortality occurs principally in females in the early stages of gestation. With modern techniques of sex determination when the sex of a conceptus can be identified shortly after fertilization, this question can and should be answered.

Sex Determination

Many attempts have been made to control the sex of offspring. These have involved various physical, chemical and mechanical factors. None has been completely successful insofar as veterinary medicine is concerned. In human medicine it is now possible to separate spermatozoa containing X or Y chromosomes and through artificial insemination or extra-uterine (or extra-oviducal to be precise) fertilization produce offspring of desired sex. At present the techniques are too delicate and expensive to be used routinely on domestic animals. However, there is a good probability that an inexpensive field procedure will eventually be developed. The search will go on, for if it becomes feasible to separate the X and Y chromosome-containing spermatozoa, it would be a great help to the livestock industry, since in many species males are relatively unwanted. Experiments with rabbit sperm indicate an apparent specific gravity difference between X and Y chromosome-containing sperm. Proper dilution, centrifugation and pipetting give a preponderance of male or female offspring. Whether this relatively simple technique can be applied to other species remains to be seen.

Recent advances in the sex determination of embryos may revise some of the above concepts. The discovery that "sex chromatin" exists in somatic cells has a significant bearing upon the determination of primary sex ratios. This substance (sex chromatin) is usually found in Feulgen-stained cells and lies close to or against the inner surface of the nuclear membrane, or next to the nucleolus. It is present only in genotype females and is thought to be derived from fused X chromosomes. A technique called amniocentesis (puncture of the amnion) has been developed to diagnose genetic defects in human embryos. Cells obtained by this technique can be used to determine the sex of embryos as small as one millimeter in length, which (in most animals) is long before grossly visible sex differentiation has taken place, or even where sex differentiation can be accomplished by tissue study. Sex chromatin appears to be present in embryonic primordia before any such structure as a genital ridge has been developed. So far, except for humans, the technique has no application to animals, except in the laboratory, and mainly on monkeys. Other studies made on freemartins have shown that prolonged exposure of the female embryo to male hormones apparently does not affect sex chromatin although somatic conformation can be greatly altered.

Intersexes

In all mammals the developing embryo contains primordia of both female and male. In some animals, both of these primordia develop and result in the condition known as intersex, wherein portions of both female and male reproduction apparatus may be present in a single individual. Intersex can be divided into two categories: true hermaphroditism and pseudohermaphroditism. In the true hermaphrodite, which occurs rarely or not at all in mammals, there is functioning testicular tissue and a penis and functioning ovarian tissue and female reproductive organs. The common intersex in mammals is the pseudohermaphrodite in which fragments of the female reproductive apparatus may be present in a more or less functional male or vice versa. Generally the male sex is dominant and will be somewhat deformed. The gonad tissue is usually testicular, but the testicles will be smaller in size and may or may not be associated with a scrotal sac. The penis of the pseudohermaphrodite is short, particularly in sheep and swine, and the prepuce is split.

A special condition called the freemartin occurs in cattle. This syndrome affects female calves born twin with a male in a single chorionic envelope. Apparently the male sex hormones develop sooner in the male twin, and override the ability of the female to form definitive reproductive apparatus. In the freemartin a number of anomalies may occur, but three are most prominent. The animal will have the external genitalia of the female, but a body conformation similar to the male. The uterus will be small or imperfectly developed. The cervix may often be absent and other internal parts of the reproductive system may be missing. The external genitalia are small, with the exception of the clitoris which is extremely large and grooved. Such animals are infertile. Freemartinism will occur in approximately 90 percent of male-female twinnings in cattle. It can be diagnosed at birth with reasonable accuracy by examination of the placenta of the twins. If there is only a single allantochorion present, the female will invariably be a freemartin. If two separate chorions are present, the female may be normal.

Genotype and Phenotype

Sex is the sum of all characteristics of structure and function that determine masculinity and femininity. This includes not only visible physical dimorphism, but the subtler differences, many of which are not visible to the unaided eye. Differences, particularly in morphology, may range from slight to extreme and can only be accounted for by the fact that each individual is composed of a nonuniform mixture of male and female qualities, one or the other of which is predominant. Thus, all males have a greater or lesser amount of female characteristics that modify their masculinity and the reverse is true for females. In selection of animals for breeding purposes, we tend to be influenced by somatic conformation, e.g., "masculine" head, shoulders, etc., and "feminine" outline, face, etc., rather than the more important qualities such as productive performance, rate of gain, intelligence, and somatic health.

Selection thus may be partially due to prejudice and partially to lack of understanding of the difference between genotype and phenotype sex. Fortunately, these latter two factors tend to go together; so efforts to promote sex dimorphism do no great harm, but nevertheless, some helpful qualities may have been lost in our efforts to produce, for example, 2,000-pound "masculine" bulls and 1,000-pound "feminine" cows.

Genotype sex is determined by the sex chromosomes an individual possesses. These in mammals are XY for the male and XX for the female. This, of course, is the classic pattern and, like most classic descriptions, is neat, simple, and not entirely correct. There can be as many as eight X-Y components in a genotypic sex. This has been shown by chromosome studies, but is fortunately more of academic than practical interest except to the individual who has the excess. Nevertheless, this multiplicity of sex chromosomes is not as rare as one might think. It does exert some proved effects in humans, and has been theorized to exert similar effects in animals. The majority of animals of a given species, however, have a genotypic character that appears to be controlled by the presence (or absence) of a single chromosome or perhaps even a single gene. The characteristic is apparently imposed upon the cell at the instant of fertilization and is one that apparently cannot be changed and is not subject to environmental modification.

Although genotypic sex determines whether an individual will be functionally male or female, there are somatic variations in sex, degrees of morphology that range across the entire spectrum of physical difference. These somatic variations are called phenotype sex. A current hypothesis states that the sex chromosomes determine the genotype (functional) sex and female somatic characteristics, and autosomes carry male somatic characteristics. Thus, if the number of "maleness" genes on the autosomes is large, they may override the "femaleness" genes on the sex chromosomes and give male morphology (phenotype) to an individual with a female sex apparatus (genotype). Conversely, lack of male autosomal genes in a male genotype would result in the individual having a female phenotype. This does occur, and there are degrees of morphologic masculinity and femininity that are not necessarily correlated with the sexual apparatus.

If the above hypothesis is correct, sex chromosomes do not establish all of the secondary sex characteristics. Since secondary sex characteristics have been known to have a relationship to androgen or estrogen production, the mechanisms that involve phenotypic sex could function to control sex hormone production rate.

Postnatal Precocity and Resistance

Postnatal or neonatal precocity is a phenomenon observed in herbivorous ungulates, and in some other species such as hares, guinea pigs, elephants, rhinoceri, hippopotami, and cetaceans. With the exception of cetaceans it is not seen in carnivores, nor is it present in primates.

Ordinarily, within minutes after birth a precocious offspring is on its feet and within a few hours will be running with its mother. This phenomenon appears to be associated with the relative size of the placenta and the fetus, the length of gestation, and the fraction of the total lifespan spent in the uterus.

A neonatally precocious mammal spends about $1/20$ to $1/25$ of its normal lifespan in the uterus, whereas those that are not precocious spend about $1/40$ to $1/90$ of their life in prenatal development. Moreover, the size and capacity of the placenta of the neonatally precocious animal is large, and the newborn are usually single rather than multiple (except in hares and guinea pigs). It has been argued that the long prenatal period and the quantitative superiority of uterine nutrition produce an offspring more mature and more capable physically. Environmental and evolutionary pressures also play their part in developing precocious offspring.

Oddly enough, postnatally precocious animals usually have little resistance to disease if taken from their mothers at birth and raised on an artificial diet. Ordinarily these animals will die of dysentery or some other systemic disease within weeks after birth. In contrast, the nonprecocious offspring of rats, rabbits, dogs, cats and other species have a surprising resistance to infections. This appears to be the result of a better transfer of antibodies across the placental barrier in these latter species. However, the resistance of both groups is improved if they receive colostrum (first milk) in the first 24 hours of their life. It is significant that the newborn developed in uteri with epitheliochorial and syndesmochorial placentas need colostrum to survive. Generally, nonprecocious offspring are born to mammals with endotheliochorial, hemochorial, and hemoendothelial placentation. These uterochorial relationships offer less resistance to the passage of maternal antibodies than do the thicker and more numerous tissues present in the uterochorial relationship of species with precocious offspring.

CONTROL OF PREGNANCY AND OVERPOPULATION

One of the more practical aspects of knowledge of the physiology of the reproductive process is the control of pregnancy. Not only is this important in animal husbandry, but it is becoming ever more important in human biology. The explosive growth of human population, if continued for another century at the present rate, will result in some 45 billion people on this planet, or about ten for every one who exists today. This will result in severe overcrowding and can be most dangerous for mankind.

J.B. Calhoun's classical work with overcrowded rats and mice, and other observations of the effects of overpopulation on animals in natural environments, and of overpopulation and confinement in zoos and laboratories have produced data that are disturbing, to say the least. If the laboratory results observed in unnaturally crowded situations can be translated to other animals and to human beings, the results become frightening. The classical adrenal stress syndrome reported by Hans Selye is a frequent finding in overcrowded animals and may contribute to the abnormal behavior, and mass suicide marches of the prolific lemmings in years of gross increase in the natural population.

Another ill of crowded animals is abnormal sex behavior. In normal rats, sex is surrounded with intricate courtship rituals. In overcrowded rat populations it can become rape, and gang rape. Such aberrant behavior patterns are never found in rats living in environments with low population densities.

In captivity and in overcrowding situations, homosexuality, especially among males, becomes common, although it is virtually unknown among mammals in the wild state. Dominant males sometimes kill and eat other males of the same species, in contrast to the wild state attitude of clemency toward a defeated foe who submits. Females tend to lose their ability to build nests and/or care for their litters. They often kill, maim, or eat their young and become increasingly unable to carry pregnancies to term. Dominant males appear in what are ordinarily monogamous species, establish harems, and attack and kill other males who invade the harem territory.

Individuals of both sexes frequently withdraw from society and emerge only when the other members of the community are asleep. With increased incidence of social withdrawal the number of behavioral aberrations increase. In their natural habitats and under normal population pressures wild animals, with virtually no exceptions, do not mutilate themselves, masturbate, attack their offspring, develop gastric and/or duodenal ulcers, form homosexual pair bonds, or commit murder. In laboratories and zoos, especially with crowding, all of these things occur. They also occur in crowded human environments. The parallel is unpleasant, particularly since we do not know what is optimum density for either animal or human population.

The only remedy for overpopulation is some form of control, and if that control is not exercised by rational and humane means, the less humane (but entirely logical) natural controls such as plague, famine, malnutrition and warfare will inevitably appear. If violent methods of population control such as war, suicide, and euthanasia are excluded, the only alternative to natural control is some sort of interference with the normal reproductive process. Considerable research is being done in this area. At present, two physiologic methods and one surgical method are available that will permit normal sex activity, but inhibit the production of offspring. The surgical method (vasectomy in males and salpingectomy in females) is most commonly employed in males, and has long been used by animal breeders and stockmen to produce infertile males that can be used to determine estrum in females. The major disadvantage of surgery is that there appears to be an increased incidence of seminoma (sperm cell tumor) in vasectomized males. Salpingectomy, because of the difficult surgery and essentially unprofitable results, is seldom performed on domestic animals and is a relatively rare operation in humans when compared with vasectomy.

Two physiological methods, one physical and one chemical, are commonly employed in females. The physical method involves introduction of a foreign body (usually a sterile plastic shape) into the uterus. This will prevent implantation of a fertilized ovum more than 85 percent of the time.

According to unconfirmed reports, this technique is not new. It was purportedly employed by horse owning communities in the Middle East for many centuries as a form of physiological warfare. Pebbles introduced into the uteri of brood mares of an enemy tribe would reduce the foal population and make the tribe more likely a victim than a victor in future wars. It was the antithesis of the other physiological warfare trick, i.e., surreptitiously impregnating enemy mares with the semen of inferior stallions, thus lowering the quality and combat effectiveness of future horse populations, which would also put the enemy at a disadvantage. It would be interesting if these reports could be confirmed, which would make the modern eugenics techniques of the intrauterine device (IUD) and artificial insemination "spinoff" from ancient biological warfare.

The chemical method, using contraceptive drugs ("the pill") has also been known for some time. Veterinarians have used progesterone to "recycle" hunting dogs and prevent estrus from occurring during hunting season, and to "adjust" reproductive cycles of sheep and cattle. However, the full effect of progesterone was not discovered until the drug was employed as an estrus inhibitor, which would prevent the formation of fertile ova. These studies have made "the pill" the most popular method of contraception in human beings. Recently some doubts have arisen as to the safety of this procedure, but the death loss of adult human females from pregnancy is about nine times that which might be expected from "the pill." Hence, relatively speaking, hormonal contraceptives can be considered to be safe.

Although these are the effective means available for mammalian population control at present, interesting new drugs are being developed or are under study. Among these is a diamine compound which blocks spermatogenesis. However, this compound seems to be contraindicated for human use since the one who uses it must completely abstain from alcohol. Ingestion of as little as one-half an ounce of ethanol by a diamine user can produce serious side effects. Another method being tried experimentally is vaccination of females in the hopes of producing antibodies against spermatozoa. So far, success has not been achieved. Other chemicals are under investigation, all aimed at either breaking the cycle of spermatogenesis or oogenesis, or making the uterus an inhospitable environment for sperm or ova. The earlier abortifacient drugs which were once used to abort pregnancies are all but passé except for heavy doses of synthetic female hormone (in dogs) to cause abortion of unwanted or unplanned litters.

At Boston University (1973), Dr. Charles Terner developed what may be a successful male contraceptive pill. In rat experiments the pill stopped spermatogenesis but did not inhibit sex drive. The pill is a mixture of androgen and progesterone, and apparently has a suppressive rather than a destructive effect upon spermatogenesis, since the experimental rats regained fertility when the drug was withdrawn. Further studies may still be in progress, but little has been published.

REPRODUCTION PATTERNS

Horses

The mare is a uniparous (monotocous) mammal and is seasonally polyestrus with a breeding cycle that normally occurs from March through July. The demands of horse breeders, however, have advanced the season to the early part of February as far as thoroughbreds are concerned. Generally, mares will not breed at this early date but are put to stud in order to conceive as early in the year as possible. This is because the rules of the Jockey Club, which govern the ages of horses, decree that all Thoroughbreds' birthdays fall on January 1 of the year in which the foal was born. Thus a foal born on December 30 would be classed as a yearling two days after birth. Since physical growth and muscular efficiency are at least partly a function of age, a foal born on January 2 would have a year's start insofar as official age is concerned over a foal born on December 30. And it is official age that determines the eligibility of horses for the stakes at the Kentucky Derby, Belmont, Saratoga, and Santa Anita. The system works to the detriment of normal breeding processes, which result in foals being born during the months from March through May. A more rational method of determining age, i.e., date of birth, has often been proposed, but has never been accepted for a number of reasons, most of which have to do with breeding tricks, the racing season and the geographical location of the largest and most influential stud farms.

The estrous cycle in mares is quite variable and is due principally to the duration of estrus or heat, which may range from as short as two to as long as 11 days. The average duration of estrus is approximately six days. Most mares have visible signs of heat, which include swelling of the external genitalia, tail switching, and postural attitudes. The period between the disappearance of one heat period and the appearance of the next is approximately 16 days, which gives the mare an average estrous cycle of 21 to 22 days with a range variation from 18 to 27. Ovulation in mares occurs one or two days before the end of estrus, which makes breeding a problem since there is so much variation in the duration of the heat period. Mares should be bred every three days during the time they are in heat, since sperm will survive in the uterus for that length of time. In the mare, a so-called foal heat occurs between four and 14 days after parturition. This period lasts for about 11 to 20 hours and has been considered by many breeders as the ideal time to rebreed the mare. Actually, this period should not be used for breeding since the uterus of the mare has not returned to normal. It is devitalized, and may contain infection resulting from mishandling or damage during the birth of the previous foal. Long term records kept on foal heat breedings indicate that a 60 percent conception rate is all that can be expected, and that of this number, a larger than normal number of foals will be aborted, small, deformed or otherwise damaged.

The ovary of the mare is a peculiar structure. As the filly grows older, the poles of the ovaries move inward and the ovaries become closely invested with a layer of connective tissue except for a narrow band or zone called the ovulation fossa. In mature mares, all ovulation takes place in this area. The size of the Graafian follicle of the mare is extremely large. Normal follicles have been recorded which contain up to 80 cubic centimeters of fluid. This may explain the frequency of diagnosis of "cystic ovaries" in the mare. Actually, the cysts are probably Graafian follicles. One case has been reported of a follicle seven inches in diameter. Upon rupture of the Graafian follicle there is usually considerable hemorrhage which fills the follicle cavity with blood and forms a large corpus hemorrhagicum.

The ovaries of the mare do not necessarily alternate in the production of ova during the breeding season. If pregnancy occurs, the corpus luteum of pregnancy will persist functionally for approximately 150 days of the 340-day gestation period. After this time the chorionic gonaotropin produced by the endometrium maintains the fetus within the uterus. Twinning is not a normal phenomenon in horses. It may occur, but a frequent result is that one or both of the fetuses will die in the uterus, or be deformed at birth.

Male equines are called studs, stallions, or colts; females are called mares or fillies, depending on their age. The suckling young, irrespective of sex, are called foals, and the act of parturition is called foaling. A castrated male is called a horse or gelding. A male with one or both testicles retained in the abdominal cavity is called a cryptorchid, rig, or ridgling. There is no specific term for a spayed (ovariectomized) female.

Popular names are almost always applied to species that are important to man and his environment. The general rule is that the more names used to identify a species, the more value (either esthetically or economically) the species is to man. Since animals such as water buffaloes, elephants and camels have never been intimately connected with life or development of the English-speaking people, there are no specific popular names for these species. However, this omission is more than adequately replaced in the languages where these animals are important factors in the economy.

Donkeys and Asses

Both females and males are considered to reach puberty at one year of age. The female is uniparous and seasonally polyestrus, with regular cycles that occur (usually) from March to August. The estrous cycle lasts from 21 to 28 days with the duration of estrus being from two to seven days. Ovulation is usually in the last third of the heat period. As in the mare, a foal heat occurs after parturition. The usual appearance of this phenomenon is two to eight days postpartum, but great variation can occur. The duration of foal heat is from two to six days. The female is uniparous. The duration of pregnancy is about 365 days but may be slightly less if the female is carrying a hinny.

Donkeys are more important for the production of horse-donkey hybrids (mules and hinnys) in Western countries than for use as a breed in farm or livestock operations. A variant of the donkey derived from the Spanish Ass is used as a pack animal in the United States and Mexico, and is called a burro. There is no essential difference between this variant and the ass.

Most observations on horses will apply to asses and donkeys.

The male is called a jack, the female is called a jenny or jennet, and the suckling young is called a foal. Parturition is called foaling. There is no specific term for a castrated male. Hybrids arising from matings of jacks and mares are called mules. Hybrids arising from matings of stallions and jennys are called hinnys. There are no specific names for male or female mules or hinnys. Male mules or hinnys are routinely castrated. Neither male nor female horse-ass hybrids are considered to be fertile, although occasional instances of mule or hinny fertility, always involving females mated to a jack, have been reported.

Cattle

The cow is a uniparous (monotocous) mammal, although multiple births (mainly twins) sometimes occur. Cows have a true polyestrus heat cycle. Puberty occurs usually at about eight to ten months of age. The heat interval averages 21 days with a maximum variation of from 18 to 24 days. The duration of heat will average between 10 and 14 hours depending upon the age of the animal. Cows show visible signs of heat, which include mounting or being mounted by other cows, swelling of the external genitalia and a vaginal discharge of clear mucus which may contain a few flecks of blood. Like the mare, the ovaries do not necessarily alternate in the production of ova. The right ovary has statistically significant increased activity over the left. Approximately one to five Graafian follicles develop during each heat period. Of these, one or more will mature and rupture the first day after heat has subsided. Breeding, therefore, should be during the latter phases of the heat period and in an emergency can be done after the subsidence of heat. Bovine sperm will survive for about 48 hours in the genital tract of the female. In cattle, rupture of the Graafian follicle does not result in hemorrhage. A corpus hemorrhagicum is not formed, and the ruptured follicle metamorphoses directly into a corpus luteum.

The corpus luteum of pregnancy persists in a functional state throughout the gestation period, which is an indication of a lack of chorionic gonadotropin activity. Ordinarily cows will show estrus 30 to 70 days after parturition. If the cow is nursing a calf, the onset of estrus will be delayed.

There has been a great deal of work done with cattle in an attempt to find a diagnosis for pregnancy other than rectal palpation of the uterus and fetus. Since the cow has among the lowest progesterone levels of any of the domestic animals, chemical tests that rely upon detection of this substance within the blood and urine have until recently been unsuccessful.

Tests have been developed that can measure progesterone levels as low as 1.0 nanogram per milliliter of substrate. These tests applied to cattle blood plasma have given accurate results in determining nonpregnancy in cattle and sheep, and 85 percent or better accuracy in detecting pregnancy. Similar tests applied to milk are better than 80 percent accurate in determining pregnancy and nonpregnancy.

Work has also been done upon vaginal and uterine epithelium and has resulted in a great deal of information, none of which, unfortunately, is of any use in practical pregnancy diagnosis. The cervical mucus has also been studied in the hope that the changes in the physical and chemical structure of this substance would offer a clue to whether or not the cow has conceived. Although there appears to be a correlation between the microscopic appearance of cervical mucus and ovarian activity, there does not appear to be a correlation between it and pregnancy. Although work has been reported on amnionic fluids in laboratory animals and humans, the techniques have little apparent application to normal livestock operations. Practical pregnancy diagnosis in cattle is still based upon rectal palpation wherein the examiner inserts the hand into the rectum of the cow and feels the ovaries, uterus and associated structures through the rectal wall to determine the presence of the fetus or conditions that are indicative of pregnancy.

Work with ovum transplants has become quite successful with survival rates of 85 percent. The technique involves surgical removal of ova from a donor cow at estrum, extrauterine fertilization, and transfer of the fertilized ovum to the uterus of a surrogate mother (usually a grade cow). In this way, high quality purebred calves can be born by low grade cows. Since a donor cow can produce up to 18 mature ova per year, which can result in 14 or more calves from surrogate mothers, the technique can rapidly improve a herd of cattle. There are obvious dangers in using this method to upgrade a herd, but the rewards probably warrant the risks and expense.

Males are called bulls; females are called cows or heifers depending on their age. The act of parturition is called calving and the suckling young are called calves. The castrated male is called a steer, a bullock or an ox. There is no specific term for an ovariectomized (spayed) female.

Sheep

The ewe is a uniparous mammal but may often bear twins or triplets. Most breeds are seasonally polyestrus with the exception of the Merino, Rambouillet, and Dorset Horn, which are true polyestrus. The breeding season for sheep occurs in the fall, usually between the first of September and the middle portion of December. This seasonal stimulus is apparently correlated with light and temperature, and perhaps with the fall alterations in the growth of plant life after the summer season. In the southern hemisphere the breeding season is also in the fall, yet this season corresponds to spring in the northern hemisphere.

Sheep have an estrous cycle of approximately 18 days with the duration of estrus averaging 36 hours. Unlike cattle and horses, sheep are capable of producing twins or triplets without reflecting or inducing abnormalities in the genital tract. Like cattle, the corpus luteum of pregnancy persists in a functional state throughout the gestation period, which lasts approximately 150 days.

It has been found that "flushing" ewes immediately prior to breeding will cause superovulation and the fertilization of more than one viable ovum. Flushing involves feeding of increased amounts of high quality feed, and in order to be effective must be associated with a gain in weight. Ordinarily, the process is not as effective in fat ewes as it is in lean. Usually the ewes are placed upon restricted diets before the beginning of the breeding season. At the start of the season they are then subjected to the "flushing" process. Thyroid extract or thyroprotein added to the diet has also produced successful superovulation.

The male is called a buck or a ram, the female is called a ewe, and the young are called lambs. The act of parturition is called lambing. Castrated males are called wethers. There is no specific term for an ovariectomized female. In English-speaking (in contrast to American-speaking) countries feeder lambs are called hoggets.

Goats

The nanny often has twins but is usually uniparous. Estrous cycles are seasonably polyetrus with a definite breeding season occurring in the fall of the year. The duration of heat, however, is longer than in the ewe, averaging about 48 hours, and the estrous cycle averages about 21 days. Four to five follicles develop within the ovary and of these, one or two may be fertilized. Twinning is normal and not uncommon in goats, otherwise the statements for sheep are generally applicable.

The male is called a buck or a billy, the female is called a doe or a nanny, and the young are called kids. The act of parturition is called kidding. There is no term for a castrated male or female.

Swine

The sow is a multiparous (polytocous) animal with a true polyestrous heat cycle occurring throughout the year, except during the nursing period when estrus may be inhibited for a variable period of time. The duration of heat is one to five days with an average of three. The estrous cycle ranges from 18 to 24 days with an average of 21. Puberty occurs in swine at about five months of age. At ovulation one or both ovaries may be involved and release 16 to 40 potentially fertile ova into the upper reproductive tract. Of these up to 16 may be carried to term although the normal litter size is about eight or nine. The remaining eggs, if they are fertilized, usually die in the early stages of embryonal development. It is not uncommon to find that embryonal or fetal death occurs throughout the pregnancy period as the uterus adjusts to the load of fetuses carried within it. Frequently there is a marked variation in the size of pigs at birth, with one or more from any given litter being "runts." Runt pigs, as a rule, will not develop normally and are an economic liability to the swine producer. Certain breeds appear to have greater fertility than others. Of these, Yorkshires are notable for their large litter size, but large litters are often accompanied by poor mothering ability of the sow, and in herds with suboptimal management the end results expressed in terms of live market pigs are about the same as for sows of other breeds.

The male is called a boar; the female is called a sow or a gilt, depending upon age. The young are called piglets up to weaning and shoats from weaning to maturity. Parturition is called farrowing. A castrated male is called a barrow if castrated young, and a stag if castrated after sexual maturity. The castrated female is sometimes called a gelding.

Dogs

The mating behavior of dogs is so well known that it is not considered extraordinary. Yet the copulatory lock or "tie" exists in relatively few species, and in dogs is a most unusual sort of union. After the male mounts and copulation begins, the penile bulb (bulbus glandis) becomes enormously distended within the bitch's vagina, producing the "tie." After a minute or two of conventional copulatory movements the dog dismounts, swings one hind leg over the bitch's back and turns away so that dog and bitch are facing in opposite directions. The two maintain this peculiar tail-to-tail stance, locked together for several minutes up to almost an hour until the distended penile bulb subsides and the animals can separate.

This act divides the mating of dogs into two distinct phases: the mounted, and the dismounted phase. Both appear to be necessary for successful breeding. The dismounted or "tie" phase is the most remarkable and unconventional of the two. In this movement, the axis of the dog's penis is bent through an arc of nearly 180°, and should be very uncomfortable to the dog although he exhibits no signs of distress.

The tail-to-tail position as yet has no completely adequate explanation. Theories ranging from helping ejaculation (which is notoriously poor in the dog) to the "popular opinion" expressed by Harrop (1960) that the stance makes dogs "less vulnerable to attack" have been presented. To date none has been experimentally confirmed, although the latter idea would argue an interest in and a dedication to sex above and beyond any reasonable level that might be attributed to dogs.

The bitch is a multiparous animal and is ordinarily diestrus, with heat periods occurring twice a year. Some individuals may have three estrous periods a year. As might be expected from the length of the period between heats, the duration of heat in dogs is quite long and is divided into two parts: a preheat or proestrum and a true heat or estrum. Proestrum lasts for approximately one week and is characterized by a swelling of the vulva, vaginal bleeding, increased excitement, and a tendency to be attractive to males. The true heat period lasts for at least a week, and it is in this period that the bitch will accept service. Ovulation usually occurs in the first or second day of the heat period. A peculiar fact about canine ovulation is that the ova are not mature at the time the follicles rupture. Neither the first nor the second polar body has been given off. It requires from two to three days residence in the oviduct before the eggs are ready for fertilization.

Fertilization in the dog is a rapid process. Spermatozoa arrive at the oviduct within 30 seconds after ejaculation, aided apparently by

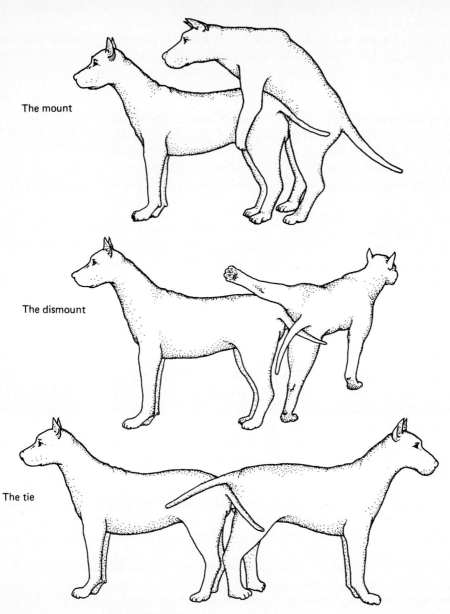

FIGURE 14-14 *Canine copulation.*

reverse peristalsis of the uterus. Ascent of the oviduct by sperm is somewhat slower, yet is fast enough that a receptive egg may be fertilized within an hour after breeding. This probably does not normally happen since the second maturation division of the canine ovum does not occur until the egg has been expelled from the ovary, and a certain amount of time would be required for the meiotic division, during which the ovum would not be receptive to sperm.

The fertilized and dividing ovum may not implant for several days. During this period the eggs tend to migrate and space themselves equally throughout the uterus. This results in an even distribution of implantations through the horns of the uterus irrespective of which ovary has been most active.

The duration of pregnancy is approximately 63 days. The corpus luteum persists throughout pregnancy and the uterus does not produce chorionic gonadotropin. Pregnancy is not noticeable during the first month. There is no appreciable change in body shape or weight of the pregnant bitch. During the second month, however, rapid changes in shape and weight occur. Increases of weight, as much as 16 or 17 pounds, have been recorded in large breeds. The weight increase is associated with increase in nutritional demand.

Mammary gland changes appear about five weeks after fertilization. These are most noticeable in bitches having their first litter and consist of enlargement of the glands and a reddening of the unpigmented skin around the nipples. Glandular growth becomes obvious after the seventh week of pregnancy and milk may appear shortly before term. This latter phenomenon is more common in bitches that have had several litters of puppies.

Litter size is variable and depends somewhat on the size of the dog and the breed. Toy breeds do not ordinarily have large litters. Of the large breeds, Irish Setters, are noted for large litters. Because of the many breeds of dogs, no average figures can be given. Certain breeds, e.g., the achondroplastic dwarf types such as Basset Hounds and Dachshunds, the megalocephalic types such as Boston Terriers, and the achondroplastic-megalocephalic types such as the English Bulldog and Pekingese, have considerable difficulties with parturition. Caesarian section is almost a requirement in English Bulldogs and may often be required among other breeds of unsual conformation.

Pseudopregnancy in the bitch is not unusual and may last for periods up to ten weeks. In prolonged pseudopregnancy, nest building, mammary enlargement, and lactation may occur. Abortions are relatively rare and are usually associated with uterine disease, trauma, or nutritional or endocrine deficiency. Postparturient eclampsia (hypocalcemic convulsion) is not uncommon, particularly in bitches that have whelped large litters. Treatment is by intravenous injection of soluble calcium, usually calcium borogluconate.

Imminent parturition is indicated by restlessness and nest-making activity. The bitch appears very conscious of her hindquarters. Abdominal distress, panting, and vomiting are also seen. The relaxation of the sacrosciatic, sacroiliac and pelvic ligaments gives the hindquarters a shrunken appearance.

Labor commences with involuntary uterine contractions and cervical relaxation and dilation. Voluntary contractions commence as the fetus is introduced into the cervix, and these are mainly the abdominal press. Puppies generally appear head first, although this position is not mandatory. Puppies are born enclosed in the amniotic membrane with the umbilical cord still attached to the placenta. The bitch removes the membrane, severs the cord, and stimulates the newborn puppy by licking it. The placenta is usually shed within 15 minutes after birth of the puppy, and under ordinary circumstances is promptly eaten by the bitch. Retained afterbirths are more frequent in toy breeds. Subsequent puppies appear at intervals ranging from 30 minutes to several hours. Total whelping time will normally range from 8 to 12 hours.

Most bitches prefer privacy; some desire human attention. In either case, the desire of the bitch at this time should be respected.

Males are called dogs, females are called bitches, parturition is called whelping, and the young are called pups or puppies. There is no specific term for the castrated male or female.

Cats

The queen is a multiparous animal and is diestrus or seasonally polyestrus with two or more breeding seasons occurring per year. The duration of heat is about four days with a two-week interval between heats. This may explain the so-called constant heat, which is seen in cats. The owners affirm that the queen is continuously in season when, in actuality, they have merely seen the separate heat periods at two-week intervals. Ovulation is apparently stimulated by breeding. Fertilization normally occurs early in the second day after mating, and implantation occurs two weeks later. As in the horse, a corpus hemorrhagicum is formed, and the corpus luteum of pregnancy does not persist for the entire duration of the 65-day pregnancy period. After the 49th day the corpus luteum becomes nonfunctional and the maintenance of the fetuses in the uterus becomes the responsibility of the chorionic gonadotropin secreted by the uterine epithelium.

Cats usually reach sexual maturity between seven and 12 months of age. Most cat breeders delay breeding the queen until she is one to one and one-half years old. Toms begin to show sexual behavior at about 11 months of age.

In the presence of a tom, estrus lasts about four days, and about 10 days if the male is not present. The cycle recurs at intervals of from 14 to 21 days.

Estrus activity in queens includes playful rolling. Excessive rubbing is observed, together with a curious low call two to three days before estrus. An "estrus crouch" is commonly seen, wherein the hindquarters are elevated and the queen makes treading movements with her legs. Sexual behavior is exaggerated in the Siamese cat and has been mistakenly considered to be nymphomania.

Queens should be bred at least twice. If not bred, they will continue to show sexual behavior for several days, and this behavior may occur over a period of several weeks as successive heats ensue.

Ovulation is stimulated by coitus and occurs 24 hours after mating. The egg reaches the uterus in five days. Gestation lasts 57-69 days (average is 63-64 days). Queens should be limited to two litters per year. Maximum litter numbers occur between two to eight years of age. Litter size is smaller in young and old females. Litters exceeding eight kittens normally cannot be nursed successfully.

Abortions are more common in cats than in dogs and usually occur after four to five weeks of gestation. Pseudopregnancy may take place after a nonfertile mating and lasts 30 to 40 days. Estrus will usually occur a week after the termination of pseudopregnancy.

Labor lasts about 45 hours, with the first stage taking 12-24 hours. First stage is characterized by a sharp drop in body temperature and restlessness. The queen seeks a secluded place to make a nest. In the second stage, birth occurs. Placentas are usually expelled with the fetuses or within 24 hours after the kittens are born. The queen will usually eat the afterbirth. Kittens are born like dogs, surrounded by the amniotic membranes which the queen removes. The queen stimulates the kittens to breathe by licking them. There is usually a 10 to 60 minute rest period between births. Kittens are commonly delivered head first, but there appears to be no anatomic or physiologic requirement for this type of presentation.

Male cats are called toms, females are called queens, and the young are kittens. There is no specific term for the male or female castrate.

Camels

There is a bit of scatologic doggerel about the sex life of the camel that is known to almost every collector of illicit verse. It begins as follows, "The sexual life of the camel is stranger than anyone thinks," and this is more true than the writer of the verse ever dreamed.

Among the domestic herbivores, which are essentially harem-keepers, the camels have an unusually tolerant attitude toward sex. Instead of being acquisitive sex tyrants and intolerantly possessive male chauvinists, bull camels have a remarkable restraint toward each other and toward a she-camel who is in season. Both sexes appear to be remarkably unbiased. If several males are present they will not fight each other for possession of sex privileges, but will instead, form a line and quietly, without fuss, each take their turn at servicing the female as long as she will permit this activity, which is usually for an hour or two.*

Moreover, the camel is the only one of man's domestic animals that breeds in a kneeling position. The entire spectrum of the camel's reproductive activity is worthy of comment from any viewpoint, and it is too bad that more is not published on this interesting animal. Unfor-

---

*For a first-hand description of this unusual aspect of a camel's life-style see *Circus Doctor* by J. Henderson DVM, Little Brown & Co., 1951.

tunately, camels do not fit particularly well into a modern economy and are on the decline throughout the world.

The female is uniparous, twins being an event of considerable rarity. She has an irregular estrous cycle of 10 to 20 days, a duration of estrus of from one to seven days, and a gestation period of approximately 370–440 days. A post-partum estrus is present which occurs from 24 to 72 hours after calving. Despite regular estrous cycles there is a spring rutting season. Both males and females have glands on the neck below the occipital region, which are active in the spring or early summer, and are apparently associated with rut.

Males are called bulls or he-camels; females are called cows or she-camels. The young are called calves. There is no term for a castrated male or female. This probably arises from the fact that camels are normal inhabitants of Moslem nations, and a Moslem does not practice spaying or castration of domestic animals.

Fur-Bearing Animals

The captive fur-bearing animals as a general rule have but one heat season a year. Most are multiparous. Foxes and mink are monestrus, and have one breeding season in the spring and a gestation period of approximately 55 days. A notable exception to this is the marten, which is a relative of the mink. It, too, has only a single breeding season, but this season occurs in the fall. The kits are produced in late spring, which gives this animal a gestation period of better than 150 days. Actually, the true gestation period of the marten is about 55 days, since the embryo develops to the morula stage in the fall and then passes into a state of suspended animation, which persists until spring when the gestation period is resumed. The phenomenon of delayed implantation or suspension of intrauterine growth is not unique with the marten. A number of other species possess this trait. Among the more notable of these are the marsupials, where suspended development of the zygote is the rule rather than the exception.

Rabbits

The male rabbit is unusual anatomically since the testicles are located anterior to the penis. Structurally this should present some copulation problems, but practically rabbits do very well in reproducing their species.

The female rabbit will breed at almost any time of the year. There is no apparent estrous cycle. Ovulation occurs only after breeding or from strong mechanical or hormonal sex stimulation. Pregnancy lasts from 30 to 32 days, and litter size may be as high as 14 to 16 although six to eight are considered normal. The corpus luteum of pregnancy lasts throughout pregnancy if conception occurs, but disappears at about 16 days if it does not. With the involution of the corpus luteum (either at 16 days or at the end of the pregnancy period) the female builds a nest and lines it with fur pulled from her breast. False pregnancy (pseudopregnancy) can only be told from true by the time lapse between breeding and nest building and, of course, by the fact that no young are produced. Newborn rabbits are blind, naked, helpless, and unusually small in size compared with the size of their parents.

Females and males come to sexual maturity (puberty) from six to nine months after birth. Rabbits born in the spring take longer to mature than those born in the fall. The best time for reproduction is from the first of May to the first of July. During this period, numbers of follicles mature—except for secretion of liquor folliculi—and then regress. This maturation occurs in waves about seven to 10 days apart. If breeding occurs, the follicles grow rapidly by secretion, and retention of follicular fluid and rupture will occur eight to 12 hours after breeding. The stimulation that produces this secretory activity is apparently nervous, but involves the brain rather than the local receptors. Mounting by another female will cause follicle maturation and rupture, and anesthetizing the lower reproductive tract prior to breeding will not prevent it.

Does nursing large litters will breed and conceive if they are bred soon after parturition, but the embryos will not survive past the blastocyst stage. Usually lactation results in suppression of breeding activity and regression of uterine and ovarian activity. After weaning, sexual potency is quickly restored. Buck rabbits seem to recognize that suckling bunnies are the source of this lack of receptivity by the doe, and in the wild or in a free association state, the buck will usually kill the bunnies if he can reach them.

Male rabbits are called bucks; female rabbits are called does; young are called bunnies; parturition is called kindling. There is no specific term for a castrated male or female.

Guinea Pigs

The guinea pig is probably the smallest mammal domesticated by man for food. In South America, guinea pigs are kept by the Quechua Indians in their dwellings as a handy source of fresh meat. Since animal protein is relatively scarce, the guinea pig is an important item in the native diet.

The female guinea pig is multiparous and is true polyestrus with a 14- to 18-day estrous cycle and a duration of estrus of about half a day. Pregnancy lasts from 59 to 72 days and from two to four large and precocious offspring are produced per litter. Ovulation is spontaneous and occurs toward the end of estrus. The corpus luteum of pregnancy is functional throughout gestation. The vagina is normally closed by a membrane which opens at heat (estrus). Puberty occurs in the male from 60 to 70 days after birth and depends somewhat upon the diet. In the female, minimum breeding age varies from 30 to 45 days. It is best to wait until about 80 days before mating young animals.

Males are called boars, females are called sows, parturition is called farrowing, and newborn young are called piglets. There are no special terms for castrated males or females, young females, or nurslings.

Rats

Rats come to puberty from about 45 days (males) to 75 days (females) of age. They are true polyestrous animals but, in a natural environment, will not breed during the winter months. The estrous cycle is complete in four to five days, with a heat period that ranges from 10 to 20 hours and ordinarily begins in the late afternoon or evening. The gestation period is usually three weeks, although longer periods have been recorded. Both ovaries are involved in the production of ova during any given heat period. Ovulation is spontaneous and commences from 8–10 hours after the appearance of heat. Heat is demonstrated by postural attitudes, rolling, and acceptance of the male—or better—by examination of vaginal smears for typical cells. About 10 ova will be shed at ovulation. Breeding should be done from one to four hours after the onset of estrus, but it is a better practice to let the male rat pick the appropriate time since certain courtship rituals are inherent in rat matings and should be observed.

Sperm are deposited in the cervix and uterus, and take about one hour to reach the uterine end of the oviduct. Fertilization, however, requires considerably longer, and occurs about seven to 10 hours after mating, in the proximal third of the oviduct. At about the 18th day of gestation the female will begin nest building. Newborn rats are small, naked and helpless, and depend upon milk for their total nourishment for the first 10 to 12 days of life. They are ordinarily weaned by the mother about three weeks after birth. Lactation interferes with estrus, but the female will return to estrus within four days after her pups are weaned or removed.

Rats in captivity are polygymous with one male serving about four females. This trait is not noticeable in wild rats with normal population densities. The only special name for rats is for the neonates, which are called pups.

Mice

The pattern for rats is generally applicable to mice with the following exceptions: the age at puberty is 35 days, heat begins late in the evening to early in the morning (10 p.m. to 1 a.m.), and about six ova are shed. The gestation period is about 20 days. Copulation time can begin immediately upon onset of heat, and fertilization occurs about two hours after mating. Estrus is usually determined by vaginal smears. There are no specific common names for mice except for neonates, which are called mouselings.

Hamsters

The golden hamster is the most common variety of the hamster group and has a reproductive pattern similar to the rat. Both the four-day estrous cycle and the 16-day gestation period are the shortest among the mammals routinely used by man. Like rats and mice, heat begins in the evening and lasts about six hours. Litter size is variable, but should average about six to eight baby hamsters. There are no specific common names for either adults or babies.

Marsupial Reproduction

The act of parturition, which is quite spectacular in mammals, can be a bit odd in some of the protomammalians, animals that seem to be the linkage between the egg-laying reptiles and the placental mammals. These evolutionary bypaths are represented by the monotremes, which lay eggs, and the marsupials, which have a uterus but are nonplacental. These latter animals usually carry their developing young in an external pouch (certain small South American marsupials do not have a pouch).

The uterine phase of the gestation period in marsupials is quite short. In the Red Kangaroo it is about 34 days. The pouch phase is about five months, giving a total gestation and maturation period of about six months before the baby kangaroo (called a "joey") becomes a completely living individual.

The kangaroo at birth is little more than an embryo with well-developed forequarters. It is less than an inch long and weighs approximately one gram. This tiny organism must climb a distance of eight to 12 inches from the urogenital opening to the mouth of the pouch, find and affix itself to a teat, and maintain an unbreakable hold upon the teat for at least 30 days if it is to remain alive.

During the last few hours of the uterine phase of the gestation period the female kangaroo cleans the inside of her pouch and assumes a sitting position, with her back against a tree or a rock and her tail extended between her outthrust hind legs. This position serves to bring the mouth of the pouch as close as possible to the urogenital opening. Some marsupials like the North American opossum lick a moist path from the genitalia to the pouch, but this act seems to be omitted by the kangaroo.

The joey appears suddenly, preceded by a small amount of fluid, and at once begins the hand-over-hand ascent up the hair stream to the pouch. The hairs point in the direction of the pouch, which may act as a guide for the embryo. This guide may be aided by the scent of the pouch as the embryo has well-developed olfactory organs, which may aid it in locating its future home.

Once inside the pouch, the infant attaches to one of the four teats. Its mouth becomes rigidly fixed to the nipple, which swells to form a tight bond. This bond between embryo and teat must remain intact for at least a month. If it is broken, the infant cannot retrieve the teat and starves to death. For this 30-day period the baby kangaroo is entirely dependent upon the teat and since it cannot suck, milk must be forced into its mouth by the muscles that surround the mammary gland.

Although the joey can separate from the teat and reattach after 30 days in the pouch, it usually remains attached for about two and a half months, and does not emerge from the pouch for about a month after that. Three and a half months after entry into the pouch the baby kangaroo begins to look at the outside world. Two months after that the youngster emerges for short periods to feed, returning for transportation and milk. It always enters the pouch head first and then turns around inside, which is probably uncomfortable for mama. About six months after initial entry (at mama's insistence) the joey leaves permanently. The mother is about ready to give birth again and needs to clean her pouch for the new boarder. Right up to the day of departure the joey continues to suck the teat to which it was originally attached.

Similar, but shorter gestations are characteristic of other marsupials.

Since kangaroos come into estrus about every 35 days, and may conceive as soon as the new joey enters the pouch, delayed embryonal development is mandatory. At conception the ovum begins to divide in normal fashion until the blastocyst stage is reached a day or two after fertilization. Development then stops and the cell mass becomes quiescent. This resting state appears to correspond with lactation and lasts as long as the embryo in the pouch continues to absorb a full supply of milk. Once the developed joey begins to eat solid food (or if death occurs) milk consumption drops, and the blastocyst in the uterus begins to develop. About 30 days later a new embryo is born.

chapter 15
The Common Integument and Its Derivatives

GENERAL CONSIDERATIONS

The common integument is the protective covering of the body and is continuous at the natural openings with the mucous surfaces of the digestive, respiratory, and urogenital tracts. It consists of skin, hair, hooves and claws, horn, feathers, and other epidermal derivatives. Associated with the common integument are sweat and sebaceous glands, and numerous sensory nerve endings. It functions as a protective envelope, as a secretory and excretory mechanism, as a sense organ and a temperature regulating device, and as a respiratory structure.

EPITHELIUM, MESOTHELIUM, AND ENDOTHELIUM

The outermost layer of the protective covering is called epithelium, mesothelium, or endothelium, depending upon its location in the body and its ontogenic derivation. Covering membranes are derived from either the so-called outer, middle, or inner of the three germ layers laid down in early embryonal development (i.e., ectoderm, mesoderm, or endoderm). External or surface epithelium is of ectodermal derivation. Epithelium of the digestive and respiratory systems is derived from mesoderm.

462
THE COMMON INTEGUMENT AND ITS DERIVATIVES

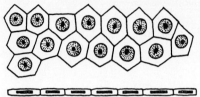

1. Squamous

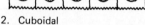
2. Cuboidal

3. Columnar

4. Ciliated columnar

5. Stratified columnar

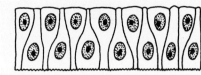
6. Pseudostratified columnar

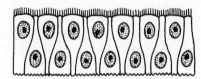
7. Ciliated pseudostratified columnar

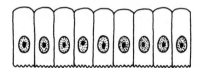
8. Three-layered columnar

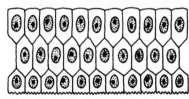
9. Transitional

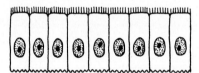
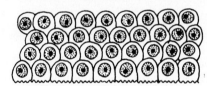
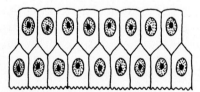
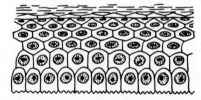
10. Stratified squamous

FIGURE 15-1   Types of epithelia.

Mesothelium and vascular endothelium comes from endoderm. Most of these structures have been discussed at some length elsewhere in these pages, but it seems wise to include them as a group. It is always a problem where to incorporate something that doesn't quite fit either structurally or functionally into a predetermined format. Covering membranes fall into this category. Whether they should be included as a part of the discussion of cells, as part of specific systems, or given a separate listing as something unique proposes a problem.

EPITHELIUM, MESOTHELIUM, AND ENDOTHELIUM

There are seven types of epithelium that are generally recognized, i.e., squamous, cuboidal, columnar, pseudostratified columnar, transitional, stratified squamous, and stratified columnar.* Two of these types, columnar and pseudostratified columnar, may bear cilia on their free surfaces.

Since the ultrastructure of cells has been discussed at length elsewhere, there is no point in repeating it here. The gross and microscopic appearance, however, is relatively untouched and can stand some further elaboration.

Squamous, or simple squamous epithelium is composed of flat scale-like cells that are extremely thin when viewed edge on, and irregularly polygonal (usually hexagonal), when viewed at right angles to their free surface. Morphologically, the squamous (flattened) form of cell is not only characteristic of certain epithelia, but is the only form in which mesothelium and endothelium occurs in adult mammals. Squamous epithelium is found covering the lens of the eye, the inner wall of the membranous labyrinth of the ear, the parietal layer of Bowman's capsule, the descending limb of the loop of Henle, and the epithelium of the rete testis.

Mesothelium, which is squamous in its developed state, covers the serous membranes of the thoracic, abdominal and pelvic cavities. Embryologically it is columnar and may bear flagella which may persist. At birth the mesothelium is already flattened and eventually becomes extremely thin with irregular interdigitated borders. As a tissue, it is somewhat primitive and destroyed adult mesothelial cells can be replaced by fibroblasts. This phenomenon is not found in other covering membranes. Mesothelium also has the protean capability of changing into fibroblasts in the event of inflammation.

Endothelium forms the lining of the vascular system and the lining of the subdural and subarachnoid spaces in the brain, the perilymphatic spaces of the ear, and the anterior chamber of the eye. The latter structures (excluding the vascular) are sometimes called mesenchymal epithelium, a term that is inherently incorrect since epithelium does not come from mesenchyme.

Cuboidal epithelium consists of a single layer of cells with a roughly cuboidal form. Ordinarily the borders of these cells appear relatively straight under light microscopy and the cells themselves appear squarish in cross section. The pigmented epithelium of the retina is the cuboidal epithelium that most closely approaches squamous in general morphology, although it bears fringe-like processes on its free surface. The less exotic forms of cuboidal epithelium are found in the secretory and excretory portions of glands, in the choroid plexus in the brain, in the organ of Corti, in the duct systems of the kidney and in the germinal

---

*Stratified columnar is often called stratified cuboidal. In some classifications stratified cuboidal is listed as an eighth category. Additionally, some texts list ciliated columnar, ciliated pseudostratified columnar, and three layered columnar epithelia as separate entities.

epithelium of the ovary. In adult male mouse kidneys, cuboidal epithelium is also found in Bowman's capsule. A two-layered cuboidal epithelium is often encountered in the uterus and in the duct system of skin glands. When cuboidal epithelium is associated with alveolar or acinar glands it can be distorted into a pyramidal or conical form that has been called pyramidal or glandular epithelium.

Columnar epithelium consists of roughly cylindrical or prismatiform cells that are taller than they are wide. The nucleus tends to be found in the deeper portions of the cell, and the free surface may bear cilia or striations which are actually microvilli. Columnar epithelium is found in the lining of the gut, in the gall bladder, the excretory ducts of the kidney, in many glands, and in the sensory epithelium of the olfactory region. Ciliated columnar epithelium is found in the upper respiratory tract, in the head of the epididymis, the oviduct, and the paranasal sinuses. Ciliated columnar cells can also be found in the ependyma of the cerebral ventricles.

Pseudostratified columnar epithelium is usually ciliated. It consists of cells that all connect to a common basement membrane, most of which (but not all) reach the free surface. The nuclei tend to lie at different levels, giving a stratified aspect to the tissue. It may be difficult at times to distinguish between stratified and pseudostratified columnar epithelium. The presence of cilia is a major basis for differentiation. Pseudostratified columnar epithelium is found in the respiratory tree from the nasal cavity through the bronchi, in the epididymis and in the oviduct.

Transitional epithelium is a structure that is relatively unique and lies somewhere between simple and stratified epithelium. The cells in the deeper layers tend to resemble a stratified low columnar epithelium, while those near the free surface appear to be highly variable in shape, ranging from spherical through cuboidal to squamous, although the squamous are never as flat as those found in either simple or stratified epithelia. There is a basement membrane present, but it is so thin and tenuous that it is frequently overlooked under conventional microscopic examination. It is, however, quite apparent in electron photomicrographs. Transitional epithelium is found in the ureters, urinary bladder and upper urethra. It is capable of marked distension and can be as few as one or two cells thick or as many as eight or ten.

Stratified squamous epithelium consists of cells organized roughly into three layers (basal, middle, superficial) that become progressively flatter as they approach the free surface. There is a distinct basement membrane upon which lie the basal layer of low columnar or cuboidal cells with oval nuclei. These basal cells may connect with deeper tissue layers via delicate cytoplasmic fibrils. The basal cells are overlaid by the middle layer cells, which are mainly polyhedral and possess spherical nuclei. These cells possess fine rodlike or thorny processes which sometimes cause them to be called "prickle cells." The processes, together with intracellular bridges, give the middle layer a homogenous aspect.

In the skin epithelium, cytoplasmic fibrils known as tonofibrils may extend through the cytoplasmic processes and connect several cells. They are thought to be supporting structures. The superficial layer becomes progressively more flattened as the cells near the free surface. The nuclei are flattened and the outermost cells are frequently cornified. The epithelium where cornification occurs is called keratinized stratified squamous epithelium. If the outer layers are shed and cornification does not occur, the epithelium is called nonkeratinized. The cornified cells scale off and are lost as new cells are pushed upward from the basement membrane. Stratified squamous epithelium is tough, flexible, and relatively impervious. It forms the epithelium of the skin, the mouth, esophagus, upper stomach, anus, vestibule, vulva, glans penis, lacrimal canal, bulbar conjunctiva, and cornea. Usually it occurs wherever there is severe mechanical or chemical wear. Squamous metaplasia or conversion of other epithelia to a squamous or stratified squamous type can occur in areas of prolonged irritation. Many investigators consider this to be a forerunner of cancer.

Stratified columnar epithelium consists of an outer layer of columnar cells, underlain by one or more layers of conical or fusiform cells whose apices are directed toward the free surface. The lowest layer adheres to a basement membrane. This epithelium is relatively rare and may be found in the excretory ducts of glands, in the palpebral conjunctiva of horses, in the nongravid uteri of some species of mammals, in the lacrimal sac and nasolacrimal duct, and in discrete areas of the distal urethra. If the cells of this epithelium are sufficiently low, the term stratified cuboidal has been used to describe it. It is doubtful if there is any difference in the two, since one form can blend into the other in a single organ or structure.

Another type of epithelium is found in the secreting areas of glands. This is glandular epithelium, and is discussed in greater detail under glands of the skin (p. 473). The cells are essentially cuboidal or low columnar, but may, depending on the surface they cover, assume a pyramidal or conical shape. The principal difference from ordinary epithelia is in physical properties. Glandular epithelium secretes rather than protects. The manner of secretion determines the type of gland according to its physiological activity, i.e., merocrine, where none of the cell contents are lost; apocrine, where part of the cell contents are lost; and holocrine, where the secretion consists of all, or virtually all, of the cell. One must differentiate between glandular epithelium and secretory cells. Glandular epithelium is, of course, composed of secretory cells, but these cells are organized into, or are a part of, a covering layer, usually enclosing a lumen, alveolus, or other secretory space. Secretory cells *per se* do not have to be organized into epithelium in order to function, but glandular epithelium must be either a covering layer or intimately associated with one.

Glandular cells are ordinarily displaced from free surfaces to deeper locations and communicate with the surface by ducts which are lined with epithelium. The glandular cells may be either unicellular or multicellular. Unicellular glandular elements are usually interspersed with covering epithelium and are typified by the so-called goblet cells found in the gut, respiratory tree, and conjunctiva. The cell is connected to the basement membrane by a tapering stem, which contains the nucleus and undifferentiated cytoplasm. The bulbous superficial portion of the cell is filled with a scant matrix that supports pale secretory granules which contain mucin. The granules are relatively unstable and in the presence of imbibed water coalesce and exude mucus at the free surface.

Multicellular glandular epithelium is more diverse and is usually aggregated into definitive glands, which may be either endocrine or exocrine. A basement membrane is present together with a number of structures such as canaliculi, crypts, pits and vesicles. Special cells are called "basket cells," and myoepithelial cells may aid secretory activity by their contractions. Generally, these glands produce serum and/or mucus. The glands that produce both serum and mucus are known as "mixed" glands. However, serum and mucus are not the sole products of glandular epithelium; other chemical substances ranging from simple compounds, such as sodium chloride to complex molecules, such as the proenzyme trypsinogen, are also produced.

Still another form of epithelium is associated with the nervous system. These are the neuroepithelial cells, which act as sensory receptors. They tend to be elongated cuboidal or fusiform cells and usually occur in clusters. The basal portions of the cells are connected to, or are part of, nerve fibers. The free surfaces end in rods or hairlike processes. Neuroepithelial cells are found in the olfactory region, in the taste buds, the retina, and the organ of Corti.

THE SKIN

The skin is the largest organ of the body. Its thickness and closeness of attachment to underlying structures varies in different animals. It is tough, resilient, and highly elastic. The skin is attached to underlying structures by subcutaneous tissue. This consists mainly of fascia (a form of fibrous connective tissue) and fatty tissue. The skin is composed of two layers; the epidermis and the corium. The epidermis is an avascular superficial layer, which is subdivided into the corneum (a hard, dead, cornified external layer of skin), and the deeper stratum germinativum, which is live, moist, and contains pigment. The corium is the deep skin layer and consists of a superficial feltwork of white and elastic fibers containing the corpus papillare (the superficial layer of corium which is folded into papillae), and the deeper tunica propria, which is a loose network of interlocking bundles of fibrous tissue. There is no clear line of demarcation of the subcutaneous tissues from the deeper layers of the tunica propria.

The skin receives blood from small arteries, which form a network in the deeper parts of the corium. Small vessels and capillaries go to the sweat and sebaceous glands, hair follicles, and papillae and form a dense network at the junction of the corium and epidermis. The veins form two plexuses, one beneath the papillae, another at the junction of the corium and subcutis. Lymph vessels form subcutaneous and subpapillary plexuses. The nerves vary greatly in number in various parts of the skin. They may end free in the epidermis or corium, or form special nerve endings of various types, such as Meissner's corpuscles. Pacinian (lamellar) corpuscles, or Golgi-Mazzoni corpuscles.

THE APPENDAGES OF THE SKIN

The appendages of the skin are modified epidermal structures and include hair, horns, hooves, claws, and other structures.

Hair covers almost the entire surface of the body of domestic animals. The hairs are directed in such a way as to form hair streams and vortices (points at which hair streams converge). Special hairs are known by special names, such as tactile hairs around nose and lips, eyelashes, vibrissae in the nostrils, and tragi in ears. A hair consists essentially of a root, bulb, and a shaft. The shaft is external; the root and its expanded portion, or bulb, is embedded in a skin depression called a follicle. The hair is supplied with oil by sebaceous glands, and some hairs (notably those along the backline of dogs) are supplied with erectile muscles known by a variety of names i.e., piloerectors, arrectores pili, arrectores pilorum. These muscles are composed of smooth muscle fibers and receive their nerve supply from the sympathetic nervous system. They function in some mammals as a defense against cold by erecting and thickening the pelage and increasing the dead air space and insulation. This effect, in man, is responsible for "goose pimples." The hairs are ineffective, but the reflex remains. In other animals, notably cats and dogs, certain hairs erect in response to emotional states, such as fear and anger. Horns, hooves, and claws consist of closely packed epidermal cells that have become cornified. They grow from a specialized corium which supplies the cells with nutrients.

It seems almost lesé majesty to leave the subject of hair with such bald treatment in these days of self-expression. Hair, actually, is a unique group of structures. Besides having specialized applications and functions, it identifies the class mammalia. No other animals have hair. Hair has long been thought to have magical powers, probably because it is an outgrowth of the body, yet is easily and painlessly separated from the body (finger and toenails have the same qualities and the same occult significance). Absence of hair, particularly on the scalps of male humans, has considerable societal and cultural and hormonal importance and has given rise to a whole field of medicine known as trichology.

In both humans and animals, hair has assumed a certain public health significance, not only for the bacteria, viruses and other minute forms of life that can be carried on it, but also (in humans at least) because of the danger of fire. The American Chemical Society's committee on laboratory safety has solemnly warned that hair fire may become a frequent and serious laboratory accident because of the flammable lacquers and dressings that are used, and the amount of hair (and beard) involved.

Oddly enough, humans have almost as much body hair as apes, but the individual hairs are so short and fine that the misleading statement "the naked ape" has come to mean a human, when a more truthful and scientifically accurate term would be "the fine-haired ape."

Man, the atypical mammal, commonly suffers from physiological baldness. The usual form, pattern baldness, may also rarely occur in other animals and appears to be associated with androgen production, i.e., the more androgen, the less hair. Baldness is a part of masculine life, and has never been particularly pleasant to those who suffer from it, even though it correlates with advantageous attributes. Egyptian tomb paintings, Mexican cave drawings, statuary, and a Roman legion marching song "honoring" Julius Caesar ("Homeward comes the bald-head lecher; Romans lock your wives away."), all indicate that the "billiard ball" look has been a part of human physiognomy since time immemorial. Women lose hair, too, but they seldom become bald.

Hair treatments have evolved from nostrums, such as the ancient Egyptian mixture of pulverized dogs' paws, asses' hooves, and dates cooked together in oil, percolated through a filter of linen cloth and rubbed into the hairless scalp, to modern hormone creams, which aren't modern after all. Paracelsus in the 16th century concocted a hair lotion from blood of a human afterbirth, blood of a murdered baby, and viper's oil. One of the latest lotions for combatting falling hair and seborrhea contains mucopolysaccharides derived from fetal structures such as umbilical cords and amniotic fluids.

Trichology, the science of hair, is now a recognized branch of medicine. (The rivalry between barbers and physicians was probably the reason why trichology is a johnny-come-lately.) With science now in the saddle, there seems to be some hope for the future. Although alopecia (falling hair) affects both man and animals and is still stubbornly resistant to research, the androgen-related baldness may well be a curiosity in a few more years.

The Hoof or Claw

The hoof or claw is the horny covering of the distal end of the digit. In the horse it is divided into four basic parts: wall, sole, frog, and periople. The wall is that part of the hoof visible when the foot is placed on the ground. It covers the front and sides of the foot and at the "heels" (or rear of the foot) is bent sharply inward and forward to form the bars.

On its inner surface the wall is lined with closely spaced plates of horn called horny lamellae. These fit into sensitive lamellae produced by the corium. Both the sensitive and horny lamellae have secondary laminae upon their surfaces, which also dovetail into one another. It is here that the nutrition of the wall takes place. The growth of the hoof begins at the coronary band (or coronet). The outer part of the coronet is covered with a thin brownish layer of horn called the periople. The horn grows outward from the coronet. On the ground surface the wall is united to the sole by horn of lighter color and softer texture. This area is known as the white line or zona lamellata. In shoeing a horse, care should be taken to drive the nails outside the white line as this marks the boundary of the sensitive lamellae of the corium, and a nail placed inside this line will penetrate the sensitive lamellae and lame the horse.

The sole forms the greater part of the ground surface of the hoof. It is somewhat domed or concave and fills the space bounded by the wall and bars. The part of the sole between the bars and wall is called the angle.

The frog is the wedge-shaped mass of soft horn which fills the space between the bars. The middle of the ground surface of the frog

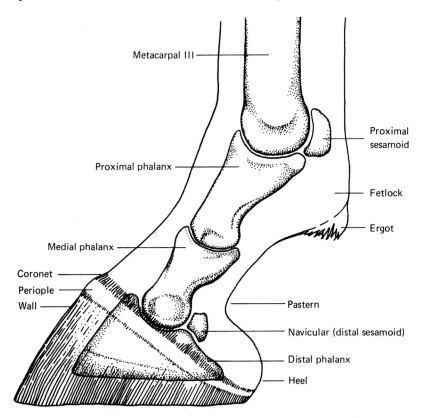

FIGURE 15-2   *Parts of the hoof of the horse.*

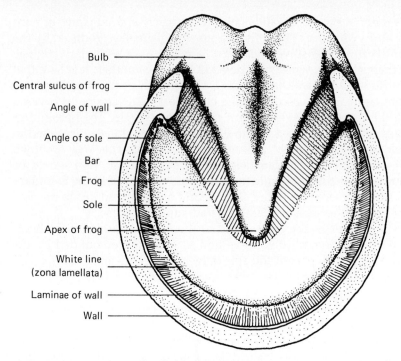

FIGURE 15-3   *Plantar surface of the hoof.*

bears a central depression called the central sulcus, which is continued through the frog dorsally as the spine of the frog (frog stay). Above and around the frog, covering the so-called heel, is a soft mass of fibrous tissue, the digital cushion.

On the lateral and medial aspects of the hoof, inside and somewhat dorsal to the coronary band, are two plates of cartilage (lateral cartilages). The structures of frog, digital cushion, and lateral cartilages form a kind of pump and are important in maintaining the blood flow to the foot region. As the horse places weight on the foot, the frog is forced into the bars, expanding the heel of the foot. The digital cushion is compressed against the frog and lateral cartilages and forces blood out of the foot and corium. Lifting the foot releases the pressure and allows blood to flow into the corium. The hoof is well supplied with blood by the digital arteries. The veins form a complex anastomosing plexus within the corium before leaving the foot as the digital veins.

The ruminants (cow, sheep, deer, etc.) do not possess such well-developed digital coverings as does the horse. In these animals the horny material is usually called the claw. In the cow, the claw is composed of the wall, sole, and periople. The periople, unlike that of the horse, is very extensive, covering the coronary border with a band about half an inch wide except at the heels, where it covers the entire surface and blends with the sole. There is no distinct line of demarcation between the periople and sole as there is in the horse, and there is no frog. Two accessory digits or "dewclaws" are also present in some ruminants. These are covered with conical capsules of horn, which are somewhat similar in structure to the horny coverings of the claws. The claws of sheep, goats, and antelope resemble those of the cow except that the dewclaws are absent. The claws of the pig resemble those of the cow except that the sole is smaller and is well defined from the bulbs, which form the greater part of the ground surface. The claws of the accessory digits are better developed than in the cow, and their parts are more recognizable.

The claws of the dog, cat, and other digitigrade animals are composed of a wall (unguis) and a sole (subunguis). The wall is compressed laterally and strongly curved in two directions (downward and forward, and from side to side). The sole blends with the wall and is indistinguishable from it except in the anterior portions of the claw.

The claws of most primates are composed of an unguis (wall) only, and are generally known as "nails."

Chestnut and Ergot

In horses and other equidae, two horny masses called the chestnut and ergot are found on each leg. The chestnut is found on the medial aspect of the forearm about halfway between the carpus and elbow, and on the medial aspect of the hindleg just below the hock. The ergot is found on all four legs, embedded in the hair on the caudal aspect of the fetlock joint. The ergot is usually small, but the chestnut, particularly on draft animals, can be quite large and prominent. There have been speculations that these horny structures are remnants of the second and fourth digits, which long ago disappeared functionally in the evolution of the equine leg. There is no horny relic of the fifth digit.

Horns

Horns are epidermal derivatives and as such are properly classed as a part of the common integument. Like hoofs, they are considered to be modified hairs bound together into a solid mass by keratin. Structurally, horns are markedly similar to hooves or claws both in composition and manner of growth. They are found on members of the ruminant ungulate genera such as cattle, buffalo, sheep, goats, and antelope.

The horn itself is composed of numerous tubules bound together into a solid mass. The horn wall is very thin at the base, but toward the apex it thickens and forms a solid structure, which is considerably harder and more dense than the basal portion. At the base, the horn is covered by a layer of softer keratinized tissue, similar to the periople of the hoof or claw. Underlying this structure is found the corium, which is relatively thick and has long slender papillae on its surface that invade the base of the horn. The horn is supported for a variable distance by the "horn core" or processus cornus, which is a bony outgrowth of the frontal bones of the skull. The horn core is generally hollow in mature animals and communicates directly with the frontal sinuses.

With the exception of one species of antelope, the American Pronghorn (*Antilocapra americana*), the following are characteristic of horns:

1. Epidermal origin, similar to hair and hooves
2. Grow from a corium
3. Possess internal nutrition and blood supply
4. Possess a bony supporting base or "horn core"
5. Not necessarily sex-linked, present in females as well as males in many species
6. Not shed annually, grow throughout life

The pronghorn does not fit this classification completely. It does possess horns, but these are shed annually, and are replaced by a new undergrowth which forms around the horn core and pushes the old horn off.

Antlers

Strictly speaking, antlers have no place in a discussion of integumental structures, since they are of dermal origin and more closely related to the skeleton than to the skin. However, for purposes of comparison, it is convenient to locate them in this section.

Antlers are characteristic of members of the deer family and possess the following characteristics:

1. Dermal origin, similar to bone
2. Grow from a germinal center
3. Possess external nutrition and blood supply while growing, via the "velvet"
4. Do not possess a "horn core"; the antler itself is a bony protuberance of the skull
5. Sex-linked, usually found only in males
6. Shed and replaced annually

The process by which antlers grow and are shed is extremely interesting and well covered by texts on mammalogy and zoology. Essentially, the antlers grow from a germinal center located on the frontal bones of the skull. This area is covered with skin which grows along with the horn and is known as the velvet. During the growth period the velvet is richly supplied with blood vessels and nerves, and both it and the underlying antler are soft, spongy, and sensitive. When the definitive annual growth is completed, a ring of connective tissue develops around the base of the antler, cutting off the circulation to the velvet, which dies and is shed. A band of columnar osteoid cells forms at the base of the antler and separates the developed structure from the skull. With the velvet dead, the antlers lose further power of growth, the cells die, and the antler becomes extremely hard and dense. The osteoid cells at the base, however, remain alive and form a firm connection between antler and skull. Later on the osteoid cells also die, forming a brittle band which breaks easily, allowing the antler to be shed. Generally, the presence of antlers is related to the male sex. Exceptions to this can be found in all species. In reindeer and caribou antlers are regularly found on females. One species, the giraffe, possesses antlers that are permanently in the velvet. Giraffe antlers do not shed and grow slowly throughout the life of the animal.

THE GLANDS OF THE SKIN

The skin of mammals is richly supplied with glands. Most of these are two kinds; sweat glands and sebaceous glands.

Sweat glands (sudoriferous glands) are usually simple tubular or coiled tubular structures. They secrete a watery fluid containing various salts and waste products of metabolism. Thus, they may be classed as organs of excretion. A second function of these glands is to cool the skin by evaporation, and aid in temperature control. These glands are merocrine, i.e., none of the cellular material is lost while the gland is performing its function.

Sebaceous glands are compound alveolar structures. They furnish an oily secretion for lubrication of the skin and hair. They are of the holocrine type in which all of the secreting cell is lost while the gland is performing its function. Modified sebaceous glands exist in the form of scent or musk glands, which are found in the anal region (skunk, mink, and civet cat) and around the hock (deer). These are present either for defense or sex attraction. A second type of modified sebaceous gland, the meibomian or tarsal glands, are found in the eyelids. These secrete an oily substance which prevents the overflow of tears across the free margins of the eyelids. A third type, the ceruminous glands, are found in the ear. These secrete a heavy waxy material.

The Mammary Glands

The mammae or milk glands are specialized secreting glands that identify the Class Mammalia. They are present in both male and female mammals. The glandular epithelium is composed of true secretory cells which utilize the bloodstream as a substrate. The structure is compound alveolar and similar to sebaceous glands, but the secretion type is apocrine (only a part of the cell is lost) and is more similar to sweat glands.

The activity of the milk glands is periodic and normally is related to parturition and occurs only in the female. Functional activity in the male (gynecomastia) is rare in all orders except monotremes. The same hormones that govern the growth and function of the uterus also affect the growth and function of the mammae. Milk production in most species is relatively brief, lasting only until either a succeeding pregnancy, or until reduced demand by the offspring results in a cessation of milk flow. However, in individuals bred for milk production, lactation can be continuous. In some species (horse, cow, rabbit), lactation tends to suppress ovulation and inhibit estrus. Ultimately, however, normal estrus cycles are restored, milk production declines, the offspring are weaned, and the glands involute. However, involution is not complete and the glands increase in size with successive pregnancies.

The mammary glands appear along epidermal ridges on either side of the median ventral line. These ridges, the so-called milk lines, extend from the pelvic to the pectoral region. The position along these ridges at which the mammae form is dependent upon the species. Humans, other primates, and elephants ordinarily have two mammae in the pectoral region. Most other mammals have mammae in the abdominal and/or pelvic region. Rodents may have mammae located along the entire length of the milk line. The number of glands is relatively species specific, but variations in number are not uncommon. Variation occurs most frequently in multipara, particularly in species which have several pair of glands located principally along the abdominal portion of the milk line.

In domestic mammals, the following locations and numbers are most often present:

Animal	Number	Location
Horse	2	Pelvic
Cow	4	Pelvic
Water Buffalo	4	Pelvic
Camel	2	Pelvic
Llama	2	Pelvic
Sheep	2	Pelvic
Goat	2	Pelvic
Pig	10-14	Abdominal
Dog	8-10	Abdominal
Cat	8-12	Abdominal
Rabbit	8-10	Abdominal
Human	2	Pectoral
Elephant	2	Pectoral

The size and shape of the mammae vary between species, from the neatly tucked glands of the mare to the grossly enlarged pendulous udder of the dairy cow. Considerable variation in size and shape may occur within a species.

The teats or nipples of the milk glands fall into two general types: true teats and false teats. The true teat occurs in primates. In this type the entire glandular area of the mammae is lifted above the general body surface and the ducts of the gland open at the tips of the nipple. The false teat occurs in carnivores, swine, horses and ruminants. In this type the glandular area is depressed and the skin is drawn out to form a nipple. A system of collecting ducts plus a milk sinus or cistern is found associated with false teats. Supernumerary or "extra" teats may be found in all animals, but the condition is most often encountered in cattle.

THE BOVINE UDDER

The bovine udder is not really typical of the udders of most domestic animals since its structures are tremendously enlarged and overdeveloped. However, this makes them easy to see on dissection, and the glandular and duct elements, other than being larger, are not markedly different from those of other animals. The bovine udder will be presented here as the type structure, since specimens are readily available from local abattoirs.

Essentially, the bovine udder consists of four parts, or quarters, each completely separate from the others and furnished with an individual duct system and teat. Small to large amounts of fatty tissue are present in each quarter. The quarters are combined to form a single mass called the udder. Extra or supernumerary teats (hyperthelia) may be present. The udder is located in the pelvic region and is suspended medially between the hind legs by two systems of ligaments (medial and lateral). The medial ligament is elastic but the lateral ligaments are not. A system of vessels and nerves ramify through the udder tissue giving it a rich blood and lymph supply and an adequate nerve supply.

Microscopic investigations have shown that the nerves to the udder consist of cerebrospinal and sympathetic fibers. There is apparently no parasympathetic nerve supply. The cerebrospinal nerves consist mainly of sensory fibers derived from the I, II, III, and IV lumbar nerves and the perineal nerve. The I and II lumbar nerves supply the anterior parts of the udder. The III and IV lumbar nerves, plus a collateral branch from II, make up the inguinal nerve, which passes through the inguinal canal together with the mammary artery and vein, and supplies the bulk of the voluntary innervation to the teats and gland substances of the udder. The perineal nerve supplies the caudal parts of the udder and sends a branch to the supramammary lymph node. It does not supply the teats. The sympathetic division of the udder's nerve supply arises from the sympathetic ganglia in the lumbosacral region and passes to the udder as postganglionic fibers.

The blood supply is derived mainly from four mammary arteries, which are branches of the two external pudic arteries (right and left). Each mammary artery supplies a quarter. A small amount of blood may be received by the cranial quarters from the caudal abdominal artery,

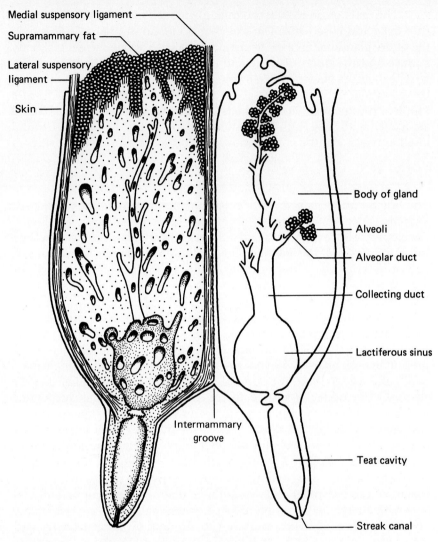

FIGURE 15-4   Schematic drawing of the udder.

and by the caudal quarters from the perineal artery (which is not always present).

The veins that drain the udder form a circle at the base of the udder and move blood toward the heart along the four main channels: right and left subcutaneous abdominal veins and the right and left external pudic veins. The subcutaneous abdominal (milk) veins are formed during a cow's first pregnancy by union of the cranial and caudal superficial epigastric veins and their subsequent enlargement under the stress of lactation. The two subcutaneous abdominal veins pass forward along the ventral line, penetrate the thoracic region at the "milk well," and empty into the cranial vena cava. The four mammary veins unite to form the two external pudic veins, which pass through the inguinal ring together with the external iliac veins, and empty either into the external iliac veins or into the caudal vena cava.

The lymphatics of the udder drain into two (right and left) very large supramammary lymph nodes, which in turn connect with the caudal abdominal lymphatics. The supramammary lymph nodes are dorsal to the udder and are located near the external opening of the inguinal canal.

The structure of the gland substance of the udder is relatively simple. A framework of fibrous tissue (stroma) supports the alveolar gland substance (parenchyma). The alveoli drain through alveolar ducts, which are aggregated to form collecting ducts. The collecting ducts empty into a large roomy space, the lactiferous sinus or cistern. The lactiferous sinus empties into the teat cavity, which in turn empties into the lactiferous duct or "streak canal."

PHYSIOLOGY OF LACTATION

Morphologically the mammary glands are derivatives of the skin, yet these structures are so closely associated physiologically with reproductive function and are so vital to the normal development of mammalian offspring that, from the physiological point of view, they should be classed as accessory reproductive organs. Functional mammaries in the male sometimes result from hormone disturbances, which usually arise from medication. The condition is called gynecomastia. Mammary glands ordinarily do not develop to any great extent in the female until puberty. At that time the glands enlarge, but most of this enlargement consists of connective tissue and fat rather than glandular tissue. With each ovulation or "heat period," there is a slight enlargement and development of the glandular and duct tissue, which recedes incompletely and is again stimulated at the following estrum. However, this enlargement is minimal and the gland does not approach a functional state until the animal becomes pregnant.

Growth of lactating tissue is dependent upon at least two hormones, estrogen from the Graafian follicle and progesterone from the corpus luteum. During pregnancy, the constant secretion of progesterone causes mammaries to develop glandular tissue, and by the time of parturition they are capable of producing milk. Milk production is a true secretory process since milk differs in structure and composition from the blood from which it is derived. In general, milk contains proteins, fat, sugar (lactose), and inorganic salts. It is slightly acid and has an osmotic pressure somewhat greater than that of blood. Ordinarily most of the milk is present in the gland prior to milking or sucking. Some secretion occurs during the milking period, but this is small compared with the total volume of lactation.

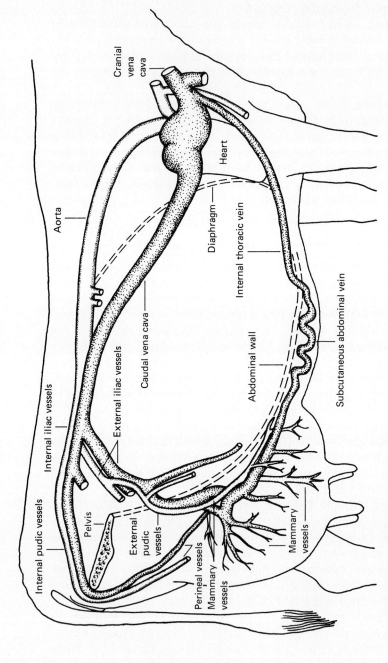

FIGURE 15-5   *Circulation of the bovine udder.*

The following table gives an approximation of normal milk composition of four common domestic animals.

Milk Composition	% Water	% Protein	% Fat	% Sugar	% Minerals
Cow	87.0	3.5	3.8	5.0	0.7
Mare	89.5	2.5	1.5	6.0	0.5
Ewe	83.0	5.4	6.5	4.3	1.0
Sow	79.5	6.2	7.7	5.6	1.0

The origin of milk fat is not definitely known. Two schools of thought exist as to its derivation: one believes that it is derived from phospholipids and the other from fatty acids transported by the blood. To date there is no explanation of why high fat levels occur late in lactation. Milk fat is mainly in the form of triglycerides composed of short chain fatty acids in the four to 12-carbon range. These short chain fatty acids are not found in the storage fat of mammals, and are probably formed from acetate in ruminants and from glucose in nonruminants. Such long chain fatty acids as are found are probably derived from fatty acids transported in the bloodstream. The protein fraction is mainly casein, a phosphoprotein secreted and synthesized from blood-borne amino acids. Albumin and globulins are present also in small amounts and are presumably derived from plasma proteins. Galactose is specifically produced by the mammary gland and is derived from blood glucose. Other constituents are taken directly from the bloodstream and include vitamins and minerals.

Milk can be a deficient dietary substance. Sow's milk lacks sufficient iron to satisfy piglets' needs, and cow's milk can be lacking in copper and cobalt. Mare's milk is low in every constituent except sugar. This appears to be a characteristic of all equidae. Cows, contrary to general opinion, do not secrete the richest milk among the domestic animals. Sow's milk contains a larger percentage of proteins, fats, and other solids. In nations where domesticated water buffalo are kept for meat, milk, and motive power, these animals produce commercial quantities of milk that is higher in nutrient levels than milk from any other domesticated mammal. Yet the 6½ to 8½ percent fat content in the milk of water buffalo is low compared to the milk fat of aquatic mammals, which is in the vicinity of 30 percent.

A number of cellular constituents will normally be found in milk. These include epithelial cells, nuclei, and leukocytes. All forms of leukocytes may be seen in normal milk. However, their number should not exceed 125,000 per cubic centimeter. Any increase in leukocytes over this amount is indicative of infection, or disturbance involving the glandular apparatus.

The act of lactation is largely under the influence of hormones, although direct mechanical stimulus acting through the sensory nerves has some effect. Although in physiological experiments, the inguinal nerves in cattle have been cut without appreciable loss in milk yield, the effect of sensory stimulation should not be discounted in the letdown of milk. The hormonal control of lactation is probably entirely due to adrenocorticotrophic hormones and oxytocin, both of which are secreted by the pituitary. There is considerable evidence that the hormone balance involved in lactation varies from one species to another.

Oxytocin apparently has the function of promoting the letdown of milk. This process is a bit more complicated than it appears at first glance, since animals under the influence of emotional stress can retard or withhold milk production. Withholding is due primarily to adrenal secretion and is the result of the hormone's effect on the vascular and nervous systems of the udder. Erection of teats and ease of milking in cattle is controlled by blood sinuses at the base of each teat, which become engorged at milking time. Under stress the lactating female secretes adrenalin which causes blood vessel vasoconstriction that inhibits engorgement of the sinuses surrounding the alveoli. Adrenalin also affects sympathetic nerve synapses which, in the udder, can result in contraction of the smooth muscle surrounding the alveolar ducts. A combination of these factors, or possibly vasoconstriction alone—which would deprive the glandular cells of oxytocin—can cause a drop, or cessation of milk production.

The duration of lactation varies between species of animals and in most breeds of domestic animals depends upon the needs of the offspring. In dairy cattle, however, the length of the lactation period and the amount of milk produced is far in excess of the requirement of any calf. Ordinarily, milk production declines as the offspring begin to consume feed, until finally the mother will reject further sucking. In dairy cattle, however, milk production can be continuous and may be very difficult to stop prior to the next parturition. In dairy cattle, in the interests of both the cow and the calf about to be born, milk flow should be interrupted for a period of at least three weeks prior to parturition. This allows the gland sufficient rest to produce colostrum which supplies essential protein fractions and antibodies the calf needs for survival. It also allows the cow some rest from the strain of lactation to prepare for the stress of parturition.

Colostrum is the first milk produced after parturition. It differs from ordinary milk in composition, color, and odor. It is necessary for the survival of those fetuses that have five or more tissues separating the fetal and maternal bloodstreams in the placenta (calves, lambs, foals, piglets). The thickness of the placental barrier prevents the fetus from absorbing antibodies and globulins from the maternal bloodstream, which are necessary to protect the newborn during the first weeks of its free-living existence until it can manufacture its own antibodies. Colostrum contains up to 25 percent protein, the greater part of which is gamma globulin. Other substances include vitamins and maternal antibodies.

During the first 12 hours of life, the intestinal tract of the newborn animal is readily permeable to globulins and other large protein molecules. After 24 hours, the intestine becomes refractory and no longer permits their passage. In the event that colostrum is not produced (a condition which occasionally occurs in dairy cattle), some of the proteins can be made up by the addition of egg white to the milk fed to the newborn calf.

chapter 16
The Organs of Special Sense

Certain specialized structures exist in all animals that function to give them an awareness of their environment and aid in survival. Mammals possess four definitive organs that fall into this category: the organs of sight, hearing (and equilibrium), smell, and taste. These are commonly lumped together as the organs of special sense. All have one characteristic in common: they are sensory extensions of the brain and are directly connected to it by means of nerve trunks. Other than that, the organs differ markedly from each other in structure and function.

LIGHT RECEPTION

Although eyes are the principal organs of light reception, sensitivity to light may not always be correlated with sight. Many blinded or eyeless animals can respond to photic stimuli, i.e., electromagnetic radiation in wavelengths between 350 and 700μ, which includes the so-called visible spectrum and all rays that stimulate phototropism in plants.

Photosensitivity may extend over much of the body and many biological scientists think that all pigmented spots on the body (particularly of lower animals) are light receptors. This "dermal light sense" has not been associated with specific sensory receptors, but has been shown to be present in a number of animal species. Some life forms with a "pineal window" and larval fish, whose spinal cords are exposed to light, give definite responses to photic stimulation. It has been shown that the pineal body in fish is involved in the pigment effector system and that if the pineal is removed the daily rhythm of color change in *Ammocoetes* is interrupted. In *Sphenodon*, a primitive lizard, the pineal has also been shown to function as a light receptor. In *Daphnia*, a tiny fresh water animal, the eyes function as a means of orientation to light, but kinetic responses depend on dermal light receptors. However, the dermal receptors are several thousand times less sensitive than the developed eye, although it appears that the photochemical systems responsible for extraocular light sensitivity are similar to those in the eye.

Studies of "dermal light sense" in man and laboratory animals have revealed some interesting data. Blind persons have abnormal water retention, and there are strong indications that vision, fluid balance, and fat deposition are interrelated in humans and other mammals, and that the dermal light sense in lower animals has a direct bearing on complex muscular activity and speed and direction of movement.

THE EYE

The eye, or organ of sight, is perhaps the most highly developed of the organs of special sense. It receives stimuli in the form of light and, by means of a complex nervous structure called the retina, changes these stimuli to nerve impulses which are sent directly to the brain along the optic nerve. The brain interprets these nerve impulses as pictures.

Two eyes are present in all normal mammals. These are situated in bony sockets or orbits, which are located in the skull at the junction of the bones of the cranium and the bones of the face. Two flaps of skin, the eyelids, serve to close or cover the eye. The free edges of these folds support special hairs, or eyelashes, which act as protective screens and dust filters. In domestic animals eyelashes are generally found only on the upper eyelids.

Specialized sebaceous glands called Meibomian (tarsal glands) are located at the base of the hair follicles of the lashes of the upper lid. These glands communicate by small ducts to the inner margin of the lid. The fatty material they secrete prevents tears from spilling over the margins of the eye.

The inner surfaces of the eyelids are covered with a thin epithelium, the conjunctiva, which also covers the anterior* surface of the eyeball and cornea. The part which covers the eye is known as the ocular conjunctiva.

---

*In modern veterinary anatomic terminology, the eye is the only structure in the animal body where the terms "anterior" and "posterior" can properly be used.

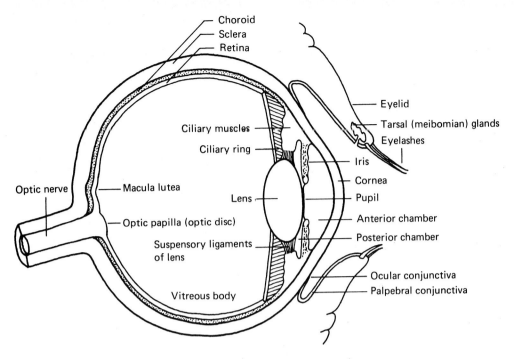

FIGURE 16-1   Schematic drawing of the eye (sagittal section).

Located in the medial canthus or corner of the eye is a transverse sheet of thin, translucent membrane, the "third eyelid" (membrana nictitans). This structure varies in extent with the species, ranging from rudimentary in the human to extensive in the horse. In most mammals the third eyelid is supported by a bar of hyaline cartilage, and bears along its base an aggregation of lymphoid tissue known as Harder's gland, which apparently functions to reduce the incidence of ocular infections.*

The lacrimal or tear glands are almond-shaped bodies located beneath the ocular conjunctiva somewhat dorsal to the lateral canthus of the eye. They secrete a serous fluid which moistens, lubricates, and flushes the conjunctiva of the lids and eyeball. The excess tears drain toward the medial canthus where they are collected by the lacrimal sac, which drains into the nasal cavity via the nasolacrimal duct.

---

*The name of the gland of the third eyelid is subject to a great deal of variation, depending to some extent upon who has written the book. The term "Harder's gland" or "Harderian gland" is essentially incorrect when applied to animals, since the Harderian gland is a human structure, and has a different cellular composition and secretion than the essentially lymphoid structure found in domestic animals. Harder's (or Harderian) gland, however, has been more or less sanctified by usage and will be used here with the implicit understanding that this gland in domestic animals differs from the structure found in humans.

The eye is located in the superficial part of the orbital cavity. It is protected in front by the eyelids and conjunctiva and in its middle by the orbital ring, and is related behind to fascia, fat, ocular muscles, and the optic nerve. It is spheroid in shape and composed of two spheres of different size. In the human the average sizes are as follows:

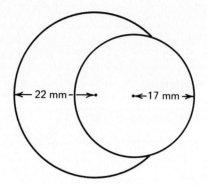

Anterior	Transparent segment	Radius 17 mm
Posterior	Opaque segment	Radius 22 mm

The eye can be divided into three major parts: the fibrous tunic, the vascular tunic, and the nervous coat or retina, which is connected directly to the optic nerve. There are three additional structures: the lens, the aqueous humor, and the vitreous humor.

The Fibrous Tunic

The fibrous tunic comprises the external coat of the eye and is composed of an opaque posterior part called the sclera and a transparent anterior part called the cornea. The sclera is a dense, fibrous membrane which forms about four-fifths of the fibrous tunic. It is generally white but may have a bluish tinge in its thinner parts. The cornea forms the anterior one-fifth of the fibrous tunic. It is transparent, colorless, and nonvascular. It is oval in outline and its outer margin blends smoothly with the sclera at the corneal-scleral junction.

The cornea is composed of five definitive layers which are, from the outside inward, the corneal epithelium, Bowman's membrane, the substantia (tunica) propria, Descemet's membrane, and the corneal endothelium. Ideally, it is of uniform thickness. It serves two purposes: an outer protective membrane for the anterior part of the eyeball and a light transmitting device.

The Vascular Tunic

The vascular tunic lies internally to the fibrous coat. It is composed of three parts and contains blood vessels, intrinsic muscles, and tissue structures of the eye. The parts are the choroid, the ciliary body and the iris.

The choroid is a thin opaque membrane that lies between the sclera and the retina. It is usually loosely attached to the sclera, but is intimately adherent at the point of entrance of the optic nerve. The inner surface is in contact with the layer of pigmented cells of the retina, which adhere so closely to the choroid that they once were regarded as a part of it.

The ciliary body is the middle portion of the vascular tunic. It is a ring of tissue that separates the transparent and opaque segments of the eye and surrounds the lens. It is roughly triangular in cross section and its base is connected to the periphery of the iris. The hypotenuse lies against the sclera, the elevation against the vitreous body, and the apex contacts the choroid. The ciliary body is divided into three parts: the ciliary ring, the ciliary processes, and the ciliary muscles.

The ciliary ring consists of a smooth band of tissue and has delicate meridional folds on its inner surface. The union of these folds forms the ciliary processes. The ciliary processes, which are very numerous, form a series of radial folds that surround the lens and act as points of attachment for the suspensory ligaments (zonular fibrils) of the lens. The ciliary muscle, or muscle of accommodation, occupies the outer part of the ciliary body, and lies between the sclera and ciliary processes. It consists of a circular band of involuntary muscles. Contractions of the

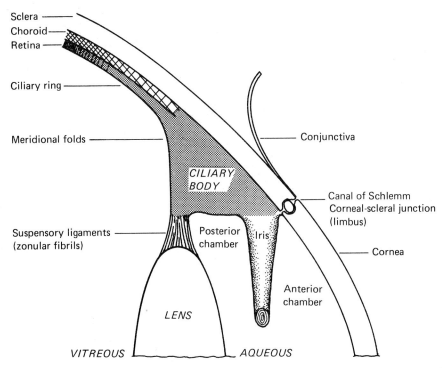

FIGURE 16-2   *Ciliary body relationships—schematic.*

muscle cause the ciliary ring and processes to be projected forward and slacken tension on the suspensory ligaments of the lens. Relaxation of the muscle tightens the suspensory ligaments.

The iris forms the anterior of the vascular coat and may be seen through the cornea. It is a muscular diaphragm that incompletely divides the space anterior to the lens into anterior and posterior chambers. It is perforated centrally by the pupil. The periphery of the iris is attached to the base of the ciliary body. The free border forms the margin of the pupil. The color of the iris when viewed from the front varies in the individual and gives the color to the eye. The iris contains smooth muscle tissue which is arranged in two distinct masses; the sphincter iridis, a band of circular muscle that surrounds the pupil; and the dilator iridis, which is composed of radially arranged muscle fibers that radiate outward from the sphincter to the periphery. Contraction of the sphincter iridis decreases the diameter of the pupil, while contraction of the dilator iridis enlarges the pupil opening. The muscles are antagonistic to each other and are respectively innervated by the parasympathetic and sympathetic parts of the autonomic nervous system.

The Nervous Coat or Retina

The retina is a thin neuroepithelium, which is closely attached to the choroid and extends from the optic disc to the border of the ciliary ring. The posterior portion is more complex than the anterior, and supports optic nerve fibers that converge at the optic disc. It gradually diminishes in thickness and loses its nerve elements as it extends toward the ciliary body. The inner surface is smooth and in some animals has a small circular yellow spot, the macula lutea, near the posterior pole of the eyeball. A similar spot in horses, the area centralis retinae, is situated above the point of entrance of the optic nerve. The optic nerve and blood vessels form a conspicuous spot of light color where they enter the vitreous chamber. This spot is the optic papilla, or optic disc. The optic disc is not covered with retina, cannot receive light stimuli, and is called the "blind spot."

The Chambers of the Eye

The lens and iris divide the transparent segment of the eye into two chambers, an anterior chamber and a posterior chamber. The posterior chamber is bounded in front by the iris and behind by the lens and ciliary processes. The anterior chamber is bounded in front by the cornea and behind by the iris. These chambers communicate by means of the pupil, and contain a watery fluid called the aqueous humor. The aqueous humor is a clear fluid consisting almost entirely of water and is almost perfectly transparent.

The Refractive Media

The refractive media are the vitreous body and the lens. The vitreous body fills the space between the lens and retina, is in close contact with the retina and acts as a support for it. Fresh vitreous is a semifluid, highly transparent, jelly-like substance composed principally of water. Anteriorly the vitreous has a cup-shaped depression which conforms to the posterior surface of the lens.

The lens is the most important part of the refractive apparatus of the eye. It is a transparent biconvex body suspended from the ciliary body by the suspensory ligament of the lens. The anterior surface is related to the iris; the posterior surface fits into the fossa on the anterior surface of the vitreous. The lens is enclosed in a transparent elastic membrane, which is called the lens capsule. The convexity of its two surfaces is not the same; that of the posterior surface being greater than the anterior. These convexities continually change in the living animal with the variations in tension upon the suspensory ligaments, giving variations in focal length to accommodate for near or distant vision.

The Muscles of the Eye

The eye is moved in its socket by seven muscles, six of which are arranged in antagonistic pairs, while the seventh serves to withdraw the eye slightly into the orbit to allow the remaining six to act with more freedom and precision. The muscles are as follows:

Muscle	Action
Dorsal Rectus	To move the eye vertically upward
Ventral Rectus	To move the eye vertically downward
Medial Rectus	To move the eye horizontally and medially
Lateral Rectus	To move the eye horizontally and laterally
Dorsal Oblique	To roll the eye axially and upward
Ventral Oblique	To roll the eye axially and downward
Retractor Oculi	To retract the eye into the orbit

The Optic Nerve

The optic nerve is purely sensory and extends from the receptors in the retina to the anterior corpora quadrigemina of the brain. Two general forms of optic transmission exist, one wherein the right optic nerve extends from the right eye to the left corpora quadrigemina and from the left eye to the right corpora quadrigemina. The nerves cross at a point slightly rostral to the tuber cinereum in most mammalian brains. This point is called the optic chiasma and nerves that cross completely are called completely decussate nerves. Animals with complete decussation of the optic nerve usually have periscopic vision or independent vision with each eye. They do not possess stereoscopic vision. The second form of optic transmission is where the optic nerves converge at the optic chiasma but do not decussate completely. Nerve fibers of the right eye, for example, go to both the right and the left corpora quadrigemina. This form of optic transmission is characteristic of animals that possess binocular and stereoscopic vision.

The optic nerve extends from the posterior aspect of the eyeball through the optic foremen in the skull and lies along the floor of the brain. It crosses the opposite nerve at the optic chiasma and disappears into the tissues of the diencephalon lateral to the chiasma. It is a large, well-formed nerve trunk. The cytons are peripheral and are located in the retina. The nerve trunk itself is composed of hundreds of thousands of sensory axons. In the human, a cross section of the optic nerve has been examined and the individual axons counted. A total of 1.2 million fibers were found to be present.

PHYSIOLOGY OF THE EYE

The eye is essentially a mechanism for the reception of light rays of the visible spectrum and for their translation into neural volleys, which are interpreted by the brain. Actually, it is the brain and not the eye that "sees," although for practical purposes the eye may be considered to do so.

Only a small part of the visible spectrum is capable of stimulating the retina of the eye. Wavelengths vary from the infrared (thermal portion of the spectrum) to the ultraviolent (actinic) portion. The actual range of the radiant spectrum is considerably greater than that which can be interpreted by the nerve cells of the retina. Only a narrow band ranging from 4,000 to 7,200 Ångstroms produces effective stimulation.

Refraction

When light travels through a transparent medium of uniform composition such as air, it moves at a constant velocity and in a straight line. But when it falls at an angle upon the surface of another transparent medium of different density, a part of the light is reflected and the rest travels through the new medium in a different direction. This bending of the light ray is called refraction. The index of refraction depends upon the density of the medium through which light passes. The medium used as a base for establishing refractive indices is air. Glass, for example, has a greater physical density than air, and consequently possesses a higher refractive index.

Lenses

A lens is a curved refractive medium. When parallel rays of light strike the curved surface, they will deviate in varying directions depending upon the degree and direction of curvature. A line perpendicular to the transverse axis of the lens and passing through the center of the curvature is known as the optic axis. Light passing along this line will not be refracted. In a lens of uniform curvature light rays parallel to the optic axis will be refracted more and more as they approach the periphery of the lens. If, for example, the refractive medium is a uniform convex curve, the parallel rays of light will be refracted so that they converge upon a point. The distance between this point and the lens is the focal length. Focal lengths vary with the curvature of the lens: The greater the curvature, the shorter the focal lengths. If the surface of the refractive medium is concave, the rays of light will be dispersed along a cone-shaped pattern whose apex is the center of the lens. In the eye, the optical formation is based upon convex refracting media and hence, measurements of the focusing power of the eye are expressed in diopters. A diopter is a focal length of one meter. A lens with a focal length of one-half meter would have the power of two diopters, or one with a focal length of two meters would have the power of one-half a diopter.

If the light from an object is made to fall upon a convex lens and a screen is placed the correct distance behind this lens, an inverted image is thrown upon the screen. Similarly, the eye, which also contains a convex lens and a screen (the retina), will produce inverted images of objects. This can be demonstrated by holding an excised eye of an animal near a candle flame. When the focal distances are properly adjusted, an inverted image of the flame can be seen upon the retina. Undoubtedly, the mammalian eye casts this inverted image during life and the ability to reinvert this upside down image to its normal upright position is a function of the brain that is acquired early in life.

Aberration

Like any optical system, the eye is subject to defects. The two commonest are spherical and chromatic aberration. In spherical aberration, the light rays falling upon the peripheral parts of the lens are refracted more than those falling upon the central parts. Consequently, the peripheral rays focus differently than the central and a blurred image results. Spherical aberration in the eye is corrected by the iris which acts to cut off peripheral rays. Chromatic aberration results from the fact that the violet end of the visible spectrum is refracted more than the red end. This tends to produce a colored outline around the borders of the image. The mammalian eye possessess little compensation for chromatic aberration, but this condition is normally not noticeable due to the fact that the ends of the spectrum which are most involved play only a minor role in retinal stimulation.

Accommodation

The above descriptions pertain to a static lens system wherein the focal distance is fixed and the curvature of the lens does not undergo change. Any mechanisms that can vary either lens curvature or focal distance is a mechanism of accommodation. Accommodation is essentially the adjustment of an optical system to compensate for distance of the observed object and to bring the object into focus. As far as the mammalian eye is concerned, objects 20 feet or greater from the eye will be seen in focus. This is because although light rays diverge somewhat from their point of origin, the angular divergence is greater than the diameter of the eyeball and only the rays that are nearly parallel will enter the eye. Objects situated at a distance of less than 20 feet will give off divergent rays, which must be brought to a focus upon the retina in order for them to be seen distinctly. There is a point, approximately 8-12 inches from the eye (the "near point"), where distinct vision ends and beyond which the normal eye cannot adjust. Objects closer than the "near point" cannot be seen distinctly since the physical limitations of the optic structures will not permit proper accommodation.

The adjustment of the optical system of the eye (accommodation) is a function of the lens. While a camera focuses through a fixed lens by moving the lens closer to or farther from the plate, the eye is incapable of such a movement. Focusing in the eye is accomplished by changing the curvature of the lens. This is effected through the action of the ciliary muscles and the pressure of the aqueous and vitreous humors upon the soft and relatively easily deformed lens, which, in the living animal, is about the consistency of hardened gelatin. The fibers of the ciliary muscles are so arranged that when they constrict, the choroid is drawn forward and the ciliary ring is reduced in diameter. Thus, tension upon the suspensory ligament is reduced and the lens, which is normally somewhat flattened, bulges anteriorly since the pressure of the aqueous humor is less than the pressure of the vitreous. Relaxation of the ciliary muscle allows suspensory ligaments to tighten and the lens to passively resume its former shape. Development of the ciliary muscle is directly associated with accommodation powers. In herbivores, ciliary muscle (and likewise accommodation) is poorly developed. With increased age there is a gradual decrease in the flexibility of the lens (presbyopia) and a concomitant decrease in the powers of accommodation. Under normal conditions a system of involuntary motor and sensory nerves automatically accommodates the eye to near or distant vision.

To assist accommodation and also to moderate the intensity of light stimulation, the pupil of the eye dilates or contracts. In close vision the pupil tends to contract. This phenomenon will also be observed in the

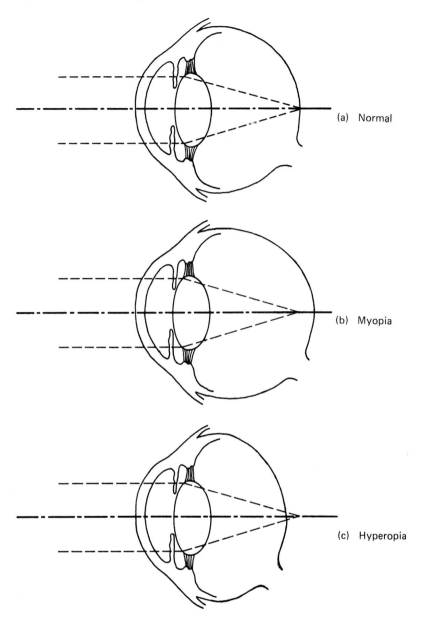

FIGURE 16-3 *Shape of the eyeball as related to normal, myopic, and hyperopic vision.*

presence of bright light. Pupillary reflexes are controlled by the autonomic nervous system. Here is the best place in the body where direct antagonism of the sympathetic and parasympathetic may be observed. Division of the nervous system may be observed since the sympathetic dilates the pupil and the parasympathetic constricts it. The pupil also may be influenced by the action of drugs that stimulate or simulate the actions of the autonomic nerves.

Myopia, Hyperopia, and Astigmatism

A normal eye wherein the eyeball, cornea, and lens are of normal shape will focus on an image 20 or more feet away without accommodation. Such an eye is known as emmetropic. Emmetropism, while theoretically normal, exists in only a few individuals. The shape of the eyeball is seldom perfect. In general, it deviates in one or two directions forming either a long or a flattened oval. In the long oval eyeball, the point of focus is anterior to the retina. Such a condition results in myopia or nearsightedness and its converse, where the eyeball is a flattened oval and focus occurs behind the retina, produces hyperopia or farsightedness. In general, myopia is more characteristic of humans than of any other species. Both domestic and wild animals tend to be hyperopic.

Another common defect is astigmatism. This is a defect of refraction and can be found in either the cornea or lens, although the cornea is most often involved. When the refractive surfaces are not segments of spheres or have areas within them which have different radii of curvature, the light rays do not uniformly converge upon the retina and the eye is incapable of sharp and accurate focusing. Erosions, local thickenings, or oval shape of the cornea produce aberrant refraction and are the principal causes of astigmatism. Variations in the shape or curvature of the lens are less frequent.

The Retina

The retina lies between the choroid and the vitreous body and functions to convert light stimuli into nerve impulses. It is a complex transparent structure composed of ten layers (Figure 16-4). The rods and cones are the visual receptors and lie close to the pigmented layer of the choroid. Nerve impulses triggered by light striking the rods and cones are passed to the first and second order neurons, which transmit the nerve impulses to the third order neurons and to the correlation neurons. Processes of the third order neurons unite to form the optic nerve. The cell bodies of the optic nerve lie in the retina and the processes extend inward to the brain where they synapse with association neurons in the corpora quadrigemina. Light, in order to strike the rods and cones which lie against the choroid, must pass through this layer of nervous tissue, which covers the innermost surface of the vitreous chamber.

That the rod and cone cells are the light receptors may be shown by anatomical arrangements, physiological response, and logical deduction. The optic disc where the optic nerve leaves the eye has no light receiving powers and is, in fact, called the "blind spot." Hence, the nerve fibers, per se, are not sensitive to light. Structures such as the

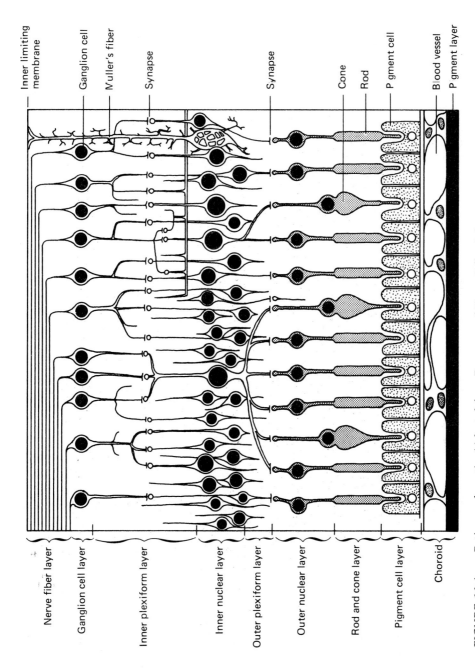

FIGURE 16-4 Retina—schematic. (Adapted from Freeman and Bracegirdle, Atlas of Histology.)

macula lutea and the fovea centralis in man have a high concentration of cone cells and known physiological properties of most acute vision. This indicates that the cones are related to visual acuity. Domestic mammals do not possess a fovea, but do have an area of concentration of cones along the visual axis of the eye. The rods and cones apparently have different functions, the rods acting as visual receptors in dim light, and the cones for color and bright light. Nocturnal animals such as cats have eyes in which rods are the predominating receptors. In most diurnal animals, cones are more prominent. There is a similar chemical factor that affects the reception of light by the rods and cones. The substance called visual purple or rhodopsin appears to be confined to the rods. A similar but not identical substance known as visual violet occurs in the cones. Rhodopsin decomposes and bleaches in the presence of light, but regenerates in the dark. Its presence in the rods and its physical characteristics indicate that it is concerned with dim light vision.

Color vision is not characteristic of most animals. Among the mammals, man has the keenest and most precise color vision. It is a normal characteristic of human vision and its lack is known as "color blindness." Color vision is apparently a function of the cone cells. The principal deficiency in human color blindness is the inability to perceive red and green. Since most colors are made from various red-green combinations, the result is that red-green vision defects result in the afflicted individual perceiving colors as various shades of gray. Specific deficiencies of red or green color perception result in some rather peculiar color perceptions similar to those gained by looking at objects through red or green glass.

Insofar as animals are concerned, most dogs and cats are color blind. Rats, mice, and rabbits have minimal color perception. Hamsters and opossums are color blind. Deer and giraffe can differentiate colors from gray, but cannot differentiate between different colors of similar intensity or albedo. Mink can detect yellow, blue, red, and green but are not able to selectively differentiate shades of color. The ability of mustelline species to detect color seems to be fairly good throughout the family. Mongooses have excellent color perception, but the best among the animals is possessed by mankind's primate cousins, apes and monkeys, with the top color vision (other than human) possessed by the chimpanzees.

Periscopic, Binocular, and Stereoscopic Vision

Physical differences in vision are exemplified in the domestic animals and man. These differences in type are classified as periscopic, binocular, and stereoscopic. Animals in which the eyes are set laterally in the head possess periscopic vision. In this type each eye observes a hemispherical segment of space on either side of the animal, and extreme muscular effort is necessary to produce converging vision. In terrestrial mammals, the condition is best exemplified in rabbits, although sheep, horses, swine and other domestic animals have it in some degree.

Binocular vision is where both eyes are set along parallel axes and view the same object simultaneously. Man is perhaps the best example of this, although other carnivores and predators possess it to a greater degree than ruminants, monogastric ungulates or rodents.

With stereoscopic vision three dimensions of space are clearly seen. It is best developed in primates, although cats and dogs whose visual axes are only slightly divergent probably have good stereoscopic vision.

For stereoscopic vision, three things are necessary. The vision must be binocular, the nerve fibers of the optic nerve must not completely decussate at the optic chiasma, and the eyes must be so constructed that the observed image will fall on corresponding points of both retinas. Many animals demonstrably lack stereoscopic vision because the optic nerves decussate completely or the other criteria are not met. Incomplete decussation is present in the dog, cat and man. Stereoscopic vision criteria, however, are open to debate and argument. It seems obvious to anyone who has watched animals other than cats, dogs and primates or who has ridden horses, that these other species possess an excellent judgment of distance, which is a function of stereoscopy. If stereoscopic vision does not exist in them, a substitute for it probably does.

THE ORGAN OF HEARING AND EQUILIBRIUM

The ear, or organ of hearing and equilibrium, is almost as highly developed as the eye. It performs two functions. It receives sonic vibrations and transforms them to nervous impulses by means of a complex structure called the organ of Corti. The nerve impulses are sent directly to the brain via the auditory nerve. The ear is also responsible for the maintenance of balance and equilibrium through nerve endings located in semicircular canals, the utricle and saccule of the inner ear, and through an aggregation of calcium concretions called otoliths, which are found in the utricle and saccule.

Two ears are present in normal mammals. These each consist of three divisions, the external ear, the middle ear, and the internal or inner ear. The principal structures of the ear are enclosed in the petrous temporal bones of the cranium.

The External Ear

The external ear consists of a funnel-shaped outer part composed of skin and elastic cartilage. This part is what is commonly referred to as "the ear," a term that is misleading and inexact. More properly, this portion of the external ear should be called the auricula or pinna. From the pinna, the external ear is continued into the skull by a membrane-lined tube, the external acoustic meatus.

The Tympanic Membrane

The eardrum, or tympanic membrane, closes the inner end of the external acoustic meatus and separates the external ear from the middle ear. The tympanic membrane is structurally interesting in that it is the only tissue in the body derived from all three of the original germ layers of the embryo (i.e., ectoderm, mesoderm, and endoderm). It functions as a vibratory structure, and converts sound waves to mechanical action.

The Middle Ear

The middle ear is a hollow chamber entirely enclosed by the petrous temporal bone. It contains the ossicles or ear bones and has four openings: the external acoustic meatus, the fenestra ovalis, the fenestra rotundum, and the eustachian tube. The first three of these openings are closed by membrane partitions. The eustachian tube connects with openings in the lateral walls of the nasopharynx. It functions to equalize the air pressure between the middle and external ear and thus prevent bulging of the eardrum and impairment of eardrum efficiency in receiving sound waves.

The ossicles are composed of three bones: the malleus (hammer), incus (anvil), and stapes (stirrup). The three bones are connected to each other and to the walls of the cavity of the middle ear by delicate ligaments. The malleus is further attached to the inner surface of the tympanic membrane, and the stirrup is attached to the outer surface of the membrane closing off the fenestra ovalis. These bones serve to transmit and amplify the vibrations of the tympanic membrane to the inner ear.

The fenestra ovalis (oval window) and the fenestra rotundum (round window) separate the middle ear from the inner ear.

The Inner Ear

The inner ear is the essential portion of the organ of hearing and equilibrium. Because of its complex shape, it is called the labyrinth. It can be divided into two main parts, the osseous labyrinth, which is within the petrous temporal bone, and the membranous labyrinth, a layer of covering for the walls of the osseous labyrinth.

The Osseous Labyrinth

The osseous labyrinth is divided into three portions: the vestibule, the semicircular canals, and the cochlea. It is filled with a watery fluid

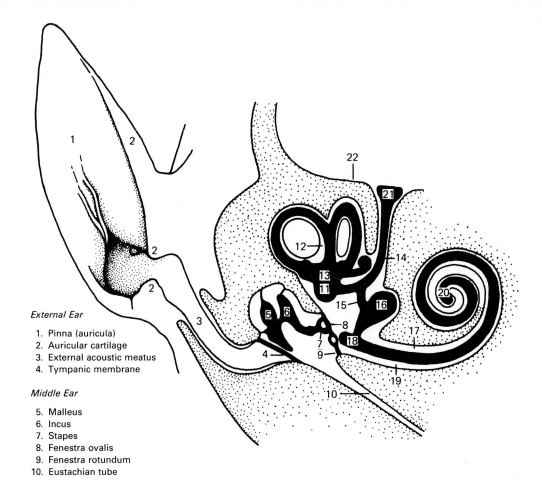

External Ear

1. Pinna (auricula)
2. Auricular cartilage
3. External acoustic meatus
4. Tympanic membrane

Middle Ear

5. Malleus
6. Incus
7. Stapes
8. Fenestra ovalis
9. Fenestra rotundum
10. Eustachian tube

Inner Ear

11. Utricle
12. Semicircular canals
13. Utricular branch of endolymphatic duct
14. Endolymphatic duct
15. Cochlear branch of endolymphatic duct
16. Saccule
17. Scala vestibuli
18. Cochlear duct
19. Scala tympani
20. Helicotrema
21. Endolymphatic sac
22. Dura mater

FIGURE 16-5   *The ear—schematic.*

called perilymph and surrounds the soft tissue of the membranous labyrinth.

The vestibule is the central part of the osseous labyrinth. It is somewhat oval in outline, communicating in front with the cochlea and behind with the semicircular canals. Its inner wall has a foramen through which pass the filaments of the vestibular branches of the auditory nerve and the endolymphatic duct. The external wall contains the "oval window" or fenestra ovalis.

The semicircular canals are three curved bony tubes, which communicate with the vestibule and form an almost complete circle. Each circle is of different size and their axes are essentially transverse, longitudinal and vertical with respect to the long axis of the body. Here are contained the organs and nerve endings concerned with balance and equilibrium.

The cochlea is a conical and helical cavity, somewhat resembling a snail shell, that forms the rostral part of the labyrinth. This is the location of the hearing organs.

The Membranous Labyrinth and Otolith Organs

The membranous labyrinth consists of a series of closed membranous sacs containing endolymph. The sacs conform more or less to the bony labyrinth, and occupy a central position within it. The vestibule contains two membranous sacs, the utricle and saccule, which communicate with each other and contain the maculae or otolith organs. The utricle is the larger of the two and receives the openings of the semicircular canals. The saccule is united with the cochlear duct, which is a spiral tube situated within the cochlea. Within the utricle and saccule are the otolith organs, which contribute to the function of the equilibratory mechanism in the inner ear. The otoliths are small calcium crystals or concretions, and are suspended together with hair cell neural receptors in a gelatinous medium called the otolith membrane. The entire organ is called the macula. The two maculae, one in the utricle, the other in the saccule, are situated roughly at right angles to each other, and the suspended otoliths, under the influence of gravity, stimulate the hair cells to send equilibratory messages to the brain over the acoustic nerve.

Fish have relatively large single otoliths, which are found in haircell-lined cavities below the brain. In fish the otoliths apparently also function to sense low frequency sounds. Since in these species, the size of the otolith nucleus is directly associated with egg size, juvenile members of virtually identical species such as rainbow trout and steelhead can be identified by the size of the otolith nucleus.

The cochlear duct contains the auditory apparatus of the organ of Corti. Both the utricle and saccule are connected to the endolymphatic duct, which proceeds to the dura mater of the brain along with the

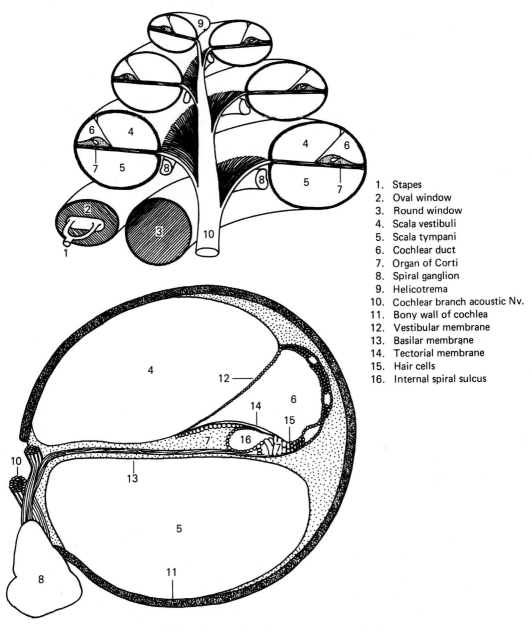

FIGURE 16-6   Cross section of the cochlea—schematic.

auditory nerve. At the dura mater the endolymphatic duct expands to form the endolymphatic sac.

The organ of Corti is made up of a series of rodlike bodies arranged in rows and connected with several rows of ciliated cells. These bodies are contacted by a vibratory portion called the tectorial membrane, which vibrates with the changes in fluid pressure in the peri and endolymph. In a sense the organ of Corti can be compared to the retina of the eye in that it receives and converts waves to nervous impulses. These nerve impulses are carried to the brain over the acoustic (auditory) nerve.

PHYSIOLOGY OF THE EAR

The ear is a remarkable organ. In performing its ordinary functions, it makes most sensing mechanisms constructed by man appear crude. Consider its accomplishments. It is so sensitive that it can hear the dropping of a pin, yet it can withstand the pounding of sound waves strong enough to shock the entire body. It is so highly selective that in the blended sounds of a symphony orchestra it can select the one instrument that is not performing properly. One measure of its sensitivity is the minuteness of vibration to which it will respond. At some sound frequencies the vibrations of the eardrum are as small as one billionth of a centimeter and the vibrations of the sensitive membrane of the inner ear, which transmit these stimuli to the auditory nerve, are about 1/100 this amplitude. This almost immeasurable smallness explains why hearing has been one of the greater physiological mysteries. Even today we do not know exactly how these minute vibrations stimulate the auditory nerve.

The ear is least sensitive in the low frequency range. This apparent insensitivity is a physical necessity because otherwise an animal would hear all the vibrations of its own body. This can be demonstrated by placing a finger in each ear, closing off airborne sounds. One should hear a low irregular tone which is produced by the contractions of the muscles of the arm and finger. It is interesting that the ear is just insensitive enough to avoid the disturbances of noises produced by muscles and body movement. On the high frequency side, the range is remarkable; bats, for instance, can hear 100,000 cycles per second, dogs about 35,000 and man about 20,000. With age, acuteness of hearing falls steadily in the high frequency range. This loss of efficiency is not difficult to understand if one considers that the elasticity of the tissues in the inner ear declines in much the same manner as the elasticity of the skin. Loss of hearing sensitivity with age may also be due to nerve deterioration, drugs such as streptomycin, or disease.

Exposure to high decibel levels also impairs hearing ability. Workers in high sound level jobs tend to become partially or totally deaf at a fairly early age. A newer and more peculiar threat to hearing lies in the high decibel output of modern musical groups, which consist essentially of a high capacity amplifier with ancillary instruments (usually an organ, a string bass, one or two guitars and a woodwind). The sheer volume of noise generated by this setup can produce irreversible damage to the ears of those who listen, and who, in their desire to "feel" sound, destroy first the sensitivity and then the capacity of their auditory apparatus. One could be philosophical about this and remark that most sounds aren't worth hearing, but there are such noises as police sirens, horns, and warning bells that may be handy to preserve one's life, and other sounds of beauty or affection that pleasantly affect one's emotions; but these can in part be restored by a high capacity hearing aid.

To understand how the ear performs, a knowledge of functional anatomy is essential. Sound waves cause vibrations of the eardrum, and these vibrations are transferred by the ossicles to the fluid of the inner ear. The stapes, fastened to the membrane covering the fenestra ovalis, acts on the fluid-filled inner ear like a piston, moving the membrane back and forth in rhythm with the sound pressure on the eardrum. These movements send vibrations through the fluid, which transmits them to a thin membrane within the cochlea called the basilar membrane. This membrane vibrates in harmony with the vibrations in the fluid and in turn transmits the vibratory stimulus to the organ of Corti, a complex-structure that contains the receptor endings of the auditory nerves.

Recently, through application of a physical principle known as the Mossbauer Effect, research workers have been able to study the mechanisms of basilar membrane vibration in living animals. The technique involves surgical exposure of the cochlea and the placement of a tiny piece of radioactive cobalt (Co^{57}) on the basilar membrane. As the membrane vibrates it moves in relation to a fixed detector, thus altering the energy of the Co^{57} gamma ray emission. The changes in energy reception are compared with known standards and computer analysis is used to determine how much and in what direction the membrane is vibrating. This is a great step forward, since prior to this time research into hearing mechanism was principally a matter of logical reasoning based on anatomic structures. Logic, unfortunately, doesn't work well with some physiological processes.

Most early thinking agreed that the basilar membrane vibrated less and less on a linearly decreasing scale for decreasing levels of sound intensity until it moved less than the width of a hydrogen atom at very low sound levels. It did pose a problem as to how the ear discriminated such tiny movements, but this was in part resolved by the theory that larger portions of the membrane and its neural connections were involved in low sound intensity reception.

It now appears that the motion of the basilar membrane decreases along a logarithmic curve rather than a linear one and thus does not decrease an equal amount for each decrease in sound level. Indeed, at extremely low sound intensities it moves relatively more than it does at high intensities. The recent work also indicates that pitch discrimination, as has been previously suggested by anatomic studies, is determined by extremely small areas of receptors along the basilar membrane.

A natural question is why is this long and complicated chain of transmission necessary? The answer is to amplify the relatively weak sound waves that strike the eardrum. Usually, when sound strikes a solid surface, most of the energy is reflected. The problem the ear has to solve is to use this energy rather than reflect it. To do so it must act as a transformer that will convert the amplitude of sound waves in the air into more forceful vibrations of smaller amplitude. The footplate of the stapes and the membrane covering the oval window act like a hydraulic press. The tiny footplate of the stirrup receives the total leverage of the malleus and incus that transforms the small pressure on the surface of the eardrum into a 22 times greater pressure on the fluid of the inner ear. But the organ of Corti needs another transformer to amplify the pressure of the fluid into a still larger force that will act upon the sensitive hair cells in which the acoustic nerves terminate. This second amplifying mechanism is based on the principle that a flexible membrane stretched to cover the opening of a tube has a lateral tension along its surface. This tension (shearing force) can be increased enormously if pressure is applied to one side of the membrane. This is the essential basis of the function of the organ of Corti. The organ is constructed in such a way that pressure upon the basilar membrane is transformed into shearing forces in the tectorial membrane many times larger than the original vibration. These enhanced forces rub upon the extremely sensitive hair cells containing the endings of the auditory nerve and produce the stimuli that the brain interprets as sound.

The eardrum, of course, is not the only means by which sound may be conducted to the inner ear. Hearing can occur through the bones

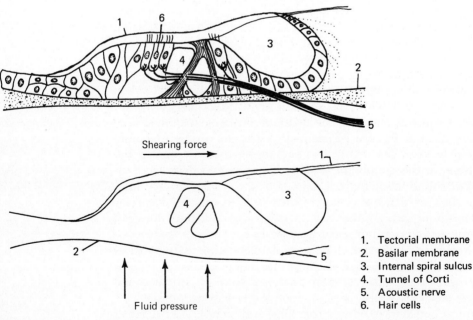

FIGURE 16-7 *Schematic of structure and function-organ of Corti.*

of the skull. Some of these vibrations can be transmitted directly to the inner ear and bypass the middle ear entirely.

The ear is highly selective and is able to ignore all sounds except the first that strikes it. Hence, it has a remarkable locating ability which is based upon the inhibition of secondary sounds or echoes. Any stimulation of the ear is translated into nervous impulses, but this stimulation varies with the frequency of the sound. In the lower range (up to 60 cycles per second) the vibration of the basilar membrane produces within the auditory nerve a series of electric spikes synchronized with the rhythm of the sound. As the frequency increases, the number of spikes packed into each part increases. Thus, two variables are transmitted to the brain: the number of spikes and the rhythm. These two variables convey the loudness and the pitch of low sounds.

A new effect appears at a vibration of about 60 cycles per second. The basilar membrane vibrates unequally over its affected area, and each tone produces a maximal vibration in a different part of the membrane. This results in local selectivity that determines pitch. Above 4,000 cycles per second, pitch is determined entirely by the location of the maximal vibration along the basilar membrane and these pitch-determining locations are extremely small in area. Apparently there is an inhibitory mechanism, which suppresses weaker stimuli and sharpens the reception around the areas of highest stimulation. The basilar membrane thus makes a mechanical frequency analysis, which is sharpened by the auditory nervous system in some manner that is not yet understood.

THE ORGAN OF SMELL

The receptor region of the olfactory (smell) apparatus lies in the caudal portion of the nasal cavity and is supplied with sensory and motor neurons derived from branches of the trigeminal nerve. This neural connection, however, has only an indirect relationship to the sense of smell, since it does not function in the act of odor reception and transmission to the brain. The trigeminal neurons in this area react mainly to irritants such as pollen, dust, or other substances by triggering the sneezing reflex.

The sense of smell (olfaction) is concentrated in a relatively small area in the caudal portion of the turbinates called the olfactory region. In living mammals this is yellowish in color and lies upon the outer surface of the ethmoid bones and along the base of the turbinates and consists of four main parts: the olfactory cells, basal cells, supporting cells, and the branched tubuloalveolar glands of Bowman.

The basal and supporting cells form the bulk of the organ and surround the olfactory cells. The olfactory cells are fat spindle shaped bipolar cells with relatively short dendrites and moderately long axons. The dendrites penetrate the spaces between the basal and supporting cells and lie free upon the surface of the olfactory epithelium, where they are continuously bathed in a mucoid exudate secreted by the glands of Bowman. The axons of the olfactory cells pass through the openings in the cribriform plates of the ethmoid bones and join together to form the paired olfactory nerves, which pass directly into the brain without synapsing. In this respect the olfactory neurons are unique, since they are the only ones in the body that do not synapse prior to entering the brain. The olfactory neurons are unique in another way, which may someday be of great value in medicine, since they possess centrioles and are capable of replication to replace cells that are damaged or destroyed. They are the only nerve cells in the body that can reproduce (cf. p. 312) after an animal is born.

Within the brain the olfactory neurons enter the olfactory bulbs, which are located in the rostroventral portion of the forebrain. There, the olfactory neurons terminate in spheroidal structures, called glomeruli, and synapse with mitral cells, which connect to the limbic system, hypothalamus, thalamus, pituitary olfactory cortex, and neocortex, and affect such diverse functions as hormone secretion, discrimination, emotion, memory, and intellect. There is also, apparently, a neural loop that returns to the olfactory epithelium and can modify responses to olfactory stimulation.

PHYSIOLOGY OF SMELL

The sense of smell is well developed in most animals and undoubtedly plays a large part in their life and habits. The sensation of odor arises through the stimulus provided by small particles of material dispersed in the inspired air. The glands of Bowman produce a mucoid fluid that constantly washes the naked nerve endings and removes odor-producing particles from them. Persistence of odor usually depends upon how refractory the stimulating particles are to being washed away. Substances like mercaptans and sulfhydryls are particularly persistent, and tend to produce subjective sensations long after the stimulus has been removed.

Nerve impulses from the olfactory receptors travel to the olfactory bulbs in the brain and then along the two olfactory tracts, which channel the impulses to interpretive centers or cells deeper within the brain. A fairly recent theory of neural function has appeared, which suggests that this nerve transmission may be a two-way street. It states that the control or interpretive areas send back impulses over the olfactory tract, instructing the relay cells in the olfactory bulb to focus upon certain odors and ignore or deemphasize others.

These studies have in part arisen from the anatomic structure of the olfactory tract and the microscopic anatomy of the olfactory nerve fibers within the tract. It has been noticed that there is a distinct difference in cross section size of the olfactory nerve fibers. The largest fibers are myelinated and are as much as 10 times greater in area than the smallest which are unmyelinated. There is, as has been previously indicated in the chapter on the nervous system, a transmission speed along myelinated axons (about 10 meters/second) which is much faster than along the nonmyelinated fibers (about 1 meter/second). These variations in nerve size and speed of conduction are speculated to have some effect upon scent discrimination. Considerable work remains to be done in this area, for while olfaction seems to be a simple process when compared to sight and hearing, relatively little investigation has been made into its precise mechanisms.

A theory has been proposed that offers a physiological explanation for the sense of smell and reduces the complexity of scent perception and discrimination to seven primary odor stimuli. The scent of a polecat, according to the theory, is distinguished from a peachblossom by the shape of its odor molecules. Every smelly substance has its particular combination of molecules that function like keys in seven different receptor areas of the olfactory epithelium and unlock the sensations that range from "aah" to "ugh." First proposed by Dr. John Amoore of the Western Regional Research Laboratory in 1952, this theory has been developed somewhat since its initial promulgation. Another theory, offered in 1963 by Dr. Robert Wright of the British Columbia Research Council, states that specific vibratory frequencies of molecules stimulate specific olfactory receptors. Since Dr. Amoore feels that pungent and putrid odors (in contrast to the other five, i.e., camphoraceous, musky, floral, pepperminty, and ethereal) have no specific shape or size but gain their effect through an excess of electrons and electrical charge, the two theories may be parts of a complete mechanism.

The five primary odor shapes are spherical (camphoraceous), rod-like (ethereal), wedge-shaped (pepperminty), disc-like (musky) and disc-with-a-tail (floral). The theory assumes that each class of olfactory nerve ends has a receptor site whose shape conforms to the scent molecules. Complex molecules incorporating more than one shape may occupy more than one site and send complex messages to the brain. Unfortunately, the specific receptor sites have yet to be discovered.

The "key-lock" theory has had some experimental confirmation, but at present it remains a theory. And, in any event, the interpretation of odors will probably always be individual and subjective, no matter how neatly science may classify the mechanism.

THE ORGAN OF TASTE

Taste sensations originate principally from specialized receptors located on the tongue. These receptors are enclosed in structures called papillae and are composed of sensory nerve endings which send fibers to the brain. Four types of papillae are found in the tongue: filiform, fungiform, vallate, and foliate. Filiform papillae are fine, thread-like projections that are found on the top and sides of the tongue near the tip. They are absent at the root of the tongue and increase in size from the tip of the tongue to its base. Fungiform papillae are larger and are found principally along the lateral portions of the tongue. Their form is like that of a toadstool, being large at the free end and supported by a slender neck. Vallate (circumvallate) papillae are relatively few in number; seldom more than four are encountered in an animal. They are circular depressed patches surrounded by an annular wall. Their free surface usually bears small, rounded secondary papillae. Foliate papillae are located just anterior to the anterior pillars of the soft palate and form rounded, leaf-shaped eminences crossed by a number of transverse fissures. Like the vallate papillae, these also bear a number of microscopic secondary papillae on their surface.

PHYSIOLOGY OF TASTE

The sensation of taste is associated with these well-defined papillary areas, which are commonly called taste buds. In the cow, fungiform papillae are numerous and distinct and are found scattered over the dorsal portion and the free edges of the tongue. Circumvallate papillae in all animals tend to form a narrow band, which may be found along the edges of the base of the tongue or in a transverse narrow row across the base of the tongue. Their number varies greatly with the species. The taste receptors consist of long thin cells with a large nucleus in their midportion. These cells end in a delicate hair-like process, which projects through the open orifices of the taste buds. Most of the sensations of taste are carried to the brain by branches of two cranial nerves, the glossopharyngeal and trigeminal. In the human, taste sensations are also carried by the facial nerve, which supplies the anterior portions of the tongue. The exact status of an animal's sensation of taste cannot be determined, but in man it has been discovered that all taste sensations can be classified under four chief headings: sweet, bitter, acid, and salt. Each of these appears to have a definite area of the tongue in which the specific sensation is most prominent. In all cases, substances must be dissolved in order to act upon the taste receptors. While animals may not have as efficient organs of taste as does mankind, there is no doubt that they have a highly developed and discriminatory taste apparatus. Substances which to a human are either bland or tasteless may often be rejected by animals. It is quite probable that the taste sensations vary between species, but further work needs to be done before adequate evaluation of the sensitivity of animal taste buds can be determined.

A group of research workers at Northwestern University in Evanston, Illinois, has succeeded in growing taste buds in tissue culture. The exact significance of this achievement is still unclear, but it is virtually certain that new discoveries in the nervous mechanism and physiology of taste will arise from this technique.

TACTILE SENSES

The senses of touch (pressure), pain, heat, and cold are not confined to any particular region of the body. The nerve endings involved in the transmission of these stimuli to the brain are connected to the sensory branches of the peripheral nervous system. The nerve endings that deal with touch sensation are chiefly aggregated in the hands, feet, and lips, although they may be found elsewhere. Pain, heat, and cold receptors have a somewhat different distribution. One of the more interesting features of this portion of the sensory apparatus are the proprioceptors, which occur in muscle-tendon junctions and are responsible for a general awareness of muscular movement, tone, and the position of the limbs and body. These sense organs are responsible, in conjunction with the semicircular canals of the middle ears, for the maintenance of balance and posture. In addition they convey messages to the brain that prevent muscular over-exertion and/or rupture of muscle-tendon junctions.

chapter 17
Anatomy and Physiology of the Fowl

While there are similarities between fowl and mammals, the differences tend to outweigh them. Arguments have been advanced that the similarities that exist are merely examples of parallel evolution. Without belaboring the point, it is obvious that so many structural and functional differences exist between birds and mammals that their relationship is remote at best.

Recent discoveries indicate that birds are probably direct descendants of dinosaurs, and since the dinosaurs were already considerably evolved in the Mesozoic era, their descendants, the birds, can only have added to their evolutionary divergence from mammals.

The solutions that the two classes of vertebrates have found for physical and environmental problems are often similar in ends, but not in means. Considering the variations in anatomy and physiology, it is probable that the study of birds should be separate from that of mammals, since attempts to coordinate two fundamentally different life forms are difficult at best. However, practical considerations of husbandry and management of domestic species dictate that they be studied together, insofar as possible. Therefore, this chapter will attempt to coordinate the two classes.

Major emphasis will be placed on differences in structure and function, and subtle or inconsequential variations will be minimized or ignored. True similarities will not be discussed. This is done to retain the nature of this book as a basic text, and because minimal variations from previously discussed data would not contribute enough to make their inclusion warranted.

The material in this chapter will be organized in the same manner as for mammals to facilitate reference to comparable mammalian structures and functions. The chicken will be used as the type species, not necessarily because it is typical, but because it is most common and easiest to obtain for postmortem and antemortem studies. However, it should not be forgotten that there are at least as many evolutionary adaptations between species of birds as there are in mammals, and in consequence there are hordes of exceptions and qualifications to any statement that can be made about chickens.

CELLS

No significant difference exists between the cells of fowl and mammals insofar as their structure and function are concerned.

BONE AND CARTILAGE

Cartilage is relatively rare in fowl and is either hyaline or fibrous. The only cartilages that persist in adult chickens are the fibrocartilages on the joint surfaces of movable bones; the fibrocartilage menisci in the spaces between the bodies of the unfused vertebrae; the hyaline cartilage in the sclera of the eye; and in the cricoarytenoid cartilages of the larynx. In the pigeon there are tracheal cartilages; in other domestic fowl, the tracheal rings are bone.

Bones in fowl are essentially of cartilagenous origin. The cartilage matrix from which the bones are formed persists for several months after hatching, and then slowly disappears except in the few permanent locations mentioned above. Cartilage persists longer in the keel than in any other bone.

Structurally, there is considerable difference between the developed bones of birds and mammals. Some avian bones are pneumatic, i.e., their medullary cavities are filled with air spaces, which communicate with the air sacs of the lungs. Bones that have these structures vary considerably between species. Small flying birds have none; large flying birds have many. In the chicken the posterior cervical vertebrae, the thoracic vertebrae (except the fifth), the pelvis, the sternum, the humerus and the distal coracoid are pneumatic. However, this list should not be carried over to other large birds, since it will not apply to other species. Some diving birds, such as penguins, have no medullary cavity at all.

Marrow is not found in the medullary cavity of most avian bones. In the limbs the nonpneumatic medullary cavities may be filled with soft medullary bone. This tissue has mistakenly been called marrow. It is most noticeable in laying hens or pullets, and is possibly a means whereby fowl can store calcium for eggshells.

The development of medullary bone is restricted to those bones which have a good blood supply and which are not pneumatic. In consequence, the femur, tibia, scapula, clavicle (furcula), radius and ulna, carpals and metacarpals can be involved in medullary bone formation. The tarsometatarsus and toe bones do not form medullary bone. The skull and cervical vertebrae are minimally involved, probably because these bones are at least partly pneumatic. The development of medullary bone appears to be under control of hormones, notably estrogens, thyroxin, and calcitonin.

Most birds do not have Haversian systems, probably for the same reason small mammals do not possess them. Large to very large birds such as emus, cassowaries, and ostriches have been reported to have these structures.

The structure of avian bone and the process of ossification differs from that of mammals. Developed bones are relatively thinner, except in penguins and other nonflying diving birds, and have less ossein and more tricalcium phosphate in their makeup. In consequence, avian bone is harder but more brittle than mammalian bone. The bones of birds tend to change in character during molting, and at least in some species there are epiphyseal nuclei, which are not found in mammalian bone. The function of these structures is unknown.

THE SKELETON

Like mammals, the avian skeleton is divided into axial, appendicular, and visceral portions. The visceral skeleton is the scleral ring in the eye. For nomenclature, Figure 17-1 should be studied.

Although serious attempts have been made to relate the bones of mammals and birds, there are fundamental differences that should be mentioned. These are as follows.

Appendicular Skeleton

PECTORAL LIMB

All three bones of the pectoral girdle are present. The coracoid is the best developed of the three. The clavicles are fused ventrally and are generally called the furcula or wishbone. An intact furcula is necessary for flight since a bird cannot spread the shoulder girdle if the

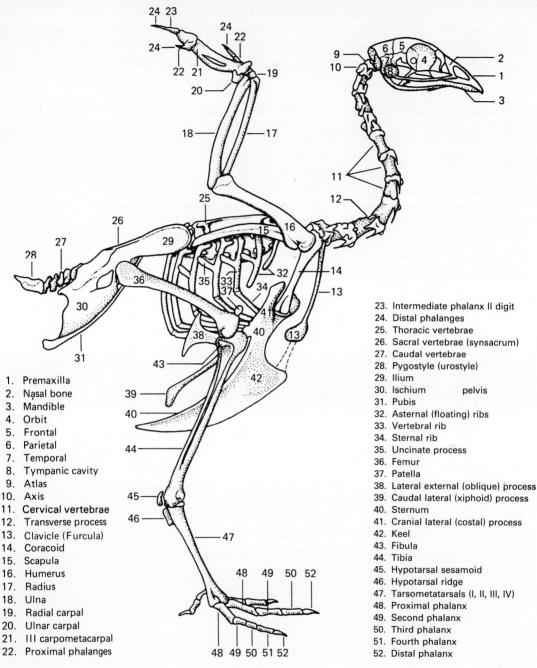

FIGURE 17-1   Skeleton of the chicken.

1. Premaxilla
2. Nasal bone
3. Mandible
4. Orbit
5. Frontal
6. Parietal
7. Temporal
8. Tympanic cavity
9. Atlas
10. Axis
11. Cervical vertebrae
12. Transverse process
13. Clavicle (Furcula)
14. Coracoid
15. Scapula
16. Humerus
17. Radius
18. Ulna
19. Radial carpal
20. Ulnar carpal
21. III carpometacarpal
22. Proximal phalanges
23. Intermediate phalanx II digit
24. Distal phalanges
25. Thoracic vertebrae
26. Sacral vertebrae (synsacrum)
27. Caudal vertebrae
28. Pygostyle (urostyle)
29. Ilium
30. Ischium   pelvis
31. Pubis
32. Asternal (floating) ribs
33. Vertebral rib
34. Sternal rib
35. Uncinate process
36. Femur
37. Patella
38. Lateral external (oblique) process
39. Caudal lateral (xiphoid) process
40. Sternum
41. Cranial lateral (costal) process
42. Keel
43. Fibula
44. Tibia
45. Hypotarsal sesamoid
46. Hypotarsal ridge
47. Tarsometatarsals (I, II, III, IV)
48. Proximal phalanx
49. Second phalanx
50. Third phalanx
51. Fourth phalanx
52. Distal phalanx

bone is broken, and this movement is essential to the functioning of the wings.

The carpals are fused into a single mass with the exception of the radial and ulnar carpal bones, and the carpus is fused to the metacarpals to form the carpometacarpal bone. The digits of the forelimb are reduced to the I, II, and III digits of which only the II digit possesses three phalanges. The phalanges of the I and III digits are rudimentary.

PELVIC LIMB

The pelvis is a broad shield-like plate covering the dorsal and dorsolateral portions of the posterior half of the body. It is united to the fused lumbar and sacral vertebrae by a strong bony union. The ventral portion of the pelvis is not fused, and the pubis lies along the ventral border of the ischium.

The femur, patella, tibia, and fibula are similar to those of mammals, although the femur is relatively short compared to the tibia. In some birds (pigeon, duck and goose) the patella is replaced with a tendinous structure. The tarsal and metatarsal bones are fused into a single structure, the tarsometatarsus, which possesses four digital articulations for the proximal phalanges of the digits. In chickens, a spur core is present on the medial site of the bone in the distal third. It is the most prominent in males. A hypotarsal sesamoid is present on the posterior surface of the joint between the tibia and the tarsometatarsal.

The digits are four in number. The I digit possesses three phalanges, the II digit has three, the III digit four, the IV digit five.

The Axial Skeleton

VERTEBRAL COLUMN

The vertebral formula for the chicken is C_{14} T_7 L-S_{14} Cy_6. The structure of the column varies considerably from that of mammals. The cervical vertebrae vary in number and tend to be more complex than in mammals. The chicken has 14 cervical vertebrae; the goose, 17; and the duck, 15. Swans may have over 20. The number is fairly constant for each species but varies from one species to another. The cervical vertebrae are arranged in an S-shaped curve, which cannot be straightened completely without rupturing the spinal cord and killing the animal.

A notarium or os dorsalis (a fusion of the thoracic and sometimes two to four caudal cervical vertebrae) exists in some birds and is absent in others. Generally the first thoracic vertebra is not fused. A thoracic notarium is present in chickens, turkeys and pigeons. Ducks and geese do not possess a notarium. The number of thoracic vertebrae varies from three to ten, depending on the species. In the chicken, the first thoracic vertebra is free, the second through the fifth are fused, the sixth is free, and the seventh is fused with the lumbosacral bone.

The lumbar and sacral vertebrae are fused to the ilium of the pelvis, and to each other to form a strong bony mass, the lumbosacral bone. The lumbosacral mass is composed of from 10 to 20 fused vertebrae, depending on the species. This type of vertebral union appears to be a prerequisite for flight.

The coccygeal vertebrae of the chicken consist of five movable vertebrae plus a sixth which is different enough to be named. The sixth coccygeal vertebra is a three-sided elongated mass called the pygostyle, and supports muscles that control movement of the tail feathers. Coccygeal vertebrae also vary in number from three to ten between species and sometimes vary within a given species. Ducks, geese and pigeons all normally have eight coccygeal vertebrae.

RIBS AND STERNUM

The chicken has seven pairs of ribs. The first two pairs are floating (asternal) and consist of only one part. The last five pairs are sternal and are composed of two parts, a vertebral part and a sternal part. The second through the sixth pair of ribs have a flat uncinate process on the posterior border of the vertebral part. This process overlays the succeeding rib and makes the rib cage a more solid and firm mass. Since the rib cage does not move appreciably, this is an advantage in supporting the large sternum and the sternal muscles. The sternal parts of the ribs unite with the lateral margins of the sternal plate in the anterior third of the sternum.

The sternum is the largest bone in the body of most birds. It may or may not be keeled and is not divided into sternal segments (sternabrae) as in mammals. The keel (carina or ventral crest) is a cartilaginous medial ventral spine that projects downward from the sternal plate. It ultimately ossifies in adults. It is prominent in the chicken, turkey, and pigeon, somewhat less prominent in the goose and duck, and may be absent in some species. In the chicken, three bony prominences, the cranial lateral (costal) process, the oblique (lateral external) process and the posterior lateral (xiphoid) process arise from the craniolateral margins of the sternal plate immediately caudal to the sternal attachments of ribs III through VI. These structures are greatly reduced in the pigeon and duck, and are absent in geese.

THE SKULL

Considerable differences exist between the skulls of birds and mammals. As in mammals, the bones of the avian skull may be divided into two groups: the bones of the cranium and the bones of the face. The bones of the cranium are similar in name and location to those in mammals, except that the interparietal bones are absent. In the bones of the face, the malar is called the zygomatic bone, and the zygomatic processes of the malar and temporal bones are named (from rostral to caudal) the jugal, the quadratojugal, and the quadrate. The quadrate bone joins with the lateral surface of the temporal and with the pterygoid.

The rostral bones of the face unite to form the beak. There are no teeth, and the maxilla is greatly reduced. The hard palate is absent, and the palatine process of the premaxilla forms the lateral margins of the upper part of the beak. The nasal foramina on each side of the upper beak are very large, the orbits are incomplete ventrally and are separated from each other by the thin perpendicular plate of the ethmoid. The vomer is functionally absent although embryonic traces remain. The nasal septum is absent. The turbinate bones in each nasal cavity are three in number, rather than two as in mammals, are located in a row, and are named rostral, middle, and caudal. The middle is the largest. They arise from the lateral walls of the nasal cavity and greatly reduce the size of the cavity in the living animal. In the middle ear the malleus, incus and stapes are replaced by a single bone, the columella. The external acoustic meatus and pinna are absent and are replaced by the tympanic cavity. The hyoid bone consists of a rod-like flattened body and two lateral processes lying between the rami of the mandible. It consists of several segments. The body of the hyoid is divided into the endoglossal segment (lingual process), the basihyoid segment, and the urohyoid segment, which rests upon the larynx. The lateral processes each consist of two segments: the basibranchial segment, which unites with the basihyoid rostrally and the ceratobranchial segment caudally. The caudal extremities of the ceratobranchial segments do not unite with the skull although they lie close to the occipital bone.

The occipital condyle is single rather than paired as in mammals.

ARTHROLOGY

Although the joints of birds differ structurally from those of mammals, there is no appreciable functional difference, except that the fleshy joint in the scapular region of coursing mammals does not occur in birds. Structurally the differences involve the synovial membranes, which extend into the joint cartilages and contain cartilaginous villi. The joint cartilages are fibrocartilage rather than hyaline as in mammals.

PHYSIOLOGY OF THE SKELETON

There is little essential difference in the physiology of the skeleton in birds and mammals, except that the skeleton is not a major aid to respiration since the rib cage has only minimal movement. Despite the fact that some bones are pneumatic and others have medullary cavities which do not contain marrow, the hemopoietic function of the skeleton is virtually the same as it is in mammals with the marrow of the ribs, sternum and thoracic vertebrae being involved in blood formation.

MUSCULATURE

Gourmets insist that there are seven different kinds of muscle in birds, but for anatomical purposes there are only three: striated, smooth, and cardiac. Lungenherz muscles do not occur in birds. The color of muscle in various body locations varies from one species to another and is principally caused by the relative amounts of fibrils and sarcoplasm, and secondarily by the amount and density of the capillary beds. Muscles with many fibrils and little sarcoplasm tend to be light colored, while muscles with few fibrils and plenty of sarcoplasm tend to be dark. Poor vascularization is associated with light-colored muscle, while extensively vascularized muscle will be darker.

Major muscle masses in birds tend to be more closely related to each other than they are in mammals since there is less connective tissue separating the muscle bundles. Long tendons in birds are ordinarily ossified, and short tendons may be ossified. A number of muscles exist in birds that do not occur in mammals: e.g., the patagial muscles in the forelimb; the pterylar muscles in the feather tracts: the protractor pterygoid, the protractor quadratus, the tensor adductor, the posterior mandibular adductor and the pseudotemporal in the head; the teres, the transverse oblique and the intertubercular muscles in the thorax; the depressor caudae in the coccygeal region; and the perforating digital flexors in the legs. Other muscles correspond roughly to similarly located mammalian muscles but may have different names.

Some muscles, such as the diaphragmatic, are rudimentary in birds. Functionally, however, muscles in birds operate in practically the same manner as they do in mammals, and their physiology and metabolism appear to be virtually identical. Gait, however, shows marked variations, owing to the fact that birds have only two running limbs, and flying birds have entirely different movements of the forelimbs than do mammals.

In the hindlimbs of running birds, the short femur is limited to what is essentially a quick up and down movement, with the tibia and the tarsometatarsals being extended rapidly forward on the upstroke of the femur and propelled abruptly backwards on the downstroke. The flexion and extension of the digits in rhythm with this movement adds greater force to the stride. The leg motion is considerably simpler in birds than mammals and the muscle masses are closer to the body, with the bones below the tibiotarsal joint having no muscle coverings at all.

Swimming, hopping, and raptorial (hawks, owls, etc.) birds have

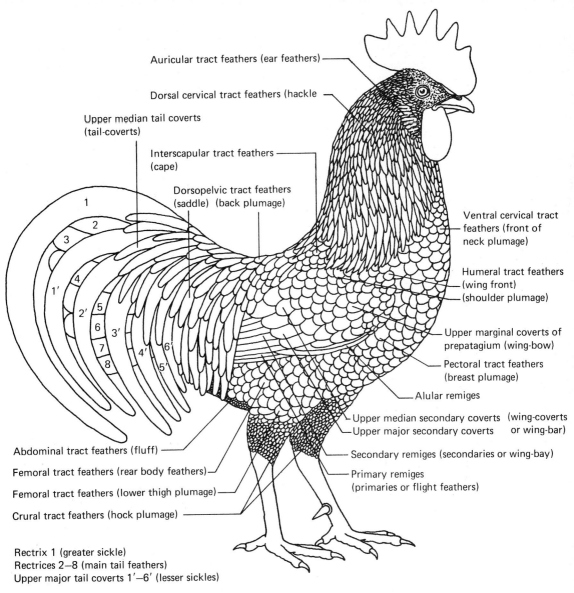

FIGURE 17-2 *Feather coat of the male Single Comb White Leghorn chicken—right lateral view. (From Lucas and Stettenheim,* Avian Anatomy—Integument.*)*

different leg functions and, in general, different leg movements. Perching birds also have a specific development of the tendons of the legs, which run in bony grooves and are capable of being locked in place with virtually no expenditure of muscular effort. Climbing birds such as woodpeckers and parrots have modified the toe and muscle arrangement to make their feet efficient grappling hooks. The grasping or seizing feet of the raptors are so designed that merely straightening the leg causes the talons to contract with enormous force, and a hawk's or owl's grip, once locked, can only be removed by amputating the leg of the bird and removing the talons one by one, unless the bird can be persuaded to let go.

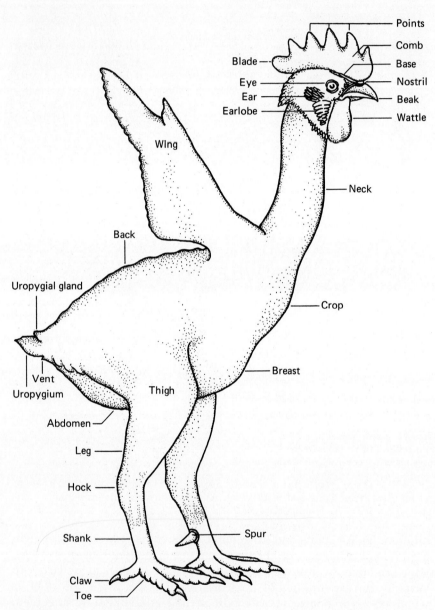

FIGURE 17-3   *External anatomy.*

Flying can be divided into a number of groups, each with somewhat different wing motions. For interest and to illustrate why the wing motions are not specifically described, various forms of flight and exemplar birds are listed below:

1. Flutterers—chicken, quail, partridge
2. Intermittent soarers—finches, sparrows, crows
3. Flappers—swifts, swallows, ducks, geese
4. Buzzers—hummingbirds, nighthawk
5. Wing dippers—cormorants, puffins, e.g., birds that fly well under water but poorly in air
6. Sail flyers—gulls, hawks, vultures, albatross, eagles (soaring birds)
7. Flightless divers—penguins, auks, grebes, certain cormorants, e.g., birds that use wings as fins and have lost their ability to fly in the air

Beak

Birds are toothless. The beak in all its varied forms replaces teeth. In general, beak forms are equated with diet and can run a gamut of shapes, from the tubular beak of the hummingbird to the massive beak of the hornbill, from the wedge-shape of the grosbeak to the cutting beak of the parrot, from the spoon-shape of the roseate spoonbill to the sword-like bill of the heron, and from the essentially simple seed-eating beak of the chicken, to the curved and notched scimitar beak of the hawk.

DIGESTIVE SYSTEM

The beak, or bill, forms the beginning of the digestive system. Considerable variation occurs between the alimentary tracts of birds and mammals. In this section, only the chicken will be discussed, but it should be understood that insectivores, raptors, fish eaters, and carrion eaters have extensive variations from gramnivorous, fruit or nut-eating, birds.

The Mouth

The beak opens directly into the mouth. In virtually all birds there are no cheeks as there are in mammals. The tongue is hard and relatively inflexible and can be considered to be bone covered by mucous membrane, which supports papillae that point posteriorly. The tongue divides the mouth into the beak cavity and the sublingual cavity. The beak cavity is partially closed off from the nasal cavity by an incomplete hard palate, which is divided by a median slit, the choanal space. Along the midline, somewhat caudal to the bifurcation of the choanal space, can be found a second longitudinal slit, the common opening of the eustachian tubes.

The choanal space is crossed by three transverse rows of caudally directed papillae. Near the pharyngeal opening at the base of the tongue are two lateral slits that contain the pharyngeal tonsils. Under each tonsil is a tonsillar salivary gland. Other salivary glands are the sublingual salivary glands, the maxillary salivary gland, the palatine salivary glands and a gland that lies near the angle of the beak, called the parotid gland. Chicken saliva is mildly amylolytic. Tongues can vary from the simple tongue of the chicken to the relatively rudimentary tongues of herons, storks, and pelicans or to the long, flexible, barbed, and hooked tongues of woodpeckers and the extensible tubular tongues of hummingbirds. The hyoid muscles, through the entoglossal segment of the hyoid, are the muscles that move the tongue.

At the base of the tongue are several rows of papillae together with a number of taste buds. These structures mark the anatomic demarcation between the mouth and pharynx.

The pharynx is open dorsally and communicates with the caudal nares via the choanal space. Ventrally the pharynx is perforated by a longitudinal medial slit, the laryngeal opening. Behind and above the laryngeal opening is the esophageal inlet.

The esophagus is a muscular membranous tube consisting of an outer three-layered sheath of striated muscle (outer longitudinal, middle circular and inner longitudinal). In some species the outer layer is absent. The muscles cover an epithelium, well-supplied with mucous glands and lymph follicles. The esophagus dilates near the thoracic inlet to form the crop.

The crop is not found in all birds. Pure insectivores (woodpeckers) and fruit-eating birds may not have crops. The crop can vary from the simple spindle-shaped food storage crop of raptorial birds to the glandular crop of gramnivorous birds.

In the chicken and other gallinaceae, the crop is located to the right of the median plane and is a single pouch. In the pigeon the crop is centrally located and has three subdivisions. In geese an anatomic variation called the "throat crop" exists. This is actually an extremely dilatable esophagus, which can be stuffed with food and which will retain it for considerable periods of time. This anatomic feature is used to advantage by makers of pâté de foie gras who literally stuff the throats of geese with food. This in turn causes increased absorption from the digestive tract, and produces large, fatty livers which are used to make the liver paste (pâté de foie gras) that is beloved by gourmets. In pigeons, sex hormone influences plus the presence of squabs, produce an exfoliation of crop lining cells, which, when mixed with mucus secretions from the crop, forms a fatty semifluid substance called "pigeon's milk" or "crop milk," that provides nourishment to the squabs.

The Body Cavity

Since birds do not have a true diaphragm, the body cavity is a single structure extending from the thoracic inlet to the cloacal vent. The cavity is not lined by serious membranes such as are found in mammals but contains seven discrete membranous sacs which apparently have no serosal relationships: two pleural sacs which cover the lungs, a pericardial sac which surrounds the heart, a dorsal hepatic sac, two ventral hepatic sacs, and a peritoneal sac, which covers all of the abdominal viscera except the kidneys.

The Stomachs

Caudal to the crop is the stomach. In most birds, including the domesticated species, the stomach is physically divided into two parts: a cranial, thin-walled, spindle-shaped glandular stomach (proventriculus) and a caudal, thick-walled, muscular, keratin-lined muscular stomach (ventriculus or gizzard). This latter structure has no homologue in mammals. The stomach structure is varied in the various species. In herons it is single. In raptors, the gizzard is thin-walled and the lining membranes are soft.

The glandular stomach produces enzymes and hydrochloric acid. The muscular stomach produces a secretion (koalin) which coats the lining membrane with a leathery yellow-brown covering that forms the grinding pads. Near the pylorus in the chicken, pigeon and goose is an area covered with villi, which supports true pyloric glands. There are neither cardiac nor pyloric sphincters.

The Small Intestine

The pyloric opening connects to the small intestine as in mammals. No major differences in function or structures exist between the small intestine of birds and mammals except that Peyer's patches are absent in birds, and the duodenum does not contain Brunner's glands. Crypts of Lieberkühn, however, are present throughout the entire intestinal tract.

The Jejunum

The jejunum commences as a funnel-shaped dilation at the end of the duodenum and is marked by the two pancreatic ducts and the bile duct(s), which empty into the duodenum near its termination. It is wound into a series of loops, and in the chicken, duck, and goose, terminates at the ileocecal ligament. In a majority of the domestic birds (chickens, 60%; ducks, 80%; geese, 90%) there is a structure in this terminal area called Meckel's diverticulum. It is a relict of the yolk sac stalk and functions as a lymphatic organ.

The Ileum

The ileum lies parallel to the duodenal loops and terminates at the cecal openings. For practical purposes there is no colon, although the band of tissue between the cecal openings has occasionally been so designated.

The Cecae

In most birds there are two cecae. In a few species (kestrels, herons) there is only one, and several species (pigeons, parrots, woodpeckers and others) have no cecae. The cecae act as terminal digestive and water absorption structures and contain scattered follicles of lymphoid tissues (cecal tonsils). The cecae are usually retrograde, i.e., point toward the cranial end of the body, and are larger at the apex than the base. They are delimited from the large intestine by the cecal valves.

The Rectum and Cloaca

The rectum is short and straight and expands rapidly to form the cloaca. The cloaca is divided into three segments—coprodeum, urodeum, proctodeum—by two annular folds. The coprodeum is an ampulla-like structure which opens into the urodeum, the shortest of the three sections. The urodeum is lined with a glandular mucous membrane and in chickens marks the transition from the columnar epithelium of the gut to the stratified squamous epithelium of the skin. Dorsally, two ureters open on papillae. Lateral to these papillae are (in the cock) the openings of the vasa deferentia or (in the hen) on the left is the slit-like opening of the vagina. Immediately caudal to the urodeum is the second annular fold, which marks the beginning of the proctodeum or cloaca. On the roof of the proctodeum is a transverse slit that opens into the sac-like bursa of Fabricius. On the floor of the proctodeum in the males of some species (ducks, geese, swans) is an erectile, lymphoid organ similar to the penis in mammals. It has a spiral groove rather than a urethra and is present in drakes, ganders, and cobs, but not in roosters or male pigeons. The proctodeum (cloaca) terminates at the vent (anus), which is closed by the two sphincter muscles, the sphincter ani internus (smooth muscle), and the sphincter cloacae (striated muscle).

The intestinal length of birds is considerably shorter on a comparative basis than the intestines of mammals. A bird's gut is about four to six times the body length compared to more than 30 times the body length in cattle and sheep.

ACCESSORY GLANDS OF THE DIGESTIVE SYSTEM

Four accessory structures can be recognized in birds: the liver, the pancreas, Meckel's diverticulum, and the bursa of Fabricius.

The Liver

The liver is a large mahogany-colored organ composed of a right and left lobe connected by a narrow isthmus of liver tissue. Each lobe is enclosed in its own serous sac. The left lobe is larger in chickens (although not in many other birds) and is divided into two parts. The liver is related cranially to the heart and lungs; cranially and dorsally to the lungs, to the proventriculus, spleen and intestinal tract; laterally and ventrally to the thoracic and abdominal walls; and caudally to the intestines and ventriculus. In the cock there is usually a testicular impression on the cranial portion of the right lobe. The right lobe also contains a ventral fossa that encloses the gall bladder. In chickens (also in geese and ducks) each lobe of the liver has its own bile duct. The duct from the left lobe empties directly into the intestine, while the duct from the right lobe empties into the gall bladder. This arrangement is not present in most birds although all domestic fowl except the pigeon have a gall bladder, and the structure is present in most birds with the exception of doves, ostriches, some parrots, and a few other species. The terminations of the bile duct from the left liver lobe, and the duct from the gall bladder, mark the approximate end of the duodenum and the beginning of the jejunum. The two pancreatic ducts empty in this area. Bile in birds is amylolytic, and weakly acid. This equates with the acid pH of the entire digestive tract.

The Pancreas

The pancreas consists of three lobes of loosely arranged glandular tissue which lie between the duodenal loops. Pancreatic secretion is essentially the same in birds and mammals although the pH in birds is less alkaline. The pancreas ordinarily empties into the terminal part of the duodenum via two pancreatic ducts, although there may be three in a significant percentage of chickens. Pigeons have three ducts 60 percent of the time. There is no apparent microscopic or functional difference between the pancreas of birds and mammals although birds have higher glucagon production than mammals, and are more insulin tolerant.

Meckel's Diverticulum and the Bursa of Fabricius

Meckel's diverticulum is found in a majority of examinations of the intestinal tracts of domestic fowl. It varies in shape from button to wormlike and is a relic of the yolk sac placenta. It contains lymphoreticular tissue and is one of the two major regions of lymphatic tissue, the other being the bursa of Fabricius.

The bursa of Fabricius is a dorsal evagination of the cloacal wall that joins to the cloaca via a transverse slit-like opening. It is covered by a well-developed layer of smooth muscle and is lined with folded mucosa containing glands and lymphoid tissue. The bursa undergoes physiologic atrophy either shortly before or at the time of sexual maturation. The structure is found in all birds and is apparently unique to them.

PHYSIOLOGY OF DIGESTION

Food recognition is almost entirely by sight. Odor and taste have little significance. Touch in some birds such as geese and ducks apparently plays a significant minor role.

Prehension is a function of the beak, although raptors use their talons to secure food and chickens and other galliformes scratch to uncover grubs, worms, and subterranean insects. The beak can crush, grind, cut, or tear food, depending upon the beak conformation and the dietary habits of its owner. Food ordinarily does not remain in the beak for any period of time, except in pelicans, which have a pocket below the beak, and puffins, and some insectivores, which can carry food in their beaks for considerable distances without swallowing it. In the oral cavity saliva is slightly acid and amylolytic. This appears to be particularly true in chickens and geese among the domestic birds. The crop, in these species, serves as a storage organ.

Swallowing does not involve peristaltic action of the oral muscles as in mammals, because of the absence of a soft palate and appropriate pharyngeal and tongue muscles. The act of passing food from the oral cavity to the esophageal opening is essentially a cramming action, but once the esophagus encloses the food, active peristaltic waves move it to the crop or stomach. The peristaltic action is slow compared to mammals, being in the neighborhood of 1.5 centimeters per second.

Drinking, since the nasal openings cannot be closed and the oral muscles for suction are not present in most birds, involves filling the mouth cavity and raising the head to allow the water to flow into the esophagus. The pigeon is an exception, since it can close off its nasal openings and establish suction.

In food acquisition, the glandular and muscular stomachs are filled first. After the stomachs are filled, additional food is passed into the crop where it is stored, and in some birds is prepared by maceration for further digestion. Fluids are supplied by the salivary glands, esophageal glands, and glands in the crop mucosa. As food is moved out of the stomach the crop replaces the food, which has been passed into the gut.

Glandular stomach contents can be transferred to the crop by reverse peristalsis. Although this does not usually occur, it is not necessarily an abnormality. Crop contents have a moderately acidic reaction (pH 5.0), which may be induced by the presence of gastric secretions, but which is more probably derived from the acid mucus produced by the salivary glands, esophageal glands, and intrinsic glandular structures. The crop is supplied with motor nerves from a branch of the left vagus nerve. The nerves apparently control the rate and quality of food passed from the crop to the stomach. The softer foods are usually passed out of the crop first. The crop also appears to be a center for hunger contractions comparable to those in the stomachs of mammals.

The secretion of the proventriculus varies in composition according to the kind of food and the quantity of gastric secretion, but is similar to the gastric juice of mammals. Pepsin concentration is relatively weak; HCl concentration is relatively high (pH 4.0). As in mammals, gastric secretion can be increased by psychic stimulation as well as the physical presence of food.

The movements of the two parts of the stomach of birds proceed at different rates and are not coordinated. The peristaltic movements of the glandular stomach are relatively slow, about one per minute in hungry chickens.

The contraction movements of the muscular stomach are considerably more rapid and vary in both rate and force with the type of food being acted upon, and its resistance to crushing and grinding. Normal contraction rate of the chicken's ventriculus is about three per minute, but can increase to more than four if the food is hard or decrease to two if it is soft. Because of the acid reaction in the gizzard the grit usually found in the ventriculus is either granite or silicate, since calcium is quickly dissolved and passed through the intestines where some of it is absorbed. Grit appears to have little useful function, although birds (particularly chickens and turkeys) will selectively take it if they have an opportunity to do so.

The enzyme and acid effect upon the food in the proventriculus is minimal in gramnivorous birds. However, after being crushed or ground in the gizzard, the enzymatic action is pronounced. Unlike mammals, birds appear to have no alkaline phase digestion, and the entire gut has an acidic reaction, although the reaction gets weaker as the food approaches the vent. However, the intestinal enzymes have no essential variation from those found in mammals.

The ceca in birds function as areas of bacterial digestion where further breakdown and absorption of nutrients takes place. They also function in water absorption. Not all the ingesta which passes into the coprodeum is taken into the ceca; an appreciable amount is passed out of the body without further processing.

The so-called large intestine of birds functions more like the rectum of mammals than as a digestive organ, and most birds require the presence of ceca for water conservation reasons if for nothing else. However, the mechanics of avian cecal function, particularly in reception and moving of ingesta, are as much of a puzzle as they are in mammals with large ceca, such as horses and rabbits. The ceca apparently selectively take the fluid portions of the ingesta. Cecal retention times are longer than the rest of the avian gut, which passes food quite rapidly (about 4-11 hours). Cecal contents are expressed once or twice a day and are semisolid, brown, and smooth in consistency. The ceca are not essential for normal functioning of the digestive tract since they can be removed or surgically occluded without apparent ill effect.

Food and water absorption takes place across the gut epithelium and the epithelium of the ceca. Large intestine absorption of food is minimal and inconsequential, but some absorption of water does occur.

Nutrition and Metabolism

Diets of birds vary as much as do their physical characteristics, but some general observations can be made that correspond with their anatomy and physiology. Protein in bird diets should be high quality, because of the relatively short gut and gut retention time. Poor quality protein will often be digested and metabolized poorly, if at all. If anything, the essential amino acids in birds are even more essential than they are in mammals, and include glycine and glutamic acid besides the 10 normally required by mammals. Occasionally cystine, tyrosine, and proline appear to be necessary in amounts greater than growing birds can supply through synthesis.

Birds digest and absorb small carbohydrate molecules (starch and lower) with facility. Digestion of crude fiber occurs mainly in the cecum. Cellulose, lignins and pentosans derived from grains are reported to be 10 to 40 percent digestible, although similar substances derived from corn fodder are almost completely indigestible. Moderate amounts of fats are well digested although there appears to be no gastric lipase.

Vitamins and Minerals

Vitamins must always be considered in fowl nutrition, particularly vitamins B_6, B_{12}, A, and E. Birds synthesize most of the B-complex and all the vitamin C and D they need.

Minerals should be watched carefully. The calcium-phosphorus ratio differs from that of mammals, being 1.6:1 in growing animals and 3.7:1 in layers, or about 1.0 percent of the weight of feed in calcium, and 0.6 percent in phosphorus. Available iron is necessary for blood formation at a level of about 1.0 gm per kg of feed, and copper at a level of about 2.0 gm per kg. Manganese deficiency causes a number of associated skeletal and metabolic disturbances, the most spectacular of which is perosis. Manganese should be supplied at a level of 55 mg/kg total dry weight of feed for chicks and 35 mg/kg total dry feed for laying hens. Iodine is needed at a level of about 0.50 mg/kg total dry feed. Selenium should be furnished as 0.1 mg Na selenate per kilogram of dry feed. Excesses of selenium, molybdenum, and cobalt can be dangerous. Selenium not only can be toxic in excess, it can inhibit enzyme production, produce anemia, liver damage, lameness, blindness, and respiratory and nervous disturbances. Excess molybdenum can produce sterility, and excess cobalt can cause polycythemia and hyperglycemia. Salt can be poisonous if given as brine or furnished free choice (ad libitum) to overheated birds. In normal birds salt may be found concentrated in discrete areas of the skin.

RESPIRATORY SYSTEM

Although there are anatomical differences, the basic function of the respiratory system is the same in birds and mammals. Structurally, however, the differences are extreme.

Commencing with the nasal cavity, the difference in the number and location of the turbinates, the lack of an epiglottis and the tracheal rings of bone rather than cartilage, the respiratory systems of birds and mammals diverge. A nasal gland, which has no homologue in mammals and allows sea birds to drink salt water is found in the anterior portion of the beak. A syrinx, which also has no mammalian homologue, contains the vocal cords and is located just cranial to the tracheal inlet into the body cavity. The two lungs (right and left) do not have lobes, do not appreciably expand or contract, and do not have alveoli. They are bright pink in color, fit between the thoracic ribs and are attached to the ribs by connective tissue. They are marked by deep costal impressions on their parietal surfaces and are relatively small and triangular in outline. The lungs extend from a pointed apex in the region of the first ribs to a wide base at the cranial border of the kidneys. A system of tubes of decreasing diameter ramify through the lung tissue and then enlarge again to form sinuses that terminate in air sacs. There are nine air sacs in most species of birds (11 in pelicans, cormorants, grebes, penguins, cranes, storks and others): two abdominal, two caudal thoracic, two cranial thoracic, two cervical, and one clavicular. Those birds with 11 sacs have two subcutaneous sacs in addition to the nine listed above.

The tubules and the air sacs form a continuous interconnecting series of passageways. Evidence indicates that the air is inspired through relaxation of the abdominal and pectoral musculature which creates a negative pressure in the air sacs, and sucks air through the air passages. The air passes rapidly through the lungs and into the large abdominal air sacs and, to a lesser extent, into the others. From here contraction of the abdominal musculature compresses the abdominal and thoracic sacs, and air is displaced forward and outward. Some of the expired air passes successively through the respiratory tubules and into the thoracic sacs, and again through the tubules and into the cervical and clavicular sacs and then to the bronchi, trachea, and the outside. In the process, the air becomes progressively more humidified and some molecules of air may pass through the entire sac linkage before they are expired. However, a certain amount of air from each air sac is directly expired after passing through the respiratory tubules (air capillaries). Most gaseous interchange between air and bloodstream takes place on expiration rather than inspiration.

The entire mechanism is still not well understood, but it is highly efficient and at a high level of evolutionary development. Birds' respiratory surface per kilogram of body weight is greater than the respiratory mechanism of any other animals. On a gm body wt/cm^2 respiratory area basis, birds will run about 1:300 and mammals about 1:15. The birds will have an average respiratory advantage (on this basis) of about 20 to 1. However, it should be remembered that mammals have a positive system of ventilation with a diaphragm and specific inspiratory and expiratory muscles, which birds do not possess. In consequence, avian inspiration is essentially passive and expiration is weak, though fairly rapid. The flow of air through the respiratory tubules is therefore reduced in quantity and low in pressure. However, the question of respiratory musculature has never been completely settled. Experiments involving transection of the pectoral muscles and the abdominal muscles (although not both at the same time) do not stop respiration. The two diaphragms of the chicken (the pulmonary diaphragm and the abdominal diaphragm) have been inactivated or sectioned without affecting respiration. However, curare, a drug that paralyzes all muscles, causes death unless artificial respiration is applied.

Further questions involve expansion and contraction of the lungs. On inspection, expansion or contraction of lung tissue appears doubtful because of the construction of the lungs, and location and tension of the pulmonary diaphragm, although small changes in size might occur with the flow of air through the smaller tubules. However, there is no question that the air sacs expand and contract. Nevertheless, destruction of the air sacs does not stop respiration although it alters the quality and character of the respiration. This is not surprising since the body cavity could (and probably does) function as an air sac in this event, and the reflux of air through the lung would still take place.

For more than 90 years it has been known that there is minimal movement of the chest region. The sternum, furcula, coracoid, and ribs can (and do) move forward and downward with relaxation of the thoracic muscles. This increases the vertical diameter of the thorax and helps expand the thoracic and abdominal air sacs while compressing the clavicular and cervical air sacs. This action results from relaxation rather than contraction of muscles. The abdomen can also relax and enlarge, but specific inspiratory muscles are lacking.

Birds easily suffocate if they "pile" as chicks and turkey poults often do when chilled. The lower layers of birds simply cannot inhale. Piglets, which also tend to "pile," virtually never suffer from this kind of activity.

Breathing frequency in resting chickens ranges from 20–40 respirations per minute. The rate is roughly proportional to size. Ducks have 60–70 respirations per minute, geese 20–22, turkeys 12–14, and pigeons 20–35. Control of respiration seems to be essentially the same in birds and mammals.

BLOOD VASCULAR SYSTEM

In birds, the blood vascular system operates at a much higher pulse rate and blood pressure level. Considering that some birds live to be more than 100 years old with such an arrangement, there should be little question of its efficiency or effectiveness.

The main difference in the vascular system lies in the heart. The right atrium of the chicken is larger than the left but the mass of the left ventricle is nearly three times that of the right. The right A-V valve is simply a muscular flap, but the left A-V valves are membranous. The aortic and pulmonary valves are similar to those in mammals.

Purkinje System

As in mammals, a specialized conducting (Purkinje) system exists in the heart muscle, but the Purkinje fibers in avian hearts also ramify through the atria. This does not occur in mammals. In birds the right fourth aortic arch becomes the definitive arch of the aorta. In mammals it is the left (if the right arch persists in mammals it results in the tetralogy of Fallot). Birds require nearly one and a half times as much heart muscle as mammals to achieve similar tissue oxygenation. This is probably due to the larger size, fewer number, and more primitive development of the erythrocytes.

Blood

It is in the blood of birds where the principal blood vascular differences exist. The red blood cells (erythrocytes) are nearly half again as large as mammalian red cells, and are oval and nucleated and there are significant differences in the white blood cells (leukocytes). Compared to the sheep a bird has about half the number of red blood cells (male, 3.3 million/mm^3; female, 2.75 million/mm^3). The sex variation in erythrocyte number in chickens is similar to that in mammals. However, many species show no sex variation. Thyroidectomy reduces the number of erythrocytes in male chickens but not in females. The erythrocytes of avian blood are capable of synthesizing hemoglobin during their life in the bloodstream, whereas mammalian erythrocytes cannot do this. Hemoglobin content expressed in milligram percent is about 11 percent in chickens and 13 percent in pigeons, which is comparable to mammals.

Blood platelets are not found in avian blood and megakaryocytes are not found in the bone marrow. The thrombocyte in avian blood is a complete cell and appears to be a cell of the erythrocyte series. It can be recognized by its different staining characteristics and the presence of one or more bright red granules in the cytoplasm.

Lymphocytes are similar to those found in mammals and so are basophils. The monocyte does not routinely have the indented nucleus of the mammalian cell and lacks the metachromatic granules in the cytoplasm. The eosinophil has smaller granules in the cytoplasm and the neutrophil is replaced by the heterophil, a cell with a lobed nucleus and bright red oval granules in its cytoplasm. Lymphocytes are by far the most numerous of the leukocytes in avian blood. Functionally, the leukocytes of birds mirror their mammalian counterparts:

	Total Leukocytes, Thousands Per mm^3	Lymphocytes	Heterophils	Percentages Eosinophils	Basophils	Monocytes	Thrombocytes, Thousands Per mm^3
Chicken	20.0	59.0	27.2	2.0	1.7	10.2	25.4
Chick	29.4	66.0	20.9	1.9	3.1	8.1	26.5
Duck	23.4	61.7	24.3	2.1	1.5	10.8	23.4

Partly from Kölb, *Lehrbuch der Physiologie der Haustiere*.

Insofar as blood coagulation is concerned the mechanism in birds is similar to that in mammals. Functionally, however, there is poor clot retraction. Chickens appear to require an exogenous source of vitamin K for normal blood coagulation.

Lymphoid Tissue

One of the interesting contrasts between the chicken and mammals is the lack of definitive lymph nodes in the chicken. Lymphoreticular tissue in lymph vessels, the cecal tonsils, the bursa of Fabricius, Meckel's diverticulum, and scattered areas of lymphoreticular tissue in the gut form the main lymphoid structures except for the spleen, which is similar in microscopic structure and in function to that of mammals. Other birds, particularly waterfowl, have true lymph nodes.

URINARY SYSTEM

The kidneys and the ureter form the urinary system. The kidneys are more primitive and relatively larger than those of mammals and are located in bony pockets formed by the fusion of the transverse processes in the lumbosacral region with the ilium of the pelvis. The cortex and medulla are not sharply defined, and two types of nephrons exist: those with Henle's loops that enter the medulla and those with structures confined entirely to the renal cortex. The kidneys vary greatly between species of birds, particularly in microscopic structure. Some differences between chickens and mammals are as follows:

> The cells of the distal convoluted tubules in chickens have no radial striations.
>
> The collecting tubules are lined with columnar epithelium.
>
> The ureters are lined with pseudostratified epithelium.
>
> The urinary output is a urate paste, which is voided directly into the cloaca.
>
> The urinary excretion of the avian kidney has a preponderance of uric acid over urea, and creatine over creatinine.
>
> There are three major arteries to each kidney: the cranial renal artery, the renal branch of the femoral artery, and the renal branch of the ischiadic artery which supply the cranial, middle, and caudal parts of the kidney, respectively.
>
> The kidney has a renal portal system that is afferent and supplies blood to the tubules via a peritubular capillary network, which ultimately joins the medullary plexus. This structure does not occur in mammals.

A valve at the junction of the renal portal and the iliac veins apparently governs the flow of blood into the kidney. Functionally, the kidneys are similar to those in mammals, although their excretory products differ.

REPRODUCTIVE SYSTEM

Male

The testicles of birds are internal and are located cranial to the kidneys. The glands are structurally and functionally similar to those of mammals, but avian sperm differs from mammalian in one important aspect. There is no dominant male sex chromosome. Avian spermatogonia contain two identical sex chromosomes (ZZ) in contrast to the recessive X and dominant Y in male mammals. On maturation, therefore, every avian spermatozoon carries a Z chromosome. This makes the female the sex determiner for the chicks. The female sex chromosome pattern is WZ. Therefore, on maturation, an avian oocyte will have either a Z or a W chromosome. Fertilization will result in the restoration of the diploid number of somatic chromosomes plus either ZZ or WZ sex chromosomes.

This lack of a dominant male sex chromosome is a solution to sex determination of offspring that is rare in the animal kingdom. Other than birds, only a few species of moths have this mechanism. It is a major and fundamental difference between birds and other vertebrates. It also explains why avian parthenogenesis (p. 539) invariably produces male offspring, which is opposite to what would happen in other vertebrates. This is one more piece of evidence that birds are the end product of parallel rather than concurrent evolution.

A copulatory organ (not a penis) is found on the floor of the cloaca in some birds. It is ordinarily composed of two papillae and a rudimentary mass of erectile tissue. There are no ampullae, prostate glands, seminal vesicles or Cowper's glands in male birds. Vascular bodies near the terminal end of the vas deferens dilute the sperm. There is no essential difference in male sex hormones and their action in birds and mammals. Lack of vitamin E in a male bird's diet will produce permanent testicular damage and infertility.

Female

Only the left ovary and oviduct persists. Since birds are oviparous, there are a number of structures in the female reproductive system that are not homologous to mammalian organs. The ovary lies to the left of the median plane, cranial to the left kidney. It is suspended from the dorsal portion of the abdominal cavity by the mesovarium. In appearance, the functioning avian ovary looks like a bunch of grapes and contains follicles in all stages of development. Although the follicle numbers are finite, breeding can increase the number of follicles that develop during a breeding season.

The domestic chicken has been greatly modified sexually from its jungle fowl ancestors that laid relatively small clutches of eggs every year. Through breeding and selection hens have been developed that can lay more than 300 eggs per year, which makes a laying hen, for all practical purposes, an egg-laying machine with a useful life of from one to two years.

In chickens developed for egg laying, the production of eggs is not ordinarily a daily event, although some hens manage to produce an egg a day for considerable periods of time. There is, regardless of the hens involved, a certain rhythmicity to egg production.

Egg Production

Type and Rating	Sun.	Mon.	Tue.	Wed.	Thurs.	Fri.	Sat.	Sun.	Mon.	Tue.	10 Day Record
1. Excellent	+	+	+	+	+	+	+	+	+	+	10
2. Very good	−	+	+	+	+	+	+	−	+	+	8
3. Good	−	+	+	+	+	+	−	+	+	+	8
4. Mediocre	−	+	+	+	−	+	+	+	−	+	7
5. (Arrhythmic)	−	+	+	−	+	+	+	−	+	+	7
6. (Irregular)	+	−	+	+	+	−	+	+	−	+	7
7. Poor	+	−	+	−	+	−	+	−	+	−	5

Poultrymen prefer birds of types 1 and 2. More often they get types 3, 4, and 5. Type 6 would probably be culled and type 7 certainly would.

The egg that is the object of all this activity is quite complex since it must contain inside the shell everything necessary for fertilization of the ovum and development of the ovum into a complete and functioning individual within 30 days.

The white (albumin) is principally proteinaceous building material. The yellow (yolk) is principally nutrient for the developing embryo. The germinal disc contains the haploid female nucleus.

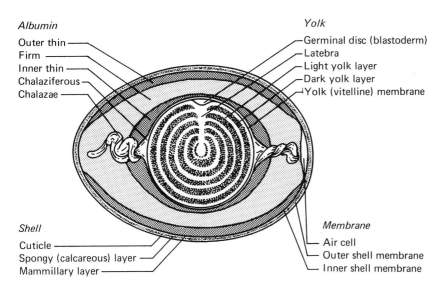

—from Egg Grading Manual USDA Ag Handbook #75

FIGURE 17-4   Structures of the egg. (From Egg Grading Manual, USDA Ag Handbook #75.)

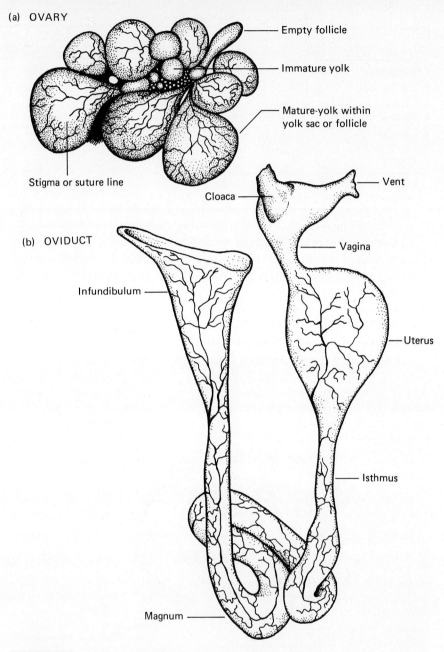

FIGURE 17-5 Ovary and oviduct of the chicken. (From Egg Grading Manual, USDA Ag Handbook #75.)

The shell membrane and the shell are protective coverings.

The egg passes successively through the ovary, oviduct, cloaca and vent to reach the outside.

To produce a structure of the complexity of an egg, a number of adaptations occur in the oviduct and are called infundibulum, magnum, isthmus, uterus, and vagina. While some of these parts have the same name as parts of female mammalian reproductive apparatus, that is about as far as the likeness goes.

The ovary produces the germinal disc and the yolk, together with its covering vitelline membrane. When the yolk reaches definitive size it is expelled from the ovary into the infundibulum. A corpus luteum is not formed from the ruptured follicle that contained the yolk. In the infundibulum of the bred female are "sperm nests," i.e., clusters of spermatozoa in epithelial crypts held in readiness to fertilize each egg as it comes from the ovary.

The inner thin albumin, the chalaza, and the outer thick albumin are secreted in the magnum and laid over the yolk and sperm. In the isthmus, the outer thin white is produced by diluting the outer portion of the thick albumin with water secreted by the isthmus, and the outer and inner shell membranes are also produced in this area.

The uterus produces the shell and shell pigment (if any) and supplies some fluid to lubricate the shell and assist the passage of the egg through the vagina and vent. The vagina is merely a connecting tube between the uterine structures and the vent. When the egg begins to travel through the oviduct it starts to move small end first, but turns end for end in the uterus and passes down the vagina and out the vent blunt end first. Ordinarily the vagina everts into the cloaca as the egg descends and the egg is thus kept from touching the cloacal mucosa as it is pushed into and through the vent by a combination of peristalsis and abdominal press.

Broodiness

In normal birds as well as highly bred laying hens, lay can be interrupted by "broodiness" or by molt. The exact mechanism of broodiness is unclear. Prolactin appears to be associated with its onset. The phenomenon is marked by a rise in body temperature, particularly in the breast region. Progesterone can halt the process. The trouble with using progesterone to stop broodiness is that the drug initiates molt. Broodiness is the method that nature uses to induce hatching of a clutch of eggs, and while it is detrimental to egg production, it is a necessary function in wild fowl that has not yet been eliminated in the domesticated egg layer.

Hatching times vary in different species. In chickens, the time is three weeks, during which the eggs must be kept warm, mildly humid, and must be turned several times a day to prevent improper development of the embryo. Incubators do this job in modern poultry operations.

Since newly hatched chicks and poults have poor temperature regulation mechanisms, they must be kept warm for several weeks. In the wild state, the female broods her young. In mechanized operations, brooders or heat lamps are used. Temperatures for newly hatched birds are about 32°C and can be reduced 2°C per week over a period of six weeks. Optimum temperature for growing chickens is about 21°C with 60–70 percent humidity. Temperatures, above 25°C or below 15°C are unfavorable for growth and development.

Molt

Molt, or feather loss, is ordinarily seasonal and occurs in the spring and fall. It is usually less severe in the spring and is, in the male, characterized by the production of brighter mating plumage. In laying hens, severe molt can be induced by administration of progesterone, injection of thyroxin, or by withholding food and water long enough to halt egg production. Prolactin injections can also (inconstantly) induce molt.

In wild birds or in meat-type or fancy chickens, the heaviest molt occurs in the fall of the year and possibly correlates with length of the day, since similar molt can be induced by keeping birds in darkness. Molt is accompanied by a rise in metabolic rate and body temperature, particularly during the loss and replacement of the flight feathers. The exact mechanisms that cause these reactions are still unclear, and severe denuding molt is thought by some to be a pathologic rather than a physiologic phenomenon, since generalized loss of plumage is not a survival trait.

Mating and Sexual Behavior

Mating behavior and sexual behavior are not the same thing, although the first leads to the second. Sexual or copulatory behavior is relatively simple. The birds couple with the male on top. The female everts her vent bringing the vaginal mucosa to the body surface. The male covers the everted vagina with his own everted vent and deposits sperm into the vaginal orifice. The sperm make their way up the female reproductive tract by rheotaxis, moving against the slow currents of oviductal and vaginal secretion, and upon reaching the oviduct are held in crypts ("sperm nests") in the epithelium. The sperm remain capable of fertilizing eggs for several weeks in chickens, two weeks in turkeys, and 10 days to two weeks in ducks.

Although it takes only about an hour after copulation for sperm to reach the oviduct, it requires about two days for fertile eggs to be laid. In chickens the highest fertility percentage occurs 10–14 days after mating (turkeys seven days).

Mating behavior, or courtship behavior, in most species is a ritual designed to attract the male and impress the female (if we can relate anthropomorphic attitudes to avian psychology), but this, while interesting, is a quagmire that is better avoided. It should be enough to state that there is ritual behavior after which the birds either pair for a season (most songbirds), for life (hawks, geese, swans), or for a while (ducks), or establish harems (chickens, turkeys).

In turkeys, the hen apparently initiates the sex act by crouching and spreading her tail. The tom then struts, raises his feathers, expands his tail, and rasps his primaries against his legs. If the female remains down, he mounts, treads on the female's back, and breeds. In some overly developed strains with excessively broad breasts, copulation cannot take place and hens must be artificially inseminated.

In chickens the breeding ritual is more involved. The cock initiates the copulating act by approaching with a peculiar waltzing gait, fluttering his wings, and repeatedly rasping his primaries against his shanks. The hen then runs: the cock pursues: the hen crouches, raises her wing fronts and spreads her scapulae. The cock mounts, grasps the hen's neck with his beak, and treads. The hen moves her tail aside and everts her cloaca. Cock and hen then bring their vents into apposition and ejaculation occurs. The cock steps off and may waltz a few steps. The hen stands and fluffs her feathers, or she may run. A cock can perform from 50 to 100 of these acts a day.

Other birds have other ways, but the mechanism for getting vent to vent is essentially the same, although birds such as the swan, goose, and duck have hemipenes which allow true coupling rather than superficial contact.

Embryonal development begins prior to laying, but can be suspended (at least for awhile) by lowered temperature. Cleavage occurs about five hours after fertilization, while the egg is still in the uterus. A laid fertile egg is in the early cell mass stage. Embryonal development of birds and mammals is much the same, except that the egg must supply all the nutrition and building material required to produce a viable free-living, organism.

Not infrequently, particularly in Beltsville small white (BSW) turkeys, and less often in other turkey breeds and chickens, unfertilized eggs will spontaneously develop embryos and a certain percentage of these will hatch. The phenomenon of parthenogenesis, was first noted at Washington State University a number of years ago. More recent investigations indicate that parthenogenesis can be stimulated by fowl pox or Rous sarcoma virus. The trait, once established, can be bred into birds and gives lines that have an incidence of 10–14 percent parthenogenates in their unfertilized eggs.

Parthenogenates are usually weak, and many have genetic defects, but some are essentially normal. All have the diploid number of chromosomes for their species, and all are male. This latter event is to be expected since the female bird determines sex (p. 534). The diploid WZ

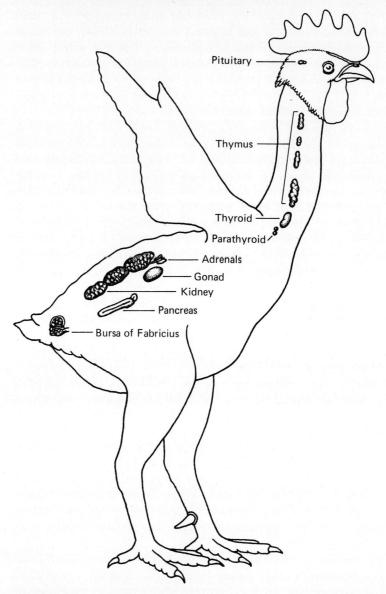

FIGURE 17-6   *Endocrine glands of the chicken.*

sex chromosome pattern becomes either W or Z after the second maturation division of the ovum. Restoration of diploidy in eggs with a W chromosome would result in a WW configuration. Such individuals would be genetic freaks and would not develop since certain vital genetic information other than sex is located on the Z chromosome. Since the diploid number of chromosomes is restored after meiosis II has occurred, a WZ pattern could not be formed. The only viable alternative left is a ZZ configuration, which is male.

NERVOUS SYSTEM

The brain of fowl is relatively small compared with the brain of mammals and has minor anatomic differences, i.e., the cerebrum is not convoluted and the olfactory tract is single. There is a corpora bigemina rather than a corpora quadrigemina. This is extraordinarily large and is often called the optic lobes. The cerebellum and medulla are disproportionately large compared to similar structures in the mammalian brain.

In the spinal cord there is no cauda equina, and there is a peculiar enlargement in the pelvic region called the sacral organ or glycogen body.

There are no gray or white rami communicantes in the thoracic region; the sympathetic trunk ganglia are fused with the dorsal root ganglia. However, commencing at the thoracolumbar junction, sympathetic trunk ganglia and rami appear.

However, the functional similarity between birds and mammals insofar as the nervous system is concerned is quite close. The only major difference is that mammals have peripheral and autonomic nerve supply to mammary glands, structures absent in birds.

The sacral organ or glycogen body is peculiar to birds. It is found in the lumbosacral region of the vertebral canal and lies dorsal to the spinal cord. It is essentially a honeycomb structure of polyhedral glial cells which contain considerable amounts of glycogen. It is thought to be an area of glycogen storage. It is resistant to glucagon treatment and to the glycogenolytic activity of injected epinephrine. The glycogen body seems to be able to store and release glycogen or glucose although the mechanism has not yet been discovered. Experiments indicate that the structure has little, if any, metabolic importance. At present, its possible function is more a matter of guesswork than knowledge. It may be a protective mechanism to ensure an adequate supply of glucose to the nervous system, or it may be a regressive structure, an evolutionary relict of the enormous pelvic ganglia found in some of the dinosaurs. Whatever it may be, its close anatomic relationship to the spinal cord makes it feasible to describe the structure as connected with the nervous system.

ENDOCRINE SYSTEM

Although there is a difference in anatomical structure and location of some of the endocrine glands of birds as compared to mammals, notably the thyroid, parathyroid, adrenals, testicle, and progesterone secreting tissue, the hormonal products are essentially the same and have the same function. Mammalian origin hormones cause essentially the same effects in birds as do natural avian hormones.

Progesterone offers an interesting study area. Since a corpus luteum is not produced in the avian ovary, the exact origin of the hormone, which can be detected in hens and cocks, is something of a mystery. It is probable that the hormone is produced in the ovary or uterus, but the location has not yet been pinpointed.

The study of endocrinology in fowl is presently an expanding field of research.

COMMON INTEGUMENT

General

The skin of fowl is virtually devoid of glands and is thin, dry, and tends to wrinkle in a lattice-like pattern. The outer surface produces a profuse amount of epidermal scales (danders). With the exception of Silkie chicken skin, which is black, and scattered melanin deposits in turkeys, the skin of most domestic fowl is a yellowish white color. In certain areas of the body the epidermis forms horn. The beak, spurs, claws, horny plates, and buttons of the shanks and toes are the specific locales for horn formation.

The corium is thin and not papilliform except in the eye region.

The pennate muscles in the skin are smooth muscle and similar to, but better developed than, the arrectores pili muscles in mammals. There are also striated "false" skin muscles, which are branches of skeletal muscle and ramify through the corium and subcutis.

There are no sweat, sebaceous, or ceruminous glands except for an oil gland (the uropygial gland) located in the pygostyle. There are no mammary glands.

The appendages of the skin are combs, wattles, ear lobes, caruncles, crests, snoods, and other structures which can be quite extensive and startling in color and conformation. They are essentially thin skin foldings well supplied with blood. Colors other than red or black are supplied by lipochrome pigments. Their purpose is unknown, although they may help synthesize vitamin D through exposure of the skin to sunlight.

Birds do not have hair, although certain filoplumes, particularly in the "beard" of turkeys look enough like tactile hairs to fool the uninitiated.

Feathers

The outstanding accessory structures of avian skin are feathers. These, like hair in mammals, are a unique development in birds and have no direct relationship to scales or any other epidermal structure. Their closest comparable structure is hair. The idea that reptilian scales are antecedent to hair is about as valid as the idea that they are antecedent to feathers. More properly, all three structures are unique evolutions of coverings that may have a common ancestor somewhere far back in evolutionary time.

There are at least five basic categories of feather: flight feathers, covert or contour feathers, down feathers, filoplumes, and bristles. Each of these categories are constructed on the same basic plan, with modifications that principally involve the vanes, barbs, and barbules.

Anatomically, a feather is composed of four basic parts: a quill (calamus), a shaft (rachis), an aftershaft (hyporachis), and vanes (vexilla). The quill or calamus lies in an epidermal sheath or follicle, the bottom layer of which is formed into an epidermal collar and a dermal papilla. The epidermal layers of the skin (the corneum and germinativum) fold backward at the base of the follicle to become the sheath, the intermediate, and the basilar layers of the quill. The basilar layer forms a layered series of cavities (pulp cavity or pith cavities). The lowest layer of the series is filled with pulp, a tissue derived from the dermal portion of the skin. A small opening (the inferior umbilicus) at the bottom of the quill permits the entrance of blood vessels to the pulp cavity and provides a continuity of tissue between the dermis and the base of the quill (Figure 17-7).

Intermediately, above the level of the skin, the quill and the basilar layer terminate in the superior umbilicus. The sheath and the intermediate layer continue as the rachis and hyporachis. The sheath layer disintegrates along the medial (ventral) side, with the bulk of the sheath layer and intermediate tissue forming the shaft (rachis) and the remainder forming the aftershaft (hyporachis).

Barbs grow from the free edges of both rachis and hyporachis forming the outer and inner vanes (vexilla). The outer vane is the smaller.

From the barbs grow barbules, and from the barbules grow the hooklets (Figure 17-8). The complete feather with all the structures present is usually the primary or secondary flight feather (remiges), but may include coverts, contour feathers, sickles (rectrices) or main tail feathers (Figure 17-2).

All the other feathers represent losses of one or more components of the complete feather. Coverts and contour feathers are complete in their terminal portions but at the base the hooklets are absent, producing a fluff of barbs and barbules.

In down feathers, the hooklets are completely absent. In filoplumes, the barbs and barbules are present only at the tip of the rachis. In bristles, the extreme modification of the feather, only the rachis remains.

Feathers grow along distinct tracts of the skin. The intervening spaces are either nude or contain a scant amount of filoplumes. The general location of the feather tracts is shown in Figure 17-9 and can be easily seen on the skin of a plucked bird.

Feathers serve similar physiologic purposes as hair, i.e., protective covering and insulation. In addition, feathers give enough resistance to

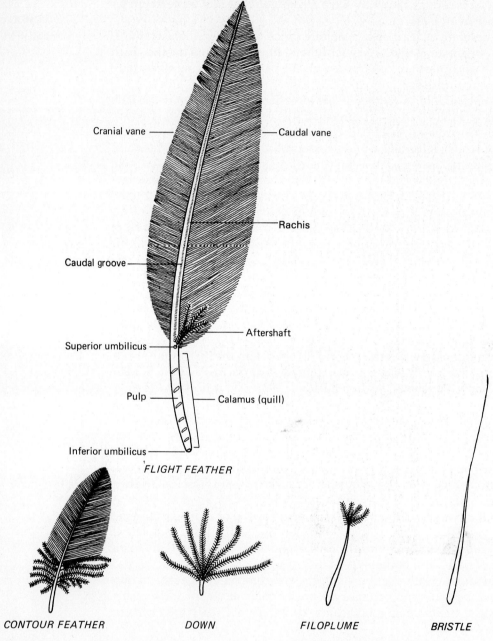

FIGURE 17-7 Types of feathers.

544

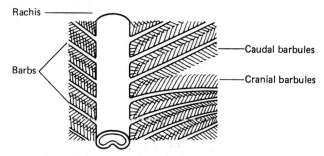

(a) Section of vane and rhachis

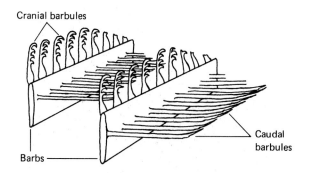

(b) Portion of two adjacent barbs

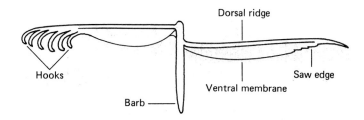

(c) Section through a barb

FIGURE 17-8   Structure of the feathers. (From Robinson, *Anatomy of the Domestic Chicken.*)

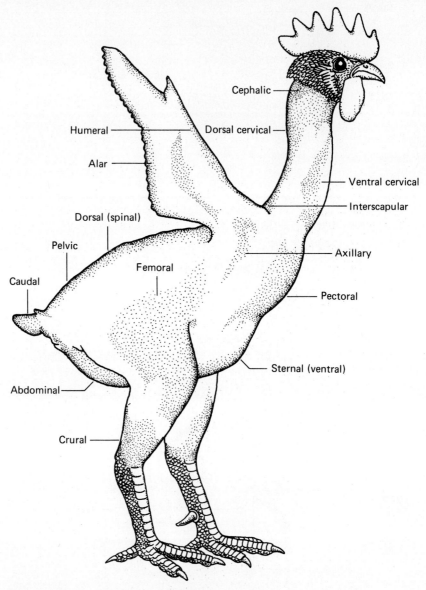

FIGURE 17-9   Feather tracts (pterylae).

air to permit flight, which, while not a unique phenomenon in birds, is brought to a high state of development in this class of animal.

Temperature Regulation

Since birds are homoiothermous and since feathers are excellent insulation, there must be some cooling mechanisms to compensate for the lack of sweat glands. There is a high radiation of heat from the unfeathered parts of the body between the feather tracts. Birds can also compensate a bit by breathing with an open beak to increase evaporation from air sacs and lungs. However, birds' temperature adaptation is not good. Ambient temperatures of 45°C (110°F) can kill through failure of the breathing and/or circulatory regulation of temperature. It is easier for birds to resist cold than heat. Bird body temperatures are high compared to mammals, and are in the neighborhood of 40–43°C (104–108°F). In domestic birds, the control of ambient temperature is a matter of great importance to poultrymen. Extremes should be avoided. Optimum ambient temperature is 21°C (72°F).

THE ORGANS OF SPECIAL SENSE

Of all the organs of special sense, the eye is the most highly developed in birds. Anatomically, the eyes and their accessory structures differ somewhat from those of mammals. Birds have three eyelids: an upper lid that is relatively fixed, a lower lid that is movable, and a third eyelid or nictitating membrane that is extensive and movable by two muscles which are not present in mammals. There are no eyelashes in birds.

The eyeball tends to be flattened or compressed along the visual axis. The lens is more nearly spherical. At the corneal-scleral junction is a bony ring, the scleral ring, composed of overlapping plates that form the visceral skeleton in birds. The scleral ring is continued backward in

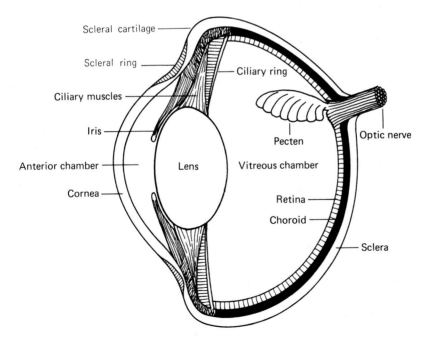

FIGURE 17-10   Schematic drawing of the avian eye.

the sclera to the optic nerve as thin overlapping plates of scleral cartilage. At the point of entrance of the optic nerve (the optic papilla) is the pecten, a strange formation composed of a folded, highly vascularized layer of tissue with a comb-like free edge. Its function is unknown although a number of speculations exist about it. These include fluid exchange, regulation of intraocular pressure, nourishment, warming, and a navigation aid for estimating the sun's angle.

The lens, in addition to being small and almost spherical, has a peculiar edge or annular pad through which it is joined to the ciliary muscles. The pad is soft and apparently acts as a cushion.

Functionally, the avian eye behaves similarly to the mammalian eye.

The Ear

With the exception of the lack of a concha (pinna) and an external acoustic meatus, and the substitution of a columella for the malleus, incus, and stapes, the avian ears are comparable to the ears of mammals. Hearing, however, should not be as acute as in mammals. The organs of balance, i.e., semicircular canals and otolith organs, are well developed.

Smell

Olfactory epithelium exists, but the sense of smell appears to be poor.

Taste

Taste buds exist, but the sense of taste appears to be rudimentary. Sight seems to be the principal means of recognizing food.

There appears to be no essential difference in the organization of either the olfactory or taste apparatus in birds and mammals. The kinesthetic senses are comparable to those of mammals and are extremely well developed, particularly in flying birds.

VALEDICTUM (L. *vale*—farewell, *dictum*—speech)

Well, I guess that's it. I hope the result of all these words and sketches has been to stimulate a desire to learn more about the fascinating way in which life operates. For the errors and omissions in this text, I accept full responsibility. For the rest, I must pay tribute to the research workers and investigators whose labors constantly expand the frontiers of knowledge and without whose contributions this book could not have been written. Their works and contributions to human knowledge are far too numerous to be specifically mentioned in a basic book, but can be found summarized in *Biological Abstracts, Quarterly Cumulative Index Medicus, Index Veterinarius, Current Contents,* and review articles in many professional journals. To all of them I owe a debt of gratitude for improving my imperfect knowledge of a complex field.

Those who go on from here will enter a world where reading is a way of life and where eyestrain is an occupational hazard. For those who stop here, the attached bibliography should be enough to enlighten, confuse, and keep one 10 years or more out of date; for, even as I write these words, many of the things that I have stated as accepted fact have been superseded by new discoveries and breakthroughs. That, unfortunately, is a fact of scientific life, and, as long as mankind has curiosity, little can be done to change it.

A Glossary of Terms Used in Anatomy and Physiology

Abaxial. Not located on the axis of the body, part, or organ
Abduct. To take away—to move outward from the median plane
Adduct. To bring toward—to move inward toward the median plane
Alveolus. A pocket—a small closed sac connected to a duct system—a tooth socket (pl. alveoli)
Anaphase. The third stage of cell division
Anatomy. The study of tissues and organs and their relationship
Androgen. Male sex hormone
Angle. A sharp bend formed by the meeting of two borders, surfaces, or lines
Ångstrom. A unit of measurement: 1/10,000,000 of a millimeter (10^{-7} mm)
Ankylosis. A bony fusion
Anterior. In front—forward—toward the head (superseded by cranial and rostral)
Apex. The top or pointed extremity of a conical mass—the top angle of a triangle
Apnea. Stoppage or lack of breathing
Apocrine. A type of secretion where part of the secreting cell is lost with the secreted substance
Appendicular. Pertaining to the limbs; legs
Articular. Pertaining to a joint
Atypical. Irregular—abnormal—not conforming to type
Axis. A line about which a revolving body turns—a center line

A GLOSSARY OF TERMS USED IN ANATOMY AND PHYSIOLOGY

Axon. A nerve fiber, usually defined as one that carries impulses away from the nerve cell

Bone. The material composing the skeleton of most vertebrates—chiefly composed of tricalcium phosphate and ossein

Bursa. A pocket or depression—usually with glandular functions

Calcaneus. The heel bone—the largest bone of the ankle (or hock) joint

Calcification. The process by which soft tissue becomes hardened by a deposition of calcium salts within its substance

Canaliculi. Minute canals or ducts which connect the lacunae in a bone

Cancellated. Lattice-like—lacelike—perforated with numerous holes or channels

Cannon bone. The metacarpal or metatarsal bones

Carpal. Of or pertaining to the wrist

Carpus. The wrist joint

Cartilage. A form of connective tissue—a precursor of bone—composed of cells embedded in a mucoprotein matrix

Caudal. Pertaining to the tail—posterior the fifth region of the spinal column

Cervical. Pertaining to the neck—the first region of the spinal column

Chondroblast. A cell which forms primitive cartilage

Chondroclast. A cell which breaks down cartilage

Chondrocyte. A cartilage cell

Clavicle. The collar bone

Cloaca. A sewer; a common opening for excrement, urine, and reproductive matter

Coccygeal. Pertaining to the tail—posterior—the fifth region of the spinal column

Coffin bone. The distal phalanx of the horse—the os pedis

Condyle. A cylindrical articular surface

Condyloid. Resembling a condyle

Connective. That which unites one part or structure to another

Coracoid. Beak-shaped

Cornified. Horny; calloused; covered with or changed into horn

Coronoid. Crown-shaped

Cotyloid. Cup-shaped

Cranial. Pertaining to the head—anterior (as used in newer terminology)

Crest. A sharp ridge

Cyton. A nerve cell

Deep. Not near the surface, opposite of superficial

Deltoid. Having a triangular shape or outline

Dens. A tooth—a tooth-shaped projection

Dendrite. A nerve fiber—usually carries impulses toward the cyton

Diaphysis. The shaft of a long bone

Digital. Pertaining to the finger or toe—the terminal part of a limb

Dissection. The cutting apart or cutting away of tissues

Distal. Farthest from the center of the body

Dorsal. Upward—toward the backline

Duodenum. The fixed or primary portion of the small intestine

Dyspnea. Difficult breathing—labored breathing
Dystocia. Difficult birth
Endocrine. Internal secretion—a ductless gland
Endosteum. A sheath of fibrous tissue lining the marrow cavity of a bone
Enteric. Pertaining to the intestines
Epicondyle. A bony prominence located above a condyle
Epidermis. The outer layer of skin
Epiphysis. The extremity of a long bone
Epithelium. A form of covering membrane
Estrogen. Female sex hormone
Estrus (estrum). Heat period (adjective: estrous)
Eupnea. Normal quiet breathing
Exocrine. External secretion—a gland with ducts
Exostosis. A bony outgrowth or projection—usually abnormal
Facet. A small articular surface
Femur. The thigh bone
Fibrous. Composed of fibers or strands
Fibula. The smaller of the two bones connecting the knee and ankle
Foramen. A hole—a perforation through or into a bone for the entrance of blood and lymph vessels and nerves
Fossa. A depression—a concavity
Fovea. A pit—a markedly concave depression
Frontal. Toward the front—pertaining to the forehead
Gastric. Pertaining to the stomach
Glenoid. A pit—a cup-shaped articular surface
Haversian. Discovered or described by Havers—the system of canals, ducts and lacunae within a bone
Head. A hemispherical articular surface
Hock. The ankle joint—the tarsus
Holocrine. A type of secretion where the entire secretory cell is lost in the process
Humerus. The upper arm bone
Hyperpnea. Increased depth and/or rate of breathing
Hypo-. Below; underneath
Ileum. The terminal portion of the small intestine
Ilium. The broad anterior or upper part of the pelvis
Incisor. Cutting teeth located in the anterior portion of the dental arch
Inferior. Under—in human anatomy, posterior or caudal
Infundibulum. Any funnel-shaped passage or cavity
Intermediate. In between
Interosseous. Situated or occurring between bones
Ischium. The lower or caudal part of the pelvis
Jejunum. The intermediate part of the small intestine
Jugal. One of the facial bones of fowl
Keratin. A binding substance found in horns, hooves and claws
Lacrimal. Pertaining to tears

A GLOSSARY OF TERMS USED IN ANATOMY AND PHYSIOLOGY

Lacunae. Small cavities within bone or cartilage usually containing cells
Laminar. Composed of layers, lamellar
Lateral. External—toward the outside—sideways
Ligament. Bands, sheets or cords of white fibrous or yellow elastic connective tissue which support or bind together parts of organs
Line (linea). A small ridge of bone—a thin elongate region of tissue of different color or composition
Lumbar. Pertaining to the loin—the third region of the spinal column
Lymph. Tissue fluid similar to blood plasma
Manus. The hand—the forefoot
Marrow. The soft material filling the cavities within cartilage bones
Medial. Median—pertaining to the midline, toward the inside, inner
Meiosis. Specialized cell division in which one-half the species number of chromosomes appear in the daughter cells
Membrane. A thin layer of tissue which covers a surface or divides a space or organ
Meninges. The three layers of membrane (dura mater, arachnoid, pia mater) which cover and enclose the brain and spinal cord
Merocrine. A type of secretion where none of the secretory cell is lost in the process
Metabolism. Process of building up or breaking down nutrient material
Metacarpal. The bone connecting the wrist and digit
Metaphase. The second stage of cell division
Metatarsal. The bone connecting the ankle and digit
Micron. One thousandth of a millimeter—approximately 1/25,000 of an inch
Microscope. An optical instrument for viewing very small objects
Mitosis. Normal cell division
Molar. Grinding teeth located at the caudal part of the dental arch
Mucoid. Similar to mucus
Mucus. A slimy fluid secreted by certain glands—a lubricant
Navicular. The distal sesamoid of the horse's foot
Neck. A constricted portion of a bone which connects the head to the shaft
Nephron. The secretory portion of the kidney tubules
Neuron. A single nerve cell and its processes
Nucleus. The central portion or life center of a cell
Obturator. A natural or artificial disc or plate closing an opening
Obturator foramen. The hole in the pelvis formed by the junction of the ilium, ischium, and pubis
Olecranon. The bulbous proximal bony prominence of the ulna—the point of the elbow
Olfactory. Pertaining to the nose or the sense of smell
Oncogenic. Cancer producing
Optic. Pertaining to the eye or the sense of sight
Ossein. A jelly-like protein found in bone
Ossification. The process of forming bone
Osteoblast. A bone forming cell
Osteoclast. A cell which breaks down bone

Pastern. A portion of the horse's foot occupied by the proximal and intermediate phalanges
Patella. The kneecap
Pelvis. The ring or basin-shaped bone connecting the hind limbs to the spinal column
Periosteum. The fibrous membrane surrounding the outer surface of a bone
Peritoneum. The serous membrane lining the abdominal cavity and covering the abdominal viscera
Pes. The foot
Phalanx. A bone of the finger or toe (pl. phalanges)
Pinna. The external ear
Plantar. The sole surface of the rear foot
Pleura. The serous membrane lining the thoracic cavity and covering the thoracic viscera (pl. pleurae)
Polypnea. Panting breath, rapid shallow breathing
Posterior. Behind—to the rear (superseded by caudal)
Premolar. Grinding teeth located in the intermediate portion of the dental arch
Process. A general term for a bony prominence
Prophase. The first phase of cell division
Protoplasm. The only known form of matter which manifests life
Proximal. Nearest the center of the body—nearest the heart
Pubis. The middle cranioventral portion of the pelvis
Radius. The medial of the two bones of the forearm
Renal. Pertaining to the kidney
Retroperitoneal. Outside the peritoneal cavity
Rostral. The anterior portions of the head; the nose; the beak
Sacral. Pertaining to the sacrum
Sacrum. The fused mass of vertebrae found in the hip region—the fourth region of the spinal column
Sagittal. Planes parallel to the median plane
Saliva. Spit, spittle
Salivary. Of or pertaining to saliva
Scapula. The shoulder blade
Segment. A piece cut off or marked off by actual or imaginary lines
Septum. A dividing wall or partition
Serous. Pertaining to or capable of secreting serum
Serum. A watery tissue fluid which acts as a lubricant—the fluid medium of the blood
Sesamoid. A small bone developed in a tendon where it moves across a bony surface
Skeletal. Pertaining to the skeleton
Spine. A pointed projection—the spinal column
Sternum. The breastbone
Stifle. The knee joint of a horse
Striated. Striped
Superficial. Pertaining to or situated near the surface

A GLOSSARY OF TERMS USED IN ANATOMY AND PHYSIOLOGY

Superior. Uppermost—in human anatomy; same as cranial in veterinary terminology

Suture. A line of junction (joint) between two membrane bones

Symphysis. A line of junction—an immovable joint between two cartilage bones

Synovia. An albumin-like fluid found in joint capsules and tendon sheaths

Tactile. Pertaining to the sense of touch

Talus. The second largest bone in the tarsus; the movable part of the ankle joint

Tarsal. Pertaining to the tarsus; the ankle; the plate of connective tissue in the eyelid

Tarsus. The ankle joint

Taxonomy. The system of biological classification; identification of life forms

Telophase. The fourth (and final) stage of cell division

Thoracic. Pertaining to the chest—the second region of the spinal column

Thorax. The chest

Tibia. The larger of the two bones connecting the knee and ankle

Tissue. An aggregation of similar cells working together to perform a particular function

Transverse. Placed crosswise to the long axis of the body or of a part or organ

Tricho-. Hair; hairlike

Trochlea. A pulley-like articular mass

Tuber. An enlargement or swelling

Tubercle. A small rounded projection

Tuberosity. A large rounded projection

Typical. Presenting the distinctive features of the type or species

Ulna. The lateral of the two bones of the forearm

Umbilicus. The navel; a depression or hole in a free surface

Vagina. The birth canal—entends from cervix uteri to vulva

Vas. A tube or duct

Vascular. Pertaining to the blood and lymph vessels

Ventral. Down—toward the belly line

Vertebra. A single bone of the spinal column (pl. vertebrae)

Vertebral. Pertaining to the vertebrae

Viscera. The large internal organs of the thoracic, abdominal and pelvic cavities of the body

Visceral. Pertaining to the viscera; the soft tissues; organs of the body

Vitamin. An essential metabolite which is not produced by the body

Volar. The sole surface of the forefoot

Vulva. The external termination of the vagina

Bibliography

Bloom, W. and Fawcett, D.W. *A Textbook of Histology*. 10th edition, Philadelphia: Saunders, 1975.

Bone, J.F. *et al.* (eds). *Equine Medicine and Surgery*. Santa Barbara: Am. Vet. Pub., 1963.

Bradley, O.C. *Topographical Anatomy of the Dog*. 5th edition. Edinburgh: Oliver and Boyd.

Chauveau. A. *The Comparative Anatomy of Domesticated Animals*. 2nd English edition, N.Y.: E. Appleton & Co., 1908.

Cotchin, E. and Roe, F.I.C. *Pathology of Laboratory Rats and Mice*. Oxford: Blackwell, 1967.

Davison, A. *Mammalian Anatomy with Special Reference to the Cat*. 4th edition, Philadelphia: Blakiston, 1955.

De Laubenfels, M.W. *Life Science*. 4th edition, N.Y.: Prentice-Hall, 1950.

Dellman, H.D. and Brown, E.M. *Textbook of Veterinary Histology*. Philadelphia: Lea & Febiger, 1976.

Dorland, W.A.N. *Dorland's Illustrated Medical Dictionary*. 25th edition, Philadelphia: Saunders, 1975.

Dukes, H.H. *The Physiology of Domestic Animals*. 7th edition, Ithaca: Comstock, 1957.

Ellenburger, W. and Baum, H. *Handbuch der Vergleichenden Anatomie der Haustiere*. Berlin: achtzehnte Auflage, 1943.

Evans, H. and Christensen, G. (eds.) *Miller's Anatomy of the Dog*. 2nd edition, Philadelphia: Saunders, 1979.

Ganong, W.F. *Review of Medical Physiology*, 5th edition, Los Altos: Lange, 1971.

Getty, R. *Atlas for Applied Veterinary Anatomy*. Ames: Iowa State Univ. Press, 1964.

Getty, R., *et al*. (eds.) Sisson and Grossman's Anatomy of the Domestic Animals, Vols 1&2. Philadelphia: Saunders, 1978.

Gray, H. *Anatomy, Descriptive and Applied*. New edition, N.Y.: Lee & Febiger, 1913.

Grollman, A. *Essentials of Endocrinology*. Philadelphia: Lippincott, 1947.

Guyton, A.C. *Textbook of Medical Physiology*. 5th edition, Philadelphia: Saunders, 1976.

Ham, A.W. *Histology*. 7th edition, Philadelphia, Lippincott, 1960.

Harrop, A.E. *Reproduction in the Dog*. Philadelphia: Williams & Wilkins, 1960.

Hegner, R. *Parade of the Animal Kingdom*. N.Y.: Macmillan, 1947.

Innes, J.R.M. and Saunders, L.Z. *Comparative Neuropathology*. N.Y.: Academic Press, 1962.

Koch, Tankred R. *Bau und Funktion des Geflügelkörpers*: Translated by Skold, B. and DeVries, L., as *Anatomy of the Chicken and Domestic Birds*. Ames: Iowa State Univ. Press, 1973.

Leach, W.J. *Functional Anatomy of the Mammal*. 2nd edition, N.Y.: McGraw-Hill, 1962.

Leisering, A. *Anatomie für Kunstler*. Berlin: achte Auflage, 1943.

Lloyd, C.W. (ed). *Endocrinology of Reproduction*. N.Y.: Academic Press, 1959.

Lucas, A.L. and Stettenheim *Avian Anatomy—Integument*, Parts I & II. USDA ARS Supt Documents WDC, 1972.

Marshal, F.H.A. and Halnan, E.T. *Physiology of Farm Animals*. 4th edition, Cambridge, 1948.

Messer, H.M. *An Introduction to Vertebrate Anatomy*. N.Y.: Macmillan, 1945.

Miller, M.E. *Guide to the Dissection of the Dog*. N.Y.: Ithaca, 1947.

Miller, M.E. *et al. Anatomy of the Dog*. Philadelphia: Saunders, 1964.

Nalbandov, A.V. *Reproductive Physiology*. San Francisco: Freeman; 1958.

Pantelouris, E.M. *A Handbook of Animal Physiology*. London: Bailliere, Tindall and Cox, 1957.

Patten, B.M. *The Embryology of the Pig*. 2nd edition, Philadelphia: Blakiston, 1944.

Phillis, J.W. (ed). *Veterinary Physiology*. Philadelphia: Saunders, 1976.

Playfair, J.H.L. *Immunology at a Glance*. Boston, Oxford: Blackwell, 1982.

Porter, K.R. and Bonneville, M.A. *Fine Structure of Cells and Tissues*. 3rd edition, Philadelphia: Lea and Febiger, 1968.

Prince, J.E. *et al. Anatomy and Histology of the ye and Orbit in Domestic Animals*. Springfield: Thomas, 1970.

Robinson, M.C. *Laboratory Anatomy of the Domestic Chicken*. Dubuque: Wm. C. Brown, 1970.

Ruch, T.C. and Fulton, J.F. *Medical Physiology and Biophysics*. 18th edition, Philadelphia: Saunders, 1946.

Seigmond, O.H. *et al*. (eds). *The Merck Veterinary Manual*. 4th edition, Rahway: Merck, 1974.

Sisson, S. and Grossman, J.D. *The Anatomy of the Domestic Animals*. 4th edition, Philadelphia: Saunders, 1952.

Smythe, R.H. *Animal Habits*. Springfield: Thomas, 1962.

Storer, T.I. *General Zoology*. N.Y.: McGraw-Hill, 1943.
Sturkie, P.D. *Avian Physiology*. Ithaca: Comstock, 1965.
Svendsen, P. *An Introduction to Animal Physiology*. Westport: Avi, 1974.
Zeman, W. and Innes, J.R.M. *Craigie's Neuroanatomy of the Rat*. N.Y.: Academic Press, 1963.

Index

A-band (anisotropic band), 102, 114
Abdominal aorta, 269, 287, 291, 302
Abdominal cavity, 86, 159, 172, 173
Abdominal muscle, 104, 214, 221
Abdominal press, 221
Abdominal respiration, 214
Abdominal ("milk") vein, 275
Abduction, 98, 104
Aberration of eye, 491–92
Abomasum, 150, 156, 157–58, 159, 188, 199
Absorption, 149, 193, 194
Accessory sex glands, 407, 427
Accommodation of eye, 492–94
Acetabular fossa, 64, 65
Acetabulum, 60, 63, 64, 65
Acetate (ruminant) metabolism, 198–201
Acetylcholine, 111, 185, 316, 331, 386
Acetylcholinesterase, 336
Acid phase, digestion, 193
Acid tide, 188
Acoustic nerve, 319, 330
Acromion, 39, 42
ACTH (adrenocorticotropic hormone), 25, 347, 348, 366, 387
Action potential, 335
Adduction, 98, 104

Adenine, 314
ADH (antidiuretic hormone), 307, 349
ADP (adenosine diphosphate), 115, 116, 371, 374, 377, 380–81, 382, 385
Adrenal
 cortex, 353–54, 357, 364, 366
 gland, 345, 353–55
 hormones, 354–55, 366
Adrenalin, 111, 355, 364
Aerobic process, 115
Afferent blood vessel, 233
Age estimation, 132–45
Agglutination, 257, 258
Agglutinin, 256
Aging, theories of, 22–25
AGP (alpha glycerophosphate), 115
Agranulocyte, 240–46
Airway, obstruction of, 222
Alar foramen, 80, 81
Albumin, 235, 248, 251
Albuminoid, 235, 248, 251
Aldosterone, 179, 308, 354
Alimentary canal, 149–66
Aliphatic molecule, 18
Alpha-type cell, 167, 355
Altitude, effect on erythrocytes, 234
Alveoli

 lung, 205–208, 211
 skin glands, 473
 udder, 477
Alveolus (dental), 131
Amino acid, 174, 177, 178, 184, 391
 excretion, 396–98
 metabolism, 391–395
Ammonia (conversion reactions), 393–95
AMP (adenosine monophosphate), 371
Amphiarthrosis, 99–100
Anabolism, 177, 371
Anaerobic process, 115, 116
Anal reflex, 221
Anatomy
 history, 1–2
 methods of study, 3–5
Androgen, 357, 402, 406, 442, 445
Anemia, 183, 217, 238, 239–40
Angiotensin, 307–08
Angles
 of bones, 37–40, 57, 60, 63, 64, 65
 of teeth, 133–34, 139, 145
Angular movement, 98
Angular process of mandible, 93
Ankylosis, 83
Annulus fibrosus, 100

Anoxia, 223–24, 237
Antagonism, muscular, 104
Antibody
  cellular, 252–53
  maternal, 442
  plasma, 234, 235
  production, 194, 235, 241–45 passim, 249, 252–58, 282
  spermatocide, 445
Antigen, 246
Antigen-antibody system, 252–53, 258–59, 352–53
Antihistamine, 252
Antitoxin, 253
Antlers, 472–73
Anus
  glands, 166
  physiology, 198
  sphincter, 104
Aorta, 268–70
Apnea, 204
Apocrine gland, 465, 474
Aponeurosis, 104
Appendicular skeleton, 33–77
  carpus, 51–53
  digits,68
  fabellae, 76–77
  femur, 66–67
  fibula, 71–72
  humerus, 41–46
  ilium, 60–63
  ischium, 63
  metacarpals, 54
  metatarsals, 75
  os acetabuli, 64
  patella, 68
  pelvic girdle, 60, 64, 65–66
  phalanges (digits), 54–57
  pubis, 63–64
  radius, 46–49
  ribs, 85
  scapula, 37–41
  sesamoids, digital, 57–60
  sternum, 86
  tarsus, 72–74
  tibia, 70–71
  ulna, 49–51
Aquatic mammals
  dentition, 146
  vertebrae, 79
Aqueduct of Silvius, 322, 323, 325
Arbor vitae, 319
Arch
  costal, 35, 85, 86
  dental, 132
  vertebral, 78
  zygomatic, 88, 91, 92
Arcuate vessel, 291, 296, 305
Area cribrosa, 289, 294, 296, 298, 301
Argentaffin cell, 155
Aristotle, 1, 5

Arterial
  branches, 268–69
  fetal circulation, 283
  flow, 266
  system, 268
Arteriole, 270
Artery
  pulmonary artery, 209
  renal artery, 291
  structure of, 269
Arthrodial joint, 99
Arthrology
  avian, 517
  mammalian, 33, 97–100
Arthus phenomenon, 353
Articular cartilage, 29, 98
Articulation, types
  Amphiarthrosis, 99
  arthrodia, 99
  enarthrosis, 99
  synarthrosis, 98
  synsarcosis, 100
Artificial respiration, in newborn foals, 216
Asternal (false) rib, 85
Astrocyte, 314
Atelectasis, 207
Atlas, 35, 79, 81, 99
ATP (adenosine triphosphate) 15, 18, 102, 115, 116, 161, 184, 371
Atrial muscle, 112, 260, 263
Atrophy of muscle, 104
Auerbach's plexus, 333–34
Auricle (atrium), 260
Autonomic nervous system, 311, 316, 330–33
A-V (atrioventricular) node, 112, 261, 266
A-V bundle of His, 112, 261
Avian skeleton, 512–13
Axial skeleton, 33, 77–95
  physiology, 84
  vertebral column, 78–79
    caudal vertebrae, 83–84
    cervical vertebrae, 79–81
    lumbar vertebrae, 83
    sacral vertebrae, 83
    thoracic vertebrae, 81
Axis, 35, 79, 81, 99
Axon, 102, 312–13, 316, 331, 333
Azygos vein, 275

Bainbridge reflex, 264, 279
Ball and socket joint, 99, 119
Banting and Best, 356
Basement membrane
  intestine, 162
  lung, 208
  urinary tract, 292, 294, 464, 466
Basioccipital bone, 89, 90
Basophil, 234, 246, 247

Beak, avian, 521, 522
Beta cell, 167, 355
Biangular table surface, 139
Bibliography, 557–59
Biceps brachii, 104, 105, 117
Bile, 167, 168, 194, 239
  duct, 153, 155, 161, 162, 168, 194, 239
Bilirubin, 194, 234, 239
Bilirubinemia, 235
Biliverdin, 194, 239
Binocular vision, 490, 496
Biological rhythms, 21
Bishoping, 139, 140, 141
Bladder, urinary, 288, 294, 299
Blastula, 433
Blind sac, 159, 160
Blind spot, 488, 494
Bloat, 200
Blood
  cell, red (see erythrocyte)
  cell, white (see leukocyte)
  clotting, 249–51
  flow/pressure, 276–77
  formation, 243
  pathology, 239–40, 251–56, 258–59
  plasma, 234, 235, 248
  pressure, 276–79
  renal, 291
  types, 256–58
Blood-vascular system, 233–79
  avian, 531–32
Body chief cell, 154, 157, 188
Body systems, 4, 5
Body water intake/removal, 179–82
Bohr effect, 225–26
Bone, 27–29, 29–95 (see also Avian skeleton; Axial skeleton; Appendicular skeleton; visceral skeleton)
  cancellous bone, 31
  cartilage bone, 28
  membrane bone, 28
  skull, 87–95
  spicules and trabeculae, 31
Bowman's capsule, 291, 293, 300–303, 464
Brachiocephalicus, 117
Bradykinin, 252
Brain, 88, 311, 318–26, 321, 322, 323, 325
Breathing, 213–15 (see also Respiration-mechanics)
  rate of, 216
  regulating mechanisms, 219
Bridle teeth, 138
Bronchiole, 205–6
Bronchus, 205–6, 208, 210
  vascular system, 209
Broodiness, avian, 537–38
Brownian movement, 19, 203
Brunner's glands, 149, 160–61, 162, 164, 193–94
Brush-bordered cells, 292–93

562

B_{12} vitamin, 184–85, 240
Buffer system–blood, 223, 225
Bursa of Fabricius, 244, 526, 532
Butyrate, 200

Calcitonin, 351, 352
Calcium ions 111, 115, 182, 189
  metabolism 351, 352, 366
Calyces, 289, 294, 295, 296
Canal alimentary, 149
  optic, 89
  vertebral, 78, 87, 100
Canaliculus, 16
  bile, 168
Cancellous bone, 31, 51, 238
Canine tooth, 131, 133, 134, 135, 138, 141, 143, 145, 146
Cannon bone
  long, 37, 54
  short, 37, 56
Capillary, 207, 208, 224, 225, 270–72, 276–77
  endothelium, 208, 224, 246
Capsule-joint, 98, 99
  renal, 287, 289, 290, 295
Carbohemoglobin, 237
Carbohydrate, 19, 174–76
  digestion, 174–76
  metabolism, 229, 230, 231
Carbon dioxide, 115, 116, 176, 225
  dissociation in blood, 228, 229
  respiration, 221, 223–32 *passim*
  transport, 223–31
Cardia, 154, 155, 156, 159, 198
Cardiac-control center, 319
  cycle, 261–63
  division-ANS, 334
  gland, 156, 158, 188
  muscle, 111–13, 260–61
  region-stomach, 150, 154, 155, 191
  sphincter, 154, 192
Cariniform cartilage, 86
Carnivore
  dentition, 119, 129, 146–47
  dietary requirements, 134, 145–46
Carotene, 234–35
Carotenoid, 235
Carotid artery, 268
Carpus, 35, 37, 38, 51–53, 118–19
Cartilage, 27, 28, 29, 30, 85–6
  avian, 512
  bars, 29, 30
  bone, 28
Caruncula sublingualis, 151
Catabolism, 177, 184, 371, 373 *ff*
Catalyst, 185
  enzyme, 185–6, 195–7
  hormone, 345–6
  vitamin, 183–85
Caudal appendage
  artery, 271 (fig.), 274 (fig.)

  pituitary, 318, 321
  vertebrae, 35, 36 (figs.) 79, 83
  vein, 273 (fig.), 275
Cecum, 150, 153 (fig.), 162–3, 165 (fig.), 197
Cell
  Anatomy of, 11–20
  interaction, 19
  membrane, 14
  organization, 20
  reproduction, 418–20
  respiration, 225
  rupture (laking), 237
Cellular immunity, 254
Cellular "soup," 18
Cellulose, 175, 198, 201
Celsus, 251
Cement, 130, 131
Central nervous system, 311, 318*ff*, 326*ff*
Centriole, 15
Centrosome, 15, 16
Cephalic vein, 273, 274, 275
Cerebellum, 317, 318, 319, 321, 322, 324
Cerebral hemisphere, 318, 319, 320, 321, 323
Cerebrospinal fluid, 318, 324
Cerebrum, 317, 322, 325
Cervical vertebra, 35, 79–81, 84
Cervix, 412, 413, 414, 415, 416
Cheek teeth, 130, 131, 132, 133, 135, 145, 146, 147
Chemistry
  of blood, 279
  of external respiration, 220
  of heart rate regulation, 263–65
  of intermediate metabolism, 371–82, 382–88, 388–89, 390–99
  of muscle contraction, 115–16
Chemoreceptor, 337
Chestnut, 56, 471
Chewing, 149
Cholecystokinin, 194, 360
Cholesterol, 178, 194, 354, 383
  metabolism, 388–89
Choline, 174, 177, 185, 360, 387
Cholinergic synapse, 102
Chondroclast, 28
Chondrology, 33
Chorion, 434, 440
Choroid plexus, 324
Chromatin (linin) filaments, 17
Chromatolysis, 312
Chromosome, 13, 17
Chyle, 193
Chyme, 191, 193
Cilia, 15
  arterial, 268–72
  cardiac, 267
  lymph, 279–81
Circulation

  fetal, 283–85
  portal, 276
  venous, 272–79
Circumduction, 99
Clavicle, 34, 37, 86
Climacteric, 437
Cloaca, 524, 537, 539
Cloning, 19, 20
Clotting (blood), 235, 249–51
Cobalt, 182
Coeliac artery, 269
Coffin
  bone, 38, 56
  joint, 119
Colliculi, 319
Colloid, 19, 351
Colon, 150, 153, 164, 165
Color blindness, 496
Colostrum, 442, 481
Complement, 253
Condyle (capitulum-trochlea)
  femoral, 67, 69
  humeral, 43, 44, 45
  mandibular, 88, 91, 92
  occipital, 79, 87, 88, 89, 90, 91, 92
  temporal, 93
  tibial, 71, 73
  zygomatic, 93
Condyloid fossa, 67, 93
Constantly erupting teeth, 129, 131
Constantly growing teeth, 129
Contraction, 114–15
  cardiac, 260–61
Copper, 183
Copulation, 432
Coracoid bone, 34, 37
Cori cycle, 373
Corner tooth, 134, 135, 137, 138, 139
Cornus (processus), 91
Coronary vessels, 212, 260, 268
Coronoid fossa, 43, 44
Coronoid process, 88, 91, 92
Corpora quadrigemina, 319, 320, 322
Corpus callosum, 320, 322
Corpus luteum, 348, 358, 412, 477
Corpus striatum, 24
Cortex
  adrenal, 354, 357, 365, 366
  cerebral, 320
  renal, 287, 289, 290, 294, 295, 296, 297, 298
Corti, 12, 15
  organ of, 463, 466, 497, 500, 501, 504
Cortin, 354
Costal cartilage, 85, 86
Costal facet, 80, 81, 82
Cotyledonary uterus, 382
Cotyloid cavity, 64
Cowper's gland, 407, 411
Craniad precocity, 433
Cranial bones, 87–93

563

Cranial nerves, 311, 329, 330
Creatine, 115, 230, 398
Creatinine PO$_4$, 115
Cribriform plate of ethmoid, 89
Crown of tooth, 129, 130, 131, 139
Crypts of Lieberkühn, 149, 160, 161, 163, 194
Cursorial leg, 123, 127
Cusp, heart valve, 260
Cutaneous receptor, 337
Cysterna chyli, 281
Cystic duct, 168, 171
Cytochrome, muscle, 102
Cyton, 312, 313, 315, 329, 330
Cytoplasm, 12, 13, 14, 15, 16, 17, 243
  intestine epithelial cell, 161, 162
  muscle, 102
Cytosine, 344

Dead space, 215, 218, 467
Deamination, 178, 391–92, 396, 398–99
Death hormone, 25
Decerebration, 341–42
Decompression, 228
Defense mechanisms (disease), 236, 240, 252–54, 256
  lack of, 209–10
Dehydration, 179–81
Dendrite, 312, 313, 316, 334
Dens, 79, 80, 81
Dental star, 135, 139
Dentine, 130, 131
Depolarization, 113, 335, 336
Dermal light sense, 483–84
Desmosome, 14, 16, 162
Dewclaw, 56, 76, 471
Diabetes, 355–57
Dialysis, 19
Diapedesis, 246, 251
Diaphragm, 86, 159, 173, 210–11, 213–14, 219, 268
Diaphragmatic respiration, 213–14
Diaphysis, 29, 31, 99
Diarthrosis, 98–99
Diastole, 263, 265, 267
Diencephalon, 317, 320
Diffuse placenta, 382
Diffusion, 18, 181, 206, 220, 223, 225
Digestive fluid, 180–82
Digestive system, 149–201, 521–29
Digestive tract muscle, 104, 110
Digit, 54–60, 76–77, 123–124, 469–71
Direct respiration, 204
Disaccharide, 175, 372
Diverticulum duodeni, 155, 161, 169
Diverticulum ventriculi, 155, 158
Divider (tooth), 133, 135, 137, 138, 139
DNA (deoxyribonucleic acid), 17, 18, 23, 343, 381
Dolphin dentition (seizing teeth), 146
Donnan equilibrium, 19

Dopamine, 24, 316
Dorsal artery, 268
Dorsal foramen, 83
DPN (diphosphopyridine nucleotide), 377, 378, 380, 381, 385
Dual origin of lymphocytes, 245
Ductless (endocrine) gland, 345
Ducts of Bellini (Brunner's), 294
Ductus arteriosus, 283
Ductus venosus, 283
Duodenalgland, 149, 160, 161, 164, 193, 194
Duodenum, 150, 154, 155, 159, 160, 162, 166, 168, 169, 170, 171, 190, 191, 193, 194
Dura mater, 325, 326, 327, 328
Dyspnea, 204, 207

Ear, 93, 497–99, 502, 504, 505, 509, 548
Eccentric contraction, 114
Ectoderm, 461
Ectoturbinate, 89
EEG (electroencephalogram), 342
Effector, 315, 317
Efferent blood, 233
EKG (electrocardiogram), 266, 267
Elastic cartilage, 28
Electrical stimulation, 342
Electrolyl balance, 236, 248, 306, 354, 366
Electron microscopy, 1, 13, 17
Elephant dentition, 146–47
Embolus, 251
Embryo development, 432–34
Emulsoid, 19
Enamel (tooth), 130, 131, 138, 140, 141, 146
Endocardium, 112, 260, 261
Endocrine system, 149, 245, 345–68
  avian, 541–2
  pathology, 355–57
Endoderm, 461, 462
Endometrium, 358, 363, 364, 431, 433
Endomysium, 102, 103
Endoneurium, 315
Endoplasmic reticulum, 15, 16, 17, 162, 292
Endosteum, 29
Endotheliochorial, 434
Endothelium, 16, 224, 246, 261, 292, 461–66
Endoturbinate, 89
Ensiform cartilage, 86
Enterocrinin, 359, 365
Enterogastrone, 190, 345, 359, 366
Enzyme, 17, 18, 23, 175, 185–86, 249, 316, 330, 356, 359, 365, 366
  digestive, 189, 193–97
  renal function, 299, 307–8
Eosinophil, 234, 246–47
Epicardium, 260
Epicondyle, 43, 45, 67, 69

Epicondyloid foramen, 45
Epidermis, 461, 466–72
Epiglottis, 187, 205, 207
Epimysium, 102, 103
Epinephrine, 252, 355
Epineurium, 328
Epiphyseal plate, 28, 29
Epiphysis, 29, 30, 31
Epitheliochorial, 434
Epithelioid cell, 161, 194
Epithelium
  gastric, 461–66, 345
  intestinal, 154, 161–3, 193, 345
Equilibrium, 18, 20
Ergot, 56, 469, 471
Eructation (belching), 200, 232
Erythrocyte (red blood cell), 14, 16, 227, 234, 236–40, 249, 251, 259, 292, 309
  nucleated, 238
Erythrolysis, 258–59
Erythropoietin, 20, 234, 238, 309, 330–31
Esophageal
  glands, 152
  region, 150, 152–53
Esophagus, 86, 150, 153–55, 156, 159, 160, 191–92, 207, 212
  muscle, 104
Essential amino acid, 178
Estrogens, 358, 359, 362–63, 363–64, 364, 365, 430, 431, 441–42, 477, 541
Estrous cycle, 362–63, 363–65, 430–32
Ethmoid, 87, 89
Euglobulin, 248
Eupnea, 204
Excretion, 293–98
Exercise, 115–16, 337
Exocrine, 154
Exostosis, 83
Extension, 98, 104, 117–19
External acoustic meatus, 88, 91, 92, 93
External respiration, 204–11, 221
Extirpation of brain, 340
Extramedullary hemopoiesis, 238
Extrinsic factor, 183–84, 239–40
Eye
  avian, 547–48
  defects, 491–94
  mammalian 484–97

Fabella, 68, 69, 70
Facet
  carpus, 44, 52
  costal, 82
  humerus, 46
  radius, 48
  ulna, 46, 48
  vertebra, 80, 81, 82
Facial bones, 93, 94, 95
  tuberosity, 91
Fallopius, 2

False (asternal) ribs, 85
False digit (dewclaw), 56, 76, 470–71
False (fixed) vertebra, 78
Fascia, 287
Fasciculus, 102, 103, 111
Fat (lipid)
　cell membrane, 14
　coronary, 290
　metabolism, 162, 174, 176–78, 185, 189, 196–97, 200, 230–33
　omental, 173
　periorbital, 290, 445
　perirenal, 287, 290, 353
Feather, 461, 519, 542–45
　tracts, 547
Femoral artery, 269
Femoral vein, 275
Femur, 35, 37, 61, 66–70, 110, 111
Fertilization, 432
Fetlock, 38, 469
　joint, 119, 125, 471
Fetus, 434–35, 436
　blood, 233, 238, 259
　blood formation, 238
　circulation, 283–85
　kidney, 289
　lung, 218
　lymph, 285
Fiber
　muscle, 101, 102, 103, 111–12
　nerve, 312
Fibrillation, 113, 266–67
Fibrin, 249, 250
Fibrinogen, 234, 235, 248, 249, 250
Fibrocartilage, 28, 98, 100
Fibrous tunic, eye, 486
Fibula, 35, 37, 70, 71–72, 73, 74
Fick's method, 264
Filtration
　cell, 19
　renal, 300, 302
Flagella, 15
Flat bone, 29
Flexion, 98–99, 104, 125
Flight (avian), 521
Floating rib, 85
Fluid, body, 179–82
　balance, 366
　intercellular, 272
Foramina (foramen)
　dorsal, 83
　epicondyloid, 45
　incisivum, 90
　infraorbital, 88, 91, 92
　intervertebral, 80, 81, 326
　lacrimal, 88, 94
　lateral, 83
　Luschka, 323, 324
　Magendie, 322, 324
　magnum, 90, 326, 328
　mental, 88, 91, 92

Monro, 322, 323
nutrient, 27, 40, 55, 57, 65, 66, 75
obturator, 62, 63, 64, 65
optic, 490
supracondyloid, 43
supraorbital, 88, 91
supratrochlear, 44, 45
transversarium, 79, 80, 81
vertebral, 78, 80, 81, 82
Forefoot, 119, 120
Forehead, 35
Forelimb (leg), 34, 116, 117, 119, 121, 124
Formula
　dental, 143, 144
　vertebral, 84
Fowl anatomy/physiology, 511–49
Freemartin, 257, 440
Frontal, 3
Frontal bone, 87–95 passim
Fructose, 175
FSH, 348, 362, 363, 364
Fundic glands, 158
Fundic region (compound stomach), 150, 158
Fundic region (simple stomach), 150, 154–55, 157, 158, 181, 192

Gait, 119, 120, 121
Galactose, 175, 479
Galen, Claudius, 1
Gall bladder, 153, 168, 169, 170, 171
Galvayne's groove, 134, 141
Ganglion, 300, 313–14, 328, 330, 331–33
Gaseous, 203–4, 206–7, 208–9, 223–32
Gastric physiology, 187–91
Gastric secretions, 155–56, 187, 196–97
　enzymes, 188, 196–97
　gland, 154–55
　hormones, 190–91, 197, 345
　juice, 189
　lipase, 189, 196–97
　secretion, 155–56, 189–90
Gastrin, 190, 191, 197, 359, 366
Gastrula, 433
Geniculate nuclei, 320
Genotype, 401, 441–43
Germ cell, 19, 418–29
Gerontology, 22–25
Gestation, 432–36
Ginglymus, 79, 93
Gland, 149–68, 345–68, 404–7, 412, 473–75
　cells, 465–66
Glenoid cavity, 38, 39, 40, 42
　notch, 38
Glial cell, 313
Gliding movement, 98, 99, 119
Globin, 223, 237
Globulin, 235, 248, 252, 253
Glomerulus, 291, 294, 302, 304, 308

Glossary, 551–56
Glossopharyngeal nerve, 187, 330
Glucagon, 167, 355, 386–87
Gluconeogenesis, 354, 391, 394–95
Glucose, 175, 176, 177, 200–1, 355, 373, 385, 395–96
　muscle, 115, 116
　resorption, 303–4
Gluteal
　line, 61, 62
　muscle, 117, 119
　surface, 60
Glyceraldehyde, 385
Glycerol, 175–76, 385–86
Glycogen, 116, 175, 386, 395–96
　liver, 116, 167, 175, 171, 372, 386
　muscle, 175, 355, 373
Glycolysis, 200, 373–75, 377
Glycoprotein, 397
Goblet cell, 161, 162, 194
Golgi apparatus, 15, 16, 167, 292, 423
Gomphosis, 98
Gonad, 345, 357, 361, 426
Gonadotropin, chorionic, 359, 362–64
Graafian follicle, 348, 358, 363, 364, 412, 417, 428–32
Grafts/organs, 255
Granulocyte, 240, 246–48
Gray commissure, 326, 327
Gray matter, 315, 318, 320, 323, 326
Gray ramus, 315
Groove
　bicipital, 41, 43
　dorsal, 57
　intercondyloid, 44, 45
　musculospiral, 41, 43, 44, 45
　vascular, 37, 40, 42
Growth hormone (GH), 347
Guanine, 344
Gum(s), 175
Gynecomastia, 477
"Gypping," 139
Gyrus, 320, 321

Hair, 467–68
Haldane's method, 264
Harder's gland, 485
Harvey, William, 5
Hassall's corpuscles, 353
Haversian system, 32–33
Hearing/equilibrium, organs, 497–505, 548
Heart, 259–67
Hederiform termination, 314
Hemal arch, 79
Hematin, 237, 239
Hemoglobin, 182, 223–27, 238, 247
Hemohistioblast, 238
Hemolysis, 237
Hemopoietic factor, 174, 183, 184, 239
Hemorrhagic syndrome, 248

565

Henle's loop, 291, 294, 300–6, 463
Henson's node, 433
Hepatic artery/plexus, 168
Hepatic duct, 168
Hepatic portal vein, 276
Hepatic vein, 275
Herbivore
  carbohydrate digestion, 174–76
  dietary habits, 145
  stomach anatomy, 156–58, 159
  stomach physiology, 187–93
Hering-Brewer reflexes, 220
Hexose, 175
Hiatus aorticus, 211, 268
Hibernation, 367
Hiccup, 222–23
Hilus, kidney, 288, 289, 295
Hinge joint, 51, 79, 98, 99
Hip bone, 34
Hippocampus, 321
Histamine, 247, 252, 316
Histidine, 349
Histology, 32
Hock joint, 119
Holocrine gland, 465, 473
Homeostasis, 251, 281, 299, 304, 384
Homothermism, 367
Hoof/claw, 467, 468–71
Hooke, Robert, 2, 12
Hormone 345–68
  adrenal cortex (androgen, corticoids), 354
  adrenal medulla (adrenalin, noradrenalin), 354
  hypothalamus, paraventricular (oxytocin, pitocin), 349
  hypothalamus, supraoptic (pitressin), 349
  Kidney (erythropoietin, urogastrone), 309, 360
  ovary (estrogens, estradione, estrone, progesterone, relaxin), 357
  pancreas (glucagon, insulin), 355–56
  parathyroid (parathormone), 352
  pineal (melatonin), 350–51
  pituitary, caudal (ADH pitocin, pitressin storage), 349
  pituitary, intermediate (melatonin), 349–50
  pituitary, rostral (growth, FSH, LH, SSH, ICSH, prolactin, TTH, ACTH), 346–48
  small intestine (secretin, enterocrinin, enterogastrone, pancreozymin, cholecystokinin), 359
  stomach (gastrin), 359
  testicle (androgens, testosterone), 357
  thymus (thymin), 352–53
  thyroid (thyroxin, calcitonin), 351
  uterus (chorionic gonadotropin), 358–59

various structures (prostaglandins), 361
Hormone relationships, 349, 361, 362–63
Hormones, regulation of
  autonomic nervous system, 326
  avian endocrines, 541
  birth, 364
  blood, 238, 279, 333, 335, 309
  body temperature, 366, 368
  broodiness (avian), 537
  cell reproduction, 20
  cholesterol, 389
  CNS hormones, 341
  death, 25
  digestion, 190, 194 passim, 197, 309, 333
  electrolyte, 354, 366
  eye, 333
  fat metabolism, 387
  glucose, 325
  glands, 331
  heart, 265, 333
  hibernation, 367
  hormone relationships, 349
  horns, hoofs, claws and hair, 468–71
  intestinal tract, 193, 365
  kidney function, 299, 306–7, 309, 365
  lactation, 319, 364
  lungs, 333
  mammary function & development, 364, 477, 480
  molt, 537
  neural transmission, 331
  neuromuscular activity, 368
  parturition, 436
  pregnancy, 363–65
  relationships, 349, 361–63
  reproduction, 20
  sex hormones, 402
  sexual activity, 445–49 passim
  somatic glands, 331
  stomach, 190–91
  synapses, 331
  water and electrolyte, 179, 366
Horns, 467, 468–71
Humerus, 35, 38, 41–46
Hunger contractions, 192
Hyaline cartilage, 28, 30, 85–86
Hydrochloric acid, 155, 157, 188, 189
Hydrogen ions, 220, 225
Hyoid bone, 93, 94, 187

I-band (isotropic), 102, 103, 114
ICSH (interstitial cell stimulating hormone), 347, 357
Icterus index, 235
Ileocecal valve, 153
Ileum, 150, 153, 159
Iliac vessel, 269, 275
Ilium, 37, 60, 61, 62, 63, 65
Immovable joint, 97, 98

Immune mechanism/reaction, 24, 235, 242, 246, 252–56, 353
Immunoglobulin, 253
Immunosuppressive, 255–56, 259
Incisivum (foramen), 88, 90, 94
Incisor (tooth), 130, 131, 133, 136–46 passim
Infection resistance, 442
Inflammation, 251–52
Infraorbital foramen, 88, 91, 92, 94
Infraspinous fossa, 37, 40, 42
Infundibulum (dental), 130, 135, 139, 140
Infundibulum (pituitary), 346, 350
Inlet, pelvic, 66
Inlet, thoracic, 86
Innervation, 102, 105, 110, 332, 333–34, 447, 488
Innominate bone, 34
Insertion, muscle, 102, 104, 110
Insulin, 167, 345, 355–57
Integument
  common, 461–81
  avian, 542–47
Intelligence (intellect), 342–44
Intercalated disc, 111
Intercarpal joint, 119
Intercondyloid fossa, 67, 69, 70, 73
Intercostal space, 85
Intercostal vessels, 268, 275
Interdigitating membrane, 162
Interferon, 254
Interlobar vessels, 291, 292, 296, 303, 305
Intermedin, 179, 349
Internal respiration, 116, 204, 225–26
Interosseus space, 46, 47, 48, 49, 69
Interparietal bone, 87, 88, 89, 92
Intersex, 440
Interstitial fluid, 19, 111, 182, 234, 248, 281–82, 303, 304
Intervertebral foramen, 78, 80, 81, 311, 326, 327, 328, 331
Intestinal
  hormones, 197
  structures, 150, 157, 158–59, 162–64
Intralobular vessels, 291–92, 296, 305
Intrapulmonic pressure, 215
Intrinsic digestion, 188
"Intrinsic factor," 155, 188, 240
Intussusception, 195
Involuntary muscle, 113, 219–20
Involuntary nervous system, 261, 263–64
Iodine, 182, 183, 351
Iris, 488
Iron, 182, 183, 194, 237
Irregular bones, 77–84, 87
Ischium, 37, 60–65
Isles of Langerhans, 167, 355, 366
Isohydric shift, 226
Isometric contraction, 114
Isotonic contraction, 112, 114

Jaw, 93, 94
Jejunum, 150, 153, 160
Joint, 97-100
    cavity, 98, 99
    movements, 98, 99, 116, 117, 118, 119-22
    structures, 98
    types, 98, 99, 100
Jugular vein, 275
Junctional fold, 102
Juxtaglomerular apparatus, 179, 209, 307, 308-9

Ketone body/Ketosis, 177, 201, 287, 288-99, 300-309, 345, 354, 360, 366 386
Kidney, 317, 331, 336
Kinetosome, 15
Knob (nerve trunk), 316
Krause's corpuscle, 314
Krebs cycle, 373-74, 376-79, 396

Lacrimal
    bone, 88, 91, 92, 93, 95
    duct, 94, 485
    foramen, 88, 94
    gland, 330, 485
    sac, 485
Lactation, 477-81
Lacteal, 161, 184, 193, 279
Lactic acid, 115, 116, 221
Lactose, 175
Laminar air flow, 215
Landsteiner groups, 256-57
Larynx, 94, 153, 187, 205, 207
Lateral cartilage, 57
Lateral foramen, 83
Lateral masses (ethmoid), 89
Law
    "all or none," 111, 334
    Boyle's, 204
    Charles', 204
    Dalton's, 204
    Henry's, 204
    of orbital symmetry, 184
    Starling's, 265
Lens (of eye), 489, 491
Leukemia, 239, 240
Leukocyte, 234, 240, 241, 246, 251-52
Leukocytosis, 240
Leukopenia, 240
Lever principle (limb movement), 117-19
Leydig cell, 328, 357
LH (luteinizing hormone), 347, 357, 362, 363, 364, 365
Ligament
    accessory, 119, 126
    interosseus, 126
    joint, 98, 99
    mesovarium, 412
    pulmonary, 209
    suspensory, 125, 126, 127, 172-73, 282, 476
Ligamentum teres, 119
Light microscopy, 1, 13-14
Light reception, 483-84
Lignin, 175, 201
Limb, muscles of, 104-05, 116-19
Lip, 154
Lipid (granule), 16
Liponeogenesis, 395-96
Lipoprotein metabolism, 383-84
Liver, 153, 167-71, 193, 194, 360
    avian, 525
Lobule, kidney, 289, 295
Locomotion, 116-17, 127
Long bone, 29-30
Lumbar artery, 269
Lumbar vertebrae, 35, 78, 81, 82, 83, 84
Lung
    air components, 529-31
    air volumes, 216-18
    avian, 205-6, 207, 208, 209-10, 211, 212
    negative pressure, 211-13
    pathology, 207, 210, 215, 218, 223-24, 228
    respiratory rate, 215
"Lungenherz" muscle, 111
Lymph
    cells, 241
    ducts, 242
    node, 241, 242, 243, 244, 245, 253, 254, 279, 280, 281
    system, 209, 210, 211, 241-45, 253, 279-83
    vessels (lacteals), 27-29
Lymphatics
    facial, 151
    fetal, 285
    immunity mechanism, 242-45, 352, 353
Lymphoblast, 242
Lymphocyte, 24, 233, 234, 241, 242, 243, 244, 245, 246, 247, 252
Lymphoid organs, 279-81
Lyophilization, 180
Lysis, 253, 257-58
Lysosome, 15, 16, 292, 293

Macroglia, 313
Macrophage, 239, 246, 254
Magnesium (ions), 115
Magnum (foramen), 87, 89, 90, 219, 318
    avian, 536, 537
Malar bone, 88, 90, 91, 92, 93, 94
Male genital system, 402-11
Malpighian corpuscle, 294, 300
Maltose, 175
Mammary gland, 362, 364, 437, 473-81
Mammillary process, 78, 81, 82
Mandible, 35, 88, 91, 92, 93
Mandibular foramen, 93
Marey's reflex, 264, 279
Marginal cartilage, 98
Margo plicatus, 154, 155, 191
Marrow, 24, 29-30, 238-45, 281
    avian, 513
    cavity, 29, 99
Mast cell, 252
Mastoid process, 93
Matrix, mucoprotein, 28
Maxilla, 88, 90, 91, 92, 93, 94
Meatus
    external acoustic, 88, 91, 92, 93
    nasal, 95
Meckel's diverticulum, 524, 525, 532
Median plane, 3, 4
Medulla (renal), 290, 294, 295, 296, 298, 302
Medulla oblongata, 192, 219, 317, 318, 321, 322
Medullary plexus, 291, 300, 303, 304
Megakaryocyte, 248
Meibomian (tarsal) gland, 484, 485
Meissner's corpuscle, 314, 467
    plexus, 334
Melatonin, 350, 351
Membrane
    bone, 28
    of cell, 12-16, 217, 218, 357-58
    nuclear, 13, 16, 17, 357-58
Meninges, 318, 325-26, 327, 328
Meniscus, 78, 98, 99
Menopause, 437
Mental foramen, 88, 91, 92, 93
Merkel's corpuscle, 314
Merocrine gland, 465, 473
Mesencephalon, 317, 318, 319
Mesenteric artery, 269
Mesentery, 159, 160, 172
Mesosternum, 86
Mesothelium, 461, 462, 463
Metabolism
    carbohydrate (intermediate), 372-82
    carbohydrate (nonruminant), 174-176
    carbohydrate (ruminant), 199-201
    catabolism and anabolism, 370
    cholesterol, 388-89
    fat, 176-77, 382-87
    intermediate, 369-99
    muscle, 115-16, 220-21
    protein, 390-99
    regulation of, 366-68
    steroid, 178-79
    vitamin, 183-85, 399
    water, 179-82
Metabolites, 182, 183, 185
Metacarpal, 36, 37, 38, 53, 55, 59, 514-15
    evolution of, 123
    movement, 118-19
Metamyelocyte, 246
Metasternum, 86, 87

Metatarsal, 35, 37, 61, 75
  evolution, 123
Metencephalon, 317, 318
Methemoglobin, 237
Microanatomy, 293
Microbody, 15, 16
Microfloral digestive phase, 188
Microglia, 313
Microscopic anatomy, 32
Microvilli, 14, 16, 161, 292–93
Milk fat, 479
Milk teeth, 129, 133, 135, 137
Milk vein, 275, 476
Mineral salts, 182–83
Mitochondria, 15, 16, 18
  in cardiac muscle cell, 112
  in intestine cell, 162
  in kidney tubule cell, 293
  in membrane, 389
  in muscle cytoplasm, 102
Mitosis, 17, 241–42
  factors affecting, 19–20
  and meiosis, 418–21, 432
Moderator band, 260
Molar tooth, 131, 132, 135, 143–45, 146
Molecular biology, 13
Molt, avian, 538
Monocyte, 234, 241, 246, 247, 251, 252, 254
Monosaccharide, 174, 175, 372
Mouth, 150–51, 153, 207
  avian, 521–22
Movable joint, 98–100
Movable (true) vertebra, 79–83, 84
MSH (melanocyte stimulating hormone), 350
Mucosa, 156–57, 161–62
Mucous
  gland, 150, 151, 155, 188
  membrane, 150–66 *passim*
  cell, 188, 194
Mucus, 150, 151
Multi-unit muscle, 110
Muscle, 101–23
  accommodation, 492–94
  avian, 518–20
  cardiac (heart), 111, 238–39, 260–61, 263–66
  diaphragm, 210–11
  esophagus, 152
  eye, 489
  fiber, 102–3
  fibril, 102
  gastro-intestinal, 153–66
  genito-urinary, 404, 409, 410, 416
  multi-unit, 99–104
  penile, 409
  pharyngeal, 152
  scrotal, 404
  smooth, 110, 153, 369–82, 436
  striated, 101, 102

  tonus, 104
  uterine, 436
  vaginal, 436
Muscularis, 154, 159–60
Mutagen, 23
Myelencephalon, 317, 318
Myelin, 315
Myelocytoma, 240
Myenteric plexus, 333–34
Myocardium, 112, 260, 261
Myofibril, 102–3
Myoglobin, 102
Myosin, 110, 115–16

Nares, 94, 205
Nasal
  bone, 88, 91, 92, 93, 94, 95
  cavity, 94, 95, 152, 205, 505, 529
  foramina, 517
  glands, 181, 529
  process (premaxilla), 94
  septum, 89, 93, 94, 205
Nasofrontal, 90
Nasolacrimal duct, 485
"Nature's surgical dressing," 173
Navicular, 57, 469
Near point, 492
Neck (avian), 515
Neck (tooth), 129
Neck chief cell, 154, 155, 188, 189
Necrosis, 251
Negative pressure, 211–13
Neoplasia, 20
Nephron, 291–93, 295–97, 298–306 *passim*
Nerve center, 315
Nerve ending, 102, 314
Nervous system, 311–43
  avian, 541
Neurofibril, 312
Neurohormonal, 364–65
Neuromuscular, 368
Neuron, 312, 313, 314–18
Neutrophil, 234, 240, 246, 247, 251–52
Niacin, 184
Nipper (tooth), 133, 135, 137, 138, 139
Nissl body, 312
Nitrogen
  balance, 306
  transport, 227–28
Node of Ranvier, 315–16, 336
Noradrenalin, 111, 316, 331, 355
Notch
  acetabular, 64
  fovea capitis, 66
  glenoid, 38–39
  semilunar, 46, 47, 48, 49, 50
Nuchal crest, 35, 88, 89, 90, 91
Nuclear membrane, 17
Nuclear sap, 17
Nucleic acid, 174, 177, 184

Nucleolus, 16, 17
Nucleoplasm, 17
Nucleus
  cardiac muscle, 104, 111
  cellular, 12, 13, 16, 17
  diploidy (restoration), 432
  kidney cell, 268
  nerve cell, 313
  ovum, 428
  smooth muscle, 110
  striated muscle, 105, 110
Nutrient foramen, 29, 37, 41, 57, 66, 71

Obturator foramen, 62, 64, 65
Occipital bone, 87, 88, 89, 90, 91, 92
Odontoid process, 79, 80, 81
Odontology, 33
Offspring identity, 254, 257
Olecranon, 46, 47, 48
  fossa, 43, 44, 45
Olfactory
  bulb, 321, 322, 325, 506
  cell, 505, 506
  nerve, 322, 506
Oligodendroglia, 313
Omasum, 150, 156, 157, 158, 199
Omnivore dentition, 129, 146
Oogenesis, 424, 427–28
Optic
  cavity, 92
  chiasma, 320, 321, 322, 489
  nerve, 320, 321, 322, 485, 489–90
  structure, 320, 484–89, 547–48
Oral cavity, 150–52, 186–87
Orbit, 35, 88, 91, 92, 484
Organ of Corti, 501–2
Organelles, 12–13, 14–19
Organization, cells, 20
Organoid, 18, 240
Organs, 19
Os
  acetabuli, 64
  compedale, 38, 56
  cordis, 33, 95, 261
  corona, 38, 56
  coronale, 38, 56
  coxa, 60, 62
  pedis, 38, 56
  penis, 33, 95, 409, 410–11
  rostri, 95
  suffraginis, 38, 56
  ungulare, 38, 56
Osmosis, 18, 181, 248
Ossein, 27
Osseous labyrinth, 498–500
Ossification, 28
Osteoblast, 28
Osteoclast, 28
Osteocyte, 32, 225
Osteology, 33
Otolith organ, 500

Ovary, 358, 362–65, 411, 412, 417, 424, 427–30
Overpopulation, 443–45
Oviduct, 412–13
Ovulation cycle, 429
Ovum, 424, 427–28, 429–30
Oxydative phosphorylation, 112, 379–80
Oxygen
  debt, 115
  dissociation curve, 226–28
  transport, 14, 223–32 *passim*
Oxyhemoglobin, 225, 237
Oxytocin, 349, 362, 364, 368, 437, 480

Pacemaker, heart, 111, 112, 113, 261
Pacinian corpuscle, 314
Packaged granule, 16, 17
Palatine bone, 88, 90, 91, 92, 93, 94
Palatine gland (tonsil), 151, 281
Palatine process
  maxilla, 90, 94
  premaxilla, 90
Pancreas, 153, 159, 162, 166–67, 181, 193–94, 345, 355, 525
Pancreatic duct, 153–4, 162, 166
Pancreozymin, 194, 359–60, 365–66
Paneth cell, 161, 194
Pantothenic acid, 184
Papilla, renal, 289
Papillary duct, 289, 294, 301
Parabiosis, kidney, 299
Paramastoid process, 89, 90, 91
Parasympathectomy, 340
Parasympathetic nervous system, 311, 331, 333
Parathormone, 179, 345, 352, 366
Parathyroid, 352–53, 366
Parietal bone, 87, 88, 89, 90, 91, 92
Parietal cell, 155, 157, 188, 189
Parotid duct, 150
Parotid lymph node, 151
Parotid salivary gland, 151
Pars distalis, 346–47, 347
Pars intermedia, 346–47, 349–50
Pars nervosa, 346–47, 349
Pars tuberalis, 346–47, 347
Parturition, 436
Pastern bone, 38, 469
Pastern joint, 119, 125
Patella, 35, 37, 61, 68
Pectoral girdle, 34–116
Pectoral limb, 34, 35
Peduncle, 290–93
Pelvic cavity, 60
Pelvic girdle, 33, 60
Pelvic limb, 34, 61
Pelvic ring, 34
Pelvis, 33, 34, 60–66
Penis, 408–11
Pentose, 176
Pepsin, 189

Pericardium, 86–87, 260, 267
Perichondrium, 28, 99
Perikaryon, 312
Perimysium, 102, 103
Periosteum, 29, 99
Peripheral nervous system, 311, 329–30
Periscopic vision, 490, 496–97
Peristalsis, 187, 191, 195
  reverse, 195
Peritoneal cavity, 172–73, 288
Peritoneum, 172–73
Permanent teeth, 137–39, 145
Petrous temporal bone, 88–93
Peyer's patches, 161, 162, 164, 194, 244, 281
pH, 183, 193, 223, 224, 225, 236
PHA (phytohemagglutinin), 242, 243
Phagocyte, 240, 241, 245, 246, 247, 252, 279, 281
Phalanx, 35, 37, 38, 53, 56, 58, 59, 61
Phenotype, 441–42
Phonation, 221
Phosphogluconate reaction, 380–82
Phospholipid, 382–83, 387, 479
Phosphoric acid, 115
Phosphorus, 182, 366
Phosphorylase, 355, 374
Phrenic nerve, 214, 291, 220
Physiology
  avian digestion, 526–28
  defined, 5
  digestive, 174–87
  ear, 502–5
  eye, 490–97
  femur, 67
  fibula, 72
  humerus, 45–46
  intestine/cecum, 193–98
  lactation, 477–81
  muscle, 113–16
  nephron, 303–9
  nervous system, 334–38
  patella, 68
  pelvis, 65–66
  radius and ulna, 50–51
  rectum/anus, 198
  renal, 298–300
  reproduction, 418–23
  respiration, 211–23
  rumination, 198–99
  scapula, 40–41
  skeleton (avian), 517–18
  stomach, 187–93
  taste, 508–9
  teeth, 145–47
  vertebral column, 84
Phytohemagglutinin, 20
Pia mater, 326, 327
Piloerector muscle, 110
Pineal body, 322, 348, 350–51
Pink muscle, 102, 110

Pinocytic vesicle, 14, 15, 16
Pinocytosis, 162
Pit, 14, 16, 17, 292
Pithing, 340
Pitocin, 345, 349, 368
Pitressin, 307, 349, 368
Pituitary, 307, 346–50, 361, 362, 363, 364, 366, 368
Pivot joint, 79, 99
Placenta, 233, 283, 415–16, 433, 434
Planes of reference, 2, 3, 4
Plasma (blood), 234, 235, 248, 251, 253
Plasmacyte, 241, 243, 245, 254
Platelet, 234, 247–48, 249, 251
Pleurae, 205, 210, 211, 212, 213
Pleural cavity, 211
Pleural membrane, 86–87
Pneumocentesis, 213
Pneumothorax, 213
Podocyte "feet," 292, 293
Poikilothermism, 367
Pole (renal), 288, 295
Polycythemia, 240
Polypeptide, 223, 237
Polypnea, 204, 220
Polysaccharide, 175, 199, 206
Pons, 317, 318, 319, 322
Portal circulation, 233, 276
Portal vein, 168, 250, 251, 276
Postganglionic fiber, 314, 315, 331, 332
Postnatal precocity, 442
Postparturition (postparturient), 436–37
Poststernum, 86
Potassium ions, 14, 111, 366
Potassium salts, 182, 183, 235, 250
Precocity, postnatal, 442
Preganglionic fiber, 314, 315, 332
Pregnancy
  control of, 443–45
  ectopic, 437–38
Premaxilla, 89, 90, 91, 92, 93, 94
Premolar (teeth), 131, 133, 135, 144, 146
Prepubic artery, 269
Presphenoid bone, 89
Presternum, 86, 87
Primitive streak, 433
Process(es) cell, 12, 13
Proenzyme, 16
Progesterone, 20, 358, 359, 362, 363, 364, 431, 445, 448, 477
Prolactin, 20, 347, 348, 364
Proline, 349
Prolymphoblast, 242
Propionate, 200
Proprioceptor, 337
Propulsion, 111–12, 122–27
Prosencephalon, 318
Prostaglandins, 316, 360–61
Prostate gland, 288, 403, 407, 408, 409
Protein, 17, 19, 174, 177–78, 195–97

569

Protein (cont.)
  metabolism, 390–96
  physical structure, 396–97
Proteolytic enzymes, 155–56, 196–97, 251–52
Prothrombin, 228, 249, 250
Protoplasm, 12, 14, 15, 18, 19, 161, 162, 163, 176, 177
Protourine, 268, 276, 277, 279, 293, 303, 306
Pseudoruminant, 140, 143, 146, 153, 156, 158
Pterygoid bone, 84, 85, 88, 89, 90, 93, 94
Pubis, 34, 57, 58, 60, 61, 62, 63–64, 65
"Puffing the glen," 130, 139
Pulmonary unit, 187, 205–6
Pulmonary vascular system, 189, 190–91, 208, 209–10, 212, 213
Pulp cavity, 120, 121, 130, 131
Purkinje system, 104, 105, 106, 111, 112–13, 238, 239, 260–61, 531
Pus, 225, 246
Pyloric gland, 143, 145, 155–56, 158, 171–72, 188–89
Pyloric region (abomasum), 138, 145, 150, 158, 188, 189, 191
Pyloric region (stomach), 138, 141, 142, 150, 154, 155–56, 158, 188, 191
Pyloric sphincter (pyloric valve), 153, 155, 158, 191
Pylorus, 155, 192
Pyridoxine, 184
Pyruvic acid, 115

Quiet breathing, 214–15

Radiation, 23
Radius, 35, 38, 46–49
Ramus communicans, 327, 328, 331
Rathke's pouch, 346
Rat/mouse eradication, 192–93
Ray, medullary, 290, 294
Reabsorption, urinary, 302–6 passim, 354
Receptor ending, 313, 314
Rectum, 150, 153, 166, 198
Red marrow, 29–30, 31, 85
Red muscle, 102
Reflex anuria, 302
Reflex arc, 317–18, 330, 333, 337–38
Refraction, 490
Regional anatomy, 4
Regnault's method, 264
Regressive structure, 154, 158, 187
Regulation, heart rate, 263–66
Reiset's method, 264
Rejection mechanism, 242–43, 244–45, 255–56
Relaxation, muscular, 115
Relaxin, 358, 362, 364
Renal column, 289–90, 295, 296

Renal function research, 272–73, 280–81, 292–93
Renal pelvis, 289–90, 293–94, 295, 296, 297, 298
Renal pyramid, 289, 294, 295
Renal secretion, 291–93
Renal threshold, 304
Renin, 279, 307, 308
Rennin, 189, 195, 196
Reproduction
  cell, 13, 20
  factors affecting, 20
Reproduction patterns
  avian, 538–40
  mammalian, 445–59
  marsupial, 459–60
Reproductive system, 369–445, 533–37
  avian, 533–37
    female, 534–37
    male, 533–34
  mammal, anatomy, 402–17
    female, 411–17
    male, 402–11
  mammal, physiology, 418–45
    contraception, 444–45
    genotype-phenotype, 441–42
    intersexes, 441
    overpopulation, 443–44
    postnatal precocity, 442–43
    sex determination, 439–40
    sex ratio, 438–39
Reserve, cardiac, 265–66
Respiration, 203–33
  CNS control center, 219–21, 319, 332
  external, 205–11, 220–21, 224
  internal, 116, 204, 225–26
  involuntary, 219–21,
Respiratory mechanics, 213–15
Respiratory quotient (RQ), 228–30
Respiratory system
  avian, 529–31
  mammalian, 203–32
Resting potential, nerve, 335
Reticular cell, 242
Reticular fiber, 102
Reticulocyte, 238
Reticuloendothelial system, 239
Reticulum, 150, 156–57, 159, 160
Retina, 485, 486, 487, 488, 547
Rh syndrome, 256, 258–59
Rhombencephalon, 318–19
Rib, 35, 36, 85
Riboflavin, 183–84
Ribonucleoprotein, 17, 312
Ribosome, 17, 312
Rigor mortis, 115
Ritual mating behavior, 539
RNA (ribonucleic acid), 17, 18, 23, 241, 343–44, 381, 397
Rod nuclear (stab) cells, 246
Romer's rule, 343

Root (tooth), 130, 131
Rootless teeth, 129
Rostral pituitary, 346, 347–49
Rotation, joint, 98, 119, 127
Ruffini's end organ, 314
Rumen, 151, 156–58, 159, 160
Ruminant dentition, 129, 142, 143
Ruminant stomach, 156–58, 159, 160
Rumination, 198–201

SA (sinoatrial) node, 111, 112, 113, 261, 263–65
Sac, rumenal, 156–57, 159, 160
Sacculated stomach, 158
Saccule, 499, 500
Sacral vertebrae, 35, 83, 84
Sacroiliac joint, 66
Sagittal plane, 2, 3
Salivary gland, 150–51, 186–87
Saphenous vessels, 274, 275
Sarcoarthrosis, 100
Sarcolemma, 105, 111
Sarcomere, 112
Sarcoplasmic reticulum, 105, 112
Scapula, 35, 36, 37, 38, 39, 40, 41, 42
  evolution, 124
Schleiden and Schwann, 2, 12
Schwann cell/sheath, 315
Scrotum, 404
Sebaceous gland, 473
Secretin, 194, 359
Sella turcica, 346
Septal cartilage, 93
Serosa, 154, 160
Serotonin, 24
Serous fluid, 150, 151, 172
Serous membrane, 210
Serum, blood, 235, 249
Sesamoid bone, 34, 35, 38, 53, 57, 58, 59, 60, 61, 68
Sex ratio/determination of, 438–39
Short bone, 29, 51, 57, 68
Sinus
  frontal, 90
  maxillary, 94
  sphenopalatine, 89
Sinusoid, liver, 233
Sisson's "intermediate zone," 290
Skeletal muscle, 101
Skeleton
  avian, 513–17
  mammalian, 33–95
Skin, 466–67
Skin appendages, 467–77
Skull, avian, 516–17
Skull, mammalian, 87–95
Smell, organ, 505–7
  avian, 546
"Smooth mouth," 143
Smooth muscle, 110–11
Sodium, 182, 183

Sodium (*cont.*)
   ions, 14, 304, 335
   salts, 183
Sol, 19
Somatic (body) cell, 20
Sounds, heart, 263
Special anatomy, 4
Special sense organs, 483–509
   avian, 547–49
Spermatic vessel, 269, 271, 273, 275, 407
Spermatogenesis, 423–26
Spermatozoa, 421, 424
Sphenoid bone, 87, 89, 90, 92
Sphincter muscle
   anal, 104, 166
   eye, 487, 488
   mouth, 104
Spicule, 31, 51–53
Spinal cord, 78, 87, 100, 311, 317, 321, 326–29
   nerve, 329–30
Spine (vertebrae), 77–84
Spinous process, 78, 79, 80, 81, 82, 83
Spleen, 241, 244, 253, 281, 282–83
Splint bone, 54
Spongy marrow bone, 31, 85
Squamous epithelium, 462–63
   occipital bone, 87, 88, 89
   temporal bone, 87, 88, 89, 90, 91, 92, 93
SSH (spermatogenesis stimulating hormone), 347, 357
Starch, 175
Stem cells, 244
Stereoscopic vision, 490, 496–97
Sternabrae, 86
Sternum, 31, 33, 85, 86
Steroid, 178–79, 354, 367
STH (somatotrophic hormone), 347, 387
Stifle, 119
Stomach
   avian, 152, 523
   compound, 152, 153, 156–58, 159, 198–201
   simple, 152, 153, 154–56, 158
Striated muscle, 101–10
Styloid cartilage, 94
Subclavian vessel, 268, 271, 273, 275
Subcutaneous muscle, 104
Subcutaneous tissues, 466
Sublingual duct, 151
Submucosa, 154, 160–61
Subscapular fossa, 37, 39, 40
Sucrose, 175
Sulcus, 320, 321, 326
Sulfur, 182, 183, 399
Supernumerary rib, 83
Supernumerary tooth, 144
Supracondyloid foramen, 44
Supracondyloid fossa, 69
Supramammary lymph node, 159

Supraorbital foramen, 88, 91
Supraorbital process, 92
Suprascapular cartilage, 37, 38, 39, 40
Supraspinous fossa, 37, 39, 40
Supratrochlear foramen, 45
Surfaces
   articular, 98
   dental, 132
   femoral, 66
   fibula, 71
   humeral, 41, 43
   ilial, 60–61
   ischial, 63
   patellar, 63
   phalangeal, 56–57
   pubic, 64
   radial, 43, 44
   scapular, 37
   tibial, 70
   ulnar, 49
   vertebral, 78
Surfactant, 206
Suspensoid, 19
Suture, 98
Swallowing, 149, 187
Sweat, 179–81, 461
   gland, 473
Sympathectomy, 340
Sympathetic nervous system, 311–18, 331–33
Symphysis, 60, 63, 64, 93, 98
Synapse adrenergic, 316–17
   cholinergic, 102, 316
Synaptic cleft, 102
Synarthrosis, 98
Synchondrosis, 98
Syncytium, 111–13
Syndesmology, 97
Syndesmosis, 98
Synovia, 100
Synovial capsule, 99
Synovial fossa, 47
Synsarcosis, 100
Systemic anatomy, 4–5
Systemic circulation, 233
Systems, 20
Systole, 261–62

Table surface (dental), 130, 131, 133, 134, 135, 139, 140, 141, 145
Tarsal glands, 484
Tarsus, 35, 37, 61, 72–75, 117
Taste, organ, 461, 507–9
Taxonomy/classification, 6–10
Teeth, 93, 94, 129–47, 150
Telencephalon, 317, 318, 320
Temperature
   ambient, 367
   body, 366–67
   regulation, 366–67
Temporal

bone, 87, 88, 89, 91, 92, 93
   external acoustic process, 93
   zygomatic process, 93
Tendon, 102, 103, 104, 105, 116, 127
Terminal bar, 162, 292
Terminal web, 161
Testicle, 358, 365, 402–7, 409, 410
Testosterone, 358
Tetany, 114
Tetrose, 176
Thalamus, 318, 320, 322
Theory
   aging, 22–25
   cell, 12
   clot formation, 227–29
   lymphocyte origin and function, 220–24
Thiamin, 184
Thoracic
   aorta, 268
   cavity, 211
   components, 211
   duct, 212
   inlet, 86
   respiration, 213–15
   vertebrae, 35, 82–83, 212
Thoracocentesis, 213
Thrombin, 249, 250
Thrombocyte, 247
Thrombokinase, 248, 249
Thromboplastin, 249, 250
Thrombus, 251
Thymus, 24, 352–53
   and lymphocytes, 241–45 *passim*
Thyroid, 345, 348, 349, 350
Thyrotropin, 348
Thyroxin, 25, 349, 351
Tibia, 35, 36, 61, 70, 73, 74, 119
Tibial veins, 275
Tight junction, 162
Tissues, 20
Tonus, 104, 212, 308, 509
Topographic anatomy, 4
Touch (tactile), 509
Toxin/antitoxin, 253
TPN (triphosphoridine nucleotide), 380
TPNH (triphosphoridine nucleotide hormone), 381
Trabeculae, 31, 53, 258
Trachea, 153, 205, 207, 208, 212
Transamination, 392–93
Transcellular fluid, 181–82
Transection, spinal cord, 338–39
Transfer mechanism, urinary, 300–9 *passim*
Transitional epithelium, 294, 300
Transplant
   autograft, 255
   heterograft, 255
   homograft, 245
Transport mechanisms, 19–20, 179–81,

Transport mechanisms (cont.)
    182, 184–85, 211, 223, 224–31, 235,
    236, 271–83, 277, 300–2
Transversarium foramen, 79, 80, 81, 82
Transverse plane, 3
Transverse process, 78, 79, 80, 81, 82, 83
Trapezius muscle, 117
TRF (thyroid releasing factor), 349
Tricalcium phosphate, 27
Triceps brachii, 117
Triglyceride, 176, 382, 383, 384, 385, 386, 479
Triose, 176
Trochanter, 66, 67, 69, 70
Trochanteric fossa, 67
Trochlea, 43, 49, 67, 74, 75
Trochoid joint, 79
"True" joint, 98
"True" (movable) vertebrae, 78
"True" (sternal) rib, 85
"True" teeth, 129, 131
Trunk, arterial, 268–69
Trunk, venous, 275
T-S (transverse membrane system), 102–3
TTH (thyrotropic hormone), 347, 348
"T.T. Hallim V.P.," 178
Tuber, 38, 39, 40, 63
Tubercle, 38
Tuberosity, 40–64 passim
Tubule, renal, 291–303 passim, 305
Tubuloalveolar gland, 166
Tunica albugenia, 428
Tunica externa, 269
Tunica intima, 269
Tunica media, 269
Tunica propria, 466
Turbinate, 89, 93, 95
Turbulent air flow, 215
Twins, 439–40
Tympanic membrane, 256–58 passim, 498

Udder, 473–75, 475–77
Ulna, 35, 44, 45, 47–48
Ultramicroscopy, 13
Umbilical vein, 233, 283, 284
Unstriped muscle, 101, 110–11
Uracil, 344
Urea, 15, 167, 196, 230
Ureter, 288, 289, 294, 295, 296, 297, 298, 300
Urethra, 288, 289, 294, 300
Uric acid, 167, 231
Urinary system, 287–309
    avian, 533
Urine formation, 300–306
Urogastrone, 309, 360, 366
Urogenital muscle, 110

Uterine anatomy (types), 413–15
Uterine contractions, 436
Uterine epithelium, 434
Uterine hormones, 364
Uterine involution, 436, 437
Uterine-ovarian vessels, 269, 271, 273, 275, 412
Uterine-placenta relationship, 347, 358–59
Uterine types, 415
Utricle, 499, 500

Vacuole, 15, 16, 17
Vagina, 412, 416
Valve
    aortic, 237, 260
    atrioventricular, 237, 260
    cardiac, 153, 155, 159
    ceco-colic, 153
    ileocecal, 153
    lymph, 279
    pulmonary, 237, 260
    pyloric, 153, 155
    veins, 276
Vascular system, 233–35
    arteries, 268–272
        flow, 278
    blood, 233–59, 276–77, 278
    fetal, 283–85
    heart, 259–69
    lymph, 279–83
    portal, 276
    pulmonary, 209–10
    renal, 291
    venous, 272–78
        flow, 277
Vascular tunic (eye), 486–88
Vascular tree, 267–68
Vater's ampulla, 162
Veins
    abdomen & hindquarters, 275
    head & forequarters, 275
    structure, 272, 276
    thorax, 275
Velum, 319, 322
Vena cava
    caudal, 167, 276, 287, 288, 291
    cranial, 261, 273, 276, 278
    foramen, 211
    fossa, 167
Ventilation, 222
Ventricle
    brain, 322–24, 325
    heart, 259–60
Ventricular muscle, 112–13
Venule, 270
Vertebrae, 77–86
Vertebral amphiarthrosis, 99

artery, 268
canal, 78
foramen, 78–84 passim
Vesalius, Andreas, 1–2
Visceral muscle, 101, 110
Visceral skeleton, 33, 95
Visceroceptor, 337
Vital phenomena, 11
Vitamin, 174, 183–85, 399
Vitreous body, 489
Volkmann's canals, 32
Voluntary muscle, 101–4
Voluntary nervous system, 311
Vomer bone, 89, 90, 93, 95
Vomition, 192–93, 213
Vulva, 416

"W" chromosome, 534
"Wasting disease," 245, 352
Water balance, 366
    body, 181, 223
    cellular, 15
    metabolism, 179–82, 354
Wave receptor, 337
White
    commissure, 326–27
    marrow, 29
    matter, 318, 326, 327
    muscle, 101–2
    ramus communicans, 327, 328, 331
Wing of atlas, 79, 80, 81
Wing
    avian, 515, 521
    ilium, 62
    sphenoid, 89

"X" chromosome, 439, 441
Xeroderma, 542
Xiphoid (ensiform) cartilage, 86
Xiphosternum, 86

"Y" chromosome, 439–41
Yawning, 222
Yellow marrow, 29

"Z" chromosome, 534
Z-band, 102, 103
Z-line, 112
Zonary placenta, 416, 434
Zygomatic
    arch, 88, 91, 92, 93
    bone, 94, 95
    process, 93
    salivary gland, 151
Zygote, 432
Zymogen granule, 155
Zymogenic cell, 155